Advances in
ATOMIC, MOLECULAR, AND OPTICAL PHYSICS

VOLUME 26

EDITORIAL BOARD

P. R. BERMAN
New York University
New York, New York

K. DOLDER
The University of Newcastle-upon-Tyne
Newcastle-upon-Tyne
England

M. GAVRILA
F.O.M. Instituut voor Atoom- en Molecuulfysica
Amsterdam
The Netherlands

M. INOKUTI
Argonne National Laboratory
Argonne, Illinois

S. J. SMITH
Joint Institute for Laboratory Astrophysics
Boulder, Colorado

ADVANCES IN

ATOMIC, MOLECULAR, AND OPTICAL PHYSICS

Edited by

Sir David Bates

DEPARTMENT OF APPLIED MATHEMATICS AND THEORETICAL PHYSICS
THE QUEEN'S UNIVERSITY OF BELFAST
BELFAST, NORTHERN IRELAND

Benjamin Bederson

DEPARTMENT OF PHYSICS
NEW YORK UNIVERSITY
NEW YORK, NEW YORK

VOLUME 26

 ACADEMIC PRESS, INC.
Harcourt Brace Jovanovich, Publishers

Boston San Diego New York
Berkeley London Sydney
Tokyo Toronto

This book is printed on acid-free paper. ∞

Copyright © 1990 by Academic Press, Inc.
All rights reserved.
No part of this publication may be reproduced or
transmitted in any form or by any means, electronic
or mechanical, including photocopy, recording, or
any information storage and retrieval system, without
permission in writing from the publisher.

ACADEMIC PRESS, INC.
1250 Sixth Avenue, San Diego, CA 92101

United Kingdom Edition published by
ACADEMIC PRESS LTD.
24–28 Oval Road, London NW1 7DX

LIBRARY OF CONGRESS CATALOG CARD NUMBER: 65-18423

ISBN 0-12-003826-9

PRINTED IN THE UNITED STATES OF AMERICA

90 91 92 93   9 8 7 6 5 4 3 2 1

Contents

CONTRIBUTORS   vii

Comparisons of Positron and Electron Scattering By Gases
Walter E. Kauppila and Talbert S. Stein

I.	Introduction	1
II.	Positron and Electron Scattering by Atoms	5
III.	Positron and Electron Scattering by Molecules	24
IV.	Summary and Concluding Remarks	41
	Acknowledgements	46
	References	46

Electron Capture at Relativistic Energies
B. L. Moiseiwitsch

I.	Introduction	51
II.	Two-State Approximation	52
III.	Second-Order Theories	57
IV.	Relativistic Continuum Distorted Wave Approximation	64
V.	Relativistic Eikonal Approximation	65
VI.	Numerical Solution of Coupled Equations	74
VII.	Experimental Data and Comparisons with Theory	74
	References	77

The Low-Energy, Heavy-Particle Collisions—A Close-Coupling Treatment
Mineo Kimura and Neal F. Lane

I.	Introduction	80
II.	General Formulation of the Close-Coupling Method	87
III.	Current Status of Theoretical and Experimental Results	113
IV.	Conclusions and Perspectives	154
	Acknowledgements	156
	References	156

Vibronic Phenomena in Collisions of Atomic and Molecular Species

V. Sidis

I.	Introduction	161
II.	Outline of the Quantum Mechanical Formulation	164
III.	Quantum Treatment of Vibronic Excitation: The IOS Approximation	170
IV.	Semi-Classical Treatment of Vibronic Excitation	175
V.	On Franck-Condon-Type Approximations	181
VI.	Studies of Vibronic Transition Processes	187
VII.	Concluding Remarks	203
	References	204

Associative Ionization: Experiments, Potentials, and Dynamics

John Weiner, Françoise Masnou-Seeuws, and Annick Giusti-Suzor

I.	Introduction	210
II.	Experiments	211
III.	The Problem of Molecular Potentials	240
IV.	Dynamics of Associative Ionization	261
V.	Summary, Conclusions, and Perspectives	289
	Acknowledgements	291
	References	292

On the β Decay of ^{187}Re: An Interface of Atomic and Nuclear Physics and of Cosmochronology

Zonghua Chen, Leonard Rosenberg, and Larry Spruch

I.	Introduction	297
II.	The Relative Production Rates of Different Isotopes of Re and of Os	303
III.	Refinements and Improvements	308
IV.	Conclusion	318
	Acknowledgements	318
	References	319

Progress in Low Pressure Mercury-Rare-Gas Discharge Research

J. Maya and R. Lagushenko

I.	Introduction	321
II.	Modeling of Low Pressure Mercury-Rare-Gas Discharge	323
III.	Altered, Low Pressure Mercury-Rare-Gas Discharge	342
IV.	Diagnostics	356
V.	Summary	369
	References	370

INDEX	375
CONTENTS OF PREVIOUS VOLUMES	385

Contributors

Numbers in parentheses refer to the pages on which the authors' contributions begin.

Zonghua Chen (297), Department of Physics, New York University, New York, New York 10003

A. Giusti-Suzor (209), Laboratoire de Photophysique Moleculaire, CNRS, Faculte des Sciences, Orsay 91405, France

W. E. Kauppila (1), Department of Physics and Astronomy, Wayne State University, Detroit, Michigan 48202

M. Kimura (79), Argonne National Laboratory, Argonne, Illinois 60439 and Department of Physics, Rice University, Houston, Texas 77251

R. Lagushenko (321), GTE Electrical Products Corporation, Danvers, Massachusetts 01923

N. F. Lane (79), Department of Physics and Rice Quantum Institute, Rice University, Houston, Texas 77251

F. Masnou-Seeuws (209), Laboratoire des Collisions Atomiques et Moleculaires, Universtite de Paris-Sud, Orsay 91405, France

J. Maya (321), GTE Electrical Products Corporation, Danvers, Massachusetts 01923

B. L. Moiseiwitsch (51), Department of Applied Mathematics and Theoretical Physics, The Queen's University of Belfast, Belfast BT7 1NN, Northern Ireland

Leonard Rosenberg (297), Department of Physics, New York University, New York, New York 10003

V. Sidis (161), Laboratoire des Collisions Atomiques et Moleculaires, Bat. 351, Universite de Paris-Sud, Orsay 91405, France

Larry Spruch (297), Department of Physics, New York University, New York, New York 10003

T. S. Stein (1), Department of Physics and Astronomy, Wayne State University, Detroit, Michigan 48202

J. Weiner (209), Department of Chemistry and Biochemistry, University of Maryland, College Park, Maryland 20742

COMPARISONS OF POSITRON AND ELECTRON SCATTERING BY GASES

WALTER E. KAUPPILA and TALBERT S. STEIN

Department of Physics and Astronomy
Wayne State University
Detroit, Michigan

I. Introduction 1
II. Positron and Electron Scattering by Atoms 5
    A. Inert Gases (He, Ne, Ar, Kr, and Xe) 5
    B. Alkali Metals (Na and K) 17
    C. Hydrogen 21
III. Positron and Electron Scattering by Molecules 24
    A. Diatomic (H_2, N_2, CO, and O_2) 24
    B. Triatomic (H_2O, CO_2, and N_2O) 31
    C. Other Polyatomic (NH_3, CH_4, Hydrocarbons, SiH_4, CF_4, SF_6) 35
IV. Summary and Concluding Remarks 41
    Acknowledgements 46
    References 46

I. Introduction

Investigations of the scattering of electrons by gas atoms and molecules has been an active area of research since the pioneering work of Ramsauer (1921a) and Townsend and Bailey (1922) in the 1920s. Meanwhile, the existence of positrons (the antiparticles of electrons) was first established in 1933 (Anderson, 1933) and the first positron–atom scattering cross sections were measured in the early 1970s by Costello *et al.* (1972). The development of the field of positron–gas scattering has in many respects followed that of the electron case. The initial positron–gas scattering experiments were primarily concerned with measuring total scattering cross sections for room-temperature gases. The 1980s have seen a blossoming of second generation positron scattering experiments (partly spurred on by the development of more intense positron beams), including measurements of (1) total cross sections for non-room-temperature gases (e.g., alkali atoms), (2) specific inelastic cross sections for excitation and ionization of atoms, and positronium formation, and (3) differential elastic scattering cross sections.

An important motivation for investigating the scattering of positrons by atoms and molecules is to possibly help to provide a better understanding of the scattering of electrons by atoms and molecules, the latter being of importance in many different fields of science and technology, such as plasma physics, laser development, gaseous electronics, astrophysics, and aeronomy. Since positrons differ from electrons only by the signs of their electric charge (and magnetic moment), comparison measurements of the scattering of positrons and electrons by the same atoms and molecules can reveal interesting similarities and differences that arise from the basic interactions (see Table I) contributing to $e^{+,-}$ scattering. The static interaction arises from the interaction between the projectile and the Coulomb field of an undistorted target atom (or molecule), while the polarization interaction results from the interaction between the projectile and the distorted (i.e., polarized) charge distribution in the target atom. The polarization interaction is important primarily at sufficiently low incident projectile energies such that the target atom has an opportunity to polarize during the time of interaction. The exchange interaction, which affects only electron scattering, arises from the indistinguishability of a projectile electron from electrons in the target atom and effects electron scattering primarily in the energy region corresponding to the kinetic energies of the atomic electrons. The static interaction contributes to $e^{+,-}$ scattering at all projectile energies. The cumulative effect of the static and polarization interactions is that for e^+ scattering there is a tendency toward cancellation, while for electrons they add, generally resulting in smaller elastic scattering cross sections for positrons than for electrons. At sufficiently high energies it is known that eventually the polarization and exchange interactions will become negligible,

TABLE I

INTERACTIONS CONTRIBUTING TO POSITRON AND ELECTRON SCATTERING

Interaction	Projectile	
	e^+	e^-
Static	repulsive	attractive
Polarization	attractive	attractive
Exchange	none	yes

leaving only the static interaction, which will result in an ultimate merging of the corresponding $e^{+,-}$-atom scattering cross sections for the various scattering channels (see Table II) that are accessible to both projectiles. Two scattering channels that exist only for e^+ scattering are annihilation and positronium (Ps) formation (both real and virtual). It is known (Massey, 1976) that annihilation is not a significant effect except at energies which are well below those (>0.2 eV) that have been used in e^+ scattering experiments. On the other hand, Ps formation has been found to play an important role in e^+-gas collision studies. It should be noted that Ps formation does result in ionization of the target atom, but in this paper the term ionization will be reserved for the process where three separate particles (e^+, e^-, and A^+) result from the associated scattering channel. Total scattering cross sections, discussed to the greatest extent in this paper, include all of the accessible scattering channels open to e^+ and e^- scattering. In many cases there have been measurements of cross sections for specific scattering channels and these will be discussed where meaningful comparisons can be made between positrons and electrons. Another aspect of e^+ scattering that has recently opened is the measurement of differential cross sections (DCSs) for the elastic scattering channel, which is particularly significant when measurements are being used to test various scattering theories because it is well known that they can provide a more sensitive test of a theory than the measurements of the total integrated cross section (TCS) for a specific scattering channel. This

TABLE II

BASIC SCATTERING CHANNELS OPEN FOR POSITRON AND ELECTRON SCATTERING BY ATOMS AND MOLECULES ("e" REFERS TO PROJECTILE POSITRON OR ELECTRON, "A" REFERS TO ATOMS AND MOLECULES, "M" REFERS TO MOLECULES ONLY, AND E_{ion} REFERS TO THE THRESHOLD ENERGY FOR IONIZATION)

Scattering Channel		Threshold Energy
$e + A \to e + A$	elastic	none
$\to e + A^*$	electronic excitation	several eV
$\to e + A^+ + e^-$	ionization	several eV (E_{ion})
$\to A^+ +$ gamma rays	annihilation (e^+ only)	none
$\to A^+ +$ Ps	positronium formation (e^+ only)	$E_{ion} - 6.8$ eV
$e + M \to e + M^*$	vibrational excitation	few $\times$ 0.1 eV
$\to e + M^*$	rotational excitation	<0.1 eV

can be illustrated using the method of partial wave analysis for low energy elastic scattering of a projectile (of momentum $\hbar k$) by an atom where

$$\text{DCS}(k, \theta) = |f(k, \theta)|^2 = \frac{1}{k^2} \Sigma (2l + 1)(2m + 1)\sin \delta_l \sin \delta_m \cos(\delta_l - \delta_m) P_l P_m \tag{1}$$

and

$$\text{TCS}(k) = \int \text{DCS}(k, \theta) d\Omega = \frac{4\pi}{k^2} \Sigma (2l + 1) \sin^2 \delta_l, \tag{2}$$

where $f(k, \theta)$ is the scattering amplitude, θ is the scattering angle, $\delta_{l,m}$ are energy-dependent and real scattering phaseshifts for each relative angular momentum $l\hbar$ of the system, $P_{l,m}$ are Legendre polynomials, and the summations are performed over all possible values of l and m. It is seen that the DCS contains interference terms between phase shifts corresponding to different angular momenta, while the TCS does not.

It is to be noted that the intent of this review paper is not to provide an exhaustive review of all aspects of the field of positron scattering, including the various experimental methods that have been used, but instead the intent is to review many of the interesting comparisons, primarily experimental measurements, that can be made between positron and electron scattering by the same atoms and molecules with the hope that these comparisons may help to stimulate a better overall understanding of electron (and positron) scattering processes. For a broader overview of the general field of positron scattering and the experimental methods that have been employed, the reader is referred to the proceedings of recent workshops pertaining to positron scattering (Humberston and McDowell, 1984; Kauppila, Stein, and Wadehra, 1986; and Humberston and Armour, 1987), and also some recent review articles (Stein and Kauppila, 1982, 1986; Charlton, 1985; Sinapius, 1988).

For most of the experimental results to be included in this paper, the e^+ and e^- cross section measurements of a given group have been made in the same experimental system with the same technique. This is significant because the relative comparisons between e^+ and e^- measurements will be more meaningful than the actual quantitative values because many of the potential experimental errors (e.g., measuring target gas number densities) would be the same for the e^+ and e^- measurements, and as a result not affect the relative comparisons. In view of the many comparisons between e^+ and e^- scattering that will be made in this paper, it is interesting to realize that an initial motivation (in our laboratory at Wayne State University) for doing the corresponding electron measurements in positron scattering experiments was merely to test the overall reliability of the experiment by comparing the electron measurements with prior measurements which were known to be quite reliable.

II. Positron and Electron Scattering by Atoms

A. INERT GASES (He, Ne, Ar, Kr, and Xe)

Historically, the inert gases have been the first atoms to be experimentally investigated for both electron and positron scattering because they exist in the atomic form as a gas at room temperature. An overall comparison of total cross sections (Q_t's) that have been measured for electrons and for positrons scattering from He, Ne, Ar, Kr, and Xe is shown in Fig. 1, where the electron and positron measurements by the Wayne State experimental group (Kauppila et al., 1976, 1977, 1981; Stein et al., 1978; and Dababneh et al., 1980, 1982) were made in the same experimental system with the same technique. The lowest energy thresholds for inelastic scattering channels that contribute meaningfully to the measured Q_t's are indicated in Fig. 1, which for electrons is atomic excitation and for positrons is the formation of positronium (Ps). Below these thresholds Q_t represents only the elastic scattering channel.

Several general observations can be made in regard to the qualitative shapes of the Q_t curves shown in Fig. 1. In the case of electron scattering, the Q_t's for Ar, Kr, and Xe are found to have very deep minima at low energies (<1 eV), where only elastic scattering occurs. These minima, which make the above gases nearly transparent to the passage of electrons at these energies, were first observed by Ramsauer (1921b, 1923), Townsend and Bailey (1922), and Ramsauer and Kollath (1929), and are referred to as Ramsauer-Townsend effects. The minima arise from quantum mechanical effects associated with a net attractive interaction between the incident projectile electron and the target atom. Another qualitative feature shared by each of these electron Q_t curves is that their maximum values occur at energies either near or below the lowest energy inelastic threshold, implying that elastic scattering dominates the electron Q_t's at the energies of the maxima.

In the case of positron scattering by the inert gases shown in Fig. 1, it is seen that Ramsauer-Townsend effects are clearly observed for He and Ne, and possibly a shallow minimum for Ar. Since a net attractive interaction between the projectile and the target atom is required in order to have a Ramsauer-Townsend effect, the observations of these minima for positron scattering suggest that at low energies the polarization interaction must be dominating the static interaction. A common qualitative feature for each of the positron–inert-gas Q_t curves is a sudden increase in Q_t as the projectile

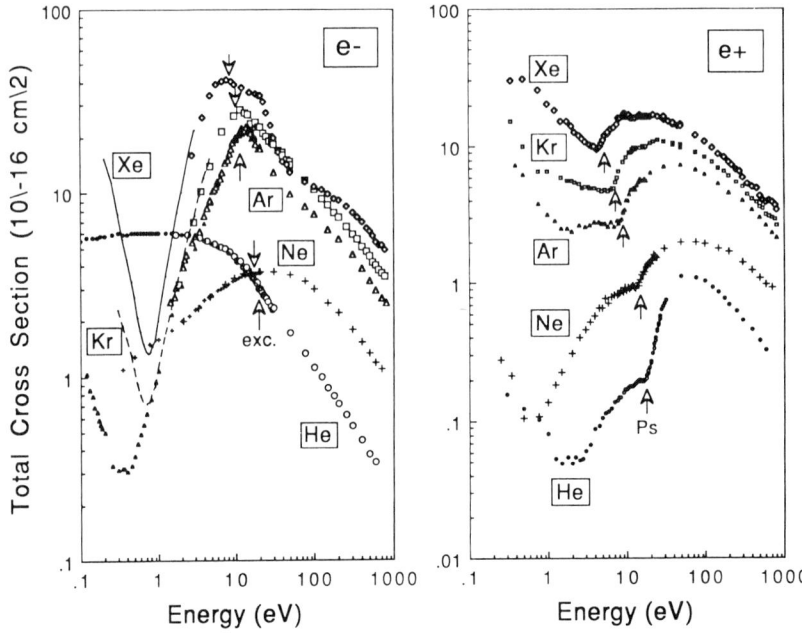

FIG. 1. Total cross section comparisons for inert gases. (a) The electron-inert gas measurements displayed at higher energies for He (○), Ne (+), Ar (△), Kr (squares), and Xe (diamonds) are those made by Kauppila et al. (1977, 1981), Stein et al. (1978), and Dababneh et al. (1980, 1982), while the lowest energy measurements are those of Buckman and Lohmann (1986a) for He (○) and Ar (△), Salop and Nakano (1970) for Ne (+), and Jost et al. (1983) for Kr (----) and Xe (———). The arrows represent the excitation thresholds for each gas. (b) The positron-inert-gas measurements for He (○), Ne (+), Ar (△), Kr (squares), and Xe (diamonds) are those made by Kauppila et al. (1976, 1981), Stein et al. (1978), and Dababneh et al. (1980, 1982). The arrows represent the thresholds for positronium (Ps) formation in each gas. It is to be noted in this and many subsequent figures that straight line segments will often be used to join discrete data points in order to generate "curves," that $10\backslash-16\,\text{cm}\backslash 2$ refers to $10^{-16}\,\text{cm}^2$, and that e− and e+ refer to electrons and positrons, respectively.

energy is increased through the Ps formation thresholds, suggesting that this channel plays an important role in positron scattering by the inert-gas atoms. All of these positron Q_t curves are found to reach relatively broad maxima in the 10–100 eV range, which are related to inelastic scattering channels. At the highest energies shown for each projectile it is seen that the measured values of the Q_t's are in order of the sizes of the target atoms.

1. Helium

Helium is the simplest atom that has been studied up to the present time in positron scattering experiments and considerable effort has been devoted to it both experimentally and theoretically. A comparison of several positron and electron ($e^{+,-}$)–He scattering cross sections is shown in Fig. 2. In the purely elastic scattering energy region it is seen that the e^- Q_t's (Kauppila et al., 1977) are everywhere more than 10 × larger than the e^+ Q_t's (Stein et al., 1978), which is what one would expect in regard to the tendency for the static and polarization interactions to add for electrons and to cancel for positrons. At higher energies the $e^{+,-}$ Q_t curves approach each other and merge (to within less than 2%) for energies above 200 eV (Kauppila et al., 1981). This

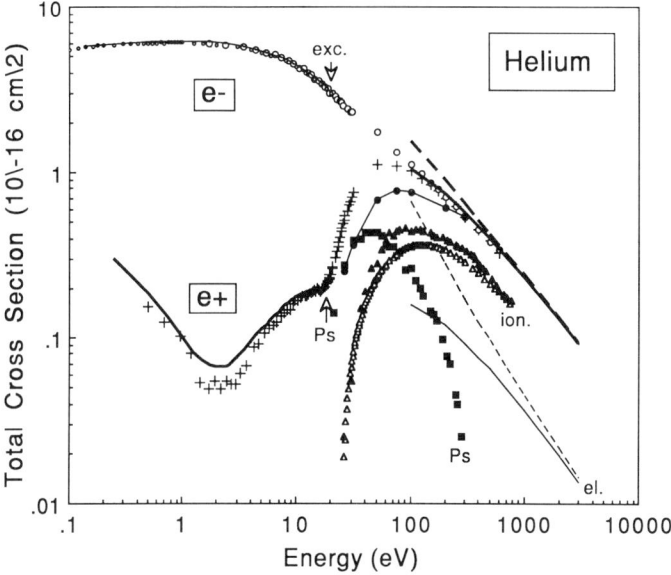

FIG. 2. Comparison of cross sections for $e^{+,-}$–He scattering. The e^+ Q_t measurements (+) are those of Stein et al. (1978) below 30 eV [which have been shifted up in energy by +0.2 eV as was suggested by Wadehra et al. (1981)], and Kauppila et al. (1981) above 50 eV. The e^- Q_t measurements shown are those of Buckman and Lohmann (1986a) below 20 eV (∘), and Kauppila et al. (1977, 1981) above 1.7 eV (○). Theoretical results for Q_{el} at energies below 20 eV are those calculated by Nesbet (1979) for electrons (———) and obtained by Wadehra et al. (1981) for positrons (heavy ▬▬▬). For energies above 100 eV the theoretical calculations of Dewangen and Walters (1977) are shown for e^+ (heavy ▬▬▬) and e^- (heavy ------)Q_t's, and for e^+ (———) and e^- (------) Q_{el}'s. The thresholds for excitation (exc.) and positronium (Ps) formation are indicated by arrows. Inelastic cross section measurements that are shown were made by Fromme et al. (1986) for Ps formation (solid squares) and e^+ impact ionization (▲), and Montague et al. (1984) for e^- impact ionization (Δ). The solid line joining the closed circles represents $Q_t - Q_{Ps}$ for positrons. It is to be noted that in most figures in this article where positron and electron results will be directly compared and the respective values may be close to each other that closed symbols and solid lines will be used to represent positrons, and open symbols and broken lines for electrons.

observed merging is particularly interesting because the theoretical distorted-wave second Born approximation (DWSBA) calculations of Dewangan and Walters (1977) do not predict this merging (to within 2%) to occur until 2000 eV. At 200 eV the theoretically predicted difference is >15%. This observed "premature" merging for the measured $e^{+,-}$ Q_t's presents an interesting puzzle. In making comparisons of $e^{+,-}$ Q_t's it should be recognized that Ps formation is a scattering channel available only to positrons. Recent Ps formation cross section (Q_{Ps}) measurements by Fromme et al. (1986) are shown in Fig. 2 and are seen to be an appreciable fraction of Q_t only between the formation threshold and 300 eV. Subtracting the Q_{Ps} measurements from the Q_t measurements gives an "adjusted" Q_t curve for positrons that does not approach the e^- Q_t curve as closely for energies below 300 eV, but still has the same degree of merging above 300 eV. Another interesting piece of information for the "puzzle" is that the "DWSBA" calculation of Dewangan and Walters (1977) predicts that the e^- elastic scattering cross section (Q_{el}) at 200 eV is more than a factor of 2 larger than the e^+ Q_{el}, which would require that the "total" inelastic cross section (sum for all of the inelastic channels) would have to be larger for positrons than it is for electrons, even after Q_{Ps} is subtracted from the e^+ Q_t. It is relevant to this discussion that the DWSBA e^- Q_{el} results agree very well (within 3%) at 200 eV with semi-empirical values that have been deduced (primarily from elastic differential cross section measurements) by de Heer and Jansen (1975), while no corresponding comparisons between theory and experiment are yet available for positrons. A clue that inelastic cross sections for positrons may be larger than for electrons was provided by the relative measurements of the e^+ and e^- ionization cross section (Q_{ion}) by Fromme et al. (1986), where they found the e^+ cross section to be larger in the energy range 35–600 eV. Normalizing their Q_{ion} measurements to the e^- measurements of Montague et al. (1984), Fromme et al. obtained the values shown in Fig. 2.

Comparing the absolute values of the experimental Q_t's with the representative calculations shown at high and low energies it is seen that the agreement is very good for positrons at high energies with the DWSBA calculation of Dewangan and Walters (1977) and for electrons at low energies with the variational calculations of Nesbet (1979). The DWSBA calculation for electrons becomes noticeably larger than the measurements at 100 eV, the lower energy limit of the calculation. The theoretical curve shown for low energy e^+ Q_{el} obtained by Wadehra et al. (1981) [using the s-wave phase shifts of Humberston (1979), the p-wave phase shifts of Humberston and Campeanu (1980), the lowest set of d-wave phase shifts of Drachman (1966), and higher order phase shifts (up to $l = 20$) from the Born approximation expression of O'Malley et al. (1961)] is somewhat higher than the experimental values in the region of the Ramsauer-Townsend minimum. An analysis by Wadehra et al. (1981) of the amount of elastic scattering at small angles in the forward direction in the vicinity of the minimum coupled with a knowledge of the inability of the experiment to completely discriminate against all small angle forward elastic scattering has shown that theory and experiment are very consistent with each other up to the Ps formation threshold. It is interesting that the low energy electron measurements (Kauppila et al., 1977) made in the same experimental system as the positron measurements and subject to similar constraints for discriminating against small-angle elastic scattering agree very well with theory because in the case of low energy e^--He scattering there is little small-angle forward scattering.

An experimental investigation by Charlton et al. (1988) of the ratio of positron and electron impact double-to-single ionization cross sections (Q^{++}/Q^+) for He has produced some very interesting results when these ratios are compared with similar ratios that have been measured for proton (Puckett and Martin, 1970; Knudsen et al. 1984; Shah and Gilbody, 1985; Andersen et al., 1987) and antiproton (Andersen et al., 1987) impact on He. The measured ratios for these four projectiles ($e^{+,-}$ and $p^{+,-}$) are shown in Fig. 3 where they are plotted versus the projectile energy divided by the projectile mass (in atomic mass units), which is equal to the projectile velocity squared. Earlier it had been found that measured Q^+'s were equal for $e^{+,-}$ impact (Fromme et al., 1986) and for $p^{+,-}$ impact (Anderson et al., 1987) at sufficiently high energies where charge transfer processes (e.g. formation of Ps and H atoms) are negligible. Furthermore, the Q^+'s for each particle–antiparticle pair were equal to each other at the same projectile velocities, which was predicted by the first Born approximation calculation of Inokuti (1971) where the collision is considered to be a single-step two-body process. In the case of double ionization it is found at higher velocities ($v^2 > 1$ MeV/amu) that the ratios of Q^{++}/Q^+ merge for each of the two projectiles having the same sign of electric charge, even though their masses are significantly different. At lower "velocities" the Q^{++}/Q^+ ratios diverge and tend toward zero for $e^{+,-}$'s (the threshold energy for double ionization is 79 eV or 0.16 MeV/amu for positrons and electrons) and become very large for $p^{+,-}$'s, which Charlton et al. (1988) attributes to the large differences in kinetic energy for these projectiles having the same velocity. Several possible physical interpretations to describe the merging of the Q^{++}/Q^+ ratios have been discussed by McGuire and Deb (1987) and Charlton et al. (1988), which all consider the double ionization process to be a more complicated process (possibly involving two steps) than for single ionization.

2. Neon

As the second simplest inert gas atom, neon has received a moderate amount of interest. Comparisons of $e^{+,-}$ cross sections are shown in Fig. 4.

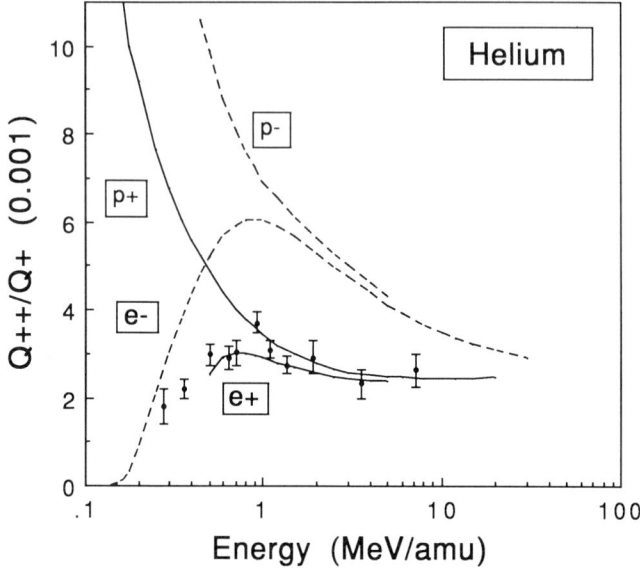

FIG. 3. Ratios of double (Q^{++}) to single ionization (Q^+) cross section measurements for positrons (e^+), electrons (e^-), protons (p^+), and antiprotons (p^-) scattering from He. The results of Charlton et al. (1988) are shown by the lower dashed curve for electrons and solid circles with error bars for positrons. The upper dashed curve represents the p^- results of Andersen et al. (1987), and the upper solid curve represents the p^+ results of Puckett and Martin (1970), Knudsen et al. (1984), Shah and Gilbody (1985), and Andersen et al. (1987). The lower solid curve is discussed in the text. (Adapted from Charlton et al., 1988.)

At the higher energies the $e^{+,-}$ Q_t measurements show a tendency toward merging, as do the DWSBA calculations of Dewangan and Walters (1977), while at lower energies the e^- Q_t results become 10 × larger than the e^+ values at the e^+ Ramsauer-Townsend minimum. The Q_{el}'s calculated by Dewangan and Walters (1977) for electrons (which are in good agreement with experiments above 200 eV) and positrons also show (as was the case for helium) a larger difference than the corresponding difference between the Q_t's clearly suggesting that the sum of the inelastic scattering cross sections for positrons is larger than for electrons with a portion of the difference being accounted for by Q_{ps} [the preliminary values of Diana et al. (1985) are displayed in Fig. 4]. In the low energy purely elastic scattering region for neon, the e^- results that are shown (Stein et al., 1978; Salop and Nakano 1970; Dasgupta and Bhatia, 1984) are in very good agreement, while the polarized-orbital approximation calculation of McEachran et al. (1978) was selected for comparison because it was one of the closest to the corresponding measurements of Stein et al. (1978). It is to be noted that the extension of the

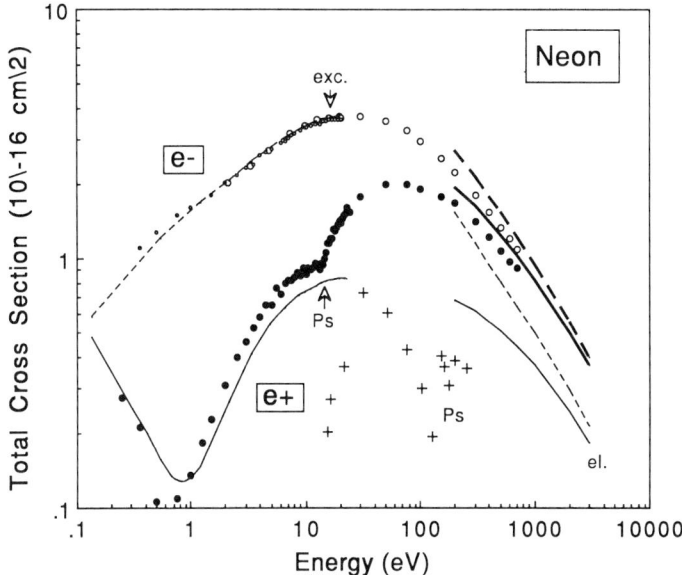

FIG. 4. Comparison of cross sections for $e^{+,-}$–Ne scattering. The Q_t measurements of Stein et al. (1978) and Kauppila et al. (1981) for positrons (●) and electrons (○) are shown, along with the e^- measurements of Salop and Nakano (1970) below 20 eV (○). Theoretical results for Q_{el}'s below 23 eV are those calculated by Dasgupta and Bhatia (1984) for electrons (------), and by McEachran et al. (1978) for positrons (———). For energies above 100 eV the theoretical calculations of Dewangen and Walters (1977) are shown for e^+ (heavy ━━━) and e^- (heavy ------) Q_t's, and for e^+ (———) and e^- (------) Q_{el}'s. The thresholds for excitation (exc.) and positronium (Ps) formation are indicated by arrows. The Q_{Ps} measurements (+) were made by Diana et al. (1985).

theoretical Q_{el} curve above the Ps formation threshold may be subject to error due to the opening up of inelastic channels which could have an effect on the elastic scattering channel.

3. Argon

Experimentally, argon has been one of the favorite inert gases to study with positrons because it is an inexpensive gas and it is a relatively large atom with correspondingly large scattering cross sections. Cross section comparisons for argon are shown in Fig. 5 where it is seen that at high energies both the $e^{+,-}$ Q_t measurements of Kauppila et al. (1981) and the optical model calculations of Joachain et al. (1977) are close but not merged for positrons and electrons. The e^- Q_{el} values of Joachain et al. agree quite well with

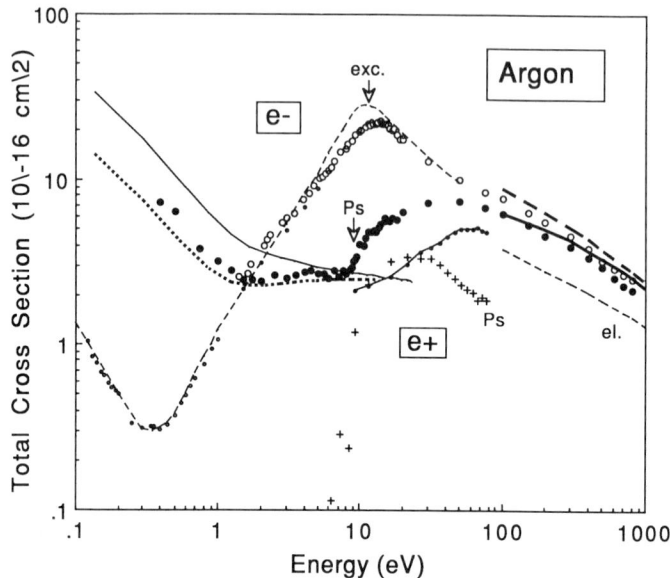

FIG. 5. Comparison of cross sections for $e^{+,-}$–Ar scattering. The Q_t measurements of Kauppila et al. (1976, 1981) for positrons (●) and electrons (○) are shown, along with the e^- measurements of Buckman and Lohmann (1986a) below 20 eV (∘). Theoretical results for Q_{el} below 50 eV are those calculated by McEachran and Stauffer (1983) for electrons (------), and Montgomery and LaBahn (1970) (·····) and McEachran et al. (1979) for positrons (———). For energies above 100 eV the "optical model method I" calculations of Joachain et al. (1977) are shown for e^+ (heavy ———) and e^- (heavy -------)Q_t's and for e^- (------)Q_{el}'s. The thresholds for excitation (exc.) and positronium (Ps) formation are indicated by arrows. The Q_{Ps} measurements (+) were made by Fornari et al. (1983). The solid line joining the closed circles represents Q_t-Q_{Ps} for positrons.

experimental values (see Joachain et al., 1977). At lower energies the e^- Q_t measurements are as much as 5 × larger than the e^+ measurements at the Ps formation threshold and then become more than 10 × smaller at the e^- Ramsauer-Townsend minimum. This latter comparison at low energies provides a nice illustration of the dramatic nature of the Ramsauer-Townsend effect where in this case the e^- Q_t decreased by two orders of magnitude from its maximum near 10 eV. The Q_{ps} measurements of Fornari et al. (1983), when subtracted from the e^+ Q_t measurements, result in a cross section curve that is noticeably smaller than the Q_t measurements at the Ps formation threshold. This discrepancy suggests the possible existence of a "cusp" in Q_{el} at the Ps formation threshold, whereby Q_{el} may change appreciably (in this case decrease) with the onset of inelastic channels. In the elastic scattering energy region the e^+ Q_t measurements of Kauppila et al. (1976) are compared with polarized orbital calculations made by Montgomery and LaBahn (1970) (their "3p-d" normalized results are shown) and by McEachran et al. (1979) with the measurements being closer to Montgomery and LaBahn at the lower energies. The low energy e^- elastic scattering calculations of McEachran and Stauffer (1983), who used an adiabatic exchange approximation with only the dipole part of the polarization potential, are in excellent agreement with the measurements of Buckman and Lohmann (1986a) in the Ramsauer-Townsend minimum region and become somewhat higher than the measurements of Kauppila et al. (1976) and Buckman and Lohmann (1986a) near the Q_t maximum.

Argon was the first (and remains the only) atom for which positron differential elastic scattering cross section measurements have been reported (Coleman and McNutt, 1979). The relative elastic $e^{+,-}$-Ar DCS measurements of Hyder et al. (1986) are shown in Fig. 6, where their e^- results are

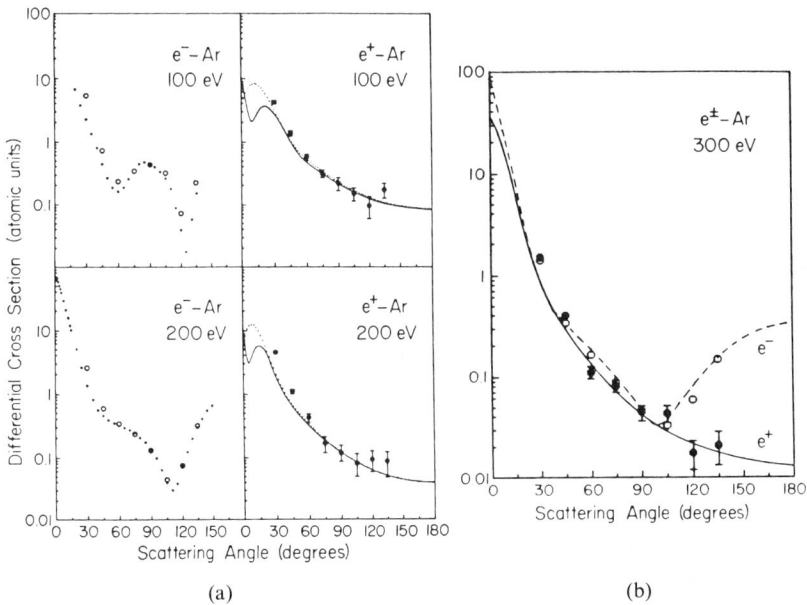

FIG. 6. Comparison of the relative elastic DCS measurements of Hyder et al. (1986) for e^- (○) and e^+ (●) scattering by Ar with (a) the e^- measurements of Srivastava et al. (1981) (∘) at 100 eV and of Dubois and Rudd (1975) (∘) at 200 eV, and the e^+ calculations of McEachran and Stauffer (1986) (———) and of Nahar and Wadehra (1987) (·····) at 100 and 200 eV, and (b) the calculations of Joachain (1978) and Joachain et al. (1977) for positrons (———) and electrons (------) at 300 eV. In each case the measurements of Hyder et al. are normalized to the comparison results at 90°.

compared with prior measurements of Srivastava et al. (1981) at 100 eV and Dubois and Rudd (1975) at 200 eV. The e^+ results at 100 and 200 eV are compared with the polarized orbital calculations of McEachran and Stauffer (1986) and the model potential calculations of Nahar and Wadehra (1987), while the $e^{+,-}$ results at 300 eV are compared with the optical model calculations of Joachain (1978) and Joachain et al. (1977). In all cases, the results of Hyder et al. (1986) have been normalized at 90° to the prior experimental or theoretical work. A comparison of the shapes of the measured $e^{+,-}$ DCS curves shows that for electrons there is considerable structure (relating to diffraction effects that vary with energy), while the e^+ curves are all monotonically decreasing with increasing scattering angle. At 300 eV, where the measured $e^{+,-}$ Q_t values are within 15% of each other (see Fig. 5), it is interesting that the $e^{+,-}$ DCSs are found both theoretically and experimentally to be very close for angles up to about 100°, and at larger angles they diverge. At higher energies the structure in the e^- DCS curve continues to diminish (see Joachain et al., 1977) and eventually the $e^{+,-}$ DCS curves will merge. As a result, at 300 eV the onset of the tendency toward merging for $e^{+,-}$-Ar elastic DCS scattering is beginning to appear. The good agreement in the shapes of the e^+ DCS measurements with the theories at 100 and 200 eV is surprising when it is considered that neither of these calculations includes effects of absorption on Q_{el} due to the presence of other accessible inelastic scattering channels. In a recent optical model calculation Joachain and Potvliege (1987) have studied the effects of absorption on the elastic differential scattering cross section and have predicted that no important structure (e.g. a minimum or maximum) should occur for small-angle positron elastic scattering by the inert gases at intermediate to high energies and that the actual differential cross sections will be reduced at medium to large angles with respect to those calculated without including absorption effects.

Some low energy elastic DCS measurements for positrons (Coleman and McNutt, 1979; Floeder et al., 1988; Smith et al., 1989) and for electrons (Williams and Willis, 1975; Srivastava et al., 1981; Smith et al., 1989) are shown in Fig. 7, where they are compared with calculations for positrons (Montgomery and LaBahn, 1970; McEachran et al., 1979; McEachran and Stauffer, 1986; Bartschat et al., 1988) and for electrons (McEachran and Stauffer, 1983; Fon et al., 1983). The 8.7 eV measurements are significant because only elastic scattering can occur at this energy (it is below the lowest inelastic thresholds for both positrons and electrons). It is interesting that at 8.7 eV structure is found in the e^+ DCS curve, which indicates that diffraction effects are also present for positron scattering at low energies. It has been suggested (Kauppila and Stein, 1987) that the lack of structure in e^+ DCS curves at higher energies and the wealth of structure for e^- DCS curves at low

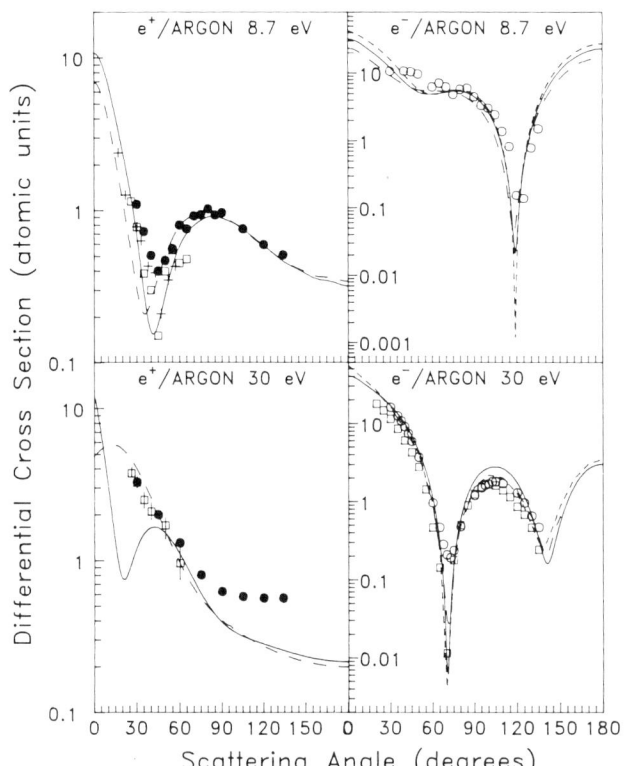

FIG. 7. Comparison of relative elastic DCS measurements for e^+-Ar scattering by Coleman and McNutt (1979) (+), Floeder et al. (1988) (open squares), and Smith et al. (1989) (●) at 8.7 and 30 eV with calculations at 8.7 eV by Montgomery and Labahn (1970) (------) and McEachran et al. (1979) (———), and with calculations at 30 eV by McEachran and Stauffer (1986) (———) and Bartschat et al. (1988) (------). Comparison and elastic DCS measurements for e^--Ar scattering by Williams (1979) (- - -), and Smith et al. (1989) (○) at 8.7 eV with calculations by McEachran and Stauffer (1983) (———) and Fon et al. (1983) (------), and of e^- measurements by Williams and Willis (1975) (- - -), Srivastava et al. (1981) (open squares), and Smith et al. (1989) (○) at 30 eV with calculations by McEachran and Stauffer (1983) (———) and Fon et al. (1983) (------). The normalization of the relative measurements to calculations or prior measurements at a particular angle does not affect the relative shape comparisons of the data due to the use of a log scale on the vertical axes.

to intermediate energies (not only for argon but also other inert gases) may relate to diffraction effects arising only when the net interaction potential is attractive, which is the case at all energies for electrons and perhaps only for low energies for positrons. The e^+ DCS measurements are in very good agreement with the polarized-orbital approximation calculations of McEachran et al. (1979) when proper consideration is given to the finite angular acceptance of the experiments. For electrons the "high precision" measurements of Srivastava et al. (1981) and Williams (1979) are in very good agreement with the adiabatic exchange calculations of McEachran and Stauffer (1983), while the "lower precision" measurements of Smith et al. (1989) are in fair agreement. At 30 eV it is very intriguing that the e^+ measurements do not show the structure indicated by the polarized orbital calculation of McEachran and Stauffer, which does not include absorption effects due to accessible inelastic channels. Instead, it is found that the general shape is in better agreement with the optical potential calculation by Bartschat et al. (1988), where consideration has been given to the absorption effects due to ten inelastic excitation channels, but not Ps formation or ionization. Meanwhile, it is curious that for e^- scattering at 30 eV the measurements, which are in good agreement with each other, are in very good agreement with the adiabatic exchange calculations of McEachran and Stauffer (1983) (which do not include absorption effects) and with the R-matrix calculations of Fon et al. (1983) (which do include absorption effects). This comparison between experiments and theories for positrons and electrons at 8.7 and 30 eV rather strongly indicates that absorption effects may be very important for positron scattering and not very important for electron scattering, which all could relate back to our earlier discussions for He, Ne, and Ar that inelastic scattering has a much more dominant effect on positron scattering than it does for electrons.

4. Krypton and Xenon

The scattering of positrons and electrons by krypton and xenon have many similarities as shown by the Q_t measurements displayed in Figs. 8 and 9. Both gases exhibit (1) pronounced Ramsauer-Townsend effects for electron scattering just below 1 eV, (2) Q_t's for positron scattering that decrease as energy increases toward the Ps formation threshold where there is a dramatic increase in Q_t, and (3) $e^{+,-}$ Q_t's that have not yet merged at the highest energies of the measurements. The Ps formation measurements (preliminary) of Diana et al. (1987) for krypton, when compared with the e^+ Q_t measurements, suggest that Ps formation may be the dominant scattering process for positrons between 10 and 20 eV (accounting for more than 50% of Q_t), which also appears to be the case for argon (see Fig. 5). For low energy e^- scattering the semi-relativistic calculations by Sin Fai Lam (1982) for elastic scattering by krypton, and the non-relativistic and relativistic calculations for xenon by McEachran and Stauffer (1984, 1987) are in quite good agreement with the experimental measurements of Jost et al. (1983) and Dababneh et al. (1980). The "knee-like" structure in the e^-–Xe Q_t curve in the vicinity of 20 eV, which seems to lead to a local anomalous crossing of the $e^{+,-}$ Q_t curves

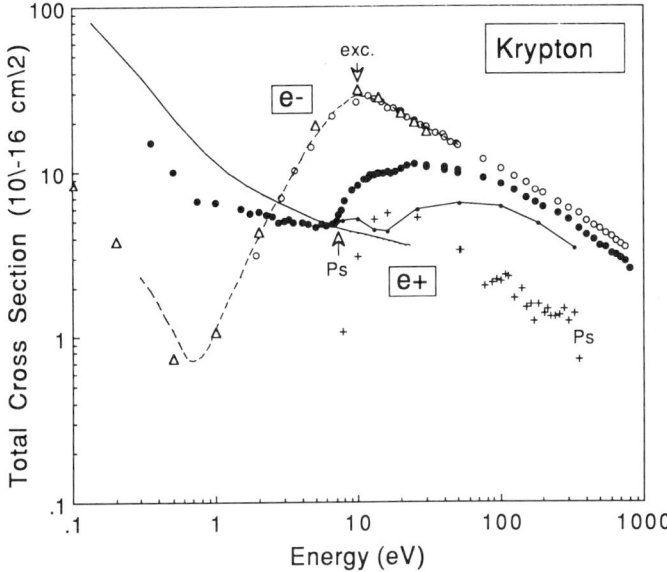

FIG. 8. Comparison of cross sections for $e^{+,-}$-Kr scattering. The Q_t measurements of Dababneh et al. (1980, 1982) for positrons (●) and electrons (○) are shown, along with the e^- measurements of Jost et al. (1983) below 50 eV (------). Theoretical results for Q_{el} below 30 eV are those of Sin Fai Lam (1982) for electrons (triangles), and McEachran et al. (1980) for positrons (———). The thresholds for excitation (exc.) and positronium (Ps) formation are indicated by arrows. The Q_{Ps} measurements (+) were made by Diana et al. (1987). The solid line joining the closed circles represents $Q_t - Q_{Ps}$ for positrons.

around 50 eV, is predicted by the calculations and related to the elastic scattering channel. Meanwhile, for low energy e^+ scattering by Kr and Xe the calculations of McEachran et al. (1980) for elastic scattering, using a frozen-core version of the polarized-orbital approximation, are in rather close agreement with the e^+ Q_t measurements of Dababneh et al. (1980) at the Ps formation thresholds, but become somewhat higher at the lower energies.

B. ALKALI METALS (Na AND K)

The alkali metal atoms potassium and sodium were the first non-room-temperature gases and are the only atoms other than the inert gases to be investigated in positron scattering experiments (Stein et al., 1985, 1987). Theoretically, alkali atoms are interesting partly because of their relatively simple structure, having a single weakly bound valence electron moving outside a core of closed shells. A unique feature regarding positron scattering

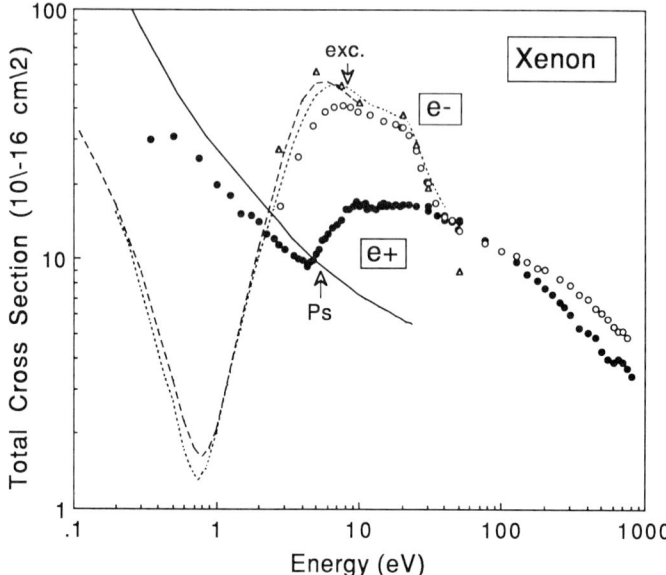

FIG. 9. Comparison of cross sections for $e^{+,-}$–Xe scattering. The Q_t measurements of Dababneh et al. (1980, 1982) for positrons (●) and electrons (○) are shown, along with the e^- measurements of Jost et al. (1983) below 40 eV (·····). Theoretical results for Q_{el} below 50 eV are the relativistic (------) and non-relativistic (△) calculations of McEachran and Stauffer (1984, 1987) for electrons, and the non-relativistic calculations of McEachran et al. (1980) for positrons (———). The thresholds for excitation (exc.) and positronium (Ps) formation are indicated by arrows.

by alkali atoms is that since their ionization thresholds are less than the 6.8 eV binding energy of Ps in its ground state, a positron with arbitrarily small kinetic energy can form Ps, in contrast with positron scattering experiments performed with the inert gases, as well as molecules, which have well-defined Ps formation thresholds. Another distinguishing feature of alkalis is that their polarizabilities are much larger than those of any other room-temperature gases that have been used in positron scattering experiments.

Comparisons of the $e^{+,-}$ Q_t measurements of Kwan et al. (1989) for Na and Stein et al. (1987) for K with determinations of Q_t from a combination of theory and experiments, as well as many of the partial cross sections, are shown in Figs. 10 and 11. The most striking feature of the Q_t measurements for both of these target atoms is the closeness of the $e^{+,-}$ Q_t values over the entire energy range that has been investigated. Furthermore, it is curious that the largest difference between the $e^{+,-}$ Q_t's is at the lowest energies where the e^+ values become larger than the corresponding e^- values, which is the

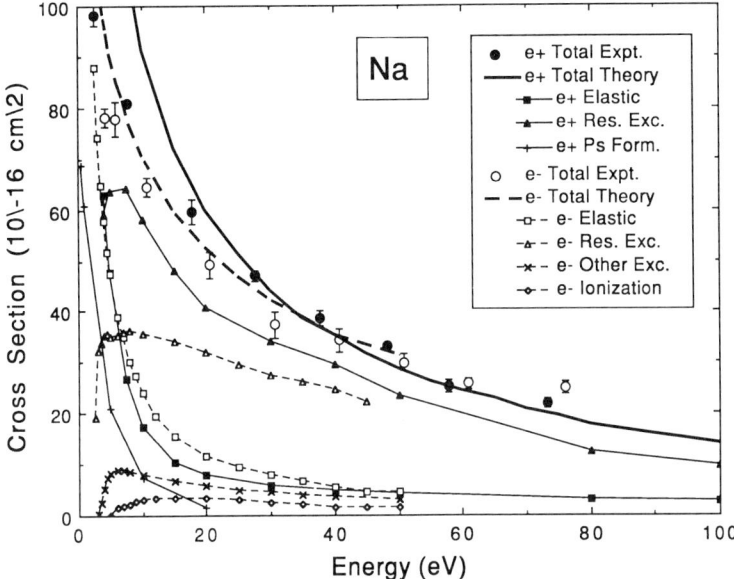

FIG. 10. Comparison of cross sections for $e^{+,-}$–Na scattering. The experimental $e^{+,-}$ Q_t values are those of Kwan et al. (1989). The e^+ cross section values for the total theory, the elastic, and resonance excitation are from the "4 state" close-coupling approximation calculations of Sarkar et al. (1988), while Q_{Ps} is from the distorted-wave approximation calculation of Guha and Mandal (1980). For e^- scattering the curve labelled "total theory" is a sum of the four separate partial cross sections that are displayed, where the elastic and ionization values are from the theoretical analysis of Walters (1976), and the resonance excitation (3s–3p) and "other excitation" (representing discrete excitations other than resonance excitation) values are from the measurements of Phelps and Lin (1981).

reverse of the normal situation (when no Ramsauer–Townsend minimum exists) observed for the inert gases.

In order to appreciate the $e^{+,-}$ Q_t comparisons it is helpful to consider the partial cross sections that contribute to Q_t. Following the lead of the theoretical analysis by Walters (1976) to deduce Q_t's for e^- scattering by Na and K, the results referred to as "total theory" for electrons are equal to the sum of the four partial cross sections shown, where Q_{el} and Q_{ion} were those selected from existing theoretical and experimental results by Walters, and the cross sections for resonance excitation (3s–3p for Na and 4s–4p for K) and discrete excitations other than resonance excitation are those measured by Phelps and Lin (1981) for Na and by Phelps et al. (1979) for K. The "total theory" results shown for e^+–Na scattering are those obtained by Sarkar et al. (1988) and consist of their close-coupling approximation calculations for Q_{el} and Q_{exc} (for 3s–3p, 3s–3d, and 3s–4p excitations), the Q_{Ps} values calculated

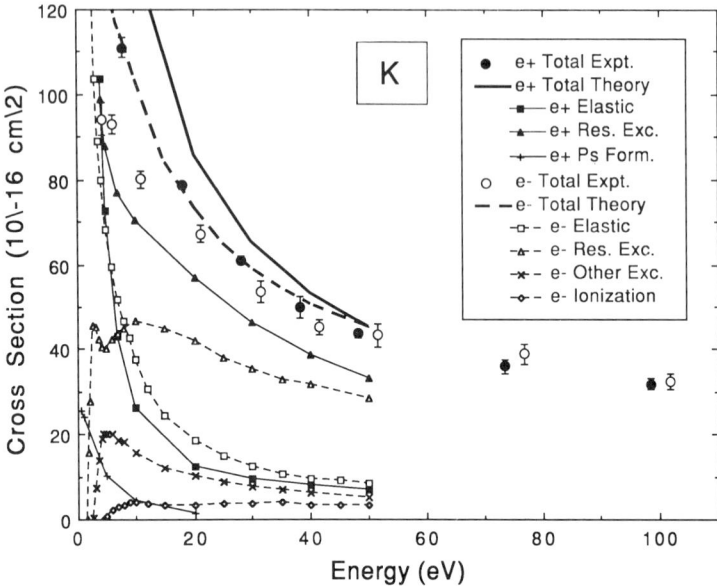

FIG. 11. Comparison of cross sections for $e^{+,-}$–K scattering. The experimental $e^{+,-}$ Q_t values are those reported by Stein et al. (1987). The e^+ cross section values for the theoretical total, the elastic, and resonance excitation are from the close-coupling approximation calculations of Ward et al. (1988), while Q_{Ps} is from the distorted-wave approximation calculation of Guha and Mandal (1980). For e^- scattering the curve labelled "total theory" is a sum of the four separate partial cross sections that are displayed, where the elastic and ionization values are from the theoretical analysis of Walters (1976), and the resonance excitation (4s–4p) and "other excitation" (representing discrete excitations other than resonance excitation) values are from the measurements of Phelps et al. (1979).

by Guha and Mandal (1980) using the distorted-wave approximation, and the first Born approximation results of Walters (1976) for Q_{ion}. For e^+–K scattering the "total theory" results are those calculated by Ward et al. (1988) using a close-coupling approximation, which included Q_{el} and Q_{exc} (for 4s-4p, 4s-5s, 4s-3d, and 4s-5p excitations), but does not include Ps formation and ionization, which would be expected to be rather small above 10 eV, as the Q_{Ps} values from the distorted-wave approximation calculation of Guha and Mandal (1980) indicate. The calculated partial cross sections for the other discrete excitations for e^+–Na,K scattering (discussed above, but not shown in Figs. 10 and 11) are close to two times larger than the e^- values below 20 eV and approach them above 20 eV.

The comparisons between the $e^{+,-}$ Q_t measurements and the "total theory" values are expected (Stein et al., 1987) to be affected appreciably at lower energies by the inability of the experiments to fully discriminate against projectiles which have elastically scattered at small angles in the forward direction, which the calculations predict to be a very pronounced effect for these alkali atoms. At energies above 20 eV, where inelastic scattering is expected to greatly dominate Q_t, the experimental measurements should be much closer to the actual Q_t's because the experiments can discriminate 100% against inelastic scattering (Stein et al., 1985). As a result, the experimental Q_t measurements of Kwan et al. (1989) and Stein et al. (1987) for $e^{+,-}$-Na,K scattering are in quite good agreement with the Q_t values expected from the "total theory" curves. Considering the various partial cross sections contributing to the $e^{+,-}$ Q_t's for Na and K shown in Figs. 10 and 11, it seems reasonable to expect that the e^+ Q_t curves may rise above the e^- curves at lower energies due to relatively small contributions of Q_{el} (which are predicted to be nearly the same for positrons and electrons) and the significant contributions of inelastic processes (which are predicted to have appreciably larger cross sections for positrons) to the overall Q_t's. In addition, Ps formation is only present for positron scattering.

A possible explanation for the observed near merging of the $e^{+,-}$ Q_t curves for K at low energies has been discussed by Stein et al. (1987), which should also apply to the Na case, if the explanation is correct. This explanation relates to a theoretical analysis by Dewangan (1980) of higher order Born amplitudes calculated in the closure approximation, which have been shown to imply (Walters, 1984; Byron et al., 1982) that if electron exchange can be ignored for electrons, and the closure approximation is valid, then a merging (or near merging) of $e^{+,-}$ Q_t's can occur at energies considerably lower than the asymptotic energies at which the first Born approximation is valid. In fact, this effect provides a possible explanation (Walters, 1984, 1988) for the observed "premature" merging of the Q_t curves for $e^{+,-}$-He scattering.

C. Hydrogen

From a theoretical point of view atomic hydrogen is one of the most interesting atoms to consider in $e^{+,-}$ scattering because of the relative simplicity of these collision systems. Experimentally, however, investigations of projectile scattering by hydrogen atoms are very difficult and up to the time of the writing of this paper there have been no reported experimental investigations of positron scattering by atomic hydrogen. For the purposes of this paper it is of definite interest to compare some of the theoretical results for $e^{+,-}$ scattering and experimental results for e^- scattering by H as shown in Fig. 12. The Q_t results displayed for e^- scattering from 0.136 to 400 eV were determined by de Heer et al. (1977), where they have deduced from theory and experiments what they consider to be the most reasonable values for the partial cross sections, Q_{el} (also shown in Fig. 12), Q_{exc}, and Q_{ion}, that contribute to Q_t. The calculations of Q_t and Q_{el} by Walters (1988) and van Wyngaarden and Walters (1986) (using a pseudostate close-coupling approximation that is supplemented by the second Born approximation) for 54.4–300 eV $e^{+,-}$–H scattering give e^- results that are in very good agreement with the determinations of these cross sections by de Heer et al (1977). These high energy calculations are also intriguing because they predict that the $e^{+,-}$ Q_t's are within 4% of each other in the 54.4–300 eV energy range while the e^- Q_{el}'s are more than a factor of 3 larger than for positrons at 54.4 eV. This is another example of a near merging of the $e^{+,-}$ Q_t's, while the cross sections for the separate accessible scattering channels

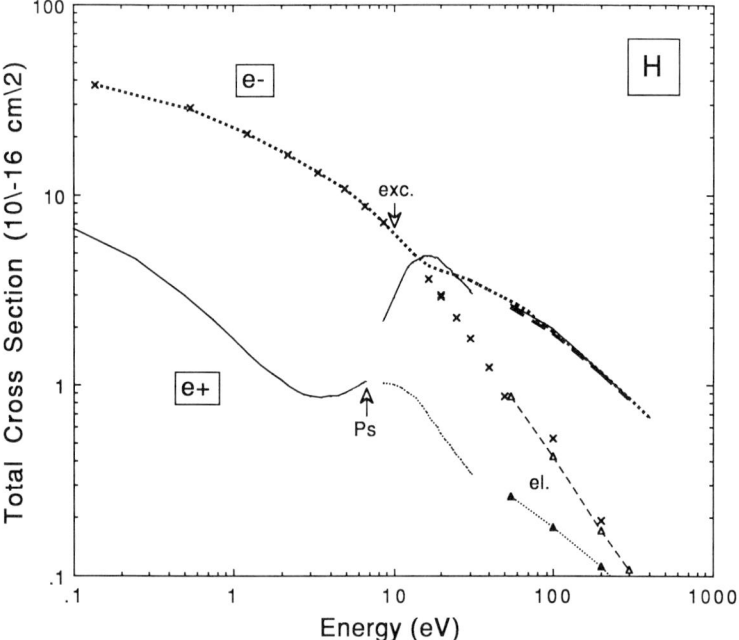

FIG. 12. Comparison of cross sections for $e^{+,-}$–H scattering. The electron Q_t (·····) and Q_{el} (×) determinations (based on a combination of experimental and theoretical results) of de Heer et al. (1977) extend from 0.1 to 400 eV. The calculations of Walters (1988) for positrons, and van Wyngaarden and Walters (1986) for electrons in the energy range 50–300 eV of elastic (el.) cross sections (▲·····▲ for positrons and △------△ for electrons) are shown clearly, while their e^+ (———) and e^- (heavy ------) Q_t's overlap with the de Heer et al. (1977) Q_t results. Theoretical results of Winick and Reinhardt (1978a, 1978b) for positron elastic cross sections from 0.1–6.8 eV (———) and 8.7–31 eV (·····) are displayed, along with positron Q_t's (———) between 8.7 and 31 eV. The threshold for excitation (exc.) and positronium (Ps) formation are indicated by arrows.

may still be significantly different, as has been observed for He, Na and K. To elaborate on this point, it is of interest to consider a comparison shown in Fig. 13 of some excitation (1s-2s and 1s-2p) cross sections with Q_t and Q_{el} for $e^{+,-}$-H scattering, as calculated by Walters (1988) and van Wyngaarden and Walters (1986). For both of the excitation processes the calculated e^+ cross sections are larger than for e^-'s, which is a necessary requirement for the total e^+ inelastic cross section to be larger than for electrons if the total cross sections are to be the same. This is additional support for the argument based on the closure approximation of Dewangan (1980) of why $e^{+,-}$ Q_t's can be equal at much lower energies than would be expected from the first Born approximation, while there are still significant differences in the elastic and inelastic scattering cross sections for positrons and electrons.

For e^+-H scattering at low (0.1-6.8 eV) and intermediate (8.7-31 eV) energy the results of the "moment T-matrix approach" calculations by Winick and Reinhardt (1978a, b) for Q_{el} and Q_t are shown in Fig. 12. Their Q_t values at intermediate energy are obtained for Q_{el} by application of the optical theorem (with the assumption of unitarity) and are considered to be a lower bound to the true Q_t. If their prediction for Q_t is correct it is interesting that the e^+ Q_t will exceed the e^- Q_t in the vicinity of 15-25 eV where e^+ scattering would be mostly inelastic and e^- scattering would be almost

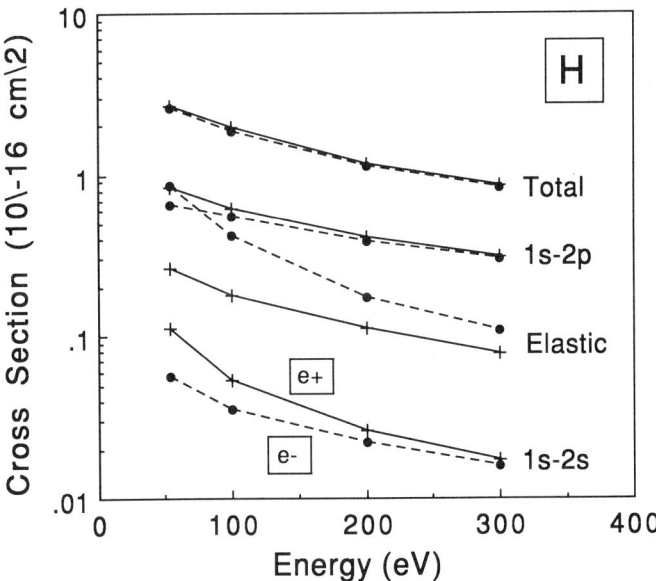

FIG. 13. Comparisons of Q_t, Q_{el}, and Q_{exc} (1s-2s and 1s-2p) for $e^{+,-}$-H scattering calculated by Walters (1988) for positrons (+ ——— +), and by van Wyngaarden and Walters (1986) for electrons (● ------ ●).

entirely elastic. Furthermore, the e^+ Q_t predictions of Winick and Reinhardt when compared with the intermediate energy e^- results suggest that a merging (or near merging) of the $e^{+,-}$ Q_t's may extend down to the vicinity of 30 eV. It is intriguing that the alkali metal atoms (Na and K), which also have a single valence e^- (and are hydrogen-like in this respect), exhibit a somewhat similar behavior in that the e^+ Q_t's are higher than the corresponding e^- Q_t's at energies lower than the region of observed merging of the $e^{+,-}$ Q_t's.

III. Positron and Electron Scattering by Molecules

A. DIATOMIC (H_2, N_2, CO, and O_2)

1. H_2

Being the simplest molecule and straightforward to study experimentally, molecular hydrogen has received considerable attention. In Fig. 14 is shown a number of cross sections that have been determined for $e^{+,-}$ scattering by H_2. The $e^{+,-}$ Q_t comparison measurements of Hoffman et al. (1982) show that the e^- Q_t is as much as 20 × larger at low energies with the e^+ Q_t increasing rapidly after the Ps formation threshold until it approaches and becomes slightly larger than the e^- Q_t from 30–100 eV. Above 100 eV the $e^{+,-}$ Q_t's are merged to within a few percent. The modified Glauber approximation calculations for the $e^{+,-}$ Q_t's above 100 eV of Jhanwar et al. (1982) are in very good agreement with the above measurements and support the observed merging. Subtracting the Q_{Ps} measurements of Fromme et al. (1988) from the e^+ Q_t measurements it is seen that the $Q_t - Q_{Ps}$ curve no longer becomes larger than the e^- Q_t curve, but they do still merge above 100 eV.

At low energies the variational R-matrix calculation of Nesbet et al. (1986) for e^- elastic scattering by H_2 is seen to agree quite well with the measurements of Jones (1985) and Hoffman et al. (1982). Meanwhile, the calculations of Morrison et al. (1984), using a model adiabatic polarization potential specifically for positrons (ADPOS), for the e^+ low energy Q_t (which includes elastic scattering plus $j_0 = 0 \rightarrow j_0 = 2$ rotational excitation) can be made to agree with the measurements of Hoffman et al. (1982) by a suitable choice of an adjustable "cutoff" parameter in the calculation. The work of Morrison et al. is of special interest to $e^{+,-}$ scattering by molecules (and atoms) because it shows that the polarization potentials for $e^{+,-}$'s, even though being attractive for both projectiles, differ due to sign dependent terms which can significantly affect the corresponding Q_t's.

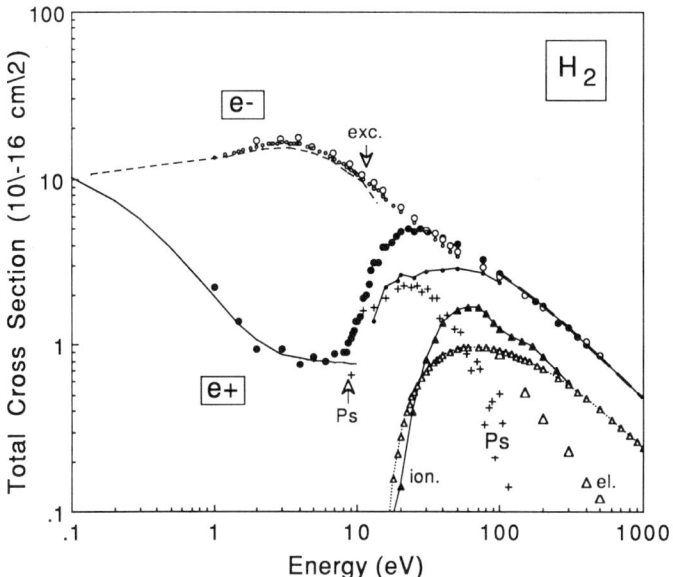

FIG. 14. Comparison of cross sections for $e^{+,-}$–H_2 scattering. The e^+ (●) and e^- (○) Q_t measurements of Hoffman et al. (1982) are shown, along with the e^- Q_T measurements (○) of Jones (1985) below 50 eV and the integrated elastic cross sections (△) above 100 eV determined by van Wingerden et al. (1977). Theoretical results shown below 20 eV are the e^- Q_{el} (------) values determined by Nesbet et al. (1986), and the "ADPOS" calculation of Morrison et al. (1984) of the e^+ total integrated cross section (———) (the sum of elastic scattering and rotational $j_0 = 0 \rightarrow j_0 = 2$ excitation). For energies above 100 eV the theoretical Q_t calculations of Jhanwar et al. (1982) for e^+ (———) and e^- (heavy ------) overlap each other and the measurements of Hoffman et al. (1982). The thresholds for electronic excitation (exc.) and positronium (Ps) formation are indicated by arrows. Inelastic cross section measurements that are shown were made by Fromme et al. (1988) for Ps formation (+) and e^+ impact ionization (▲———▲), and by Rapp and Englander-Golden (1965) for e^- impact ionization (△······△). The (●———●) curve between 13 and 100 eV represents $Q_t - Q_{Ps}$ for positrons.

Above the respective $e^{+,-}$ inelastic thresholds for Ps formation and electronic excitation it is interesting to consider the relative values of the cross sections for $e^{+,-}$ elastic and total inelastic scattering. The integrated elastic differential cross section measurements of van Wingerden et al. (1977) for e^-'s above 100 eV, if extrapolated to the electronic excitation threshold where it would meet the Q_{el} calculations of Nesbet et al., indicate an e^- total inelastic cross section that is much smaller than that for positrons (including Ps formation) at intermediate energies (if one assumes that the slope of the e^+ "elastic plus rotational and vibrational excitation" cross section does not change appreciably as it passes through the Ps formation threshold). The relative e^+ Q_{ion} measurements of Fromme et al. (1988) shown in Fig. 14 along with the e^- Q_{ion} measurements of Rapp and Englander-Golden (1965) (to which the results of Fromme et al. are normalized at the highest energies of overlap) provide additional evidence that cross sections for specific inelastic channels are larger for positrons than for electrons.

In the near-threshold energy region it can be seen in Fig. 14 that the e^+ Q_{ion} measurements become noticeably smaller than the corresponding e^- values. Fromme et al. (1988) have indicated that this may be related to differing near threshold behavior of the Q_{ion}'s for $e^{+,-}$ scattering, which could be in accord with the theoretical predictions by Wannier (1953) of a threshold law of the form

$$Q_{ion} \propto (E-E_{ion})^n, \tag{3}$$

where E_{ion} is the ionization threshold energy, and values of $n = 2.651$ for positrons and $n = 1.127$ for electrons have been predicted by Klar (1981, 1984). The measurements of Fromme et al. (1988) are unable to quantitatively determine the exponent n for positrons. It is noteworthy that similar threshold behavior has been observed by Fromme et al. (1986) for the Q_{ion}'s measured for $e^{+,-}$-He scattering (not shown in sufficient detail to be easily seen in Fig. 2).

An intriguing comparison is shown in Fig. 15 between the measured Q_t's for $e^{+,-}$-H_2 scattering (Hoffman et al., 1982) and the Q_t values for $e^{+,-}$-H scattering that were displayed in Fig. 12. It is seen that when allowance is made for the difference in the Ps threshold energies for H and H_2 that there are remarkable similarities between the respective e^+ and e^- Q_t curves. The shapes of the curves for each projectile are almost identical, except at the lowest energies for electrons. For both targets the $e^{+,-}$ Q_t curves approach each other rapidly after the Ps formation thresholds, appear to temporarily cross (where the e^+ curve is slightly higher) and then are merged at higher energies. Above 30 eV the H_2 Q_t curves average about 40–50% higher than the H Q_t curves, which in a simplistic view seems quite reasonable since one would expect a total cross section for H_2 that is between one and two times that for H.

2. N_2 and CO

It is appropriate to consider N_2 and CO together because they are isoelectronic and in many respects behave similarly in $e^{+,-}$ scattering studies as can be seen in Figs. 16 and 17 from comparisons of their respective measured Q_t's (Kennerly, 1980; Hoffman et al., 1982; Kwan et al., 1983; Buckman and Lohmann, 1986b), which have very similar shapes and magnitudes for each projectile. Both gases exhibit $^2\Pi$ shape resonances for low energy (1–5 eV) e^- scattering significant increases in the e^+ Q_t's above

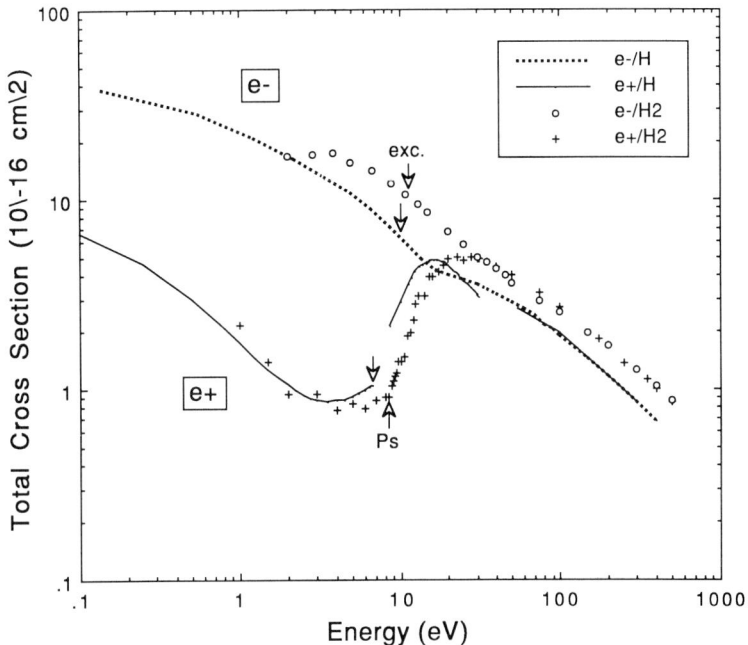

FIG. 15. Comparisons of Q_t's for $e^{+,-}$ scattering by H and H_2. The $e^{+,-}$-H_2 results were measured by Hoffman et al. (1982), the e^--H results were determined by de Heer et al. (1977), and the e^+-H results below 7 eV and from 9 to 31 eV were calculated by Winick and Reinhardt (1978a, 1978b), and from 50 to 300 eV were calculated by Walters (1988).

the Ps formation thresholds, and a tendency towards merging of the $e^{+,-}$ Q_t's at the higher energies.

At lower energies it is seen that the "boomerang model" calculation of Dube and Herzenberg (1979) for electrons is quite good at reproducing the observed vibrational structure of the temporary N_2^- ion that is responsible for the shape resonance, although it does overestimate the magnitude of Q_t below the resonance. For low energy e^+-N_2 scattering the fixed-nuclei approximation calculation of Darewych (1982) for the integrated Q_{el} is in good agreement with the measurements of Hoffman et al. (1982). For low energy e^- scattering by CO, a calculation by Jain and Norcross (1985) is quite good at predicting the position and width of the shape resonance, but overestimates the magnitude of Q_t, while the results of an adiabatic-nuclei approximation calculation by Jain (1986a) for the low energy e^+ Q_t (vibrationally elastic but summed over all possible final rotational states) are lower than the measurements of Hoffman et al. (1982) at the lower energies.

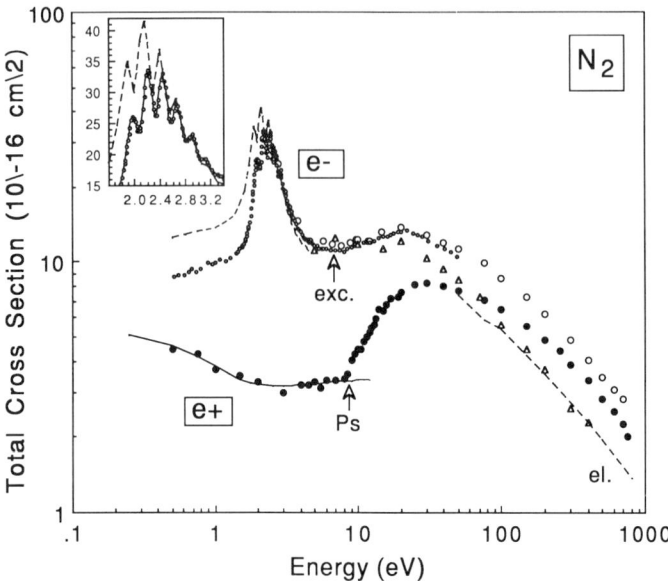

FIG. 16. Comparison of cross sections for $e^{+,-}$-N_2 scattering. The e^+ (●) and e^- (○) Q_t measurements of Hoffman et al. (1982) are shown along with the e^- Q_t measurements (○) of Kennerly (1980) up to 50 eV and the integrated elastic cross sections (△) determined by Shyn and Carignan (1980). Theoretical results shown below 20 eV are the e^- Q_t's (------) determined by Dube and Herzenberg (1979), and a calculation by Darewych (1982) of the e^+ integrated elastic cross section (———). Above 40 eV the theoretical results for integrated e^- elastic cross sections (-------) by Jain (1982) are shown. The thresholds for electronic excitation (exc.) and positronium (Ps) formation are indicated by arrows. The inset shows more detail in the region of the $^2\Pi_g$ shape resonance where the Kennerly results are joined by a solid line.

These two target gases appear to be consistent with most of the room-temperature gases already discussed in regard to comparisons of the $e^{+,-}$ cross sections for elastic and inelastic scattering. At low energies Q_{el} is larger for electrons than positrons. Extrapolations of the shapes of the low energy $e^{+,-}$ measurements of Q_t (where is consists almost entirely of elastic scattering) smoothly through the Ps formation and electronic excitation thresholds (a definite assumption), coupled with the e^- integrated elastic scattering cross sections measured by Shyn and Carignan (1980) for N_2 and calculated by Jain (1982) for N_2 and Jain et al. (1984) for CO, suggest that the intermediate energy Q_t's are dominated by inelastic scattering for positrons and dominated by elastic scattering for electrons with the e^- Q_{el} being larger than for positrons and the total inelastic scattering cross sections being larger for positrons.

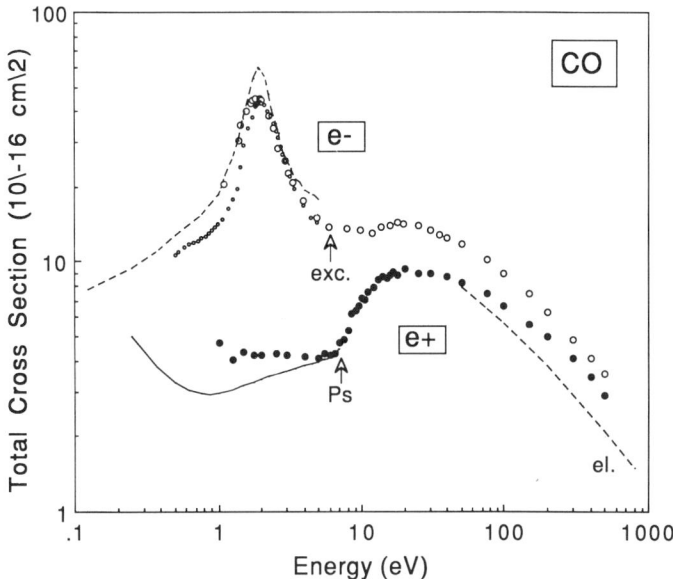

FIG. 17. Comparisons of cross sections for $e^{+,-}$–CO scattering. The e^+ (●) and e^- (○) Q_t measurements of Kwan et al. (1983) are shown along with the 0.5–5 eV e^- Q_t measurements (∘) of Buckman and Lohmann (1986b). Theoretical results shown below 10 eV are the e^- integrated cross sections (------) determined by Jain and Norcross (1985), and a calculation by Jain (1986a) of the e^+ total (includes rotational but not vibrational excitation) cross section (———). Above 50 eV the theoretical results of Jain et al. (1984) for integrated e^- elastic cross sections (------) are shown. The thresholds for electronic excitation (exc.) and positronium (Ps) formation are indicated by arrows.

3. O_2

Of all the room-temperature gases that have been studied experimentally with positrons and electrons, the measured Q_t's for $e^{+,-}$–O_2 scattering (Dababneh et al., 1988; Zecca et al., 1986) shown in Fig. 18 are the most similar in shape between 1 and 500 eV (having broad intermediate energy maxima with Q_t's that gradually approach each other as the projectile energy increases) and have among the least overall structure between them. The determination of Q_{el} (from different cross section measurements) for electrons by Shyn and Sharp (1982) indicates that below 10 eV the e^- Q_t is primarily elastic scattering. It is of interest that e^+ Q_t measurements reported by Katayama et al. (1987), which agree very well with Dababneh et al. (1988) above 1 eV, indicate an increase in Q_t to 2.6×10^{-16} cm^2 at 0.7 eV, thereby suggesting the possible existence of a Ramsauer-Townsend minimum. The lack of any dramatic increase in the e^+ Q_t after the Ps formation threshold

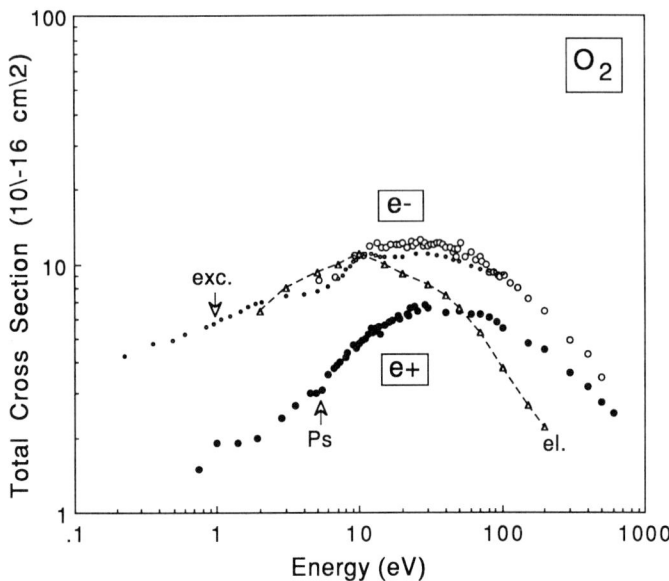

FIG. 18. Comparison of cross sections for $e^{+,-}$-O_2 scattering. The e^+ (●) and e^- (○) Q_t measurements of Dababneh et al. (1988) are shown along with the 0.2–100 eV "Trento" measurements of e^- Q_t's (○) by Zecca et al. (1986). Q_{el}'s (△ ------ △) obtained from differential cross section measurements by Shyn and sharp (1982) are also displayed. The thresholds for electronic excitation (exc.) and positronium (Ps) formation are indicated by arrows.

(although there may be a small increase) may be consistent with the Q_{Ps} measurements reported by Griffith (1983), where the measured Q_{Ps} increases from zero at the Ps formation threshold to a maximum of about 1×10^{-16} cm^2 in the vicinity of 8 eV, although it should be realized that Q_{Ps} measurements from this laboratory for other gases have been found to be lower and of different shape than measurements made elsewhere (Fromme et al., 1986, 1988). Katayama et al. (1987) have made some estimates of inelastic cross sections (based on measured time-of-flight energy loss spectra) for e^+-O_2 scattering and found that for excitation of the Schumann-Runge continuum (7.1–9.7 eV) the estimated cross sections have an energy-dependent shape and magnitudes (reaching a maximum of about 1.2×10^{-16} cm^2 near 12 eV) similar to those reported by Griffith for Q_{Ps}, while their "minimum estimate" measurements of Q_{ion} (increase from zero at the ionization threshold of 12.1 eV to about 1.2×10^{-16} cm^2 at 30 eV) are somewhat higher than the e^- Q_{ion} measurements of Rapp and Englander-Golden (1965).

B. TRIATOMIC (H_2O, CO_2, AND N_2O)

1. H_2O

The Q_t's that have been measured by Sueoka et al. (1986) for $e^{+,-}$ and by Zecca et al. (1987) for e^- scattering by H_2O are shown in Fig. 19, along with the Q_{el} measurements of Katase et al. (1986) and calculations of Q_{el} by Jain et al. (1988). The comparison $e^{+,-}$ Q_t curves are very interesting because of their relative lack of structure, their quite similar shapes, and their closeness to each other at all energies, appearing to be nearly merged at the highest and also the lowest energies of comparison with a maximum separation being a factor of two near 10 eV. It is curious that the lack of structure in the $e^{+,-}$ Q_t curves for H_2O (e.g., no appreciable change at the Ps formation threshold for positrons and no shape resonances for electrons) is similar to the situation for O_2. The observed merging of the high energy $e^{+,-}$ Q_t's for H_2O is curious because it contains two H atoms and mergings of the $e^{+,-}$ Q_t's for H_2 have been observed and for H have been suggested (see Secs. II,A. and III,A,1.).

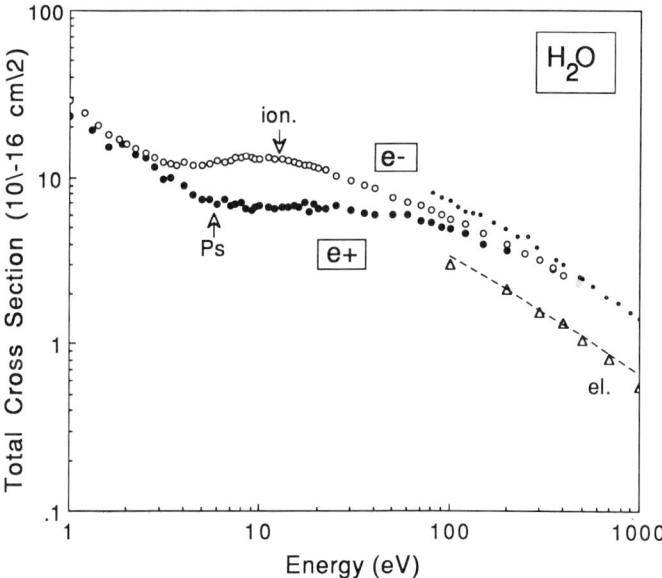

FIG. 19. Comparison of cross sections for $e^{+,-}$-H_2O scattering. The e^+ (●) and e^- (○) Q_t measurements of Sueoka et al. (1986) are shown along with the 80–1000 eV e^- Q_t measurements (○) of Zecca et al. (1987). Integrated elastic cross sections (△) obtained by Katase et al. (1986) from their elastic differential cross section measurements are also displayed along with a calculation of the e^- elastic cross section (------) by Jain et al. (1988). The thresholds for ionization (ion.) and positronium (Ps) formation are indicated by arrows.

The closeness of the $e^{+,-}$ Q_t curves is also intriguing to consider because the only other target gases for which the comparison curves are as close or closer to each other (at all energies of overlap) are the alkali atoms Na and K, which have very large polarizabilities and also large contributions of inelastic scattering to Q_t. Meanwhile, H_2O does have a permanent dipole moment, but the e^- Q_t results do not exhibit any strong evidence of appreciable inelastic scattering since the e^- Q_{el} results shown in Fig. 19 account for almost one-half of the higher energy Q_t measurements of Zecca et al. (1987).

2. CO_2 and N_2O

Just as in the case of $e^{+,-}$ scattering by N_2 and CO, the measured Q_t's for $e^{+,-}$ scattering by CO_2 and N_2O (Hoffman et al., 1982; Kwan et al., 1983, 1984; Szmytkowski et al., 1987) are remarkably similar in shapes and magnitudes for each given projectile scattering from each of these two molecules, as shown in Figs. 20 and 21. Both of these molecules exhibit

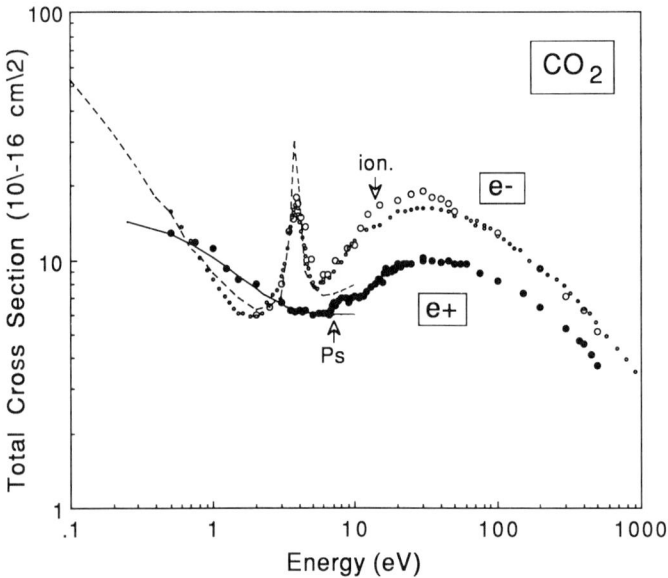

FIG. 20. Comparison of cross sections for $e^{+,-}$-CO_2 scattering. The e^+ (●) and e^- (○) Q_t measurements of Hoffman et al. (1982) below 60 eV and Kwan et al. (1983) above 30 eV are shown along with the e^- Q_t measurements (∘) of Szmytkowski et al. (1987). Theoretical results shown below 10 eV are the "SEP" model calculations by Morrison et al. (1977) of the e^- total integrated cross section (------), and a calculation by Horbatsch and Darewych (1983) of the e^+ integrated elastic cross section (———). The thresholds for ionization (ion.) and positronium (Ps) formation are indicated by arrows.

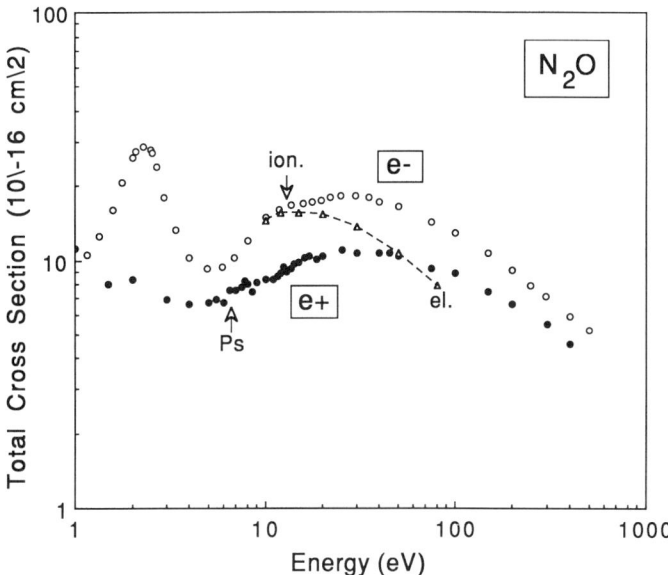

FIG. 21. Comparison of cross sections for $e^{+,-}$-N_2O scattering. The e^+ (●) and e^- (○) Q_t measurements of Kwan et al. (1984) are shown along with the integrated elastic (el.) cross sections (△ ------ △) obtained by Marinkovic et al. (1986) from their elastic differential cross section measurements. The thresholds for ionization (ion.) and positronium (Ps) formation are indicated by arrows.

prominent shape resonances at low energies for electrons (of $^2\Pi_u$ symmetry at 3.8 eV for CO_2 and of $^2\Sigma^+$ symmetry at 2.3 eV for N_2O), but only small increases in Q_t at the Ps formation thresholds for positrons. At low energies for CO_2, the fixed-nuclei approximation calculation by Morrison et al. (1977) of the e^- total integrated cross section, shown in Fig. 20, does quite well at predicting the location and width of the shape resonance, and also the magnitude of Q_t below the resonance. The fixed-nuclei approximation calculation by Horbatsch and Darewych (1983) of the integrated Q_{el} for e^+-CO_2 scattering agrees quite well with the measurements of Hoffman et al. (1982).

It is interesting that for both gases the $e^{+,-}$ Q_t's are quite close to each other just above the low energy e^- shape resonances, then separate with the e^- curve becoming about 2 × larger in the vicinity of 30 eV (where both the $e^{+,-}$ curves have broad maxima), and then become closer to each other at the highest measured energies. The Q_{el} results for e^--N_2O scattering obtained by Marinkovic et al. (1986) (by integrating their elastic DCS measurements) merge with the Q_t measurements around 10 eV suggesting that the broad maximum for electrons is to a great extent associated with the elastic scattering channel and not inelastic processes, as is generally the case for positron maxima at intermediate energies.

An interesting comparison of measured Q_t's for $e^{+,-}$ scattering by the isoelectronic pairs of molecules N_2 and CO, and N_2O and CO_2 (Hoffman et al., 1982; Kwan et al., 1983, 1984) is shown in Fig. 22, where the remarkable similarities in the shapes and the magnitudes of the Q_t's for each pair of isoelectronic molecules are displayed. It is intriguing, however, to focus on the small differences in the Q_t curves because both the polar molecules, CO and N_2O, have noticeably larger Q_t's for positron scattering than their non-polar pairs, while for electrons the Q_t's for each pair are much closer to each other than for positrons except at the low energy shape resonances, which are observed to be more pronounced (height and width) also for the polar molecules. For this discussion it is pertinent to realize that the N_2 and CO pair become the N_2O and CO_2 pair by merely adding an oxygen atom to each diatomic molecule, and that the polar CO becomes the non-polar CO_2 and vice versa for N_2 and N_2O. It is also curious that the apparent role of Ps formation, based on the behavior of Q_t after the respective Ps formation thresholds, is greatly diminished for N_2O and CO_2 (as compared to N_2 and

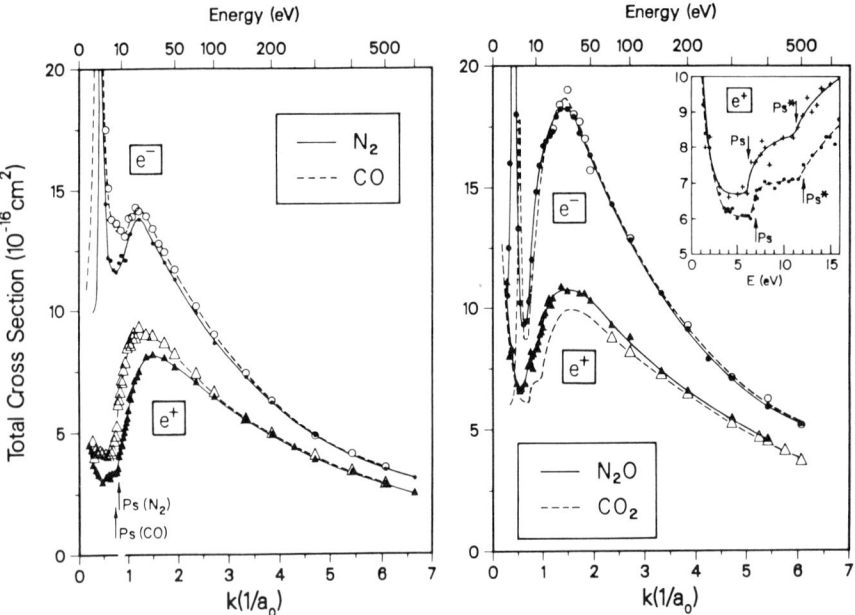

FIG. 22. Comparison of Q_t's measured for $e^{+,-}$ scattering by N_2 and CO and by N_2O and CO_2 (Hoffman et al., 1982; Kwan et al., 1983, 1984). The threshold energies are indicated by arrows for formation of Ps in the ground state (Ps) and the first excited state (Ps*). The curves have been drawn to guide the eye through the Q_t measurements. (From Kwan et al., 1984.)

CO) and more similar to the case for O_2 and H_2O, which also contain oxygen atoms. Kwan et al. (1984) speculated that an additional possible increase in the e^+ Q_t's at an energy of about 5.1 eV above the Ps formation threshold (at Ps* in the inset in Fig. 22) might correspond to formation of Ps in the first excited state. However, Laricchia et al. (1988) (using an ultraviolet-sensitive photomultiplier plus a gamma-ray detector coincidence method) have concluded that these latter features are due to ground-state Ps formation and simultaneous excitation of the resulting molecular ion. The relatively large increases in the Q_t's for electrons at the intermediate energy maxima for N_2O (mentioned above) and CO_2, and the smaller maxima for N_2 and CO are also interesting because they may relate to broad (width > 5 eV) intermediate energy shape resonances, which have been theoretically identified and discussed by Dehmer, Dill, and coworkers (Dill et al., 1979; Dehmer and Dill, 1980; Dehmer et al., 1980). These intermediate energy shape resonances, located between 15–35 eV for N_2 and 10–40 eV for CO_2 (which coincide with the energy regions of the Q_t maxima), have been found to be quite prominent and to enhance the electron impact vibrational excitation cross sections for these molecules, while they tend to be more difficult to detect in measurements of Q_{el} and Q_t.

C. OTHER POLYATOMIC (NH_3, CH_4, HYDROCARBONS, SiH_4, CF_4 AND SF_6)

1. NH_3

The measurements of Q_t's for $e^{+,-}$ scattering by NH_3 (Sueoka et al., 1987) are shown in Fig. 23. In many respects these comparison curves are quite similar to those for H_2O (see Fig. 19) in that they have the same general shapes, they are nearly within a factor of two in magnitude at all energies, and they tend to merge at the highest energies. The biggest difference between the Q_t curves of these molecules is the more pronounced maximum around 10 eV for $e^- - NH_3$ scattering.

2. CH_4

In many respects the measured Q_t's for $e^{+,-}$ scattering by CH_4 (Dababneh et al., 1988; Lohmann and Buckman, 1986), shown in Fig. 24, exhibit many shape similarities to the Q_t curves for argon (see Fig. 5), such as a Ramsauer-Townsend minimum below 1 eV and a Q_t maximum around 10 eV for electrons, an appreciable increase in the e^+ Q_t curve at the Ps formation threshold, and a tendency for the $e^{+,-}$ Q_t curves to become close to each

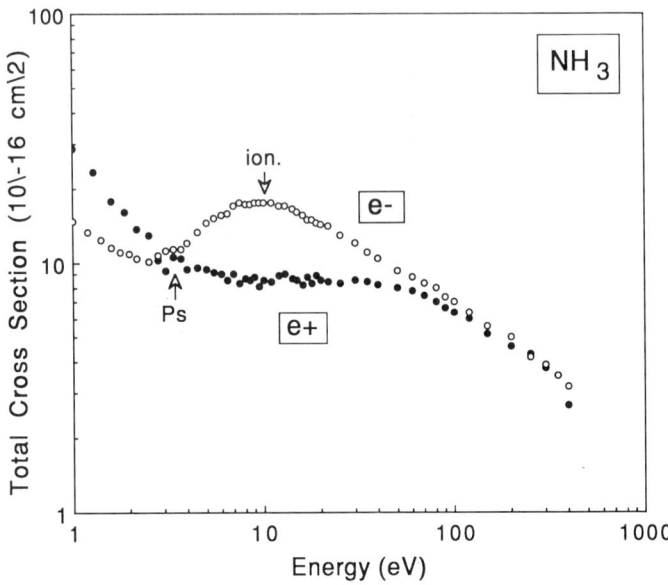

FIG. 23. Comparison of Q_t measurements of Sueoka et al. (1987) for $e^{+,-}$–NH_3 scattering. The thresholds for ionization (ion.) and positronium (Ps) formation are indicated by arrows.

other (but not merge) above 100 eV. The results of calculations by Jain (1986b, 1987) for Q_{el} and Q_t for $e^{+,-}$–CH_4 scattering, displayed in Fig. 24, are in reasonably good agreement with the corresponding Q_t measurements and predict that at higher energies inelastic scattering dominates the e^+ Q_t while elastic scattering plays a more important role for electrons.

3. Hydrocarbons

The first survey of $e^{+,-}$ scattering by hydrocarbon molecules, other than methane, consisted of the Q_t measurements of Floeder et al. (1985), which are shown in Fig. 25. These measurements have found that the Q_t's for each isocarbonic molecule (with 2, 3, and 4 carbon atoms, respectively) are quite similar in magnitude and shape for each given projectile, that the Q_t's at any given higher projectile energy increase with the number of carbon atoms in the molecule, that the $e^{+,-}$ Q_t's for each target molecule (excluding CH_4) are within a factor of two of each other at these energies, and that the $e^{+,-}$ Q_t's for each molecule are nearly merged (within 15%) at 400 eV. It was also observed by Floeder et al. that comparison of the $e^{+,-}$ Q_t's for the alkane series (CH_4, C_2H_6, C_3H_8, and nC_4H_{10}) and for the alkene series (C_2H_4, C_3H_6, and C_4H_8) also revealed additional similarities in the detailed shapes of the Q_t curves within each series.

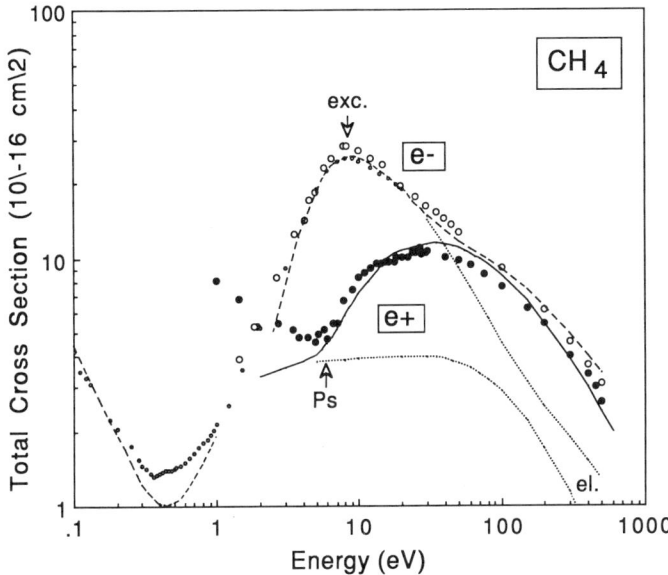

FIG. 24. Comparison of cross sections for $e^{+,-}$–CH_4 scattering. The e^+ (●) and e^- (○) Q_t measurements of Dababneh et al. (1988) are shown along with the e^- Q_t measurements (∘) of Lohmann and Buckman (1986) below 20 eV. Theoretical Q_t results shown are those obtained by Jain (1987) for positrons (———) from 2 to 600 eV, and by Jain (1986b) for electrons (------) for the energy ranges of 0.1–1.0 eV ("SEPGT" calculation), 2.5–30 eV ("SEAPJT" calculation), and 30–500 eV ("SEAPJTal" calculation). Calculations of Q_{el}'s at higher energies for electrons by Jain (1986b) ("SEAPJTal" calculation) and for positrons by Jain (1987) are represented by the upper and lower (....) curves, respectively. The thresholds for electronic excitation (exc.) and positronium (Ps) formation are indicated by arrows.

Measurements of Q_t's for 0.7–400 eV $e^{+,-}$ scattering by CH_4, C_2H_6, C_2H_4, and C_6H_6 by Sueoka and Mori (1986) and Sueoka (1988), displayed in Fig. 26, indicate that the $e^{+,-}$ curves for each molecule approach to within 10% at the highest energies. A comparison of the $e^{+,-}$ Q_t curves for the alkanes CH_4 and C_2H_6 shows that they have nearly identical shapes for each separate projectile and that the magnitudes of the Q_t's for C_2H_6 are everywhere in this energy range close to 2 × larger than for CH_4. The remarkable similarities of these curves would seem to suggest that a Ramsauer–Townsend minimum may also exist for e^- scattering by C_2H_6, as is the case for CH_4.

The measured Q_t's for $e^{+,-}$ scattering by C_2H_4 and C_6H_6 in Fig. 26 show some interesting features at low energies as pointed out by Sueoka and Mori (1986) and Sueoka (1988). Ethylene (C_2H_4) exhibits a pronounced shape resonance near 2 eV for electrons, while the small increase around 1.5 eV for benzene (C_6H_6) may be a shape resonance first observed by Sanche and

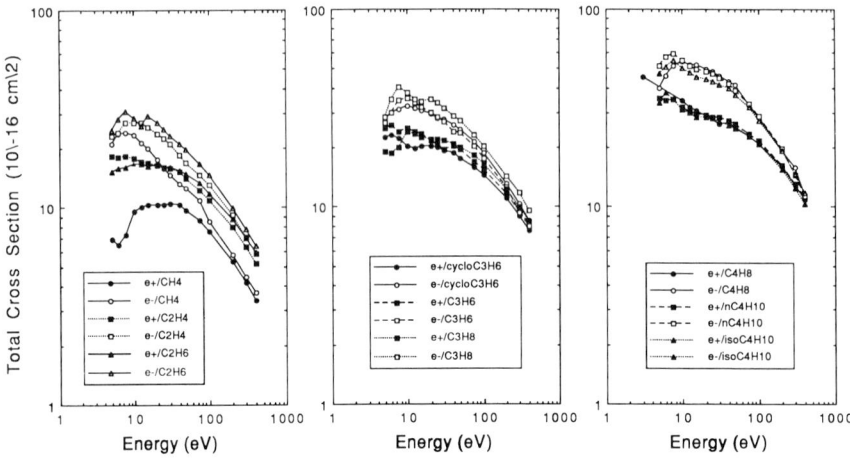

FIG. 25. Comparison of Q_t measurements of Floeder et al. (1985) for $e^{+,-}$–CH_4, C_2H_4, C_2H_6, C_3H_6, cycloC_3H_6, C_3H_8, C_4H_8, isoC_4H_{10}, and nC_4H_{10} scattering.

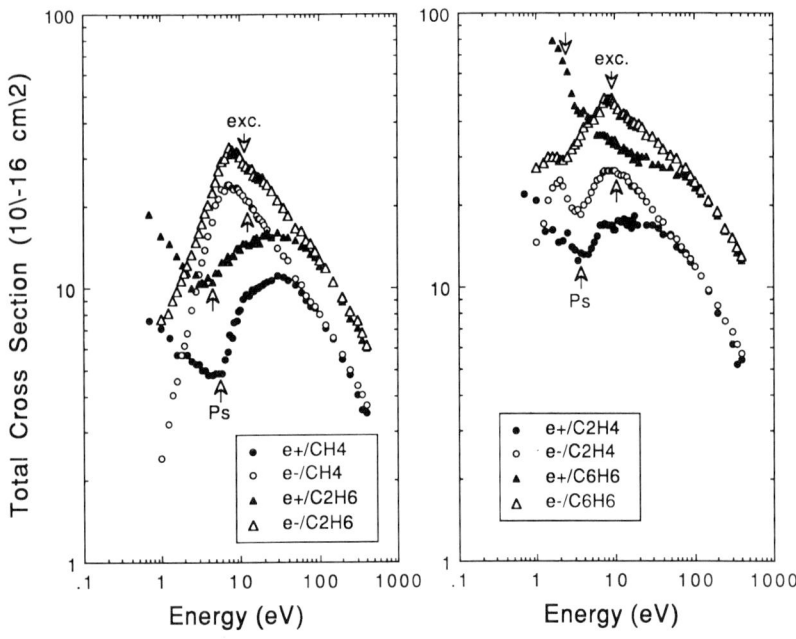

FIG. 26. Comparison of Q_t measurements Sueoka and Mori (1986), and Sueoka (1988) for $e^{+,-}$–CH_4, C_2H_6, C_2H_4, and C_6H_6 scattering. The thresholds for electronic excitation (exc.) and positronium (Ps) formation are indicated by arrows for the e^- and e^+ curves, respectively.

Schulz (1973) in an electron transmission experiment. The e^+-C_6H_6 Q_t curve is somewhat unique because of its significant increase at the lowest energies (becoming $156 \pm 59 \times 10^{-16}$ cm^2 at 0.7 eV) and because it continues decreasing through the Ps formation threshold, suggesting that some scattering process is having a very significant effect on the e^+ Q_t at very low energies. The only other gas targets that have exhibited e^+ Q_t's anywhere near this large are the alkali atoms (see Figs. 10 and 11), which are predicted to have appreciable inelastic scattering that is possibly related to their large polarizabilities.

4. SiH_4

The Q_t measurements for $e^{+,-}$-SH_4 scattering made by Mori et al. (1985) and Sueoka (1987) are shown in Fig. 27 along with the calculations of Q_{el} and Q_t by Jain (1986c) and Jain and Thompson (1987). The general shapes of the $e^{+,-}$ Q_t curves are very similar to those for CH_4 with there being a tendency

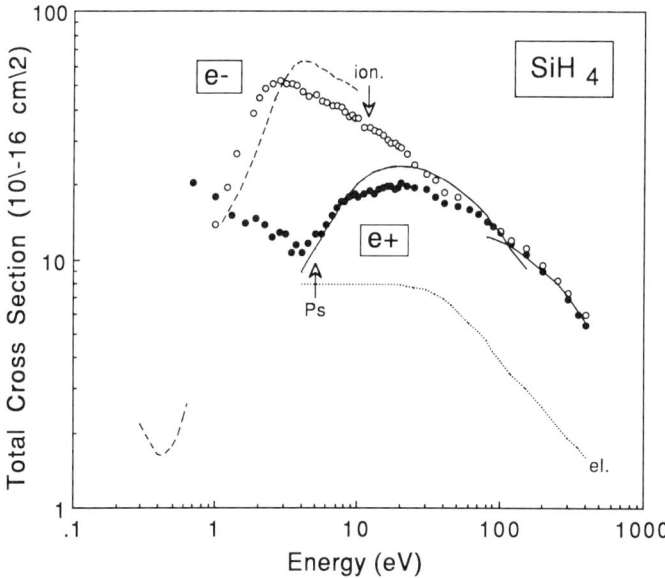

FIG. 27. Comparison of cross sections for $e^{+,-}$-SiH_4 scattering. The e^+ (●) and e^- (○) Q_t measurements of Mori et al. (1985) and Sueoka (1987) are shown along with the e^+ calculations by Jain (1986c) for Q_t (———) in the energy ranges for 4–150 eV (using a mean excitation energy parameter $\Delta = 3.0$ eV) and 80–400 eV (using $\Delta = 21.1$ eV), and for Q_{el} (·····) extending up to 400 eV. Calculations by Jain and Thompson (1987) for e^- Q_{el} (------) are shown for the energy ranges of 0.3–0.65 eV (GT model) and 1–10 eV (JT model). The thresholds for ionization (ion.) and positronium (Ps) formation are indicated by arrows.

towards merging at higher energies, a noticeable increase in the e^+ Q_t after the Ps formation threshold, and a predicted Ramsauer-Townsend minimum for electrons that appears to be consistent with an extrapolation of the lowest energy e^- measurements.

5. CF_4 and SF_6

Low energy measurements of Q_t's for $e^{+,-}$–CF_4 scattering made by Mori et al. (1985) are shown in Fig. 28 where it is seen that there appears to be a rather prominent shape resonance for e^- scattering and a change in the slope of the e^+ Q_t curve at the Ps formation threshold. At the lowest energy both the $e^{+,-}$ Q_t's are decreasing, which is a behavior that is unique to CF_4 and possibly O_2 for the gases that have been studied in $e^{+,-}$ comparison measurements.

The most massive molecule that has been investigated in $e^{+,-}$ scattering is SF_6 for which Q_t's have been measured by Dababneh et al. (1988). These measurements along with the e^- Q_t measurements of Kennerly et al. (1979) and the multiple-scattering model calculations for Q_{el} of Dehmer et al. (1978) are shown in Fig. 29. The e^- Q_t curve for SF_6 is of special interest due to the existence of several shape resonances, which have been identified by Dehmer et al. as being associated with the following symmetries (and predicted energy

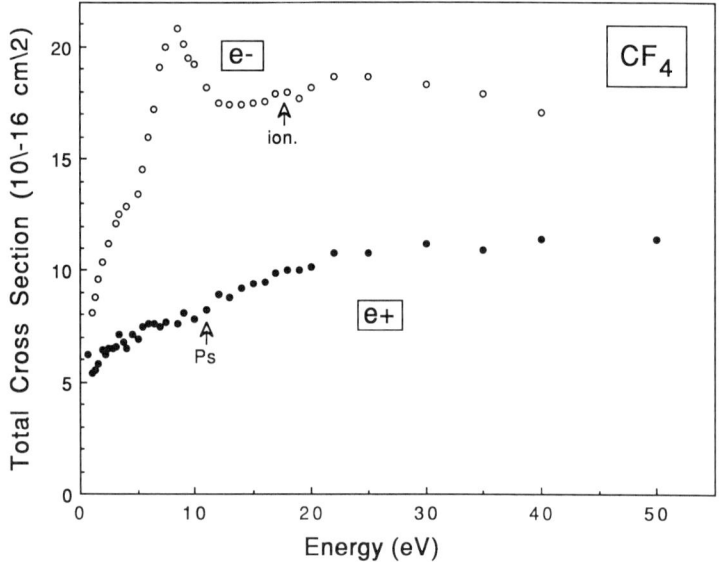

FIG. 28. Q_t measurements by Mori et al. (1985) for $e^{+,-}$ scattering by CF_4.

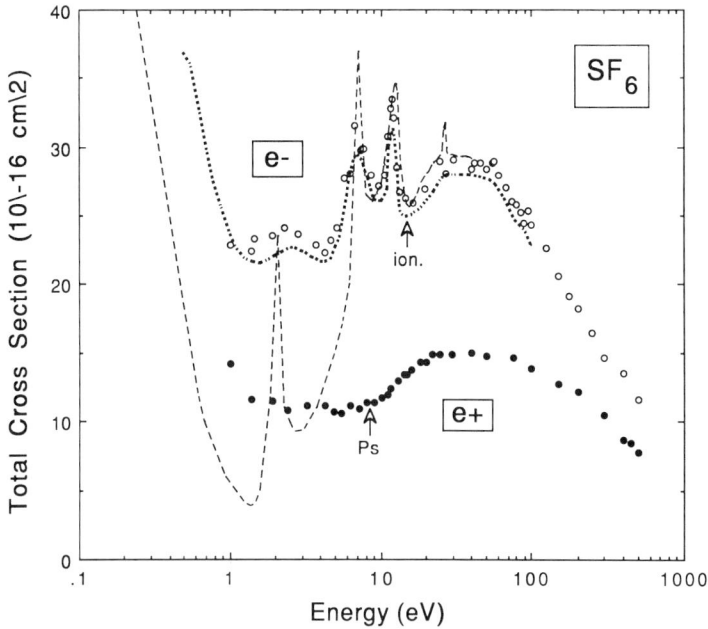

FIG. 29. Comparison of cross sections for $e^{+,-}$–SF_6 scattering. The e^+ (●) and e^- (○) Q_t measurements of Dababneh et al. (1988) are shown along with the e^- Q_t measurements (·····) of Kennerly et al. (1979). The theoretical results for e^- Q_{el} (------) were obtained by Dehmer et al. (1978). The thresholds for ionization (ion.) and positronium (Ps) formation are indicated by arrows.

locations): a_{1g} (2.1 eV), t_{1u} (7.2 eV), and t_{2g} (12.7 eV). An additional resonance predicted to occur at 27.0 eV has not been observed in the Q_t measurements. The e^+ Q_t curve exhibits a slow increase after the Ps formation threshold, which may be partially associated with Ps formation. At the highest energies the $e^{+,-}$ Q_t curves are gradually tending to approach each other but at 500 eV the e^- Q_t is still nearly 50% larger than the e^+ value.

IV. Summary and Concluding Remarks

Many of the general features of the Q_t curves for $e^{+,-}$ scattering by atoms and molecules are summarized in Tables III and IV. Most of these features have already been discussed and only a few additional comments will be made here. The ratio $R_t = Q_t^-/Q_t^+$ just below the Ps formation threshold energy shows that the largest differences between the $e^{+,-}$ elastic cross

TABLE III

SUMMARY OF $e^{+,-}$–ATOM Q_t MEASUREMENTS[a]

Atom	Fig. #	R-T effect		E_{Ps} (eV)	E_{exc} (eV)	$R_t(<E_{Ps})$	$E_{max}(Q_t)$		Degree of merging (E)
		e^+ (eV)	e^- (eV)				e^+ (eV)	e^- (eV)	
He	2	2	no	17.8	19.8	17	50	1	<2% (≥200 eV)
Ne	4	0.6	no	14.8	16.6	3.8	75	30	20% (700 eV)
Ar	5	2?	0.35	9.0	11.5	6	50	14	15% (800 eV)
Kr	8	no	0.7	7.2	9.9	3.5	30	12	21% (750 eV)
Xe	9	no	0.8	5.3	8.3	3.4	10–25	8	33% (750 eV)
Na	10	no	no	0	2.1	0.9 @10 eV	no	no	<10% (>10 eV)
K	11	no	no	0	1.6	0.75 @10 eV	no	no	<10% (>20 eV)
H	12	3*	no	6.8	10.2	8*	15*	no	<2% (≥200 eV)*

[a] The information in this table is obtained from the figures listed. "R-T Effect" refers to the approximate energies (in eV) of a Ramsauer-Townsend minimum, if it exists for either projectile. E_{Ps} and E_{exc} are the threshold energies (in eV) for Ps formation by e^+ impact and atomic excitation by e^- impact. $R_t(<E_{Ps})$ represents the ratio of Q_t^-/Q_t^+ (Q_t for electrons to Q_t for positrons) at an energy just below E_{Ps}. $E_{max}(Q_t)$ is the energy (in eV) where the highest energy maximum in Q_t is observed for each projectile. The "Degree of merging (E)" is equal to the average value for $[(Q_t^-/Q_t^+) - 1] \times 100\%$ for the energy range listed or the value at the highest energy that has been studied. "*" refers to results based on theory. "?" indicates considerable uncertainty.

TABLE IV

SUMMARY OF $e^{+,-}$–MOLECULE Q_t MEASUREMENTS.[a]

Molecule	Fig. #	e^- SR (eV)	E_{Ps} (eV)	Q_t increase After E_{Ps}	$R_t(<E_{Ps})$	$E_{max}(Q_t)$ e^+	$E_{max}(Q_t)$ e^-	Degree of merging (E)
H_2	14	no	8.6	large	15	20 eV	4 eV	<2% (≥200 eV)
N_2	16, 22	2.4	8.8	large	3.4	30	20	24% (700 eV)
CO	17, 22	2	7.2	large	3.2	25	20	20% (500 eV)
O_2	18	no	5.3	none?	2.8	30	25	30% (500 eV)
H_2O	19	no	5.8	none?	1.6	no?	10	<2% (≥200 eV)
CO_2	20, 22	4	7.0	small	1.4	30	30	40% (500 eV)
N_2O	21, 22	2	6.1	small	1.4	30	30	30% (400 eV)
NH_3	23	no	3.4	none?	1.0	no	10	<5% (>200 eV)
CH_4	24	no	5.8	large	5	30	8	<15% (≥200 eV)
C_2H_4	26	2	3.7	large	1.5	15	8	<5% (≥50 eV)
C_2H_6	26	no	4.7	medium	2.1	30	8	<5% (≥90 eV)
C_3H_6	25	—	2.9	—	—	6	10	<2% (≥300 eV)
cycloC_3H_6	25	—	3.3	—	—	6	10	13% (400 eV)
C_3H_8	25	—	4.3	—	—	10	7.5	15% (400 eV)
C_4H_8	25	—	2.8	—	—	no	10	5% (400 eV)
nC_4H_{10}	25	—	3.8	—	—	6?	7.5	2% (400 eV)
isoC_4H_{10}	25	—	3.8	—	—	6	7.5	13% (400 eV)
C_6H_6	26	2 eV?	2.4	decrease?	0.5	no	8	<5% (>80 eV)
SiH_4	27	no	5.0	large	4	20	3	<5% (≥100 eV)
CF_4	28	8.5 eV	11.0	small?	2	40?	25	—
SF_6	29	three	8.5	small?	2	40	35	50% (500 eV)

[a] The information in this table is the same as for Table III with some deletions and additions. "e^- SR" is the approximate energy location (in eV) of an e^- shape resonance if one exists. "Q_t increase after E_{Ps}" refers to the nature of the increase in Q_t as the energy is increased through E_{Ps}.

sections occur for the smallest targets (H, He, and H_2) implying that the cancellation between the static and polarization interactions for positrons is greatest for systems having only a single shell containing electrons. The target gases Ar, CH_4, and SiH_4 with the next largest values for R_t have remarkably similar overall Q_t shapes (as mentioned earlier) and their $e^{+,-}$ Q_t comparison curves are shown in Fig. 30, along with the Q_t curves for C_2H_6, which is an alkane family member with CH_4 and has Q_t shapes quite similar to the other gases shown in Fig. 30. As mentioned earlier, Ar and CH_4 are known to have Ramsauer-Townsend minima, SiH_4 is predicted to have one, and it may be anticipated on the basis of these overall Q_t shape comparisons that C_2H_6 may also have one. Another interesting observation that can be made from Fig. 30 is that the $e^{+,-}$ Q_t's for Ar and CH_4 not only exhibit a tendency toward merging of their separate $e^{+,-}$ Q_t curves, but also with each other, while the isoelectronic molecules C_2H_6 and SiH_4 exhibit an even greater tendency toward mergings of their respective $e^{+,-}$ Q_t curves above 100 eV. Recall that similar behavior was discussed earlier (in relation to Fig. 22) for the isoelectronic molecules CO and N_2, and CO_2 and N_2O, although in this earlier comparison the isoelectronic molecule pairs also had the same number

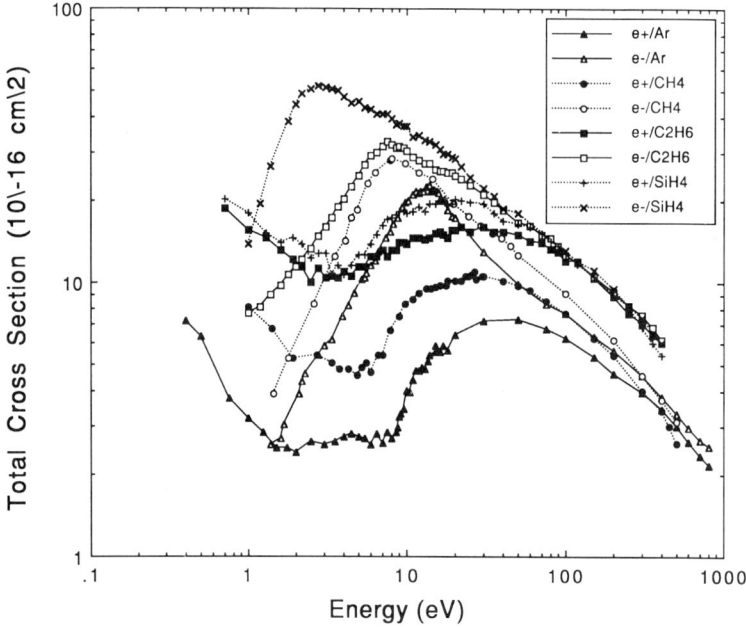

FIG. 30. Comparisons of Q_t's for $e^{+,-}$ scattering by Ar, CH_4, C_2H_6, and SiH_4, as measured by Kauppila et al. (1976, 1981) for Ar, Dababneh et al. (1988) for CH_4, Sueoka and Mori (1986) for C_2H_6, and Mori et al. (1985) for SiH_4.

of atoms and the same mass number, while for C_2H_6 and SiH_4 the number of atoms in each molecule is different as is the mass number of each molecule.

In comparing the degree of merging of the $e^{+,-}$ Q_t curves for the various target gases it is intriguing that the H atom and all of the molecules containing H atoms (that have been investigated) are either merged or nearly merged ($\leq 15\%$) at the highest energy of study. Meanwhile, with the exception of He (which is close as one can get to H), and Na and K (which are unique due to their large polarizabilities and large inelastic contributions to Q_t even at low energies), all of the other target gases that have been studied have a degree of merging that is $\geq 15\%$ at the highest energies of investigation. As a result of the R_t comparisons and the degree of merging comparisons it is curious that the smallest targets (H, He, and H_2) not only have the largest values for R_t (mentioned above), but also the greatest degree of merging at the highest energies, clearly showing that they exhibit the greatest extremes in the differences and similarities between $e^{+,-}$ scattering as one goes from low to high projectile energy.

It has been discussed earlier (Sec. II) that for $e^{+,-}$ scattering by the inert gas atoms, the maxima in the Q_t curves occur at lower energies for electrons than for positrons and appear to be related to elastic scattering for electrons and to inelastic scattering for positrons. From Tables III and IV and Figs. 1 and 30 it is also found that there seems to be a general tendency for the maxima of Q_t curves (having the same general shape for a given projectile) to shift to lower energy with increasing size (Q_t) of the target gas. Molecules which exhibit shape resonances, however, do not seem to follow this trend.

The subject of resonances is an interesting one when considering comparisons between $e^{+,-}$ scattering by atoms and molecules. In the various figures presented in this paper it is seen that low energy shape resonances are observed (see Table IV) in the Q_t measurements for e^- scattering by several molecules. Meanwhile, for e^- scattering by atoms, much narrower Feshbach resonances (temporary bound states between the e^- and atom) have been experimentally observed (using high resolution electron spectroscopy) in many cases for elastic scattering cross sections just below excitation thresholds (for reviews see Schulz, 1973a, b; Biondi et al., 1979). In the case of e^+ scattering by atoms and molecules no shape or Feshbach resonances have yet been observed experimentally, although Feshbach resonances have been theoretically predicted to occur for e^+-H scattering just below the $n = 2$ excitation threshold for H (Choo et al. 1978; Doolen et al., 1978; Pelikan and Klar, 1983), and associated with formation of Ps in its $n = 2$ state (Doolen, 1978; Ho and Greene, 1987) and in its $n = 3$ and $n = 4$ states (Ho, 1988).

Another aspect of comparisons of $e^{+,-}$ scattering by atoms that has received attention concerns tests of the validity of the sum rule, which is based on the forward dispersion relations of Gerjuoy and Krall (1960). For $e^{+,-}$

scattering by the inert gases the sum rule has the form (Bransden and McDowell, 1969)

$$-A - f_B^D + f_B^E = \frac{1}{2\pi} \int_0^\infty Q_t(k)dk, \qquad (4)$$

where A is the scattering length, f_B^D and f_B^E are the first Born elastic scattering amplitudes in the forward direction for direct and exchange scattering ($f_B^E = 0$ for positrons), k is the projectile wave number, all in atomic units, and Q_t is in units of πa_0^2. By using available Q_t measurements to evaluate the integral and theoretical results for the other terms in the above equation, it has been determined (Bransden and Hutt, 1975; Byron et al., 1975; de Heer et al., 1976; Hutt et al., 1976; Tsai et al., 1976; Brenton et al., 1977; Griffith et al., 1979; Kauppila et al., 1981) that the sum rule is valid when applied to e^+ scattering by He, Ne, and Ar, and not valid for e^- scattering by He and Ne (with the situation for e^- scattering by Ar unknown). The apparent lack of validity of the sum rule for e^--atom scattering is understood (Byron et al., 1975; de Heer et al., 1976; Hutt et al., 1976; Tip, 1977) to arise from the nature of singularities in the exchange amplitude, and since there are no exchange effects for e^+ scattering the sum rule is found to be valid for this projectile.

ACKNOWLEDGEMENTS

We would like to thank Professor J. M. Wadehra for helpful discussions, and Dr. C. K. Kwan, Mr. Steven J. Smith, and Mr. James Klemic for helpful assistance. We acknowledge, with gratitude, the support of the National Science Foundation for our research program.

REFERENCES

Anderson, C. D. (1933). *Phys. Rev.* **43**, 491.
Andersen, L. H., Hvelplund, P., Knudsen, H., Moller, S. P., Sorensen, A. H., Elsener, K., Rensfelt, K. G., and Uggerhoj, E. (1987). *Phys. Rev. A* **36**, 3612.
Bartschat, K., McEachran, R. P., and Stauffer, A. D. (1988). *J. Phys. B* **21**, 2789.
Biondi, M. F., Herzenberg, A., and Kuyatt, C. E. (1979). *Phys. Today* **32** (10), 44.
Bransden, B. H., and Hutt, P. K. (1975). *J. Phys. B* **8**, 603.
Bransden, B. H., and McDowell, M. R. C. (1969). *J. Phys. B* **2**, 1187.
Brenton, A. G., Dutton, J., Harris, F. M., Jones, R. A., and Lewis, D. M. (1977). *J. Phys. B* **10**, 2699.
Buckman, S. J., and Lohmann, B. (1986a). *J. Phys. B* **19**, 2547.

Buckman, S. J., and Lohmann, B. (1986b). *Phys. Rev. A* **34**, 1561.
Byron, F. W., Jr., de Heer, F. J., and Joachain, C. J. (1975). *Phys. Rev. Lett.* **35**, 1147.
Byron, F. W., Jr., Joachain, C. J., and Potvliege, R. M. (1982). *J. Phys. B* **15**, 3915.
Charlton, M. (1985). *Rep. Prog. Phys.* **48**, 737.
Charlton, M., Andersen, L. H., Brun-Nielsen, L., Deutch, B. I., Hvelplund, P., Jacobsen, F. M., Knudsen, H., Laricchia, G., Poulsen, M. R., and Pedersen, J. O. (1988). *J. Phys. B* **21**, L545.
Choo, L. T., Crocker, M. C., and Nuttall, J. (1978). *J. Phys. B* **11**, 1313.
Coleman, P. G., and McNutt, J. D. (1979). *Phys. Rev. Lett.* **42**, 1130.
Costello, D. G. Groce, D. E., Herring, D. F., and McGowan, J. Wm. (1972). *Can. J. Phys.* **50**, 23.
Dababneh, M. S., Kauppila, W. E., Downing, J. P., Laperriere, F., Pol, V., Smart, J. H., and Stein, T. S. (1980). *Phys. Rev. A* **22**, 1872.
Dababneh, M. S., Hsieh, Y.-F., Kauppila, W. E., Pol V., and Stein, T. S. (1982). *Phys. Rev. A* **26**, 1252.
Dababneh, M. S., Hsieh, Y.-F., Kauppila, W. E., Kwan, C. K., Smith, Steven J., Stein, T. S., and Uddin, M. N. (1988). *Phys. Rev. A* **38**, 1207.
Darewych, J. W. (1982). *J. Phys. B* **15**, L415.
Dasgupta, A., and Bhatia, A. K. (1984). *Phys. Rev. A* **30**, 1241.
de Heer, F. J., and Jansen R. H. J. (1975). *FOM Institute for Atomic and Molecular Physics, Amsterdam*, Report No. 37173.
de Heer, F. J., Wagenaar, R. W., Blaauw, H. J., and Tip, A. (1976). *J. Phys. B* **9**, L269.
de Heer, F. J., McDowell, M. R. C., and Wagenaar, R. W. (1977). *J. Phys. B* **10**, 1945.
Dehmer, J. L., and Dill, D. (1980). *In* "Electronic and Atomic Collisions" (N. Oda and K. Takayanagi, eds.), pp. 195–208. North-Holland, Amsterdam.
Dehmer, J. L., Siegel, J., and Dill, D. (1978). *J. Chem. Phys.* **69**, 5205.
Dehmer, J. L., Siegel, J., Welch, J., and Dill, D. (1980). *Phys. Rev. A* **21**, 101.
Dewangan, D. P. (1980). *J. Phys. B* **13**, L595.
Dewangan, D. P., and Walters, H. R. J. (1977). *J. Phys. B* **10**, 637.
Diana, L. M., Sharma, S. C., Fornari, L. S., Coleman, P. G., Pendleton, P. K., Brooks, D. L., and Seay, B. E. (1985). *In* "Positron Annihilation" (P. C. Jain, R. M. Singru, and K. P. Gopinathan, eds.) pp. 428–430. World Scientific, Singapore.
Diana, L. M., Coleman, P. G., Brooks, D. L., and Chaplin, R. L. (1987). *In* "Atomic Physics with Positrons" (J. W. Humberston and E. A. G. Armour, eds.) NATO ASI Series B, Vol. 169, pp. 55–69. Plenum Press, New York and London.
Dill, D., Welch, J., Dehmer, J. L., and Siegel, J. (1979). *Phys. Rev. Lett.* **43**, 1236.
Doolen, G. D. (1978). *Int J. Quantum Chem.* **14**, 523.
Doolen, G. D., Nuttall, J., and Wherry, C. (1978). *Phys. Rev. Lett.* **40**, 313.
Drachman, R. J. (1966). *Phys. Rev.* **144**, 25.
Dube, L., and Herzenberg, A. (1979). *Phys. Rev. A* **20**, 194.
Dubois, R. D., and Rudd, M. E. (1975). *J. Phys. B* **8**, 1474.
Floeder, K., Fromme, D., Raith, W., Schwab, A., and Sinapius, G. (1985). *J. Phys. B* **18**, 3347.
Floeder, K., Honer, P., Raith, W., Schwab, A., Sinapius, G., and Spicher. G. (1988). *Phys. Rev. Lett.* **60**, 2363.
Fon, W. C., Berrington, K. A., Burke, P. G., and Hibbert, A. (1983). *J. Phys. B* **16**, 307.
Fornari, L. S., Diana, L. M., and Coleman, P. G. (1983). *Phys. Rev. Lett.* **51**, 2276.
Fromme, D., Kruse, G., Raith, W., and Sinapius, G. (1986). *Phys. Rev. Lett.* **57**, 3031.
Fromme. D., Kruse, G., Raith, W., and Sinapius, G. (1988). *J. Phys. B* **21**, L261.
Gerjuoy, E., and Krall, N. A. (1960). *Phys. Rev.* **119**, 705.
Griffith, T. C. (1983). *In* "Positron Scattering in Gases" (J. W. Humberston and M. R. C. McDowell eds.), NATO ASI Series B, Vol. 107, pp. 53–63. Plenum Press, New York and London.

Griffith, T. C., Heyland, G. R., Lines, K. S., and Twomey, T. R. (1979). *Appl. Phys.* **19**, 431.
Guha, S., and Mandal, P. (1980). *J. Phys. B* **13**, 1919.
Ho, Y. K. (1988). *Phys. Rev. A* **38**, 6424.
Ho, Y. K., and Greene, C. H. (1987). *Phys. Rev. A* **35**, 3169.
Hoffman, K. R., Dababneh, M. S., Hsieh, Y.-F., Kauppila, W. E., Pol, V., Smart, J. H., and Stein, T. S. (1982). *Phys. Rev. A* **25**, 1393.
Horbatsch, M., and Darewych, J. W. (1983). *J. Phys. B* **16**, 4059.
Humberston, J. W. (1979). In "Advances In Atomic and Molecular Physics" (D. R. Bates and B. Bederson, eds.), Vol. 15, pp. 101-133. Academic Press, New York.
Humberston, J. W. and Armour, E. A. G., eds. (1987). "Atomic Physics with Positrons" NATO ASI Series B, Vol. 169. Plenum Press, New York and London.
Humberston, J. W., and Campeanu, R. I. (1980). *J. Phys. B.* **13**, 4907.
Humberston, J. W., and McDowell, M. R. C., eds. (1984). "Positron Scattering in Gases" NATO ASI Series B, Vol. 107. Plenum Press, New York and London.
Hutt, P. K., Islam, M. M., Rabheru, A., and McDowell, M. R. C. (1976). *J. Phys. B* **9**, 2447.
Hyder, G. M. A., Dababneh, M. S., Hsieh, Y.-F., Kauppila, W. E., Kwan, C. K., Mahdavi-Hezaveh, M., and Stein, T. S. (1986). *Phys. Rev. Lett.* **57**, 2252.
Inokuti, M. (1971). *Rev. Mod. Phys.* **43**, 297.
Jain, Ashok (1982). *J. Phys. B* **15**, 1533.
Jain, Ashok (1986a). *J. Phys. B* **19**, L105.
Jain, Ashok (1986b). *Phys. Rev. A* **34**, 3707.
Jain, Ashok (1986c). *J. Phys. B.* **19**, L807.
Jain, Ashok (1987). *Phys. Rev. A* **35**, 4826.
Jain, Ashok and Norcross, D. W. (1985). *Proc. Int. Conf. Phys. Electron. At. Collisions, 14th* Abstr., p. 214; and private communication.
Jain, Ashok, and Thompson, D. G. (1987). *J. Phys. B* **20**, 2861.
Jain, Ashok, Freitas, L. C. G., Mu-Tao, L., and Tayal, S. S. (1984). *J. Phys. B* **17**, L29.
Jain, Arvind K., Tripathi, A. N., and Jain, Ashok (1988). *Phys. Rev. A* **37**, 2893.
Jhanwar, B. L., Khare, S. P., and Sharma, M. K. (1982). *Phys. Rev. A* **26**, 1392.
Joachain, C. J. (1978). In "Electronic and Atomic Collisions" (G. Watel, ed.), pp. 71-93. North-Holland, Amsterdam.
Joachain, C. J., and Potvliege, R. M. (1987). *Phys. Rev. A* **35**, 4873.
Joachain, C. J., Vanderpoorten, R., Winters, K. H., and Byron, F. W., Jr. (1977). *J. Phys. B* **10**, 227.
Jones, R. K. (1985). *Phys. Rev. A* **31**, 2898.
Jost, K., Bisling, P. G. F., Eschen, F., Felsmann, M., and Walther, L. (1983). *Proc. Int. Conf. Phys. Electron. At. Collisions*, 13th Abstr., p. 91.
Katase, A., Ishibashi, K., Matsumoto, Y., Sakae, T., Maezono, S., Murakami, E., Watanabe, K., and Maki, H. (1986). *J. Phys. B* **19**, 2715.
Katayama, Y., Sueoka, O., and Mori, S. (1987). *J. Phys. B* **20**, 1645.
Kauppila, W. E., and Stein, T. S. (1987). In "Atomic Physics with Positrons" (J. W. Humberston and E. A. G. Armour, eds.), NATO ASI Series B, Vol. 169, pp. 27-39. Plenum Press, New York and London.
Kauppila, W. E., Stein, T. S., and Jesion, G. (1976). *Phys. Rev. Lett.* **36**, 580.
Kauppila, W. E., Stein, T. S., Jesion, G., Dababneh, M. S., and Pol, V. (1977). *Rev. Sci. Instrum.* **48**, 822.
Kauppila, W. E., Stein, T. S., Smart, J. H., Dababneh, M. S., Ho, Y. K., Downing, J. P., and Pol, V. (1981). *Phys. Rev. A* **24**, 725.
Kauppila, W. E., Stein, T. S., and Wadehra, J. M., eds. (1986). "Positron (Electron) -Gas Scattering". World Scientific, Singapore.

Kennerly, R. E. (1980). *Phys. Rev. A* **21**, 1876.
Kennerly, R. E., Bonham, R. A., and McMillan, M. (1979). *J. Chem. Phys.* **70**, 2039.
Klar, H. (1981). *J. Phys. B* **14**, 4165.
Klar, H. (1984). In "Electronic and Atomic Collisions" (J. Eichler, I. V. Hertel, and N. Stolterfoht, eds.), pp. 767–775. North-Holland, Amsterdam.
Knudsen, H., Andersen, L. H., Hvelplund, P., Astner, G., Cederquist, H., Danared, H., Liljeby, L., and Rensfelt, K. G. (1984). *J. Phys. B* **17**, 3545.
Kwan, C. K., Hsieh, Y.-F., Kauppila, W. E., Smith, Steven J., Stein, T. S., Uddin, M. N., and Dabebneh, M. S. (1983). *Phys. Rev. A* **27**, 1328.
Kwan, C. K., Hsieh, Y.-F., Kauppila, W. E., Smith, Steven J., Stein, T. S., Uddin, M. N., and Dababneh, M. S. (1984). *Phys. Rev. Lett.* **52**, 1417.
Kwan, C. K., Kauppila, W. E., Lukaszew, R. A., Parikh, S. P., Stein, T. S., Wan, Y. J., and Dababneh, M. S. (1989). Private communication (to be published). The absolute $e^{+,-}$ Q_t values shown in Figs. 10 and 11 are subject to change due to possible reassessment of vapor pressure data in the literature.
Laricchia, G., Charlton, M., and Griffith, T. C. (1988). *J. Phys. B* **21**, L227.
Lohmann, B., and Buckman, S. J. (1986). *J. Phys. B* **19**, 2565.
McEachran, R. P., and Stauffer, A. D. (1983). *J. Phys. B* **16**, 4023.
McEachran, R. P., and Stauffer, A. D. (1984). *J. Phys. B* **17**, 2507.
McEachran, R. P., and Stauffer, A. D. (1986). In "Positron (Electron)-Gas Scattering" (W. E. Kauppila, T. S. Stein, and J. M. Wadehra, eds.), pp. 122–130. World Scientific, Singapore.
McEachran, R. P., and Stauffer, A. D. (1987). *J. Phys. B* **20**, 3483.
McEachran, R. P., Ryman, A. G., and Stauffer, A. D. (1978). *J. Phys. B* **11**, 551.
McEachran, R. P., Ryman, A. G., and Stauffer, A. D. (1979). *J. Phys. B* **12**, 1031.
McEachran, R. P., Stauffer, A. D., and Campbell, L. E. M. (1980). *J. Phys. B* **13**, 1281.
McGuire, J. H., and Deb, N. C. (1987). In "Atomic Physics with Positrons" (J. W. Humberston and E. A. G. Armour, eds.), NATO ASI Series B, Vol. 169, pp. 83–94. Plenum Press, New York and London.
Marinkovic, B., Szmytkowski, C., Pejcev, V., Filipovic, D., and Vuskovic, L. (1986). *J. Phys. B* **19**, 2365.
Massey, H. S. W. (1976). *Phys. Today* **29** 3, 42.
Montague, R. G., Harrison, M. F. A., and Smith, A. C. H. (1984). *J. Phys. B* **17**, 3295.
Montgomery, R. E., and LaBahn, R. W. (1970). *Can. J. Phys.* **48**, 1288; and private communication.
Mori, S., Katayama, Y., and Sueoka, O. (1985). *At. Collisions Res. Jpn.* **11**, 19.
Morrison, M. A., Lane, N. F., and Collins, L. A. (1977). *Phys. Rev. A* **15**, 2186.
Morrison, M. A., Gibson, T. L., and Austin, D. (1984). *J. Phys. B* **17**, 2725.
Nahar, S. N., and Wadehra, J. M. (1987). *Phys. Rev A* **35**, 2051.
Nesbet, R. K. (1979). *Phys. Rev A* **20**, 58.
Nesbet, R. K., Noble, C. J., and Morgan, L. A. (1986). *Phys. Rev. A* **34**, 2798.
O'Malley, T. F., Spruch, L., and Rosenberg, L. (1961). *J. Math. Phys.* **2**, 491.
Pelikan, E., and Klar, H. (1983). *Z. Phys. A* **310**, 153.
Phelps, J. O., and Lin, C. C. (1981). *Phys. Rev. A* **24**, 1299.
Phelps, J. O., Solomon, J. E., Korff, D. F., Lin, C. C., and Lee, E. T. P. (1979). *Phys. Rev. A* **20**, 1418.
Puckett, L. J., and Martin, D. W. (1970). *Phys. Rev. A* **1**, 1432.
Ramsauer, C. (1921a). *Ann. Phys. (Leipzig)* **64**, 513.
Ramsauer, C. (1921b). *Ann. Phys. (Leipzig)* **66**, 546.
Ramsauer, C. (1923). *Ann. Phys. (Leipzig)* **72**, 345.
Ramsauer, C., and Kollath, R. (1929). *Ann. Phys. (Leipzig)* **3**, 536.

Rapp, D., and Englander-Golden, P. (1965). *J. Chem. Phys.* **43**, 1464.
Salop, A., and Nakano, H. H. (1970). *Phys. Rev. A* **2**, 127.
Sanche, L., and Schulz, G. J. (1973). *J. Chem. Phys.* **58**, 479.
Sarkar, K. P., Basu, M., and Ghosh, A. S. (1988). *J. Phys. B* **21**, 1649.
Schulz, G. J. (1973a). *Rev. Mod. Phys.* **45**, 378.
Schulz, G. J. (1973b). *Rev. Mod. Phys.* **45**, 423.
Shah, M. B., and Gilbody, H. B. (1985). *J. Phys. B* **18**, 899.
Shyn, T. W., and Carignan, G. R. (1980). *Phys. Rev. A* **22**, 923.
Shyn, T. W., and Sharp, W. E. (1982). *Phys. Rev. A* **26**, 1369.
Sin Fai Lam, L. T. (1982). *J. Phys. B* **15**, 119.
Sinapius, G. (1988). In "Electronic and Atomic Collisions" (H. B. Gilbody, W. R. Newell, F. H. Read, and A. C. H. Smith, eds.), pp. 73–91. North-Holland, Amsterdam.
Smith, Steven J., Hyder, G. M. A., Kauppila, W. E., Kwan, C. K., and Stein, T. S. (1989). *Proc. Int. Conf. Phys. Electron. At. Collisions*, 16th Abstr. (to be published).
Srivastava, S. K., Tanaka, H., Chutjian, A., and Trajmar, S. (1981). *Phys. Rev. A* **23**, 2156.
Stein, T. S., and Kauppila, W. E. (1982). In "Advances in Atomic and Molecular Physics" (D. Bates and B. Bederson, eds.), Vol. 18, p. 53, Academic Press, New York.
Stein, T. S., and Kauppila, W. E. (1986). In "Electronic and Atomic Collisions" (D. C. Lorents, W. E. Meyerhof, and J. R. Peterson, eds.), pp. 105–123. North-Holland, Amsterdam.
Stein, T. S., Kauppila, W. E., Pol, V., Smart, J. H., and Jesion, G. (1978). *Phys. Rev. A* **17**, 1600.
Stein, T. S., Gomez, R. D., Hsieh, Y.-F., Kauppila, W. E., Kwan, C. K., and Wan, Y. J. (1985). *Phys. Rev. Lett.* **55**, 488.
Stein, T. S., Dababneh, M. S., Kauppila, W. E., Kwan, C. K., and Wan, Y. J. (1987). In "Atomic Physics with Positrons" (J. W. Humberston and E. A. G. Armour, eds.), NATO ASI Series B, Vol. 169, pp. 251–263. Plenum Press, New York and London.
Sueoka, O. (1987). In "Atomic Physics with Positrons" (J. W. Humberston and E. A. G. Armour, eds.), NATO ASI Series B, Vol. 169, pp. 41–54. Plenum Press, New York and London.
Sueoka, O. (1988). *J. Phys. B* **21**, L631.
Sueoka, O., and Mori, S. (1986). *J. Phys. B* **19**, 4035.
Sueoka, O., Mori, S., and Katayama, Y. (1986). *J. Phys. B* **19**, L373.
Sueoka, O., Mori, S., and Katayama, Y. (1987). *J. Phys. B* **20**, 3237.
Szmytkowski, C., Zecca, A., Karwasz, G., Oss, S., Maciag, K., Marinkovic, B., Brusa, R. S., and Grisenti, R. (1987). *J. Phys. B* **20**, 5817.
Tip, A. (1977). *J. Phys. B* **10**, L11.
Townsend, J. S., and Bailey, V. A. (1922). *Philos. Mag.* **43**, 593.
Tsai, J.-S., Lebow, L., and Paul, D. A. L. (1976). *Can. J. Phys.* **54**, 1741.
van Wingerden, B., Weigold, E., de Heer, F. J., and Nygaard, K. J. (1977). *J. Phys. B* **10**, 1345.
van Wyngaarden, W. L., and Walters, H. R. J. (1986). *J. Phys. B* **19**, 929.
Wadehra, J. M., Stein, T. S., and Kauppila, W. E. (1981). *J. Phys. B* **14**, L783.
Walters, H. R. J. (1976). *J. Phys. B* **9**, 227.
Walters, H. R. J. (1984). *Phys. Rep.* **116**, 1.
Walters, H. R. J. (1988). *J. Phys. B* **21**, 1893.
Wannier, G. H. (1953). *Phys. Rev.* **90**, 817.
Ward, S. J., Horbatsch, M., McEachran, R. P., and Stauffer, A. D. (1988). *J. Phys. B* **21**, L611.
Williams, J. F. (1979). *J. Phys. B* **12**, 265.
Williams, J. F., and Willis, B. A. (1975). *J. Phys. B* **8**, 1670.
Winick, J. R., and Reinhardt, W. P. (1978a) *Phys. Rev. A* **18**, 910.
Winick, J. R., and Reinhardt, W. P. (1978b). *Phys. Rev. a* **18**, 925.
Zecca, A., Brusa, R. S., Grisenti, R., Oss, S., and Szmytkowski, C. (1986). *J. Phys. B* **19**, 3353.
Zecca, A., Karwasz, G., Oss, S., Grisenti, R., and Brusa, R. S. (1987). *J. Phys. B* **20**, L133.

ELECTRON CAPTURE AT RELATIVISTIC ENERGIES

B. L. MOISEIWITSCH

Department of Applied Mathematics and Theoretical Physics
The Queen's University of Belfast
Belfast, Northern Ireland

I. Introduction 51
II. Two-State Approximation 52
    A. The First-Order Relativistic OBK Approximation 53
III. Second-Order Theories 57
    A. Classical Double Scattering 57
    B. The Second-Order Relativistic OBK Approximation 58
IV. Relativistic Continuum Distorted Wave Approximation 64
V. Relativistic Eikonal Approximation 65
    A. Relativistic Symmetric Eikonal Approximation 66
    B. Relativistic Eikonal Phase Factors 70
    C. First-Order Born Approximation with Coulomb Boundary Conditions . 72
VI. Numerical Solution of Coupled Equations 74
VII. Experimental Data and Comparisons with Theory 74
    References 77

I. Introduction

The first investigation of electron capture at relativistic energies was carried out by Mittleman (1964) who used a relativistic generalization of the first-order Oppenheimer-Brinkman-Kramers (OBK) approximation, finding that the asymptotic behaviour of the capture cross section was E^{-1} in the limit of high relativistic energies E.

Subsequently much more detailed studies were made using the first-order relativistic OBK approximation by Shakeshaft (1979) and by Moiseiwitsch and Stockman (1980). Also a considerable amount of experimental data became available as a consequence of the work of Crawford (1979), later subjected to detailed analysis by Anholt (1985), which showed that the relativistic OBK first-order aproximation produced capture cross sections which were much too large. This stimulated the use of various higher-order approximations including the second-order relativistic OBK approximation (Humphries and Moiseiwitsch, 1984; Moiseiwitsch, 1988, 1989), the relativistic eikonal approximation (Eichler, 1985), the relativistic continuum distorted wave approximation (Deco and Rivarola, 1987c), and the relativistic first-order Born approximation with Coulomb boundary conditions (Eichler, 1987), all of which led to much better agreement with the experimental data. We shall be discussing all of these approximations, as well as others, in the following pages.

II. Two-State Approximation

Let us consider the capture of an electron from a target hydrogenic ion T by an incident bare projectile ion P moving in a straight line with constant velocity $\mathbf{v}$ and impact parameter ρ referred to the nucleus of the target. We suppose that initially the electron is bound in the state i of the target system having the Dirac time-independent four-spinor atomic eigenvector $\psi_i^T(\mathbf{r}_T)$ with position vector $\mathbf{r}_T$ and energy eigenvalue E_i^T referred to the inertial frame S_T of the target nucleus. After capture the electron is attached to P in the final state f having the Dirac four-spinor eigenvector $\psi_f^{P'}(\mathbf{r}_P')$ with position vector $\mathbf{r}_P'$ and energy eigenvalue $E_f^{P'}$ referred to the inertial frame S_P' of the projectile ion P.

In the two-state approximation the electron involved in the capture is represented by the total wave function

$$\Psi = a_i^T(\rho, t)\psi_i^T(\mathbf{r}_T)\exp(-iE_i^T t/\hbar) \\ + a_f^P(\rho, t)S\psi_f^{P'}(\mathbf{r}_P')\exp(-iE_f^{P'} t'/\hbar) \tag{1}$$

referred to the inertial frame S_T of the target nucleus, where S is an operator which Lorentz transforms the Dirac eigenvector $\psi_f^{P'}$ from the S_P' inertial frame to the S_T inertial frame and is given by

$$S = \left(\frac{\gamma + 1}{2}\right)^{1/2}\left[1 + \left(\frac{\gamma - 1}{\gamma + 1}\right)^{1/2}\boldsymbol{\alpha}\cdot\hat{\mathbf{v}}\right]. \tag{2}$$

Here $\gamma = (1 - v^2/c^2)^{-1/2}$ is the Lorentz factor, $\boldsymbol{\alpha}$ is the vector composed of Dirac matrices, and $\hat{\mathbf{v}}$ is the unit vector in the direction of the velocity $\mathbf{v}$ of the projectile ion referred to S_T.

The times t and t' are referred to the inertial frames S_T and S_P' respectively. They are related by the Lorentz transformation formula

$$t' = \gamma(t - \mathbf{v}\cdot\mathbf{r}_T/c^2) \tag{3}$$

and so we may write

$$E_f^{P'} t'/\hbar = \gamma E_f^{P'} t/\hbar - \mathbf{k}\cdot\mathbf{r}_T, \tag{4}$$

where $\mathbf{k} = \gamma E_f^{P'}\mathbf{v}/\hbar c^2$ is the propagation vector of the electron attached to P.

It is very interesting to see what happens in the nonrelativistic limit $c \to \infty$. In this limit we obtain

$$\mathbf{k} \to m\mathbf{v}/\hbar, \qquad \gamma E_f^{P'} \to mc^2 + \varepsilon_f^P + \tfrac{1}{2}mv^2 \tag{5}$$

where mc^2 is the rest mass energy of the electron and ε_f^P is the nonrelativistic Schrödinger energy eigenvalue of the final state f so that (Moiseiwitsch and Stockman, 1980)

$$\exp(-iE_f^{P'}t'/\hbar) \to \exp[-i(mc^2 + \varepsilon_f^P)t]f(t), \tag{6}$$

where

$$f(t) = \exp[-i(\tfrac{1}{2}mv^2 t - m\mathbf{v}\cdot\mathbf{r}_T)/\hbar] \tag{7}$$

is the translation factor introduced by Bates and McCarroll (1958) and corresponds to making a Galilean transformation from S'_P to S_T.

A. The First-Order Relativistic OBK Approximation

In the impact parameter formulation the capture cross section is given by

$$\sigma_{fi} = 2\pi \int_0^\infty |a_f^P(\rho, t = \infty)|^2 \rho \, d\rho, \tag{8}$$

where, to first order, the probability amplitude takes the form

$$a_f^P(\rho, t = \infty) = -\frac{i}{\hbar}\int_{-\infty}^\infty dt \int d\mathbf{r}_T \exp[i(E_f^{P'}t' - E_i^T t)/\hbar]$$
$$\times S\psi_f^{P'}(\mathbf{r}'_P)^\dagger L V'_P(\mathbf{r}'_P, \mathbf{r}'_T)\psi_i^T(\mathbf{r}_T). \tag{9}$$

Here $V'_P(\mathbf{r}'_P, \mathbf{r}'_T)$ is the interaction potential between the incident projectile ion P and the target hydrogenic ion viewed from the projectile frame, and $L = S^{-1}S^{-1}$ is the operator which Lorentz transforms the potential V'_P referred to S'_P to a potential referred to S_T.

To obtain simple analytical formulas for the scattering amplitudes and capture cross sections it is useful to change to the wave formulation. Let $\mathbf{k}_i$, $\mathbf{k}_f$ be the wave vectors of the incident ion P, and the ion P together with the captured electron after the collision has occurred, respectively. Setting $\mathbf{q} = \mathbf{k}_i - \mathbf{k}_f$ we have

$$\mathbf{q}\cdot\hat{\mathbf{v}} = k_i - k_f = (E_f^{P'} - \gamma E_i^T)/\hbar c(\gamma^2 - 1)^{1/2} \tag{10}$$

by conservation of energy, neglecting terms of order m/M_P where M_P is the mass of the projectile ion. Next we introduce the vector $\mathbf{q}'$ given by

$$\mathbf{q}'\cdot\hat{\mathbf{v}} = (E_i^T - \gamma E_f^{P'})/\hbar c(\gamma^2 - 1)^{1/2} \tag{11}$$

and $(\mathbf{q} + \mathbf{q}') \cdot \boldsymbol{\rho} = 0$ where $\boldsymbol{\rho}$ is the position vector of the point of closest approach of P relative to the target nucleus. Then it can be readily shown that

$$(E_f^{P'} t' - E_i^T t)/\hbar + (\mathbf{k}_i - \mathbf{k}_f) \cdot \boldsymbol{\rho} = \mathbf{q} \cdot \mathbf{r}_T + \mathbf{q}' \cdot \mathbf{r}'_P \tag{12}$$

which enables us to write the capture cross section in the form

$$\sigma_{fi} = \frac{2\pi}{k_i^2} \int_{q_{\min}}^{\infty} |g_{fi}(q)|^2 q \, dq, \tag{13}$$

where the scattering amplitude for capture from state i to state f is given by

$$g_{fi} = -\frac{1}{4\pi} \frac{2M_P}{\hbar^2} \int d\mathbf{r}'_P \int d\mathbf{r}_T \exp[i(\mathbf{q} \cdot \mathbf{r}_T + \mathbf{q}' \cdot \mathbf{r}'_P)]$$
$$\times S\psi_f^{P'}(\mathbf{r}'_P)^\dagger L V'_P(\mathbf{r}'_P, \mathbf{r}'_T) \psi_i^T(\mathbf{r}_T) \tag{14}$$

and $q_{\min} = \mathbf{q} \cdot \hat{\mathbf{v}}$ is the least value of q.

We now make the Oppenheimer-Brinkman-Kramers (OBK) approximation which neglects the nucleus-nucleus interaction $e^2 Z_T Z_P/|\mathbf{r}'_P - \mathbf{r}'_T|$ between the projectile ion and the nucleus of the target and set

$$V'_P = -e^2 Z_P/r'_P \tag{15}$$

in (14). This is the *prior* form of the relativistic OBK first order approximation or ROBK1 approximation to the scattering amplitude for capture.

The *post* form may be obtained by replacing LV'_P in (14) by

$$V_T = -e^2 Z_T/r_T \tag{16}$$

It is not difficult to demonstrate that the prior and post forms of the scattering amplitude are precisely equivalent. They can be evaluated in exact closed analytical form and have been given explicitly for ground state to ground state capture (Moiseiwitsch and Stockman, 1980; Stockman, 1981; Moiseiwitsch, 1985).

It was shown by Moiseiwitsch and Stockman (1980) that very simple approximate analytical forms can be obtained by neglecting quantities of the second order of smallness in the fine-structure constant $\alpha = e^2/\hbar c \simeq \frac{1}{137}$. Thus for capture from an s state with principal quantum number n_i to an s state with principal quantum number n_f it is found that

(1) *without spin flip* of captured electron,

$$g_{fi}^{(1)}(q) \simeq 16 \frac{M_P}{m} a_0 \alpha \alpha^2 Z_T Z_P \left(\frac{\alpha^2 Z_T Z_P}{n_i n_f} \right)^{3/2}$$
$$\times [\tfrac{1}{2}(\gamma + 1)]^{1/2} (\alpha^2 Z_T^2/n_i^2 + q^2)^{-3}$$
$$\times [1 - \tfrac{1}{2}\delta^2 + \tfrac{1}{8}\pi\alpha(Z_T + Z_P)q]; \tag{17}$$

(2) *with spin flip* of captured electron,

$$g_{fi}^{(2)}(q) \simeq 8 \frac{M_P}{m} a_0 \alpha \alpha^2 Z_T Z_P \left(\frac{\alpha^2 Z_T Z_P}{n_i n_f}\right)^{3/2}$$
$$\times [\tfrac{1}{2}(\gamma + 1)]^{1/2} (\alpha^2 Z_T^2/n_i^2 + q^2)^{-3}$$
$$\times \delta(q^2 - q_{min}^2)^{1/2} \exp(i\phi_q), \tag{18}$$

where q is expressed in units of $(\alpha a_0)^{-1}$,

$$\delta = \left(\frac{\gamma - 1}{\gamma + 1}\right)^{1/2}, \tag{19}$$

and ϕ_q is the azimuthal angle of $\mathbf{q}$ about the polar axis $\hat{\mathbf{v}}$.

Then the relativistic capture cross sections, per electron captured, are given by

(1) *without spin flip*

$$\sigma_{ROBK1}^{(1)} = \frac{2\pi}{k_i^2} \int_{q_{min}}^{\infty} |g_{fi}^{(1)}(q)|^2 q \, dq$$
$$\simeq \frac{128}{5} \pi (a_0 \alpha)^2 \frac{(\alpha^2 Z_T Z_P)^2}{\gamma - 1} \left(\frac{\alpha^2 Z_T Z_P}{n_i n_f}\right)^3$$
$$\times [(\alpha Z_T/n_i)^2 + q_{min}^2]^{-5}$$
$$\times [(1 - \tfrac{1}{2}\delta^2)^2 + \tfrac{5}{18}\pi\alpha(Z_T + Z_P)\delta(1 - \tfrac{1}{2}\delta^2)]; \tag{20}$$

(2) *with spin flip*

$$\sigma_{ROBK1}^{(2)} \simeq \tfrac{8}{5}\pi(a_0\alpha)^2 \frac{(\alpha^2 Z_T Z_P)^2}{\gamma - 1}\left(\frac{\alpha^2 Z_T Z_P}{n_i n_f}\right)^3$$
$$\times \delta^4 [(\alpha Z_T/n_i)^2 + q_{min}^2]^{-5}, \tag{21}$$

giving for the total relativistic capture cross section, per electron captured from state i to state f, without and with spin flip,

$$\sigma_{fi}^{ROBK1} = \sigma_{ROBK1}^{(1)} + \sigma_{ROBK1}^{(2)}$$
$$\simeq \tfrac{128}{5}\pi(a_0\alpha)^2 \frac{(\alpha^2 Z_T Z_P)^2}{\gamma - 1}\left(\frac{\alpha^2 Z_T Z_P}{n_i n_f}\right)^3$$
$$\times [(\alpha Z_T/n_i)^2 + q_{min}^2]^{-5}$$
$$\times [1 - \delta^2 + \tfrac{5}{16}\delta^4 + \tfrac{5}{18}\pi\alpha(Z_T + Z_P)\delta(1 - \tfrac{1}{2}\delta^2)]. \tag{22}$$

These are symmetric with respect to T and P since

$$(\alpha Z_T/n_i)^2 + q_{min}^2 = (\alpha Z_P/n_f)^2 + q_{min}'^2, \tag{23}$$

where
$$q_{\min} = (P_f - \gamma P_i)/(\gamma^2 - 1)^{1/2},$$
$$q'_{\min} = (P_i - \gamma P_f)/(\gamma^2 - 1)^{1/2}, \tag{24}$$
with
$$P_i = (1 - \alpha^2 Z_T^2/n_i^2)^{1/2},$$
$$P_f = (1 - \alpha^2 Z_P^2/n_f^2)^{1/2}. \tag{25}$$

The formulas (20), (21) and (22) for the capture cross sections were derived using Dirac relativistic atomic wave functions. The terms $-\delta^2 + 5\delta^4/16$ in (22) which are independent of the nuclear charges arise from the interaction between the magnetic field produced by the electric current of the moving projectile ion and the magnetic moment of the Dirac electron in the target hydrogenic ion. If nonrelativistic Schrödinger atomic wave functions are used instead, these terms are no longer present and the capture cross section has the wrong outside factor in the high relativistic energy $E \to \infty$ limit.

Let us examine the high relativistic energy limit more closely. Since $E = M_P c^2(\gamma - 1)$ we see that in the $E \to \infty$ or $\gamma \to \infty$ limit, to the lowest order in the fine-structure constant α,

$$\sigma_{fi}^{\text{ROBK1}} \to 8\pi(a_0\alpha)^2(\alpha^2 Z_T Z_P)^2(\alpha^2 Z_T Z_P/n_i n_f)^3 E^{-1}, \tag{26}$$

TABLE I

ROBK1 ELECTRON CAPTURE CROSS SECTIONS FOR
$H^+ + H(1s) \to H(1s) + H^+$

Proton energy (10^3 MeV)	Capture cross sections (barns)			
	Without spin flip		With spin flip	
	N^a	F^b	N^a	F^c
1.0(−2)[†]	1.09(−3)	1.09(−3)	1.94(−9)	1.93(−9)
5.0(−2)	7.91(−8)	7.93(−8)	3.42(−12)	3.42(−12)
1.0(−1)	1.38(−9)	1.38(−9)	2.32(−13)	2.32(−13)
5.0(−1)	1.88(−13)	1.89(−13)	6.46(−16)	6.48(−16)
1.0(0)	6.54(−15)	6.55(−15)	7.17(−17)	7.18(−17)
5.0(0)	1.95(−17)	1.96(−17)	1.57(−18)	1.57(−18)
1.0(1)	3.90(−18)	3.90(−18)	5.05(−19)	5.06(−19)

[†] Numbers in brackets denote the powers of 10.
[a] N = numerical integration of exact formulas (Moiseiwitsch and Stockman, 1980).
[b] F = formula (20) without spin flip.
[c] F = formula (21) with spin flip.

expressing E in units of the rest mass energy $M_P c^2$ of the projectile ion. Thus in the high relativistic energy limit, the ROBK1 capture cross section decays as E^{-1}. This is essentially very different from the asymptotic behaviour obtained with the nonrelativistic first-order OBK1 approximation which gives a capture cross section decaying as E^{-6} in the high nonrelativistic energy limit.

On the other hand both the relativistic and nonrelativistic capture cross sections fall off as n^{-3} with principal quantum number n at high energies, neglecting quantities of the second order in the fine-structure constant α.

Good agreement was obtained by Moiseiwitsch and Stockman (1980) between the approximate formulas (20) and (21) and their respective exact ROBK1 cross-section calculations for 1s − 1s electron capture as can be seen from Table I where cross sections for protons incident on hydrogen atoms, without and with spin flip, are displayed.

III. Second-Order Theories

In the previous section we were concerned with a first order theory involving just the initial and final states. However it is to be expected that the presence of other states might have an important effect on capture and this has been investigated nonrelativistically as well as, more recently, using relativistic approaches.

To begin with, a classical viewpoint was adopted and we shall discuss this next.

A. Classical Double Scattering

A nonrelativistic double scattering theory for capture was developed by Thomas (1927) which produced an angular distribution with a peak at the angle $\theta_T = (\sqrt{3}/2) m/M_P$ in the target frame corresponding to conservation of linear momentum in the first encounter between the incident ion and the target electron. Further he obtained a total capture cross section decaying as $E^{-11/2}$ at high energies which dominates the OBK1 cross section asymptotically.

Relativistic generalizations of the Thomas theory have been given by Shakeshaft (1979) and Moiseiwitsch and Stockman (1979) who showed that the double scattering peak occurs at the angle

$$\theta_{RT} = \frac{(2\gamma + 1)^{1/2}}{\gamma + 1} \frac{m}{M_P} \qquad (27)$$

in the frame of the target nucleus corresponding to conservation of linear momentum in the first encounter. This angle tends to the nonrelativistic Thomas angle θ_T in the limit $\gamma \to 1$ as it should. Also they obtained total cross sections falling off as E^{-3} in the asymptotic region of high relativistic energies. This is a much more rapid decay with energy E than obtained with the ROBK1 approximation and so classical double scattering does not dominate the total first-order capture cross section in a relativistic theory.

B. THE SECOND-ORDER RELATIVISTIC OBK APPROXIMATION

The first quantum-mechanical investigation of electron capture using a second-order theory was carried out by Drisco (1955) at nonrelativistic energies. He employed the OBK2 approximation and demonstrated that there is a peak in the differential cross section at the Thomas angle θ_T corresponding to conservation of linear momentum in the first encounter. Moreover he obtained a total capture cross section which falls off as $E^{-11/2}$ as E approaches very high nonrelativistic energies, in agreement with the classical theory of Thomas.

A relativistic generalization of the second order OBK approximation has been made by Humphries and Moiseiwitsch (1984, 1985) neglecting all higher order terms in the fine-structure constant α. They found that the differential cross section has a sharp double scattering peak at the relativistic Thomas classical scattering angle θ_{RT} corresponding to conservation of relativistic linear momentum in the first encounter between the projectile ion and the target electron. However as the energy is increased they found that another broader peak, not present in nonrelativistic theory, begins to form at smaller angles of scattering and eventually dominates the Thomas peak (see Figs. 1-4). This broad peak arises from the interference between the first-order and second-order terms in the ROBK2 scattering amplitude and at ultrahigh relativistic energies its maximum occurs at $q^2 = \tfrac{5}{3}$, that is at the angle

$$\theta_B \simeq (2/3)^{1/2}(\gamma^2 - 1)^{-1/2} m/M_P \tag{28}$$

which approaches zero as $\gamma \to \infty$. The main contribution to the total capture cross section now comes from this broad peak, and the Thomas peak, unlike the nonrelativistic theory, provides a negligible contribution.

Humphries and Moiseiwitsch (1984) obtained the following formulas for the ROBK2 scattering amplitudes for ground state to ground state electron capture in the limit of ultrahigh relativistic energies:

(1) *without spin flip*

$$g^{(1)}_{1s-1s}(q) \to 8\frac{M_P}{m} a_0\alpha(\alpha^2 Z_T Z_P)^{5/2}\left(\frac{\gamma}{2}\right)^{1/2}\left(\frac{1}{q^6} - \frac{1}{q^4}\right); \quad (29)$$

(2) *with spin flip*

$$g^{(2)}_{1s-1s}(q) \to 8\frac{M_P}{m} a_0\alpha(\alpha^2 Z_T Z_P)^{5/2}\left(\frac{\gamma}{2}\right)^{1/2}$$

$$\times (q^2 - 1)^{1/2}\left(\frac{1}{q^6} - \frac{1}{q^4}\right). \quad (30)$$

Now using the cross-section formula

$$\sigma = \frac{2\pi(m/M_P)^2}{\gamma^2 - 1}\int_{q_{\min}}^{\infty}(|g^{(1)}(q)|^2 + |g^{(2)}(q)|^2)q\,dq, \quad (31)$$

it can be seen that as $\gamma \to \infty$

$$\sigma^{\text{ROBK2}}_{1s-1s} \to \frac{32\pi(a_0\alpha)^2(\alpha^2 Z_T Z_P)^5}{\gamma}\int_1^{\infty}\frac{(y-1)^2}{y^5}\,dy, \quad (32)$$

where we have set $y = q^2$ and noted that $q_{\min} \sim -\delta \to -1$ as $\gamma \to \infty$.

Thus the limiting form of the ROBK2 capture cross section is given by

$$\sigma^{\text{ROBK2}}_{1s-1s} \to 8\pi(a_0\alpha)^2(\alpha^2 Z_T Z_P)^5/3\gamma \quad (33)$$

as $\gamma \to \infty$.

It follows that the ROBK2 capture cross section falls off as E^{-1} at ultrahigh relativistic energies which is the same decay law as that obtained with the ROBK1 approximation. However the outside factor given by the second-order approximation is 1/3 times that given by the first-order approximation.

In the evaluation of the ROBK2 amplitudes Humphries and Moiseiwitsch (1984) neglected all higher-order terms in α. This is entirely satisfactory for very small atomic numbers, e.g., $Z_P = 1$, $Z_T = 1$ corresponding to incident protons on target hydrogen atoms. However even for not very high atomic numbers of the projectile and target, their formulas lead to a considerable overestimation of the capture cross sections. The reason for this is the application of the peaking approximation which is invalid when Z_P and Z_T are not small.

It has been shown by Moiseiwitsch (1988) that much improvement is achieved by using an averaging approximation in place of the peaking approximation. This replaces q^{-4} by $\{(\alpha Z_T/n_i)^2 + q^2\}^{-2}$ in the second-order term of the ROBK2 scattering amplitude for electron capture giving

(1) *without spin flip*

$$g_{fi}^{(1)}(q) \simeq 16 \frac{M_P}{m} a_0 \alpha \alpha^2 Z_T Z_P \left(\frac{\alpha^2 Z_T Z_P}{n_i n_f}\right)^{3/2}$$
$$\times \left(\frac{\gamma + 1}{2}\right)^{1/2} (\alpha^2 Z_T^2/n_i^2 + q^2)^{-3}$$
$$\times \{1 - \tfrac{1}{2}\delta^2 + (\gamma + \tfrac{1}{2}\delta^2)(\alpha^2 Z_T^2/n_i^2 + q^2)D^{-1}\}; \quad (34)$$

(2) *with spin flip*

$$g_{fi}^{(2)}(q) \simeq 8 \frac{M_P}{m} a_0 \alpha \alpha^2 Z_T Z_P \left(\frac{\alpha^2 Z_T Z_P}{n_i n_f}\right)^{3/2}$$
$$\times \left(\frac{\gamma + 1}{2}\right)^{1/2} (\alpha^2 Z_T^2/n_i^2 + q^2)^{-3} \delta (q^2 - q_{\min}^2)^{1/2}$$
$$\times \{1 + (2\gamma + 1)(\alpha^2 Z_T^2/n_i^2 + q^2)D^{-1}\}, \quad (35)$$

where

$$D = q^2 + \alpha^2 Z_T^2/n_i^2 - 2(\gamma - 1) - 2i\alpha(Z_T/n_i + Z_P/n_f)(\gamma^2 - 1)^{1/2}, \quad (36)$$

retaining only leading-order terms in α except when they arise in $q^2 + \alpha^2 Z_T^2/n_i^2$. It can be seen that there is a resonance in the second-order term at $q^2 \simeq 2(\gamma - 1)$ corresponding to the relativistic Thomas angle θ_{RT} given by (27).

These formulas for the scattering amplitudes enable us to obtain a simple analytical closed expression for the total capture cross section given by the ROBK2 approximation per electron captured from state i to state f:

$$\sigma_{fi}^{ROBK2} \simeq \tfrac{128}{5} \pi (a_0 \alpha)^2 \frac{(\alpha^2 Z_T Z_P)^2}{\gamma - 1} \left(\frac{\alpha^2 Z_T Z_P}{n_i n_f}\right)^3 [(\alpha Z_T/n_i)^2 + q_{\min}^2]^{-5}$$
$$\times \Bigg\{ 1 - \delta^2 + \tfrac{5}{16}\delta^4 + 5c_1 a^{-2}\bigg[\tfrac{1}{4} + \tfrac{2}{3}a^{-1} + \tfrac{3}{2}a^{-2} + 4a^{-3}$$
$$+ a^{-4}\left(\frac{\pi}{\zeta} - \frac{2(\gamma + 1)}{2\gamma + 1} + 5\ln(2\gamma + 1)\right)\bigg]$$
$$+ 5c_2 a^{-2}\bigg[\tfrac{1}{3} + a^{-1} + 3a^{-2} + a^{-3}\left(\frac{\pi}{\zeta} - \frac{2(\gamma + 1)}{2\gamma + 1} + 4\ln(2\gamma + 1)\right)\bigg]$$
$$+ 5c_3 a^{-2}\bigg[\tfrac{1}{2} + 2a^{-1} + a^{-2}\left(\frac{\pi}{\zeta} - \frac{2(\gamma + 1)}{2\gamma + 1} + 3\ln(2\gamma + 1)\right)\bigg]\Bigg\},$$

(37)

where

$$a = 2(\gamma + 1),$$
$$\zeta = \alpha(Z_T/n_i + Z_P/n_f)/\delta,$$
$$c_1 = a[\tfrac{1}{2}\delta^4 Q - 2(1 - \tfrac{1}{2}\delta^2)P],$$
$$c_2 = P^2 + 2(1 - \tfrac{1}{2}\delta^2)P - \tfrac{1}{4}\delta^4(Q^2 + 2Q) - \tfrac{1}{2}a\delta^4 Q,$$
$$c_3 = \tfrac{1}{4}\delta^4(Q^2 + 2Q),$$
$$P = \gamma + \tfrac{1}{2}\delta^2,$$
$$Q = 2\gamma + 1. \tag{38}$$

This formula (Moiseivitsch, 1989) for the capture cross section is symmetric with respect to T and P.

The results of calculations carried out using the ROBK2 approximation given by (37) and (38) for 1s-1s electron capture by incident protons from hydrogen atoms, corresponding to taking $Z_P = Z_T = 1$ and $n_i = n_f = 1$, are displayed in Table II. Here they are compared with the values obtained using the matrix continuum distorted wave (MCDW) approximation, the relativistic eikonal (RE), and the relativistic first-order Born (R1B) approximations,

TABLE II

Electron Capture Cross Sections for $H^+ + H(1s) \rightarrow H(1s) + H^+$ Calculated Using Various Approximations

Proton energy (10^3 MeV)	Capture cross sections (barns)				
	ROBK1[a]	RE[b]	R1B[c]	MCDW[d]	ROBK2[e]
0.1	1.4(−9)[†]	5.9(−10)	8.4(−10)	6.7(−10)	7.1(−10)
0.5	1.9(−13)	6.4(−14)	9.2(−14)	1.7(−13)	1.6(−13)
1	6.6(−15)	1.7(−15)	—	7.6(−15)	7.4(−15)
5	2.1(−17)	2.9(−18)	—	2.9(−17)	2.9(−17)
10	4.4(−18)	7.5(−19)	—	3.9(−18)	4.1(−18)
50	3.9(−19)	1.0(−19)	—	1.4(−19)	1.4(−19)

[†] Numbers in brackets denote the powers of 10.
[a] Moiseiwitsch and Stockman (1980).
[b] Eichler (1985).
[c] Eichler (1987, 1988).
[d] Deco and Rivarola (1987b).
[e] Moiseiwitsch (1988).

to be discussed in the following sections, as well as the ROBK1 approximation discussed earlier in Sec. II.

We conclude this section by looking at the ROBK2 differential cross section for electron capture, including collisions without and with spin flip, evaluated using

$$\frac{d\sigma_{fi}^{ROBK2}}{d\Omega} = |g_{fi}^{(1)}(q)|^2 + |g_{fi}^{(2)}(q)|^2 \tag{39}$$

where the scattering amplitudes $g_{fi}^{(1)}$ and $g_{fi}^{(2)}$ have the analytical forms (34) and (35), respectively.

The results of calculations of angular distributions for 1s-1s electron capture by incident protons from target hydrogen atoms are shown in Figs. 1, 2, 3, and 4 for 100, 500, 1000, and 5000 MeV proton impact energies respectively. The relativistic Thomas peak at the angle θ_{RT} given by (27) and the relativistic broad peak at the angle θ_B given by (28) in the differential cross section curves are manifest.

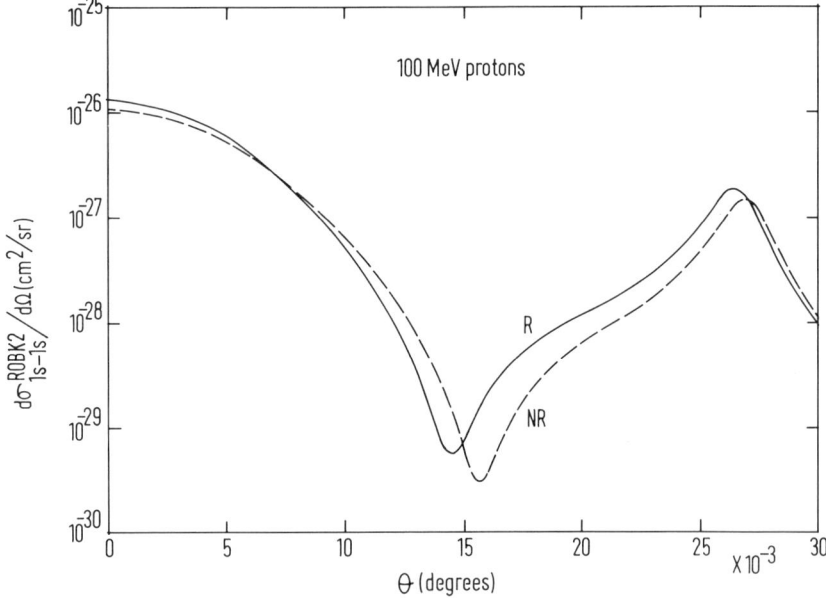

FIG. 1. Differential capture cross sections $d\sigma_{1s-1s}^{ROBK2}/d\Omega$ for $H^+ + H(1s) \to H(1s) + H^+$ given by the relativistic (R) and nonrelativistic (NR) second-order OBK approximations as functions of the scattering angle θ in the laboratory frame for 100 MeV protons.

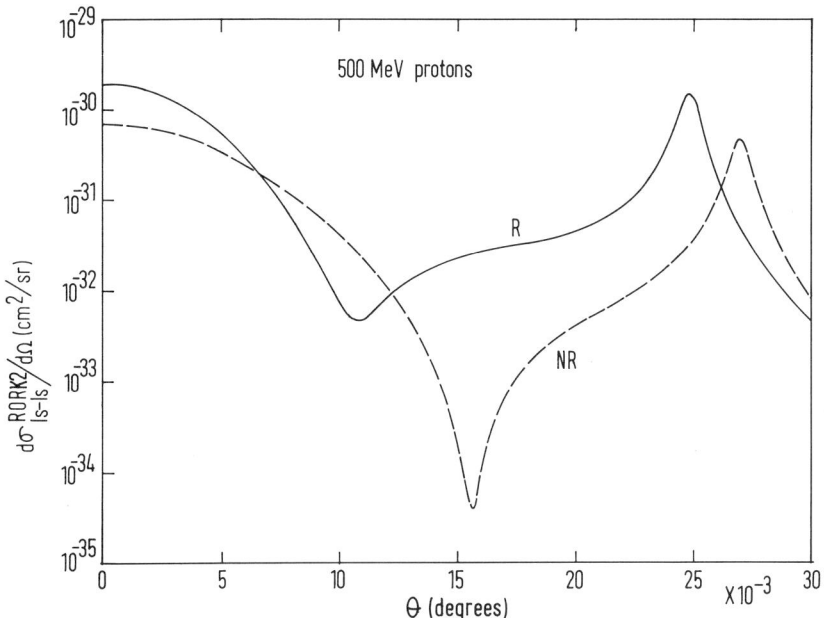

FIG. 2. Differential capture cross sections $d\sigma^{ROBK2}_{1s-1s}/d\Omega$ for $H^+ + H(1s) \to H(1s) + H^+$ given by the relativistic (R) and nonrelativistic (NR) second-order OBK approximations as functions of the scattering angle θ in the laboratory frame for 500 MeV protons.

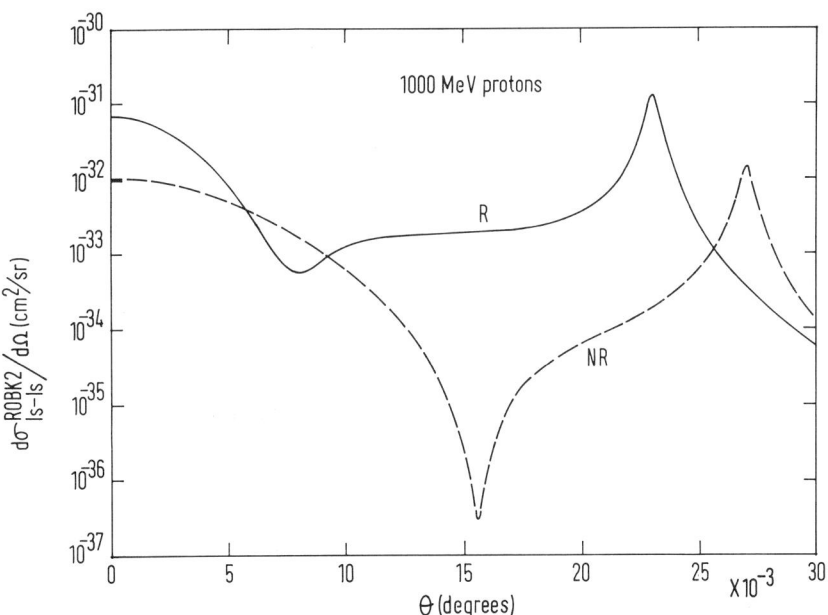

FIG. 3. Differential capture cross sections $d\sigma^{ROBK2}_{1s-1s}/d\Omega$ for $H^+ + H(1s) \to H(1s) + H^+$ given by the relativistic (R) and nonrelativistic (NR) second-order OBK approximations as functions of the scattering angle θ in the laboratory frame for 1000 MeV protons.

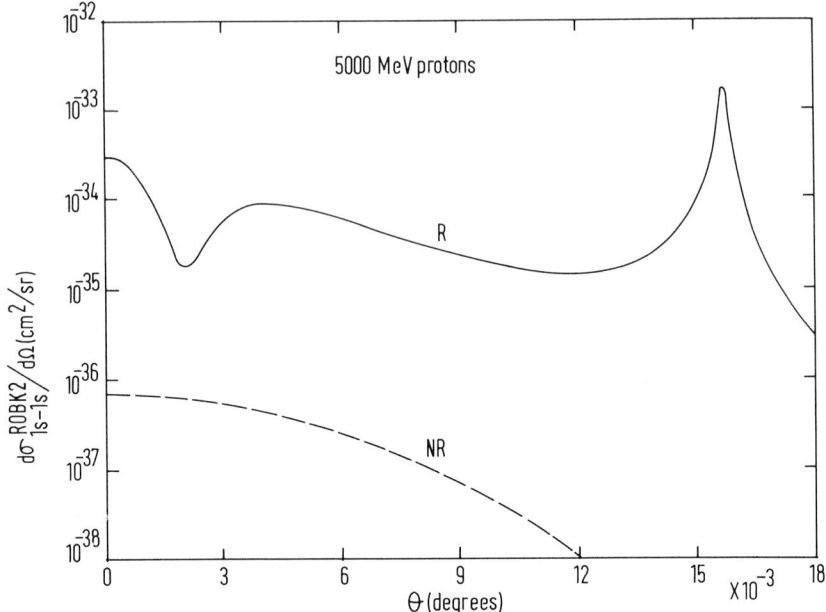

FIG. 4. Differential capture cross sections $d\sigma_{1s-1s}^{ROBK2}/d\Omega$ for $H^+ + H(1s) \to H(1s) + H^+$ given by the relativistic (R) and nonrelativistic (NR) second-order OBK approximations as functions of the scattering angle θ in the laboratory frame for 5000 MeV protons.

IV. Relativistic Continuum Distorted Wave Approximation

McCann (1985) and Deco and Rivarola (1986, 1987b, c) have developed relativistic theories of electron capture based on spinor continuum distorted waves in the entry and exit channels. Their approach regards the electron as being in a continuum state of the projectile ion in the entry channel and in a continuum state of the target ion in the exit channel. The distortions produced by the electron-nucleus Coulomb potentials $V'_P(r'_P)$ and $V_T(r_T)$ are taken into account by using the hypergeometric functions

$$_1F_1[iZ_P c/v, 1, i\gamma(vr'_P + \mathbf{v} \cdot \mathbf{r}'_P)]$$

and

$$_1F_1[iZ_T c/v, 1, i\gamma(vr_T + \mathbf{v} \cdot \mathbf{r}_T)]$$

respectively. Because matrix operators are used in a relativistic theory, Deco and Rivarola (1987c) refer to their method as the matrix continuum distorted wave (MCDW) model.

The angular distributions obtained by Deco and Rivarola (1987b) using the MCDW approximation are very close to those obtained by Humphries and Moiseiwitsch (1984) using the ROBK2 approximation. Both the relativistic Thomas peak at θ_{RT} and the broad peak at θ_B are found. However the first order MCDW model produces a spurious catastrophic dip at the relativistic Thomas peak in the same way as happens in the nonrelativistic first-order CDW theory. This catastrophic dip is due to the omission of a second order term (Crothers and McCann, 1984).

The MCDW model also leads to an E^{-1} decay in the capture cross section at ultrahigh relativistic energies in agreement with the ROBK2 approximation. Moreover there is close agreement between the total capture cross sections produced by the two approximations over a wide range of relativistic energies E for protons incident on hydrogen atoms as can be seen from Table II.

V. Relativistic Eikonal Approximation

A relativistic generalization of the eikonal (RE) approximation has been introduced by Eichler (1985) for electron capture. If $Z_T > Z_P$ the distortion arising from the target nucleus is greater than that due to the projectile ion and the prior form of the probability amplitude should be used. This is given by

$$a_{fi}^{RE}(\rho) = -\frac{i}{\hbar} \int_{-\infty}^{\infty} dt \int d\mathbf{r}_T \exp[i(E_f^{P'}t' - E_i^T t)/\hbar] \\ \times [S\psi_f^{P'}(\mathbf{r}_P')U_f]^\dagger L V_P'(r_P')\psi_i^T(\mathbf{r}_T), \quad (40)$$

where

$$U_f = \exp\left[\frac{i}{\hbar}\int_t^\infty V_T(r_T)d\tau\right] \quad (41)$$

or equivalently

$$U_f = \exp\left[i\alpha Z_T \frac{c}{v}\ln(vr_T + \mathbf{v}\cdot\mathbf{r}_T)\right] \quad (42)$$

Using this formula (40) for the probability amplitude and expanding in powers of the fine-structure constant α, retaining only the lowest-order terms, Eichler (1985) has derived a simple analytical closed formula for the 1s-1s electron capture relativistic eikonal cross section summed over non spin flip and spin flip collisions. He found that the total cross section per electron captured was given by

$$\sigma_{1s\text{-}1s}^{RE} \simeq \tfrac{128}{5}\pi(a_0\alpha)^2 \frac{(\alpha^2 Z_T Z_P)^5}{\gamma - 1}[(\alpha Z_T)^2 + q_{min}^2]^{-5}$$

$$\times \frac{v_T \pi}{\sinh v_T \pi} \exp[-2v_T \tan^{-1}(|q_{min}|/\alpha Z_T)]I_{1s\text{-}1s}^{RE}, \qquad (43)$$

where $v_T = \alpha Z_T c/v$ and

$$I_{1s\text{-}1s}^{RE} = I_0^{RE} + I_{1T}^{RE} + I_{1P}^{RE} \qquad (44)$$

with

$$I_0^{RE} = 1 - \delta^2 + \tfrac{5}{16}\delta^4 + \tfrac{5}{4}(1 - \tfrac{1}{2}\delta^2)\eta_T q_{min} + \tfrac{5}{12}(\eta_T q_{min})^2, \qquad (45)$$

$$I_{1T}^{RE} = \tfrac{5}{2}\pi\alpha Z_T[\tfrac{1}{9}\delta(1 - \tfrac{1}{2}\delta^2) - \tfrac{1}{14}\delta^2\eta_T], \qquad (46)$$

$$I_{1P}^{RE} = \tfrac{5}{2}\pi\alpha Z_P[\tfrac{1}{9}\delta(1 - \tfrac{1}{2}\delta^2) - \tfrac{1}{9}\delta^2\eta_T$$
$$- \tfrac{1}{14}\delta^2(1 - \delta^2)\eta_T + \tfrac{1}{14}\delta^3\eta_T^2]. \qquad (47)$$

where $\eta_T = v_T/\alpha Z_T = c/v$.

Eichler (1985) obtained good agreement, except for very large Z_T, between the closed analytical form (43), with (44) to (47), and his exact numerical calculations carried out using the RE approximation without explicitly making an expansion in α.

The relativistic eikonal formula (43) approaches, as it should, the relativistic first order OBK formula (22) in the limit $v_T \to 0$ which corresponds to $U_f \to 1$, that is no allowance made for distortion in the exit channel as well as the entry channel.

The RE approximation for the capture cross section is not symmetric with respect to T and P, unlike the ROBK1 and ROBK2 approximations. Only distortion in the exit channel is allowed for, distortion in the entry channel being neglected since $Z_T > Z_P$. This is a serious defect in the RE approximation even though it gives quite good agreement with the experimental data for electron capture. One would have anticipated that the RE approximation should only be valid for $Z_T \gg Z_P$.

A. Relativistic Symmetric Eikonal Approximation

A symmetric form of the relativistic eikonal (RSE) approximation for the probability amplitude for electron capture has been derived by Moiseiwitsch

(1986). He finds that the *post* form is given by

$$a_{fi}^{\text{RSE}}(\rho) = -\frac{i}{\hbar}\int_{-\infty}^{\infty} dt \int d\mathbf{r}_T \exp[i(E_f^{P'}t' - E_i^T t)/\hbar]$$
$$\times [S\psi_f^{P'}(\mathbf{r}_P')U_f]^\dagger W_T(\mathbf{r}_T)U_i\psi_i^T(\mathbf{r}_T), \qquad (48)$$

where

$$W_T(\mathbf{r}_T) = \left[1 - \frac{c}{v}\frac{\boldsymbol{\alpha}\cdot(\hat{\mathbf{r}}_T + \hat{\mathbf{v}})}{1 + \hat{\mathbf{v}}\cdot\hat{\mathbf{r}}_T}\right]V_T(r_T). \qquad (49)$$

U_f is given by (41) or (42) while

$$U_i = \exp\left[-\frac{i}{\hbar}\int_{-\infty}^{t'} V'_P(r_P')d\tau'\right] \qquad (50)$$

or equivalently

$$U_i = \exp\left[-i\alpha Z_P \frac{c}{v}\ln(vr_P' + \mathbf{v}\cdot\mathbf{r}_P')\right] \qquad (51)$$

The *prior* form is given by

$$a_{fi}^{\text{RSE}}(\rho) = -\frac{i}{\hbar}\int_{-\infty}^{\infty} dt \int d\mathbf{r}_T \exp[i(E_f^{P'}t' - E_i^T t)/\hbar]$$
$$\times [S\psi_f^{P'}(\mathbf{r}_P')U_f]^\dagger S^{-1} W_P'(\mathbf{r}_P')S^{-1}U_i\psi_i^T(\mathbf{r}_T), \qquad (52)$$

where

$$W_P'(\mathbf{r}_P') = \left[1 - \frac{c}{v}\frac{\boldsymbol{\alpha}\cdot(\hat{\mathbf{r}}_P' + \hat{\mathbf{v}})}{1 + \hat{\mathbf{v}}\cdot\hat{\mathbf{r}}_P'}\right]V_P'(r_P') \qquad (53)$$

and is equivalent to the post form (48).

By expanding in powers of α and retaining only the lowest-order terms, Moiseiwitsch (1987a) has derived a closed analytical formula for the 1s-1s electron capture cross section using the RSE approximation. He obtained for the total cross section per electron captured without and with spin flip:

$$\sigma_{1s\text{-}1s}^{\text{RSE}} \simeq \tfrac{128}{5}\pi(a_0\alpha)^2 \frac{(\alpha^2 Z_T Z_P)^5}{\gamma - 1}[(\alpha Z_T)^2 + q_{\min}^2]^{-5}$$
$$\times \frac{v_T\pi}{\sinh v_T\pi}\exp[-2v_T\tan^{-1}(|q_{\min}|/\alpha Z_T)]$$
$$\times \frac{v_P\pi}{\sinh v_P\pi}\exp[-2v_P\tan^{-1}(|q'_{\min}|/\alpha Z_P)]I_{1s\text{-}1s}^{\text{RSE}}, \qquad (54)$$

where $v_P = \alpha Z_P c/v$ and

$$I^{\text{RSE}}_{1s\text{-}1s} = I^{\text{RSE}}_0 + I^{\text{RSE}}_{1T} + I^{\text{RSE}}_{1P} \tag{55}$$

with

$$\begin{aligned} I^{\text{RSE}}_0 = {} & 1 - \delta^2 + \tfrac{5}{16}\delta^4 + \tfrac{5}{4}(1 - \tfrac{1}{2}\delta^2)(\eta_T q_{\min} + \eta_P q'_{\min}) \\ & + \tfrac{5}{12}(\eta_T q_{\min} + \eta_P q'_{\min})^2 \\ & + \tfrac{5}{12}(3 - 2\delta^2)\eta_T q_{\min}\eta_P q'_{\min} \\ & + \tfrac{5}{6}\eta_T q_{\min}\eta_P q'_{\min}(\eta_T q_{\min} + \eta_P q'_{\min}) \\ & + \tfrac{5}{8}(\eta_T q_{\min}\eta_P q'_{\min})^2, \end{aligned} \tag{56}$$

$$\begin{aligned} I^{\text{RSE}}_{1T} = {} & \tfrac{5}{2}\pi\alpha Z_T[\tfrac{1}{9}\delta(1 - \tfrac{1}{2}\delta^2) - \tfrac{1}{14}\delta^2\eta_T - \tfrac{1}{9}\delta^2\eta_P \\ & - \tfrac{1}{14}\delta^2(1 - \delta^2)\eta_P + \tfrac{1}{7}\delta^3\eta_T\eta_P \\ & + \tfrac{1}{14}\delta^3\eta_P^2 - \tfrac{1}{10}\delta^4\eta_T\eta_P^2], \end{aligned} \tag{57}$$

$$\begin{aligned} I^{\text{RSE}}_{1P} = {} & \tfrac{5}{2}\pi\alpha Z_P[\tfrac{1}{9}\delta(1 - \tfrac{1}{2}\delta^2) - \tfrac{1}{14}\delta^2\eta_P - \tfrac{1}{9}\delta^2\eta_T \\ & - \tfrac{1}{14}\delta^2(1 - \delta^2)\eta_T + \tfrac{1}{7}\delta^3\eta_P\eta_T \\ & + \tfrac{1}{14}\delta^3\eta_T^2 - \tfrac{1}{10}\delta^4\eta_P\eta_T^2], \end{aligned} \tag{58}$$

where $\eta_P = v_P/\alpha Z_P = c/v$.

The capture cross section $\sigma^{\text{RSE}}_{1s\text{-}1s}$ given by (54) and (55) to (58) is symmetric with respect to T and P. Setting $v_P = 0$ and $\eta_P = 0$ in these formulas produces the non-symmetric eikonal approximation for the capture cross section given by (43) and (44) to (47).

It is interesting to see that I^{RE}_0 and I^{RSE}_0 approach the same value 5/48 in the ultra-high relativistic limit $\gamma \to \infty$, since $\eta_T \to 1$, $\eta_P \to 1$, $\delta \to 1$ and $q_{\min} \to -1$, $q'_{\min} \to -1$ in this limit. However the scattering amplitudes $g^{\text{RSE}}(q)$ and $g^{\text{RE}}(q)$ possess different distributions with respect to momentum change q. Supposing that αZ_T and αZ_P are sufficiently small for the eikonal prefactors to be approximated by 1 and allowing both for non-spin-flip and spin-flip electron capture, we obtain in the limit $\gamma \to \infty$

$$g^{\text{RSE}}(q) \to \frac{8M_P}{m} a_0 \alpha (\alpha^2 Z_T Z_P)^{5/2} \left(\frac{\gamma}{2}\right)^{1/2} \frac{q^2 - 1}{q^5}, \tag{59}$$

$$g^{\text{RE}}(q) \to \frac{8M_P}{m} a_0 \alpha (\alpha^2 Z_T Z_P)^{5/2} \left(\frac{\gamma}{2}\right)^{1/2} \frac{(q^2 - 1)^{1/2}}{q^5}. \tag{60}$$

The limiting form of $g^{\text{RSE}}(q)$ given by (59) is the same as that given by the ROBK2 approximation after a sum over non spin flip and spin flip capture has been carried out whereas $g^{\text{RE}}(q)$ given by (60) is different.

Nevertheless the limiting RSE and RE capture cross sections are equal since, as $\gamma \to \infty$,

$$\sigma_{1s\text{-}1s}^{RSE} \to \frac{32\pi(a_0\alpha)^2(\alpha^2 Z_T Z_P)^5}{\gamma} \int_1^\infty \frac{(y-1)^2}{y^5} dy, \tag{61}$$

$$\sigma_{1s\text{-}1s}^{RE} \to \frac{32\pi(a_0\alpha)^2(\alpha^2 Z_T Z_P)^5}{\gamma} \int_1^\infty \frac{y-1}{y^5} dy, \tag{62}$$

where $y = q^2$, and fortuitously these are both equal to $8\pi(a_0\alpha)^2(\alpha^2 Z_T Z_P)^5/3\gamma$ which is the same as that given by the ROBK2 approximation (33) in the limit $\gamma \to \infty$ and is 1/3 of the limiting value of $\sigma_{1s\text{-}1s}^{ROBK1}$ given by (26).

Deco and Rivarola (1987a) have also developed a relativistic symmetric eikonal theory. Although their analysis is based on a similar approach to that of Moiseiwitsch (1986, 1987a) their formulas are quite different and rather more complicated. However the numerical results given by the two symmetric eikonal theories are very close for protons incident on hydrogen atoms but considerably smaller than given by the nonsymmetric eikonal approximation except at very high energies as can be seen from Table III.

TABLE III

CROSS SECTIONS FOR 1s-1s ELECTRON CAPTURE CALCULATED USING RELATIVISTIC EIKONAL AND SYMMETRIC EIKONAL APPROXIMATIONS

Proton energy (10^3 MeV)	RE	RSE		RE	RSE	
		DR[a]	M[b]		DR[a]	M[b]
	$Z_T = 1$ $Z_P = 1$			$Z_T = 5$ $Z_P = 1$		
0.5	6.4(−14)[†]	3.7(−14)	3.7(−14)	1.8(−10)	—	1.0(−10)
1	1.7(−15)	9.6(−16)	1.0(−15)	5.0(−12)	2.7(−12)	2.8(−12)
5	2.9(−18)	2.5(−18)	2.8(−18)	8.6(−15)	7.0(−15)	7.7(−15)
10	7.5(−19)	7.5(−19)	7.8(−19)	2.1(−15)	2.2(−15)	2.2(−15)
50	1.0(−19)	1.1(−19)	1.0(−19)	2.9(−16)	3.0(−16)	2.9(−16)
100	5.1(−20)	5.0(−20)	5.0(−20)	1.4(−16)	1.4(−16)	1.4(−16)

Cross section (barns) per electron captured — Approximations

[†] Numbers in brackets denote powers of 10.
[a] DR = Deco and Rivarola (1987a).
[b] M = Moiseiwitsch (1987a).

B. Relativistic Eikonal Phase Factors

The relativistic eikonal phase factors U_f given by (41) and U_i given by (50) arise from multiple "soft" elastic collisions in the exit and entry channels produced by the electron-nucleus potentials $V_T(r_T)$ and $V'_P(r'_P)$, respectively.

An instructive way of understanding this can be achieved by summing the relativistic Born series for electron capture assuming that "soft" elastic collisions occur before and after electron capture takes place (Moiseiwitsch, 1987b). The amplitude for electron capture from an initial state i of the target system T to the final state f of the projectile ion P plus electron is given by the Born expansion:

$$a_{fi}(\rho, t = \infty) = \sum_{n=1}^{\infty} a_f^n(\infty) \tag{63}$$

where, assuming that only one electron transfer from the target T to the projectile P takes place per collision and that "soft" elastic collisions occur before and after electron transfer, we have for the nth-order term

$$a_f^n(\infty) = \left(\frac{-i}{\hbar}\right)^n \sum_{m=0}^{n-1} \sum_{s_1,\ldots,s_{n-1}} \int_{-\infty}^{\infty} dt'_n \int_{-\infty}^{t'_n} dt'_{n-1} \cdots \int_{-\infty}^{t'_{m+3}} dt'_{m+2}$$

$$\times \int_{-\infty}^{t_{m+2}} dt_{m+1} \int_{-\infty}^{t_{m+1}} dt_m \cdots \int_{-\infty}^{t_2} dt_1$$

$$\times \exp[i(E_f^{P'} t'_{m+1} - E_i^T t_{m+1})/\hbar]$$

$$\times V_{fs_{n-1}}^{PP}(t'_n) V_{s_{n-1}s_{n-2}}^{PP}(t'_{n-1}) \cdots V_{s_{m+2}s_{m+1}}^{PP}(t'_{m+2})$$

$$\times W_{s_{m+1}s_m}^{PT}(t_{m+1}) V_{s_m s_{m-1}}^{TT}(t_m) \cdots V_{s_1 i}^{TT}(t_1). \tag{64}$$

Here the states $s_1, \ldots, s_m$ associated with the target T are all taken to have the same energy eigenvalue as that of the initial state E_i^T while the states $s_{m+1}, \ldots, s_{n-1}$ associated with the projectile P are all taken to have the same energy eigenvalue as that of the final state $E_f^{P'}$. All possible values of m from 0 to $n-1$ are summed. There are n integrations over time, the times $t_1, \ldots, t_{m+1}$ being measured with respect to the target frame and the times $t'_{m+1}, t'_{m+2}, \ldots, t'_n$ being measured with respect to the projectile frame.

The n transition matrix elements occurring in (64) are given by

$$V_{s_{l+1}s_l}^{TT}(t_{l+1}) = \int \psi_{s_{l+1}}^T(\mathbf{r}_T)^\dagger S^{-1} V'_P(r'_P) S^{-1} \psi_{s_l}^T(\mathbf{r}_T) d\mathbf{r}_T \tag{65}$$

for $l = 0, \ldots, m-1$ with $s_0 = i$;

$$V_{s_{l+1}s_l}^{PP}(t'_{l+1}) = \int [S\psi_{s_{l+1}}^{P'}(\mathbf{r}'_P)]^\dagger V_T(r_T) S\psi_{s_l}^{P'}(\mathbf{r}'_P) d\mathbf{r}'_P \tag{66}$$

for $l = m + 1, \ldots, n - 1$ with $s_n = f$; and for the post interaction

$$W^{PT}_{s_{m+1}s_m}(t_{m+1}) = \int [S\psi^{P'}_{s_{m+1}}(\mathbf{r}'_P)]^\dagger W_T(\mathbf{r}_T)\psi^T_{s_m}(\mathbf{r}_T) d\mathbf{r}_T, \tag{67}$$

while for the prior interaction $W_T(\mathbf{r}_T)$ is replaced by $S^{-1}W'_P(\mathbf{r}'_P)S^{-1}$ in (67) where $W_T(\mathbf{r}_T)$ and $W'_P(\mathbf{r}'_P)$ are given by (49) and (53) respectively.

Now we use the closure relations

$$\sum_s \psi^T_s(\mathbf{r}_1)\psi^T_s(\mathbf{r}_2)^\dagger = \delta(\mathbf{r}_1 - \mathbf{r}_2) \tag{68}$$

$$\sum_s \psi^{P'}_s(\mathbf{r}'_1)\psi^{P'}_s(\mathbf{r}'_2)^\dagger = \delta(\mathbf{r}'_1 - \mathbf{r}'_2), \tag{69}$$

and then for the post-interaction case we find that

$$a^n_f(\infty) = \left(\frac{-i}{\hbar}\right)^n \sum_{m=0}^{n-1} \int_{-\infty}^{\infty} dt'_n \int_{-\infty}^{t'_n} dt'_{n-1} \cdots \int_{-\infty}^{t'_{m+3}} dt'_{m+2}$$
$$\times \int_{-\infty}^{t_{m+2}} dt_{m+1} \int_{-\infty}^{t_{m+1}} dt_m \cdots \int_{-\infty}^{t_2} dt_1$$
$$\times \exp[i(E^{P'}_f t'_{m+1} - E^T_i t_{m+1})/\hbar]$$
$$\times \int [S\psi^{P'}_f(\mathbf{r}'_n)]^\dagger SV_T(r_n)SSV_T(r_{n-1})S \cdots SV_T(r_{m+2})S$$
$$\times W_T(\mathbf{r}_{m+1})S^{-1}V'_P(r'_m)S^{-1} \cdots S^{-1}V'_P(r'_1)S^{-1}\psi^T_i(\mathbf{r}_1)d\mathbf{r}'. \tag{70}$$

On carrying out repeated integrations by parts we can write

$$a^n_f(\infty) = \frac{-i}{\hbar} \sum_{m=0}^{n-1} \frac{1}{(n-m-1)!\,m!} \int_{-\infty}^{\infty} dt' \exp[i(E^{P'}_f t' - E^T_i t)/\hbar]$$
$$\times \int [S\psi^{P'}_f(\mathbf{r}')]^\dagger \left(\frac{-i}{\hbar}\int_{t'}^{\infty} dt'_1 SV_T(r_1)S\right)^{n-m-1}$$
$$\times W_T(\mathbf{r})$$
$$\times \left(\frac{-i}{\hbar}\int_{-\infty}^{t} dt_2 S^{-1}V'_P(r'_2)S^{-1}\right)^m \psi^T_i(\mathbf{r})d\mathbf{r}' \tag{71}$$

and since

$$\sum_{n=1}^{\infty}\sum_{m=0}^{n-1} \frac{X^{n-m-1}Y^m}{(n-m-1)!\,m!} = \exp(X+Y) \tag{72}$$

it follows that the probability amplitude for electron capture (63) is given by (48), on changing from the projectile frame to the target frame coordinates, and vice versa, where

$$U_f = \exp\left[\frac{i}{\hbar}\frac{1}{\gamma}\int_t^\infty dt_1 S V_T(r_1) S\right], \qquad (73)$$

$$U_i = \exp\left[\frac{-i}{\hbar}\frac{1}{\gamma}\int_{-\infty}^{t'} dt'_2 S^{-1} V'_P(r'_2) S^{-1}\right]. \qquad (74)$$

The presence of the S and S^{-1} matrix operators is a complicating feature here. However it is well known that

$$\int \psi_s^T(\mathbf{r})^\dagger \boldsymbol{\alpha} \psi_n^T(\mathbf{r}) d\mathbf{r} = \tfrac{1}{2} i\alpha (E_s - E_n) \int \psi_s^{T\dagger} \mathbf{r} \psi_n^T \, d\mathbf{r}, \qquad (75)$$

where $\boldsymbol{\alpha}$ is the vector of the Dirac matrices and α is the fine structure constant, and since we have been assuming that $E_s = E_n$ corresponding to "soft" elastic collisions prior to electron transfer we may take $\int \psi_s^{T\dagger} \boldsymbol{\alpha} \psi_n^T d\mathbf{r} = \mathbf{0}$ which enables us to replace $\gamma^{-1} SS$ by unity in (73). Similarly we may replace $\gamma^{-1} S^{-1} S^{-1}$ by unity in (74) and so U_f and U_i may be approximated by (41) and (50) respectively as was done in the relativistic eikonal and symmetric eikonal approximations discussed in Sec. V.

C. First-Order Born Approximation With Coulomb Boundary Conditions

The importance of satisfying Coulomb boundary conditions in electron capture was first emphasised by Cheshire (1964) in a nonrelativistic theory involving a gauge transformation. Later boundary corrected nonrelativistic first order approximation formulas were obtained by Belkić et al. (1979). These were generalized in a relativistic theory by Humphries (1985) who showed that the post form of the first order probability amplitude is given by the symmetric formula

$$a_{fi}(\rho) = \frac{-i}{\hbar}\int_{-\infty}^\infty dt \int d\mathbf{r}_T (S\psi_f^{P'} U_f)^\dagger [V_T(r_T) - V_T(R)] U_i \psi_i^T \qquad (76)$$

with

$$U_i = \exp\left[-i\alpha Z_P \frac{c}{v} \ln(vR' - \mathbf{v}\cdot\mathbf{R}')\right], \qquad (77)$$

$$U_f = \exp\left[i\alpha Z_T \frac{c}{v} \ln(vR + \mathbf{v}\cdot\mathbf{R})\right], \qquad (78)$$

where

$$\mathbf{R} = \mathbf{r}_T - \mathbf{r}_P, \qquad R = (\rho^2 + v^2 t^2)^{1/2}$$

and with corresponding expressions for $\mathbf{R}'$ and R'. The prior form of the first-order probability amplitude can be derived from (76) by replacing $V_T(r_T) - V_T(R)$ by $S^{-1}[V'_P(r'_P) - V'_P(R')]S^{-1}$ and can be shown to be equivalent to it.

We note that it is not the interaction $e^2 Z_T Z_P / R$ between the nuclei which is present in these formulas but $V_T(R) = -e^2 Z_T / R$ and $V'_P(R') = -e^2 Z_P / R'$.

The relativistic first order Born approximation with Coulomb boundary conditions given by (76) or its prior equivalent form is called the R1B approximation by Eichler (1987, 1988).

The results of calculations using the R1B approximation are given in Table II for $H^+ + H(1s) \to H(1s) + H^+$ electron capture collisions, and in Table IV for 1s-1s electron capture by bare neon ions ($Z_P = 10$) having 1050 MeV/amu energy incident on various target atoms.

It can be seen that there is quite good agreement between the R1B, RE, and ROBK2 approximations although they involve quite different approaches. Of these only the ROBK2 (and the MCDW) approximation produces a Thomas peak in the angular distribution arising from double encounters.

TABLE IV

Cross Sections for 1s-1s Electron Capture by Bare Neon Ions ($Z_P = 10$) Having 1050 MeV/amu Energy Calculated Using Various Approximations

Atomic number of target Z_T	Cross section (barns) per captured electron			
	Approximations			
	ROBK1[a]	RE[b]	R1B[c]	ROBK2[d]
13	2.0(−4)[†]	4.0(−5)	5.5(−5)	4.7(−5)
30	1.2(−2)	1.8(−3)	2.3(−3)	2.2(−3)
47	8.8(−2)	1.1(−2)	1.3(−2)	1.4(−2)
73	4.1(−1)	4.9(−2)	4.9(−2)	5.7(−2)
92	6.3(−1)	8.5(−2)	7.2(−2)	8.1(−2)

[†] Numbers in brackets denote powers of 10.
[a] Moiseiwitsch and Stockman (1980).
[b] Eichler (1985).
[c] Eichler (1987, 1988).
[d] Moiseiwitsch (1988).

However the prominence of the Thomas peak is reduced for large atomic numbers and the π/ζ terms in (37) produced by its presence become smaller as $Z_T + Z_P$ is increased. Thus the neglect of the Thomas peak by the R1B approximation may not lead to much inaccuracy for the cross sections considered in Table IV.

VI. Numerical Solution of Coupled Equations

Toshima and Eichler (1988a) have solved the time-dependent Dirac equation using a coupled channels approach for arbitrary impact parameters to calculate electron capture cross sections for relativistic collisions between the Uranium ions U^{92+} and U^{91+} at 1000 and 500 MeV/amu impact energies. They used the Dirac atomic eigenvectors for the $1s_{1/2}$, $2s_{1/2}$, $2p_{1/2}$, $2p_{3/2}$ states attached to the target and projectile ions corresponding to a basis set of 20 states.

They found that capture cross sections calculated using the R1B approximation deviated considerably from the coupled channels calculations in some cases.

However for low atomic numbers they indicate that the results obtained with the R1B approximation (Eichler 1987, 1988) were reproduced by the coupled channels calculations.

Recently, Toshima and Eichler (1988b) have extended their calculations to Xe^{54+} projectile ions on Ag and Au.

VII. Experimental Data and Comparisons with Theory

Finally we shall compare the capture cross sections calculated using the ROBK2 approximation with the experimental data obtained by Crawford (1979), carefully analysed by Anholt (1985), in Fig. 5, and with some of the other approximations described in this review in Fig. 6. These comparisons are made for incident C($Z_P = 6$), Ne($Z_P = 10$) and Ar($Z_P = 18$) bare projectile ions having 140, 250, 400, 1050, and 2100 MeV/amu energies.

The data points refer to the capture of electrons from the target atom to the K shell of the projectile ion and were obtained by taking the most probable values of the capture cross sections given by Anholt (1985) in his Table I and dividing by $\sum_1^\infty n^{-3} \simeq 1.202$, corresponding to the approximate n_P^{-3} scaling law produced by the ROBK2 approximation.

The theoretical ROBK2 capture cross section curves were obtained by using the closed analytical expression (37) for σ_{fi}^{ROBK2}. For capture from closed K, L, and M shells of the target atom the screened values $Z_T^* = Z_T - 0.3$, $Z_T^* = Z_T - 4.15$, $Z_T^* = Z_T - 11.25$, respectively, given by Slater's rules, were chosen for the target charges. The cross sections $\sigma_{1s\text{-}1s}^{ROBK2}$, $\sigma_{1s\text{-}2s}^{ROBK2}$, $\sigma_{1s\text{-}3s}^{ROBK2}$ were all multiplied by the factor 2 to take account of the two electrons in the K, L_1, and M_1 shells. The contribution of capture from the other L and M shells was neglected because the capture cross sections for target 2p and 3p electrons are considerably smaller, except for large Z_T, than the capture cross sections for target 2s and 3s electrons respectively (Stockman, 1981), being of order $(\alpha Z_T)^2$ higher.

For capture from the K shell only, it can be seen from Fig. 5 that the ROBK2 cross sections rise to a maximum and then, for the lower energies

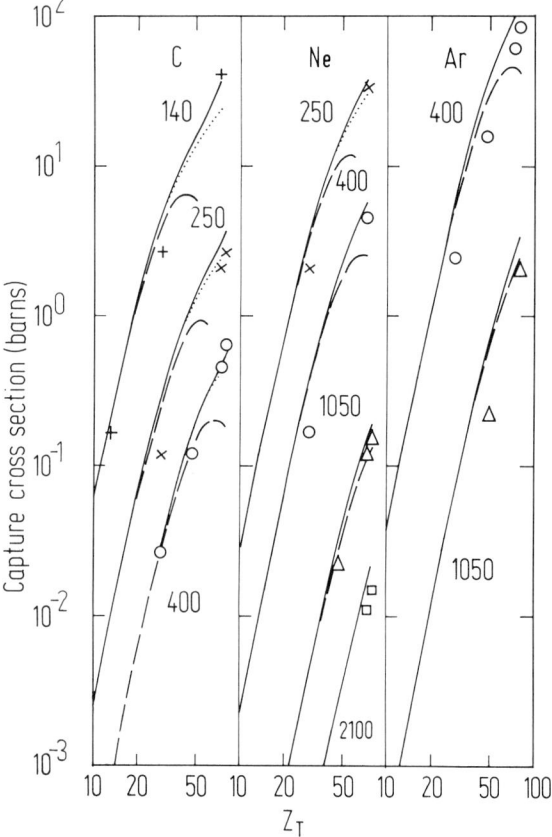

FIG. 5. Total capture cross sections for bare $C(Z_P = 6)$, $Ne(Z_p = 10)$ and $Ar(Z_P = 18)$ projectile ions incident on target atoms with atomic numbers Z_T. Theoretical curves: ———, ROBK2 cross section for capture from K, L_1 and M_1 shells of target to K shell of projectile, calculated using (37); ············, ROBK2 cross section for capture from K and L_1 shells of target to K shell of projectile; -------, ROBK2 cross section for capture from K shell of target to K shell of projectile (Moiseiwitsch, 1988). Experimental data (Crawford, 1979; Anholt, 1985): +, 140; ×250; ○, 400; △, 1050; □, 2100 MeV/amu energies.

considered, bend over as Z_T becomes larger. On including capture from the L_1 shell, the cross section curves no longer exhibit a maximum and better accordance with the experimental data points are found. The same kind of behaviour was discovered by Deco and Rivarola (1987c) using the MCDW approximation. The addition of capture from the M_1 shell has a smaller but not insignificant effect for large values of Z_T. Capture from still higher shells may also be non-negligible although they are not easy to estimate accurately.

In Fig. 6 we have compared the ROBK2 cross sections for capture from the K and L_1 shells with the calculations of Eichler (1985) using the RE

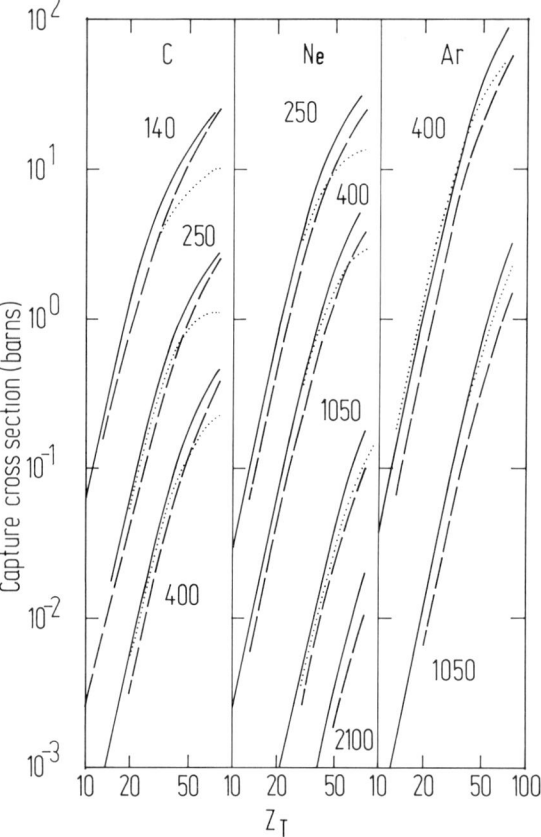

FIG. 6. Total capture cross sections for bare $C(Z_P = 6)$, $Ne(Z_P = 10)$ and $Ar(Z_P = 18)$ projectile ions incident on target atoms with atomic numbers Z_T. Theoretical curves: ———, ROBK2 cross section for capture from the K and L_1 shells of target to K shell of projectile -------, MCDW cross section for capture from K and L shells of target to K shell of projectile (Deco and Rivarola, 1987c); ......, RE cross section for capture from K and L shells of target to K shell of projectile (Eichler, 1985).

approximation and the calculations of Deco and Rivarola (1987c) using the MCDW approximation for capture from the K and L shells. Whereas the ROBK2 and MCDW curves both rise upwards quite sharply, the RE curves tend to flatten out for large Z_T for the lower energies.

The accord between the various theoretical curves and the experimental data obtained by Crawford (1979) and analysed by Anholt (1985) is reasonably satisfactory bearing in mind the approximate nature of the theories.

Better comparisons will be possible when the detailed numerical solutions of the appropriate coupled equations are forthcoming.

REFERENCES

Anholt, R. (1985). *Phys. Rev. A* **31**, 3579.
Bates, D. R., and McCarroll, R. (1958). *Proc. Roy. Soc. A* **245**, 175.
Belkić, Dz., Gayet, R., and Salin, A. (1979). *Phys. Rev.* **56**, 279.
Cheshire, I. M. (1964). *Proc. Phys. Soc., London* **84**, 89.
Crawford, H. J. (1979). Ph.D. thesis, University of California, Lawrence Berkley Laboratory Report No. 8807.
Crothers, D. S. F., and McCann, J. F. (1984). *J. Phys. B* **17**, L177.
Deco, G. R., and Rivarola, R. D. (1986). *J. Phys. B* **19**, 1759.
Deco, G. R., and Rivarola, R. D. (1987a). *J. Phys. B* **20**, 317.
Deco, G. R., and Rivarola, R. D. (1987b). *J. Phys. B* **20**, 3853.
Deco, G. R., and Rivarola, R. D. (1987c). *J. Phys. B* **20**, 5117.
Drisco, R. M. (1955). Ph.D. thesis, Carnegie Institute of Technology, Pittsburg.
Eichler, J. (1985). *Phys. Rev. A* **32**, 112.
Eichler, J. (1987). *Phys. Rev. A* **35**, 3248; Erratum (1988). *Phys. Rev. A* **37**, 287.
Humphries, W. J. (1985). Ph.D. thesis, Queen's University, Belfast.
Humphries, W. J., and Moiseiwitsch, B. L. (1984). *J. Phys. B* **17**, 2655.
Humphries, W. J., and Moiseiwitsch, B. L. (1985). *J. Phys. B* **18**, 2295.
McCann, J. F. (1985). *J. Phys. B* **18**, L569.
Mittleman, M. H. (1964). *Proc. Phys. Soc., London* **84**, 453.
Moiseiwitsch, B. L. (1985). *Phys. Rep.* **118**, 133.
Moiseiwitsch, B. L. (1986). *J. Phys. B* **19**, 3733.
Moiseiwitsch, B. L. (1987a). *J. Phys. B* **20**, L171.
Moiseiwitsch, B. L. (1987b). *J. Phys. B* **20**, 4111.
Moiseiwitsch, B. L. (1988). *J. Phys. B* **21**, 603.
Moiseiwitsch, B. L. (1989). *Phys. Rev. A* **39**, 5609.
Moiseiwitsch, B. L., and Stockman, S. G. (1979). *J. Phys. B* **12**, L695.
Moiseiwitsch, B. L., and Stockman, S. G. (1980). *J. Phys. B* **13**, 2975.
Shakeshaft, R. (1979). *Phys. Rev. A* **20**, 779.
Stockman, S. (1981). Ph.D. thesis, Queen's University, Belfast.
Thomas, L. H. (1927). *Proc. Roy. Soc. A* **114**, 561.
Toshima, N., and Eichler, J. (1988a). *Phys. Rev. ett.* **60**, 573.
Toshima, N., and Eichler, J. (1988b). *Phys. Rev. A* **38**, 2305.

THE LOW-ENERGY, HEAVY-PARTICLE COLLISIONS—A CLOSE-COUPLING TREATMENT*

MINEO KIMURA

Argonne National Laboratory
Argonne, Illinois
and Department of Physics
Rice University
Houston, Texas

and

NEAL F. LANE

Department of Physics and Rice Quantum Institute
Rice University
Houston, Texas

I. Introduction 80
II. General Formulation of the Close-Coupling Method 87
   A. Expansion Methods (Coupled Equations). 87
   B. Molecular States Expansion Method 92
   C. Atomic States Expansion Method 99
   D. Unified AO–MO Matching Method 104
   E. Analytical Model Representations of the MO Expansion Method . . . 106
   F. Relationships of Close-Coupling to Perturbation Methods 109
   G. Direct Numerical Solution of the Time-Dependent Schrödinger Equation 111
   H. Observables 111
III. Current Status of Theoretical and Experimental Results 113
   A. Ion–Atom and Atom–Atom Collisions 114
   B. Ion–Molecule and Atom–Molecule Collisions 138
   C. Ion–Surface Collisions 146
IV. Conclusions and Perspectives 154
   Acknowledgements 156
   References 156

* Work supported by the U.S. Department of Energy, Office of Health and Environmental Research, under Contract W-31-109-Eng-38 (MK), Office of Basic Energy Sciences, Divison of Chemical Sciences, and Robert A. Welch Foundation (NL).

I. Introduction

The conditions most frequently encountered in the terrestrial environment and for the most part throughout the universe are such that ions, atoms, molecules, and aggregates of atoms are the important constituents. Therefore, the improved understanding of basic atomic and molecular physics remains an important research goal. The development of intense lasers and advanced detection schemes has made possible experiments that probe fundamental microscopic phenomena in a manner not thought possible, or at least probable, a decade ago. Fundamental questions regarding the quantum suppression of chaos; the interaction of intense electromagnetic fields with atoms and molecules; subtle symmetries in highly correlated, many-electron systems; precise QED tests; and other issues of interest continue to capture the attention of atomic, molecular, and optical physicists and make the field a particularly lively one.

Atomic and molecular collision physics is a particularly important and singularly challenging field. Such collisions are ubiquitous in nature. In addition to the fundamentally interesting questions they pose, atomic and molecular collisions are also very much at the heart of a large number of applications in such areas as magnetically or intertially confined fusion plasmas, laser systems, partially ionized gases and plasmas, chemical systems, surface interactions, channeling and energy loss in solids, energy and ionization balance in the earth's atmosphere, astrophysical studies on a broad front, and a host of other examples.

Among the various subfields of atomic, molecular, and optical physics, the understanding of inelastic atomic collision processes involving heavy particles, particularly at low-to-intermediate velocities, has seen extraordinary progress, both experimentally and theoretically, in the past decade. [See a series of invited lecture notes for the International Conference on the Physics of Electronic and Atomic Collisions (ICPEAC): Olson, 1980; Salin, 1982; Fritsch and Lin, 1984a, b; Kimura, 1986; and Harel and Salin, 1988.] The purpose of this review is to attempt to describe that progress.

As just one illustration, we compare in Fig. 1 theoretical and experimental cross sections for electronic excitation and electron capture to the H(2s) state in $H^+ + H(1s)$ collisions in the energy range 1 keV to 100 keV. Figures 1a and 1b show results published before 1970, and Figures 1c and 1d show those after 1970. (We include only results that are commonly cited in the literature.) The $H^+ + H$ collision system is the simplest example of a heavy-particle collision where inelastic processes are possible. Even so, both theoretical and experimental studies of inelastic scattering for this system have proven to be difficult. This system is now so well studied that it has become a testing

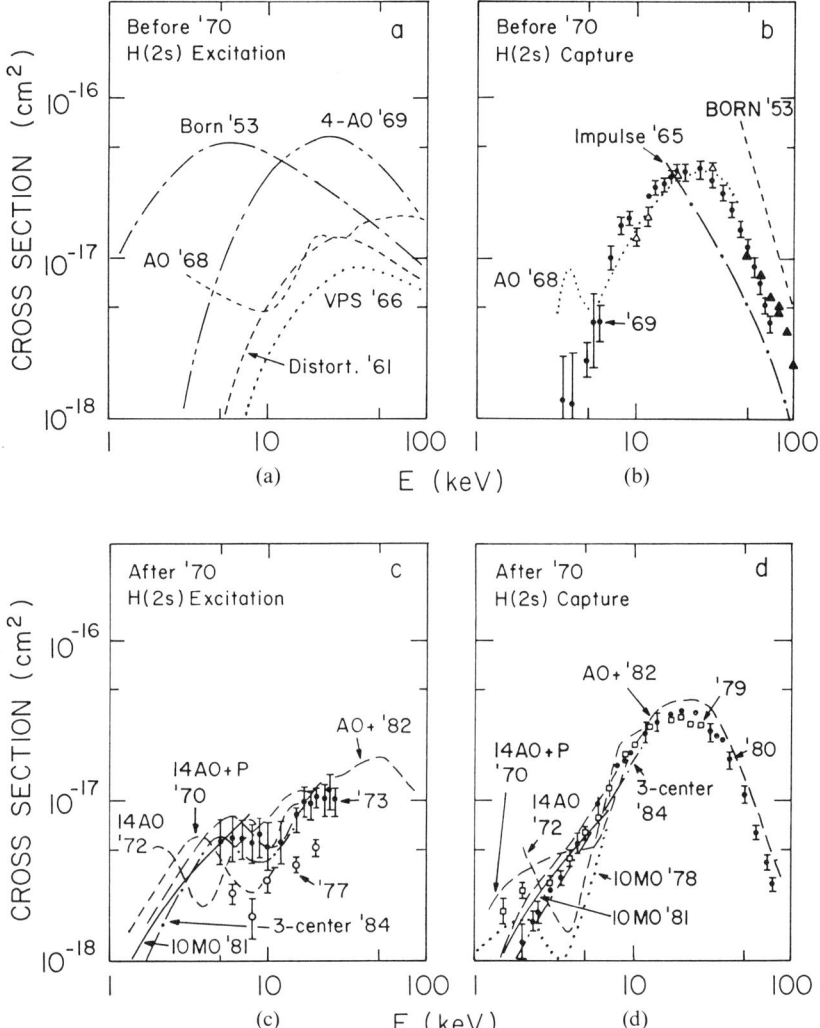

FIG. 1. Theoretical and experimental results prior to 1970 of H(2s) capture (a) and excitation (b) in $H^+ + H$ collisions. Theory: Born '53 (from Bates, 1953); Impulse '65 (from Coleman and McDowell, 1965); VPS '66 (from McCarroll and Salin, 1966); Distort. '61 (from Bates and Dalgarno, 1961); AO '68 (from Gallaher and Wilets, 1968); 4-AO '69 (from Flannery, 1969). Experiment: '69 (from Bayfield, 1969). Results after 1970 for capture and excitation are given in (c) and (d), respectively. Theory 14AO + P '70 (from Cheshire et al., 1970); 14AO '70 (from Rapp and Dinwiddie, 1972); 10MO '78 (from Crothers and Hughes, 1978); 10MO '81 (from Kimura and Thorson, 1981); AO+ '82 (from Fritsch and Lin, 1982); 3-center (from Winter and Lin, 1984a). Experiment: '73 (from Morgan et al., 1973); '77 (from Chong and Fite, 1977); '79 (from Hill et al., 1979); '80 (from Morgan et al., 1980).

ground for any new theoretical method. Two general features are evident in these figures. (i) Before 1970, the perturbative technique was the dominant theoretical tool for studying inelastic scattering in atomic collisions. Large discrepancies among such theories are seen, e.g., in order-of-magnitude differences in cross sections below 20 keV. At the same time, early experimental attempts to measure the cross sections in this energy regime were few. (ii) After 1970 it became the practice to apply large-scale, close-coupling methods; that trend has continued into the 1980s, in part because of the availability of supercomputers to the theoretical atomic and molecular scientific community. Several benchmark experiments based on advanced technologies were also performed during this latter period. Although the agreement between the results of experiment and theory has been greatly improved and the interpretation of the collisions dynamics for many systems is now satisfactory, there still remain questions to be resolved and physical mechanisms to be explained, particularly as the measurement techniques become more refined and consequently the experimental features are more precisely defined and more challenging to theoretical description.

This rapid development of the field owes much both to the advanced technology of high-speed computing, which has benefited both theory and experiment, and to modern ion sources, accelerators, and advanced electronics, which have made possible a high level of sophistication in the current measurements of inelastic processes. Moreover, the theoretical and experimental advances have kept pace with one another, permitting the close interactions that have allowed researchers to obtain an unusually good insight into the underlying dynamics in atomic collisions. Specifically, it is now possible to observe experimentally the final electronic state with a specific n, l, and m designation (at least for one-electron and two-electron systems) resulting from electron capture or excitation in an ion–atom collision by using methods such as energy gain and energy loss spectroscopy and photon emission spectroscopy. Moreover, within the last one or two years, several coincidence measurements have been reported (Hippler, 1988) of the polarization of photons emitted from an excited electronic state formed by electron capture or excitation, simultaneously with the measurement of the deflection of the scattered particle. Such results provide detailed information on how and when electron capture and excitation take place and what the transient electronic state is during the collision, which is the ultimate test of the theory and our goal in attempting to understand the collision dynamics. Such measurements offer an extremely important challenge to even the most refined theoretical models. These models, in turn, pose new questions to the experimentalist that demand even more accurate measurements. We are now in a new era in the sense that the earlier qualitative dialogue is being replaced by a far more quantitative discourse.

Historically, the theoretical tool box used to study heavy-particle atomic collisions was divided into two broad compartments, (i) close-coupling methods based on an expansion of the wavefunction in terms of a set of functions chosen to describe the electronic coordinates of the colliding atomic systems, and (ii) perturbative methods such as a truncated Born series. The first approach was generally believed to be valid for *low-energy collisions*, and the second for *high-energy collisions*. However, the terms "low" and "high" are quite ambiguous in this context and have often been misused, leaving the boundary of validity vague. The ratio of the projectile velocity, v, to that of the orbital electron of interest in the target, v_{el}, provides some guidance in categorizing low- and high-energy collisions (where low means that $v/v_{el} \ll 1$, and high means that $v/v_{el} \gg 1$). Although very interesting questions of physics are yet to be understood in high-energy collisions, we have chosen to focus this review on the study of low-energy, heavy-particle atomic collisions with approaches based on the close-coupling method.

In high-energy, heavy-particle collisions, ionization generally is the dominant channel, followed by target excitation processes. In contrast, in low- to intermediate-energy heavy-particle collisions it is not possible to single out a dominant channel in general, because often many inelastic channels strongly couple with one another, exchanging flux and phase in a complex manner. Thus, without the simultaneous inclusion of all important channels, an accurate determination of the transition probabilities is impossible. This is where a non-perturbative scheme such as the close-coupling method is indispensable. Returning to the earlier example, we illustrate how different channel cross sections behave in the low- to intermediate-energy range in Fig. 2, where representative cross sections for several inelastic scattering processes in $H^+ + H$ collisions are shown. It is apparent that certain electron capture, excitation, and ionization cross sections have the same magnitude at $E \gtrsim 5$ keV, so that one process can be expected to significantly influence the others. Consideration of all important channels on equal footing is necessary in any reliable theoretical treatment.[1]

In the version of the close-coupling method that is based on a molecular (electronic) orbital function expansion, the electrons are regarded as nearly independent particles moving in the field of the nuclei and other electrons. (Actually, in the expansion, the electronic wave function is represented by a superposition of several configurations—Slater determinants—each having a

---

[1] When multistep processes are known to be important in any dynamical process in physics, some sort of close-coupling treatment is necessary in a form appropriate for the system under investigation. One example outside atomic physics is a basis expansion method for nucleons in nucleus-nucleus collisions (Imanishi and von Oertzen, 1987).

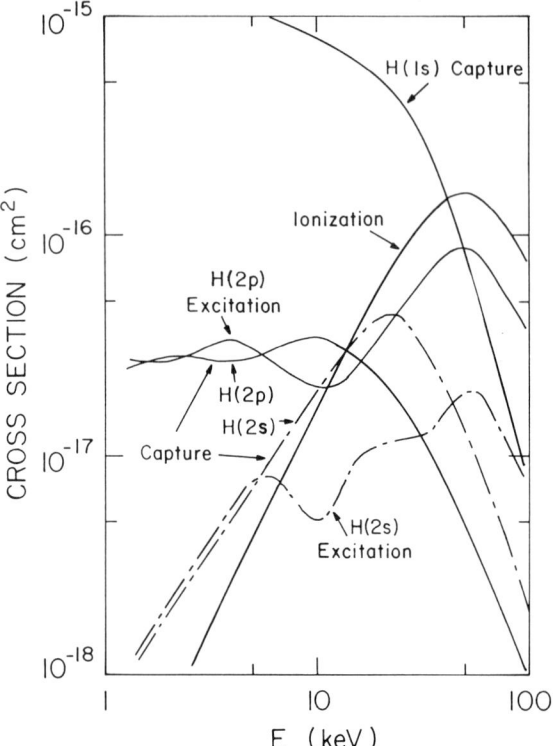

FIG. 2. Major inelastic cross sections in $H^+ + H$ collisions.

particular set of molecular orbitals.) The molecular orbital (MO) model, originally proposed within the molecular structure context by Hund (1927) and Mulliken (1928), is taken to be an appropriate description of the system when the nuclei are close. Collision phenomena for which the MO expansion method is supposed to be particularly suited include the special cases of (i) small-impact-parameter collisions (*close collisions*) and (ii) slow ($v \ll v_{el}$) collisions. The latter implies that the MO method may work well up to kinetic energies of MeV or even higher if v_{el} is sufficiently large, e.g., if molecular orbital x-rays and inner shell ionization processes (Macek and Briggs, 1972, 1973, 1974) are the subjects of study. The extensive study of the MO method over the last decade by several groups has demonstrated its applicability to a wide class of atomic collisions. The method also has been very successful in interpreting qualitatively various experimental findings. Particularly important among the many examples of the latter is the electron promotion model of Fano and Lichten (1965), which explains electron promotion from inner shell to outer shell during an atomic collision. Lichten (1963) also introduced the important concept of the diabatic (versus adiabatic) state, a concept that has helped researchers over the past two decades in interpreting experimental observations of many different excitation and electron-capture processes.

The MO method, however, possesses an intrinsic problem in that the system scattering wavefunction expanded in terms of adiabatic MOs does not satisfy the correct asymptotic boundary conditions, which specify that in the separated-atom limit the bound electron is moving either with the projectile or with the target. This translational motion of the electron is not accounted for in the conventional MO method [the perturbed stationary state (PPS) method originally proposed by Massey and Smith (1933)]. Further, the resulting coupled equations are not Galilean invariant; i.e., they depend on the origin of the coordinates. To avoid these defects in the conventional MO method, an atomic orbital electron translation factor (AO-ETF) was first introduced by Bates and McCarroll (1958). In their description of the electron translation factor (ETF), each MO is assumed to translate exactly the same way as does its corresponding atomic orbital (AO) in the separated-atom limit. This choice of ETF, while removing the difficulties associated with asymptotic boundary conditions, introduces undesirable new physical constraints on the translational motion of the electron in each MO. In a collision where the MO description is appropriate (i.e., a slow or close collision), each electron is shared by both nuclei; thus, this "molecular aspect" should be incorporated somehow into the ETF. Although agreement between MO methods that employ different versions of the MO-ETF is generally good in the case of the total cross sections for one-electron systems and for some two-electron systems, that is not true for all partial cross sections, and the decisive method for the determination of the MO-ETF is still subject to controversy. The extensive studies of alternative determinations of suitable state-dependent MO-ETF using the optimization method (Lebeda *et al.*, 1971; Rankin and Thorson, 1979) and applications to many collision processes show that it is one candidate that can provide accurate descriptions of the MO-ETF in a close-coupling treatment. However, the procedure is rather time consuming.

Another electronic representation, mentioned earlier, is the two-center AO method proposed by Bates (1958). The AO method is considered to provide an appropriate description of intermediate-energy collisions in which large impact parameters are important (*distant collisions*). In the AO method each atomic orbital is constrained to travel either with the target or with the projectile, retaining an atomic character throughout the collision. In contrast to the MO method, here there is no ambiguity in the determination of the

ETF, the AO-ETF (plane-wave ETF) of Bates and McCarroll (1958) being the definitive form. On the other hand, in molecular structure applications the two-center AO expansion, known as the linear combination of atomic orbitals (LCAO) in quantum chemistry, is known to be incapable of reproducing accurate molecular orbitals at small internuclear separations (see, for example, Schaefer, 1972). Thus the AO expansion cannot be expected, even qualitatively, to provide an accurate representation of slow collisions, where molecular effects are important, or of collisions in which close encounters are dominant. One method that has been devised to overcome the weakness in the conventional two-center AO expansion method extends the region of validity of the method by incorporating the united-atom orbitals in either the two-center (Fritsch and Lin, 1982) or multicenter (Winter and Lin, 1984a,b) expansion formalism. In principle, these "new" AO expansion approaches cure the defects of the conventional two-center AO expansion method, but at the expense of using a large basis set and at the risk of overcompleteness of the basis set.

The strengths and weaknesses, briefly mentioned above, of both the MO and AO expansion methods imply that none of these representations can adequately represent electronic wave functions in the entire space. These observations have naturally led to the idea of adopting the best features of both methods while avoiding their major defects in a new approach called the AO-MO matching method. This method uses different representations at small and large distances (Kimura and Lin, 1985 a, b; Winter and Lane, 1985) and is analogous to the R-matrix method (see, for example, Burke and Robb, 1975) used in time-independent treatments of electron–atom, electron–ion, and electron–molecule collisions and photoionization of atomic and molecular systems. The AO-MO matching method is particularly promising for treating heavy-particle collisions involving complex atomic or molecular collision partners.

Over the past three decades, a number of approximate methods that treat heavy-particle atomic collisions have been proposed that take advantage of some particular characteristics of a collision system, for example, a strongly coupled two-channel case. Such methods have provided considerable insight into the collision dynamics and could be handled with the computation technology of the day; therefore they dominated theoretical research for quite a long time. However, in view of the major advances in computer technology, and the supercomputer in particular, the significance of many of these approximate methods is likely to shift in time to the role of interpreting or predicting, qualitatively, certain observable features for special systems rather than providing quantitative results. It is not our intention to discuss the details of these approximate treatments here, but rather to refer readers to excellent review articles on the topic (Janev and Winter, 1985; Barat, 1986).

In the present review, we will focus particularly on "quantitatively reliable" close-coupling treatments of slow, heavy-particle atomic collisions. In particular, we will attempt to show what we now understand about the dynamics of such collisions, to identify remaining problems with the theory, and to suggest possible extensions of ion–atom collision models to the study of ion–molecule and ion–surface collisions.

II. General Formulation of the Close-Coupling Method

In the following brief summary of the general formalism, we will avoid details of the derivations while attempting to stress the relationships between the MO and AO representations, the quantum-mechanical and semiclassical formalisms, and the close-coupling and the perturbative methods. In a heavy-particle atomic collision at an energy above $\simeq 100$ eV, the following conditions are usually met:

(i) the de Broglie wavelength of the relative motion of the heavy particles (nuclei) is small compared with atomic dimensions;

(ii) the relative momentum of the nuclei, k, satisifes the relation $k^2/2\mu \gg \Delta E$, where μ is the reduced mass of the colliding system and ΔE is the inelasticity (energy defect of relevant states) in the collision;

(iii) $\cos^{-1}(\mathbf{k}_i \cdot \mathbf{k}_f) = \theta \ll 1$, i.e., most scattering of the projectile occurs at "small" scattering angles.

Under such circumstances, the nuclei can be assumed to move classically along some trajectory. The electrons experience an intrinsically time-dependent force field due to the moving nuclei, and hence the electronic wave function must satisfy a time-dependent Schrödinger equation. Although this semiclassical picture is an approximation and misses quantum effects such as tunneling, which can be important at low energies, the merit of using the semiclassical representation is to provide an intrinsically simpler picture of the collision dynamics and to simplify the computations. Semiclassical methods can provide quantitatively accurate results under conditions of interest in the present review.

A. EXPANSION METHODS (COUPLED EQUATIONS)

1. Quantum Mechanical Theory

Before proceeding further, we will briefly describe the derivation of the relevant system of coupled equations in the quantum mechanical formulation and in the semiclassical version. In order to provide a simple, and we hope clear, relation between the approaches that permits an easy interchange from one description to another, we choose a route that explicitly relates the quantum and semiclassical representations and includes the *ETF-modified MO* method as well as the conventional MO method (PSS method), but unfortunately with a loss of some rigor in the mathematical derivation. For a more general and extensive review, readers are advised to consult the articles by Delos (1981) and McDowell and Coleman (1970).

As the simplest example, the formulation is given for a one-electron, two-nuclei system, for which the coordinates are shown in Fig. 3. This system is described by the full, stationary Schrödinger equation

$$H(\mathbf{R}, \mathbf{r})\psi(\mathbf{R}, \mathbf{r}) = E\psi(\mathbf{R}, \mathbf{r}), \qquad (1)$$

where $\mathbf{R}$ and $\mathbf{r}$ represent nuclear and electronic coordinates, respectively (see Fig. 3). The total Hamiltonian H has the form

$$H(\mathbf{R}, \mathbf{r}) = T_\mathbf{R} + h_{el}, \qquad (2)$$

where $T_\mathbf{R}$ and h_{el} are the kinetic energy of the relative nuclear motion and the electronic Hamiltonian, respectively. In general, the solution of Eq. (1) is obtained by expanding the system wavefunction in terms of some known electronic basis functions with the inclusion of an ETF,

$$\psi(\mathbf{R}, \mathbf{r}) = \sum_i X_i(\mathbf{R})\phi_i(\mathbf{R}, \mathbf{r})F_i(\mathbf{R}, \mathbf{r}), \qquad (3)$$

where $X_i(\mathbf{R})$, $\phi_i(\mathbf{R}, \mathbf{r})$, and $F_i(\mathbf{R}, \mathbf{r})$ denote (continuum) wave functions of nuclear motion, electronic wave functions, and the ETF, respectively, the latter of which has the general form

$$F_i(\mathbf{R}, \mathbf{r}) = \exp\left(\frac{i}{2}\mathbf{v}\cdot\mathbf{r}f_i(\mathbf{R}, \mathbf{r})\right). \qquad (4)$$

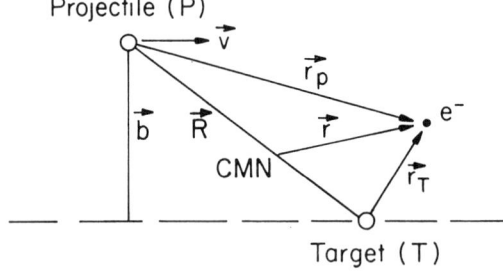

FIG. 3. Coordinates of (projectile + target + one electron) system.

This form for the ETF is intrinsically semiclassical in nature, so that including it in an otherwise quantum mechanical formalism is not rigorously correct. [A detailed derivation of a fully quantum mechanical representation based on a coordinate transformation which does not explicitly introduce the ETF can be found in Delos (1981).] In Eq. (4), $f_i(\mathbf{R}, \mathbf{r})$ represents the so-called "switching function," which serves to incorporate molecular effects into the ETF when necessary. The $f_i(\mathbf{R}, \mathbf{r})$ is fixed only in the limit $R \to \infty$, having the value ± 1 depending on the partner to which the electron attaches. By substituting Eq. (3) into Eq. (1) and projecting onto any basis function $\phi_j(\mathbf{R}, \mathbf{r})$, one obtains the set of coupled equations (in matrix form)

$$\{(1/2\mu)[-i\mathbf{S}\nabla + (\mathbf{P} + \mathbf{A})]^2 + \mathbf{h} - E\mathbf{I}\}\mathbf{X}(\mathbf{R}) = 0, \quad (5)$$

where

$$\mathbf{S}_{ji} = \langle \phi_j F_j | \phi_i F_i \rangle, \quad (6a)$$

$$\mathbf{P}_{ji} = \langle \phi_j | -i\nabla_R | \phi_i \rangle, \quad (6b)$$

$$\mathbf{A}_{ji} = \frac{-i}{2} \langle \phi_j | [h_{\text{el}}, f_i(\mathbf{R}, \mathbf{r})\mathbf{r}] | \phi_i \rangle, \quad (6c)$$

$$\mathbf{h}_{ji} = \langle \phi_j | h_{\text{el}} | \phi_i \rangle, \quad (6d)$$

and $\mathbf{X}(\mathbf{R})$ is a column matrix whose elements correspond to the various nuclear wavefunctions for the possible collision channels. Note that we have expanded the ETF and retained only the first-order term in $\mathbf{v}$ to derive Eq. (5); this is a valid approximation for slow collisions. In Eqs. (5) and (6), $\mathbf{S}$ represents the overlap matrix; $\mathbf{P}$ represents the nonadiabatic coupling, which causes electronic transitions due to the nuclear motion; h is the electronic Hamiltonian matrix; and $\mathbf{A}$ arises from the ETF.

If the chosen basis set ϕ_n does not depend on R explicitly, e.g., *atomic states*, then the $\mathbf{P}$ coupling vanishes except for the angular-coupling matrix elements. Of course, in this case h is not diagonal; this potential coupling also causes transitions. The matrix $\mathbf{A}$ denotes the ETF correction to the $\mathbf{P}$ matrix;

If the chosen basis set ϕ_n does not depend on R explicitly, e.g., *atomic states*, then the $\mathbf{P}$ coupling vanishes except for the angular-coupling matrix elements. Of course, in this case h is not diagonal; this potential coupling also causes transitions. The matrix $\mathbf{A}$ denotes the ETF correction to the $\mathbf{P}$ matrix; as a whole, $(\mathbf{P} + \mathbf{A})$ should be considered to represent the correct nonadiabatic coupling in that it possesses no origin dependency and no fictitious long-range coupling. Hence, the scattering wave function satisfies the correct boundary condition and Galilean invariance is recovered.

In the rotating coordinate frame, the $\mathbf{P}$ and $\mathbf{A}$ matrices can be expressed in terms of two separate contributions, viz., radial and angular (rotational) couplings and the corresponding ETF correction terms. The radial coupling causes transitions because of the relative radial motion of the nuclei, while the angular coupling is due to rotation of the molecular (internuclear) axis. The radial coupling couples states with the same angular symmetry (e.g., Σ with Σ or Π with Π), while the angular coupling couples states with angular momenta differing by one (e.g., Σ with Π). In molecular structure and spectroscopy, radial and angular couplings represent the breakdown of the Born-Oppenheimer approximation and explain various "perturbations" of the observed spectra. The angular coupling causes Λ doubling in molecules.

It may be appropriate here to remark on the importance of spin-orbit coupling, which arises in the leading relativistic corrections to the theory given above. For small (low-Z) atomic systems, however, spin-orbit coupling is a rather weak effect compared to the radial and angular coupling described here. [See Delos (1981) for details.]

2. Semiclassical Theory

In a semiclassical formalism, which is generally valid under conditions (i)–(iii), it is assumed that the nuclei move along a classical path specified by the internuclear separation vector $\mathbf{R}(t)$, and that the electronic wavefunction $\psi(\mathbf{r}, t)$ satisfies a time-dependent Schrödinger equation,

$$i \frac{\partial \psi(\mathbf{r}, t)}{\partial t} = h_{\text{el}} \psi(\mathbf{r}, t). \tag{7}$$

The solution of Eq. (7) can be obtained by expanding the total wave function $\psi(\mathbf{r}, t)$ in terms of a known basis set of electronic functions and the ETF, $\phi_i F_i$ in the form

$$\psi(\mathbf{r}, t) = \sum_i a_i(t) \phi_i [\mathbf{R}(t), \mathbf{r}] F_i [\mathbf{R}(t), \mathbf{r}] \exp(x), \tag{8}$$

where

$$\exp(x) = \exp\left(-i \int^t E_i[\mathbf{R}(t')] dt' - \frac{1}{2} \int^t v^2 \, dt' \right).$$

Employing the procedure used earlier to derive the coupled equation in the fully quantum mechanical version, we can reach the coupled, time-dependent equations for the semiclassical version (to first order in $\mathbf{v}$),

$$i\mathbf{S} \frac{d\mathbf{a}}{dt} = [\mathbf{h} + \mathbf{v} \cdot (\mathbf{P} + \mathbf{A})] \mathbf{a}, \tag{9}$$

where the matrices $\mathbf{S}$, $\mathbf{h}$, $\mathbf{P}$, and $\mathbf{A}$ are defined as in Eq. (6). Regardless of the precise choice of basis functions ϕ_i and ETF F_i, the respective forms of the coupled Eqs. (5) and (9) are the same.

The relation between the coupled equations of the fully quantum mechanical formulation, Eq. (5), and the semiclassical formulation, Eq. (9), is made transparent by introducing into the quantum mechanical formulation the well-known semiclassical approximation based on the limit of large reduced nuclear mass at a fixed relative nuclear velocity. This procedure reduces the quantum mechanical coupled equations [i.e., second-order coupled differential Eq. (5)] to the semiclassical equations [i.e., the time-dependent, first-order coupled differential Eq. (9)]. In the simplest derivation, we assume that the heavy-particle path is well defined as a straight line by $\mathbf{R}[t]$ and write the nuclear wave function in the form

$$X_i(\mathbf{R}) = a_i(\mathbf{R})\exp[i\mu\mathbf{v}\cdot\mathbf{R}], \qquad (10)$$

where $\mathbf{v}$ is a constant velocity vector. Inserting Eq. (10) into Eq. (5), one can easily obtain

$$[-i S\mathbf{v}\cdot\nabla_R + h + \mathbf{v}\cdot(\mathbf{P} + \mathbf{A})]\mathbf{a}(\mathbf{R}) + \theta(\mu^{-1}) = 0, \qquad (11)$$

where $\theta(\mu^{-1})$ represents higher order terms in μ^{-1}. Neglecting the latter term, which is by far the smallest, we can recover the semiclassical coupled Eq. (9), where we redefine $\mathbf{a}(\mathbf{R})$ as $\mathbf{a}(t)$. As a pictoral summary, we show schematically in Fig. 4 the relation of the fully quantum mechanical and semiclassical representations.

Up to this point, we have described a general theoretical framework to describe heavy-particle atomic collisions without introducing any specific information about the basis set $\{\phi(\mathbf{R}, \mathbf{r})\}$. However, the choice of the basis set is extremely important in attempting to describe the collision physics correctly. Put another way, an unwise choice of basis set will either give

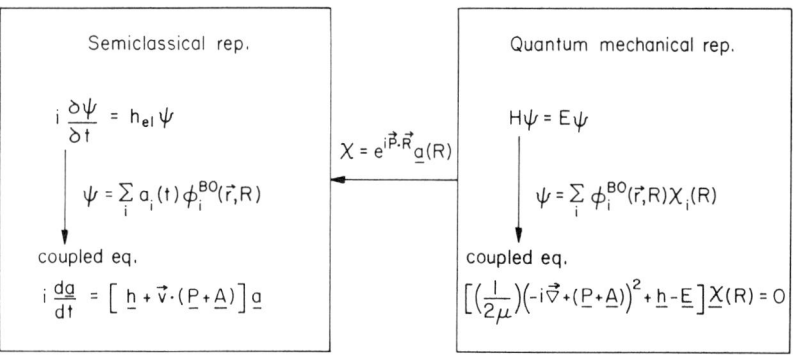

FIG. 4. Schematic diagram of relation of quantum mechanical representation and semiclassical representation.

nonphysical results or make the computational task unnecessarily time consuming and expensive.

For an extremely slow collision, $E \ll \Delta E(R)$ at all important R (say $E \lesssim 10^{-3}$ eV), the electrons completely adjust to the slowly changing potential field and the electronic state behaves adiabatically, with its energy staying always on a single Born-Oppenheimer energy surface, which corresponds to an eigenstate of the electronic Hamiltonian; in this case, no transition takes place and the collision is purely elastic. For a somewhat faster collision, in which the relative nuclear velocity is still small compared with the classical velocities of the bound electrons, the wave function is simply described as a superposition of a small number of molecular orbitals. Under these conditions, i.e., near-adiabatic collisions, the electronic wave functions have time to partially adjust to the changing nuclear potential field. The incompleteness of this adjustment causes mixing of the states leading to electronic transitions, i.e., inelastic collisions.

If the relative velocity of the nuclei is greater than the average velocity of the bound electrons in the target, i.e., in intermediate-to-fast collisions, then the electronic wave function does not have enough time to adjust to the changing nuclear potential field and may even retain "rigid" atomic characteristics during an entire collision event. In this situation, a collision system may be well described by using a set of atomic orbitals.

In the following section, we attempt to give a detailed discussion of the advantages and disadvantages of each of several basis sets commonly used in close-coupling treatments. A diagram showing the connections between theoretical methods discussed below is given in Fig. 5.

B. Molecular States Expansion Method

Even in relatively slow collisions ($v \ll v_{el}$), an inelastic transition can take place with high probability at those particular internuclear separations where near-degeneracies occur among adiabatic electronic states. Under these circumstances, it is quite natural and also appealing to adopt molecular orbitals as the expansion basis set (Massey and Smith, 1933). However, as was noted earlier, the conventional MO method (the perturbed stationary state, PSS, in the more general context) is known to have some mathematical defects, namely the violation of the Galilean invariance and the incorrect scattering boundary conditions.

1. Galilean Invariance and Boundary Conditions

As a simple illustration of the problem associated with the conventional MO (PSS) method, recall the coupled Eq. (9) with the total scattering wave

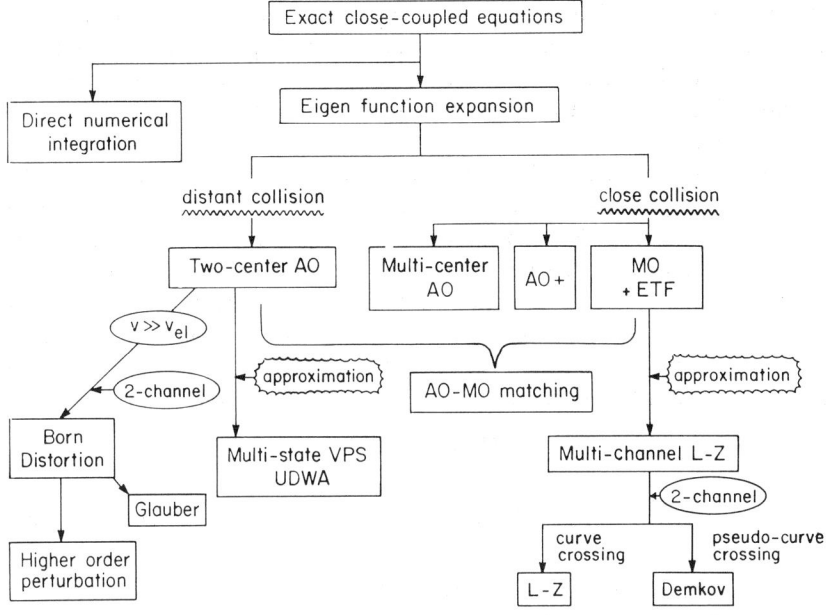

FIG. 5. Schematic outline illustrating relations among major theoretical models discussed in the text.

function expanded as in Eq. (8). In the conventional MO representation (without ETF), the coupled Eq. (9) can be recast in a simpler form by using the expansion

$$\psi = \sum_i a_i \phi_i^{MO}(\mathbf{R}, \mathbf{r}) \exp\left(-i \int^t E_i[R(t')]dt'\right), \quad (12)$$

where the coupled equations take the form

$$i\dot{\mathbf{a}} = \mathbf{v} \cdot \mathbf{P} \mathbf{a}, \quad (13)$$

with the usual form for the nonadiabatic coupling matrix

$$\mathbf{P} = \langle \phi_j^{MO} | (-i\nabla_R)_r | \phi_i^{MO} \rangle. \quad (14)$$

Here the symbol $(-i\nabla_R)_r$ signifies that the derivative with respect to $\mathbf{R}$ be carried out while holding $\mathbf{r}$ constant. The adiabatic (Born-Oppenheimer) molecular wave function $\phi_n^{MO}(\mathbf{R}, \mathbf{r})$ satisfies the electronic Schrödinger equation for fixed R,

$$h_{el}\phi_n^{MO}(\mathbf{R}, \mathbf{r}) = E_n(R)\phi_n^{MO}(\mathbf{R}, \mathbf{r}). \quad (15)$$

In Eq. (13), $\mathbf{P}$ again provides the nonadiabatic coupling that gives rise to inelastic collisions. If the basis set $\{\phi_i^{MO}\}$ is complete, then the resulting coupled Eq. (13) and its quantum mechanical counterparts appear to be exact. But for practical reasons, the expansion in Eq. (12) is truncated to a finite basis so that the set of coupled Eq. (13) to be solved is finite. This inevitably introduces errors into the solutions and the calculated collision cross sections.

As is clear from Eq. (14), the nonadiabatic coupling matrix explicitly exhibits an origin dependence with respect to the electron coordinate $\mathbf{r}$ because the derivative $\mathbf{V}_R$ is performed with the electron coordinates fixed. This means that the values of the coupling matrix elements will be different depending on the choice of the origin of the electron coordinates. Furthermore, in the asymptotic region ($R \to \infty$), for example in an asymmetric collision system, the molecular wave function $\phi_i^{MO}(\mathbf{R}, \mathbf{r})$ approaches its atomic counterpart centered either on projectile (P) or on target (T)

$$\phi_i^{MO}(\mathbf{R}, \mathbf{r}) \underset{R \to \infty}{\sim} \phi_i^{AO}(\mathbf{r}_P) \tag{16a}$$

$$\phi_j^{MO}(\mathbf{R}, \mathbf{r}) \underset{R \to \infty}{\sim} \phi_j^{AO}(\mathbf{r}_T). \tag{16b}$$

For two different states i and k that correlate to different atomic states, but dipole-coupled states, of the same atom, i.e., dipole-coupled states either on projectile or on target, the nonadiabatic coupling becomes

$$\langle \phi_k^{MO} | (-i\mathbf{V}_R)_r | \phi_i^{MO} \rangle \underset{R \to \infty}{\sim} \langle \phi_k^{AO}(\mathbf{r}_T) | \tfrac{1}{2}\mathbf{v} \cdot \mathbf{V}_{r_T} | \phi_i^{AO}(\mathbf{r}_T) \rangle. \tag{17}$$

Equation (17) can possess a nonzero value at infinite nuclear separation (the dipole coupling term). To cite an extreme example, this means that a proton on the earth still couples with a hydrogen atom on the moon! An example of nonvanishing coupling is illustrated in Fig. 6 for $1s\sigma_g \to 2s\sigma_g$ radial coupling in H_2^+; the ETF-corrected coupling is also shown. In addition to the asymptotically incorrect coupling, other problems include (i) unrealistic, symmetry-violating coupling between gerade and ungerade states of a symmetric system, and (ii) fictitious origin-dependent coupling. All of these defects in the conventional MO (PSS) method originate because the wave function does not satisfy the correct scattering boundary conditions, i.e., it is not Galilean invariant.

2. Atomic Orbital (Plane Wave) Electron Translation Factor (AO-ETF)

Let us consider an arbitrary reference frame of origin $0'$ with coordinates $\mathbf{r}'$, translating with constant velocity $\mathbf{v}$ with respect to the origin of some inertial frame 0 with coordinates $\mathbf{r}$. Then the Galilean invariance requires that

$$\psi(\mathbf{r}, t) = U\psi'(\mathbf{r}', t), \tag{18}$$

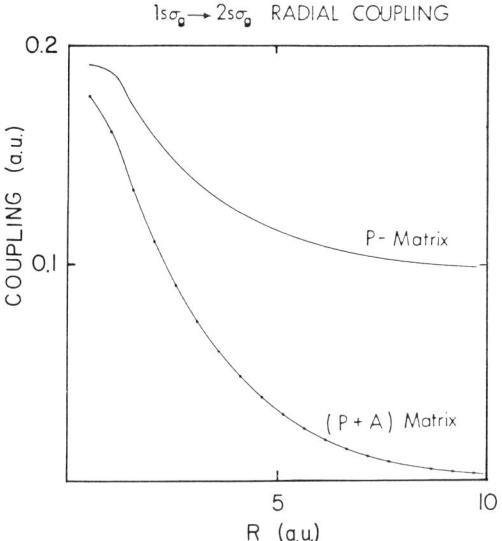

FIG. 6. $1s\sigma_g$–$2s\sigma_g$ radial couping in H_2^+ system. Note the nonvanishing feature of the P coupling and the corrected (P + A) coupling at $R \to \infty$.

where U represents some unitary operator. By substituting Eq. (18) into the time-dependent Schrödinger equation, Eq. (7), one can easily show that the unitary operator should have the form

$$U = \exp\left(i\mathbf{v}\cdot\mathbf{r} - \tfrac{1}{2}\int^t v^2\, dt'\right). \tag{19}$$

This unitary operator has been designated the *plane-wave electron translation factor* (AO-ETF) (Bates and McCarroll, 1958); it accounts for the linear momentum and kinetic energy associated with the translational motion of an electron fixed in the frame having origin 0' with respect to the origin 0. Physically, the inclusion of this form of the ETF in the MO expansion makes some sense; if we sit as a spectator on the center of mass of the collision system, we see the electron moving along with the nucleus to which it attaches. This *traveling effect* is completely missing in the conventional MO (PSS) method and results in the defects we have described.

Thus, the introduction of the AO-ETF removes the difficulties associated with the conventional MO (PSS) method, at least asymptotically; all coupling matrix elements vanish in the limit $R \to \infty$, being origin independent, and all fictitious coupling is removed. The ETF-modified MO expansion method leads to the coupled equations given in Eq. (5) for the fully quantal, and Eq. (9) for the semiclassical versions.

3. Molecular Orbital Electron Translation Factor (MO-ETF)

While the introduction of the AO-ETF into the MO method removes the obvious defects in the PSS coupling matrix elements, giving at least unique transition probabilities, it nevertheless does not necessarily provide the most rapidly convergent expansion. In the range of small to intermediate internuclear separations, the electron belongs to neither the projectile nucleus nor the target nucleus, but is shared by both as in a molecule. Because the AO-ETF is incapable of accounting for this molecular effect, a number of workers have proposed forms of the ETF that are "relaxed" by including a switching function that incorporates some molecular character into the ETF. In this way, the molecular orbital ETF (MO-ETF) was designed to have the form

$$F_i(\mathbf{R}, \mathbf{r}) = \exp\left(\frac{i}{2} f_i(\mathbf{R}, \mathbf{r}) \mathbf{v} \cdot \mathbf{r}\right), \tag{20}$$

where $f_i \to \pm 1$ as $R \to \infty$ is defined so as to recover the correct AO-ETF form as $R \to \infty$; the precise form of the switching function $f_i(\mathbf{R}, \mathbf{r})$ is not defined. Many forms of the MO-ETF have been proposed by various groups. Forms most frequently used are shown in Table I. Because no basic physical principle has yet been established for determining the precise form of the MO-ETF, the various forms have been based either on some intuitive physical picture (Schneiderman and Russek, 1969; Levy and Thorson, 1969; Rankin and Thorson, 1979; Mittleman and Tai, 1973; Vaaben and Taulbjerg, 1981; and Errea et al., 1982), or on a variational argument (Riley and Green, 1971; Crothers and Hughes, 1978; Ponce, 1979). Depending upon the form of the MO-ETF chosen, the ETF-corrected nonadiabatic coupling matrix elements show quite different structure. An example of the dependence of the radial coupling term on the form of the ETF used is shown in Fig. 7. for $2p\sigma$-$3d\sigma$ coupling in HeH^{2+}. For a large-scale close-coupling calculation (200-300 channels), which is becoming increasingly feasible, the results do not depend on the precise form of the ETF. Thus, in the near future, calculations with "semicomplete" sets may become the general rule, in which case the problems associated with the nonuniqueness of the ETF will disappear.

In contrast to the MO-ETF, the AO-ETF can be defined without ambiguity because an electron always attaches either to the target or to the projectile (except ionization event). The form is given, as before, simply by

$$F_i(\mathbf{R}, \mathbf{r}) = \exp\left(\pm \frac{i}{2} (\mathbf{v} \cdot \mathbf{r})\right), \tag{21}$$

where the origin of the electron coordinates is chosen at the midpoint of two nuclei.

TABLE I

Various Forms of the Proposed MO-ETF: $F_i(\mathbf{R}, \mathbf{r}) = \exp[\frac{1}{2}\mathbf{v} \cdot \mathbf{r} f_i(\mathbf{R}, \mathbf{r})]$

Group	Switching Function: $f(\mathbf{R}, \mathbf{r})$		Remark
Schneiderman and Russek (1969)	$\dfrac{\cos \theta}{1 + (a/R)^2}$;	θ: angle of electron to internuclear separation of projectile and target a: parameter	state independent
Levy and Thorson (1969)	$\dfrac{r_T^2 - r_P^2}{r_T^2 + r_P^2}$		state independent
Lebeda et al. (1971)	$\tanh R\beta_n \eta$;	$\eta = \dfrac{r_T - r_P}{R}$ β_n: state-dependent parameter	state dependent
Mittleman and Tai (1973)	$\dfrac{(1 - S^2)(1 - e^{-2\alpha\eta})}{(1 + S^2)(1 + e^{-2\alpha\eta}) - 4Se^{-\alpha\eta}}$	S: overlap of LCAO α: Slater exponent $\eta = \|\mathbf{r} + \frac{1}{2}\mathbf{R}\| - \|\mathbf{r} - \frac{1}{2}\mathbf{R}\|$	state dependent
Rankin and Thorson (1979)	$\tanh R\{\frac{1}{2}\beta_n[(Z_T + Z_P) + (Z_T - Z_P)] + \alpha_n \ln(Z_P/Z_T)\}$ α_n: state-dependent parameter		state dependent
Vaaben and Taulbjerg (1981)	$\dfrac{1}{2}\left(\dfrac{Z_P r_T^3 - Z_T r_P^3}{Z_P r_T^3 + Z_T r_P^3} + \dfrac{Z_T - Z_P}{Z_T + Z_P}\right)$		state independent, similar to Levy and Thorson (1969)
Errea et al. (1982)	$\dfrac{R^2}{R^2 + \beta^2} \dfrac{\tilde{Z}}{R + \alpha}$;	$\tilde{Z}$: Z direction of electron coordinate α, β: parameter	state independent, similar to Schneiderman and Russek (1969)

4. Adiabatic Representation versus Diabatic Representation

The concept of the diabatic representation was first introduced by Lichten (1963) to interpret the experimental results for electron capture in $He^+ + He$ collisions. Since then several formal definitions have been proposed by different researchers; however, a unique definition still does not exist. By recasting the coupled Eq. (9) into nonvector forms (and ignoring the ETF term for simplicity), we obtain

$$i\frac{da_i}{dt} = \sum_j \left\langle \phi_i \left| \left(h_{el} - i\frac{\partial}{\partial t}\right) \right| \phi_j \right\rangle a_j \exp\left(-i \int^t (E_i - E_j) dt'\right). \quad (22)$$

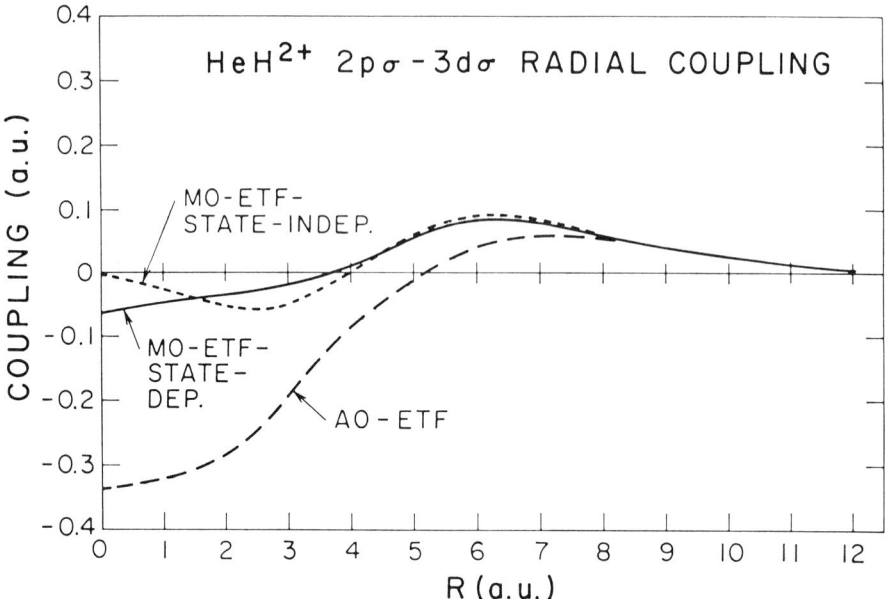

FIG. 7. 2pσ–3dσ radial coupling in HeH^{2+}. The different ETFs used are designated in the figure. (From Kimura and Lin, 1986.)

From the definition of Smith (1969) [the correct definition with inclusion of the ETF is given by Delos and Thorson (1979)], we note the following.

(i) The adiabatic representation is the representation that diagonalizes the h_{el} operator; hence, the radial part of the operator $\partial/\partial t$ ($\equiv v\, \partial/\partial R$) term is regarded as the perturbation. In the adiabatic representation, potential curves that possess the same symmetry exhibit avoided crossings.

(ii) The diabatic representation is the representation that diagonalizes the radial part of the operator d/dR; hence, the h_{el} term is regarded as the perturbation. In this representation, any potential curves can cross.

Because the radial and angular (rotational) couplings have different physical origins and different characteristics, it is often useful to treat them separately. Details of these classifications are given by Crothers (1971) and Delos (1981). The adiabatic and the diabatic representations are connected by a unitary transformation (Heil *et al.*, 1981). As a practical matter, either representation may be used. The differential equations that result from the quantum mechanical treatment with a diabatic representation are somewhat easier to solve numerically than those that result from the adiabatic representation (See Sec,IIE).

C. ATOMIC STATES EXPANSION METHOD

As the collision velocity is increased to a magnitude comparable to that of the bound electrons, the effective collision time is sufficiently short so that the electronic wave function does not have time to adjust to the changing potential field; i.e., the collision becomes nearly impulsive. Alternatively, at any collision velocity, if the important coupling takes place at large R, then the electronic wavefunction similarly is not distorted much. In either circumstance, the AO expansion method, which represents the stiffer nature of the colliding system, should clearly be adequate.

In order to treat the intermediate-energy regime ($v \simeq v_{el}$), Bates (1958) introduced the two-center AO expansion method with a basis set of traveling atomic orbitals. The troublesome ETF nonuniqueness problem seen in the MO-ETF approach does not exist in the AO method because the AO-ETF can be uniquely defined as in Eq. (21). Since Bates' pioneering study using the two-center AO expansion method, several applications of the method have been used to study electron capture and excitation in systems such as $H^+ + H$. However, not all the results are consistent and agreement with measurements has been variable. In the late 1960s and early 1970s, the conclusion seemed clear that, on the basis of studies of convergence of the cross section, atomic continuum states would have to be included in order to make basis sets more nearly complete. Hence, several AO expansion methods were developed by using bases that include the continuum in some sense, e.g., the Sturmian basis and pseudostate basis. However, the improvement with respect to the conventional two-center AO method was found to be marginal. In retrospect, this is not so surprising because in quantum chemical structure calculations, a simple LCAO method has long been known to be incapable of reproducing molecular orbitals, even qualitatively, at small internuclear separations.

When a distant collision, i.e., a collision of large impact parameter, is important, the two-center AO method works reasonably well. This class includes some examples of dominant-channel excitation and electron capture in ion-atom collisions. However, when a close collision, i.e., a collision of small impact parameter, is important, then the two-center AO method fails to predict transition probabilities even qualitatively. Processes that fall into this class include ionization and other examples of weak-channel excitation and electron capture. These observations led naturally to the multicenter AO method (Lin et al., 1982) and the AO + method (Fritsch and Lin, 1982). In the following section we describe the formal structure of the AO method and discuss the basis sets frequently encountered in the literatures.

1. One-Center AO Method (OC-AO)

When a collision takes place at high energy, i.e., $v/v_{el} \gtrsim 1$, or in cases where a distant collision causes predominantly direct excitation or ionization of target electrons, then the electron charge distribution that is initially on the target remains on the target, with little distortion toward the projectile. In this circumstance, it may be appropriate to use an expansion of atomic orbitals around the target only. This one-center AO (OC-AO) method has two advantages: (i) no ETF is necessary because the OC-AO expansion satisfies the correct scattering boundary conditions, and (ii) the coupled equations are simpler because the basis functions are orthonormal and time independent; hence, the solution of the equations is computationally simple and economical. Thus, Eqs. (8) and (9) can be reduced to the simpler forms

$$\psi(\mathbf{r}, t) = \sum_i a_i(t)\phi_i^{AO}(\mathbf{r}_T)e^{-i\varepsilon_i t} \tag{23}$$

and

$$i\dot{a}_i = \sum_j V_{ij} a_j e^{-i(\varepsilon_j - \varepsilon_i)t}, \tag{24}$$

where

$$V_{ji} = \langle \phi_j^{AO}(\mathbf{r}_T) | V_P(\mathbf{r}_P) | \phi_i^{AO}(\mathbf{r}_T) \rangle \tag{25}$$

and ε_i and ε_j are atomic energies of the ith and jth states. This method has been applied to study direct electronic excitation in $H^+ + H$ collisions (Flannery, 1969) and $H^+ + He$ collisions (Flannery and McCann, 1974) and direct ionization (Reading et al., 1976; Ford et al., 1977). Success of the method, however, is limited. Physically, this is understandable because the region of validity of the OC-AO method is quite narrow, being limited to circumstances in which the electron capture channels are *not* equal in importance to direct excitation at intermediate energies. If, for any reason, the electron cloud is pulled toward the projectile during the collision, the OC-AO expansion is ineffective in describing the electron distribution.

2. Two-Center AO Method (TC-AO)

In order to allow for the movement of the electron cloud between the target and projectile, we expand the wave function in terms of a small set of functions on the target and the projectile in the form

$$\psi(\mathbf{r}, t) = \sum_i a_i(t)\phi_i^{AO}(\mathbf{r}_T)F_i(\mathbf{r})e^{-i\varepsilon_i t} + \sum_j b_j(t)\phi_j^{AO}(\mathbf{r}_P)F_j(\mathbf{r})e^{-i\varepsilon_j t}, \tag{26}$$

where $F_n(\mathbf{r})$ again is the appropriate AO-ETF (Bransden and Noble, 1981, 1982; Shingal and Bransden, 1987; Fritsch and Lin, 1984a).

Generally, the electron capture cross section peaks at $v \simeq v_{el}$ because of the large overlap of the electron cloud between the projectile and the target that occurs under conditions of velocity matching. Further, in this energy domain electron capture primarily takes place at large impact parameters. Lin *et al.*, (1978) pointed out that in such a case a two-state TC-AO (an orbital on target and projectile) can reproduce reasonably accurate electron capture cross sections for various multicharged ion–atom collisions. However, as the energy departs from the $v \simeq v_{el}$ matching region, a close collision becomes increasingly important in the inelastic collision; hence, this simple two-state TC-AO treatment becomes inadequate (Lin *et al.*, 1978). Indeed, even a TC-AO calculation with a large basis set showed poor convergence, suggesting that some fundamental physics is missing from the method. (Note that generally a calculation with a small basis set overestimates the cross section because of the reflection of the flux within the given space.)

Many theorists believed that the failure of the TC-AO was due to the use of incomplete basis sets or the neglect of continuum states. Therefore, various methods were reported that included continuum states by using special basic sets. Among these basis functions are the following.

a. Sturmian functions are regular solutions of the hydrogenic radial equation

$$\left(-\frac{1}{2}\frac{d^2}{dr^2} + \frac{l(l+1)}{2r^2} - \frac{\alpha_{nl}(\varepsilon)}{r}\right)S_{nl}(r,\varepsilon) = \varepsilon S_{nl}(r;\varepsilon), \qquad (27)$$

where $\alpha_{nl}(l) = n(-2\varepsilon)^{1/2}$. The Sturmian radial functions are explicitly given by

$$S_{nl}(r,\varepsilon) = [\alpha_{nl}(\varepsilon)]^{1/2} R_{nl}(\alpha_{nl}(\varepsilon)r), \qquad (28)$$

where $R_{nl}(r)$ is the usual hydrogenic radial function (Gallaher and Wilets, 1968; Shakeshaft, 1975, 1976; Winter, 1986, 1987). The Sturmian functions form a complete set within an infinity of discrete states. These states span a more compact space than do the hydrogenic states and thus offer advantages as an expansion basis. However, because the Sturmian functions are not eigenstates of the atomic Hamiltonian, the transition probabilities oscillate indefinitely as t → ∞. Of course, by projecting the Sturmian functions onto true hydrogenic functions, these difficulties can be removed. Still, comparison of the results with measurements suggest that large Sturmian basis sets are required for accurate results at lower energies.

b. Pseudostates functions. (Cheshire *et al.*, 1970) were constructed to meet following requirements: (i) to avoid the problems that arise in the asymptotic region with the Sturmian basis, and (ii) to increase the electronic charge distribution in the small-*r* region while avoiding linear dependence problems by imposing the constraint of orthogonality with respect to other functions used in the expansion. Radial functions are chosen of the form

$$P_m(r) \propto e^{-\alpha r} \sum_n C_n r^{n-1}, \tag{29}$$

where the parameters α and C_n are selected to represent the continuum and to ensure orthogonality to all other functions included in the bases. Although Cheshire et al. (1970) obtained impressive results that agree quite well with more recent, large-scale calculations for excitation in $H^+ + H$ collisions, no systematic or extended studies have been carried out since this early work. Unfortunately, many unanswered questions remain with respect to the method, and more work is needed.

3. Multicenter AO Method (MC-AO) and AO+ Method

As was discussed previously, the TC-AO method fails to describe a close collision correctly because a simple LCAO is incapable of reproducing the correct electron charge distribution at small internuclear separations. For example, in the united-atom limit of the $H^+ + H$ collision system, the electronic wave function of the system is exactly that of the He^+ (1s) wavefunction, while the simple two-center LCAO $(1s_a + 1s_b)$ or $(1s_a - 1s_b)$ cannot represent it adequately. This recognition led Anderson et al. (1974) and Lin et al. (1982) to adopt the idea of a three-center AO method. In their work, a three-center AO is written as

$$\psi(\mathbf{r}, t) = \sum_i a_i(t)\phi_i^{AO}(\mathbf{r}_T)F_i(\mathbf{r})e^{-i\varepsilon_i t} + \sum_j b_j(t)\phi_j^{AO}(\mathbf{r}_P)F_j(\mathbf{r})e^{-i\varepsilon_j t}$$
$$+ \sum_k c_k(t)\phi_k^{AO}(\mathbf{r}_c)F_k(\mathbf{r})e^{-i\varepsilon_k t}, \tag{30}$$

where $\phi_k^{AO}(\mathbf{r}_c)$ is the united-atom wave function, which is placed at the center of charge. This method was tested extensively for $H^+ + H$ collisions and was shown to reproduce the MO cross sections, including the smooth trend exhibited by the transition probabilities in going from the MO picture at low energies to the AO picture at intermediate collision energies (Winter and Lin, 1984a, b, and Winter, 1988). Unfortunately, because of the need to evaluate three-center matrix elements, this three-center AO method was exceedingly expensive at the time it was developed.

Another idea emerged within this context, in which united-atom orbitals are placed on the two centers instead of at the center of charge. In this way, one can still take advantage of the ability of the united-atom functions to reproduce the molecular character at small R and still avoid the excessive computation involved in the three-center AO. This approach was called the

AO+ method by Fritsch and Lin (1982), who first proposed and applied it. The AO+ method, while still involving a lengthy computation as compared to the TC-AO method, has been extensively applied with great success to studies of excitation, electron capture, and ionization for many multiply charged ion–atom collisions at low to intermediate energies. Furthermore, because of the presence of united-atom orbitals in the expansion and their ability to represent molecular effects, the region of validity of the AO+ and the MC-AO methods is considerably broader than that of the AO methods, particularly at lower energies down to several tens of eV. It is interesting that the ETF-corrected MO method and the corrected AO methods (AO+ and MC-AO) were developed nearly simultaneously in the early 1980s.

4. Approximate AO Methods: Highly Asymmetric Systems

When the projectile charge Z_P and the target charge Z_T are nearly the same ($Z_P \simeq Z_T$, indicating a nearly symmetric collision), it is essential to treat atomic orbitals on the projectile and the target on equal footing. In this situation the TC-AO and MC-AO methods should be applied to calculate excitation and electron capture probabilities. However, in asymmetric collisions, where the projectile (target) charge Z_P (Z_T) is much smaller than the target (projectile) charge Z_T (Z_P), the coupled equations arising from the TC-AO method, for example, can be significantly simplified by recognizing important differences in the basic physics of the asymmetric collision.

a. *Case of $Z_P \ll Z_T$ (e.g., $H^+ + Si$ collision).* In a highly asymmetric collision where $Z_P \ll Z_T$, the coupled equations can be simplified by retaining only a few projectile AOs but keeping many target AOs. This simplification is justified by the small electron capture probability expected because of the large differences in binding energy in such collisions. Thus, conditions are good for applications of the ordinary OC-AO expansion. However, a small number of AOs placed on the projectile in the later stage of the collision improves the results considerably. This is the idea of the so-called "one- and one-half-center" AO expansion method originated by Reading et al. (1981). These researchers used the expansion

$$\psi(\mathbf{r}, t) = \sum_i a_i(t)\phi_i^{AO}(\mathbf{r}_T)F_i(\mathbf{r})e^{-i\varepsilon_i t} + \sum_j b_j(t)\beta_j(t)\phi_j^{AO}(\mathbf{r}_P)\mathbf{F}_j(\mathbf{r})e^{-i\varepsilon_j t}, \quad (31)$$

where $F_n(\mathbf{r})$ is the AO-EFT and $\phi^{AO}[\mathbf{r}_{P(T)}]$ represents an AO placed on the projectile (target). If the function $\beta_j(t)$ is chosen to satisfy the conditions $\beta_j(\infty) = 1$ and $\beta_j(-\infty) = 0$, a set of the coupled equations is derived that accounts for the influence on the excitation amplitudes of flux loss to electron capture channels. In fact, for cases of electron capture from inner shells of atoms by light ions, this method predicts reasonably accurate cross sections. However, the method is only valid when the electron capture probability is small in asymmetric collisions.

b. *Case of $Z_P \gg Z_T$ (e.g., $O^{8+} + H$ collision)*. In this type of asymmetric collision, electrons are captured into highly excited Rydberg states of the projectile. Hence, contrary to the previous case, a large number of atomic states on the projectile take part in the collision and must be included in the close-coupling method. Simplification can be achieved by neglecting a portion or, in certain cases, all of the coupling between the various exit channels (by neglecting virtual intermediate couplings). This simplification can be shown to be valid in the higher collision energy regime (e.g., $E > 100$ keV/amu).

A unitarized distorted wave approximation (UDWA) was constructed by Ryufuku and Watanabe (1978, 1979a, b) and Ryufuku (1982) and applied to various multiply charged ion-atom collisions in the $E \gtrsim 50$ keV/amu regime. The method gives reasonably accurate results for total and dominant-channel cross sections in the energy range from 50 keV/amu to 500 keV/amu. A similar approximation, called the multichannel-Vainstein-Presnyakov-Sobel'man (M-VPS) method, was developed by Presnyakov *et al.* (1975) and Vainstein *et al.* (1964).

D. UNIFIED AO-MO MATCHING METHOD

As was described above, the ETF-corrected MO and the AO+ or MC-AO methods have been partially successful in resolving difficulties with the theory and in predicting reasonably accurate transition probabilities and cross sections for various heavy-particle collision systems at low to intermediate energies. We believe that most of the difficulties associated with these methods will be removed in the near future. The unified AO-MO matching method was proposed as an alternative for dealing with the fundamental difficulties associated with the conventional TC-AO and MO expansion methods. Again, as was discussed previously, the TC-AO method cannot appropriately account for the relaxation of the electronic orbitals during slow collisions, and the MO method has difficulty in incorporating the translational motion of the electron. This knowledge led naturally to the idea of a unified AO-MO matching method that adopts the best features of the two approaches while avoiding their defects. In this method, the time-dependent electronic wave function is expanded in terms of two-center traveling AOs (i.e., with AO-ETF) in the outer region and in terms of MOs (without ETFs) in the inner region. Solutions in the two regions are matched at two internuclear separations, one on the incoming classical trajectory of the collision and the other on the outgoing trajectory. A schematic illustration of the regions and the appropriate choice of basis in each region is shown in Fig. 8. This AO-MO matching method is somewhat analogous to the R-matrix method (Burke and Robb, 1975) frequently used in the quantum mechanical treatment of electron–atom, –ion, and –molecule collisions.

For the collision system, the electronic eigenstates in the asymptotic region are atomic states traveling either with the target or with the projectile, under the assumption that ionization events are not important. Thus, in the asymptotic region, the physical wave function can be expanded in terms of these traveling AOs. Each atomic state in this region has a well-defined AO-ETF in the sense that the electron is moving unambiguously with a particular atomic nucleus. In the inner region, the electronic eigenstates have a molecular character and in general must be obtained by variational or equivalent methods. For slow collisions, the MO (PSS) model provides the qualitative description of the collision dynamics in terms of the MO adiabatic potential curves and nonadiabatic coupling terms. Thus, in the inner region the wave function is expanded in terms of MOs. Since the MO-ETFs are not known *a priori* in the inner region, no ETFs are incorporated in the MO expansion. Of course, this does not create a problem with the boundary conditions because the MOs are not used in the asymptotic region.

This matching method avoids many difficulties associated with the TC-AO and MO methods. However, it does introduce one new complication. The choice of the matching radius R_0 is not well defined. For a reasonable basis set, the result of the calculation should be relatively insensitive to the R_0 chosen. Obviously, if R_0 is chosen to be very large, a large MO basis set will be necessary. If R_0 is chosen to be very small, then a large AO basis set will be

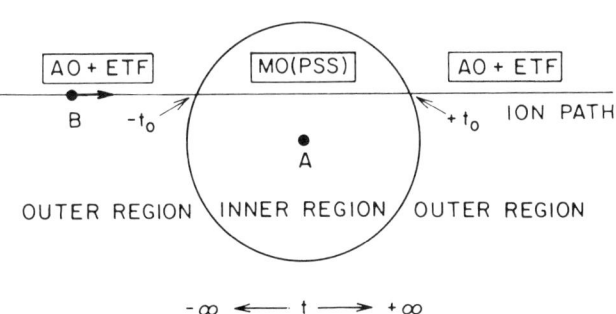

FIG. 8. Diagram illustrating inner and outer regions and corresponding representations used for each region in the AO–MO matching method. (From Kimura and Lin, 1986.)

required. From a practical standpoint, it may be more economical to use a larger MO basis set, i.e., to choose R_0 to be as large as possible, because the AO coupling matrix elements include the AO-ETF terms, which give rise to lengthy computations.

This AO-MO matching method has been applied to study electronic excitation and electron capture in ion–atom collisions for a number of systems (Kimura and Lin, 1985a, b, 1986; Winter and Lane, 1985; Toepfer *et al.*, 1987). In particular, the method successfully provided an explanation for the experimental measurements of alignment and orientation of the 2p electron cloud excited in $H^+ + He$ collisions (Kimura and Lin, 1986), a particularly stringent test of the theory. Although this method has proved its usefulness and established its range of validity, all applications thus far have included only discrete states. The next step should be the inclusion of ionization channels so that the method can be extended to higher energy collisions.

E. ANALYTICAL MODEL REPRESENTATIONS OF THE MO EXPANSION METHOD

In a low-energy collision, an inelastic transition is probable only under very special circumstances, viz., for an exact or near degeneracy of the adiabatic potentials. In such a case, the inclusion of only two (or at most three, e.g., two sigma states and one pi state) states in a close-coupling treatment is usually sufficient for an accurate description of the collision dynamics relevant to the largest cross sections. Depending on the forms of the coupling matrix elements, it is sometimes possible to approximate them in such a way that the two-state close-coupling equations can be solved analytically (Child, 1974). These analytic solutions can offer deeper insight into the collision dynamics through analytical information on, for example, the dependence of transition probabilities on various collision parameters. Schematic diagrams of two interactions that are frequently encountered are depicted in Figs. 9a and 9b.

Figure 9a illustrates a system with a strong (i.e., weakly avoided) curve crossing with a corresponding near-degeneracy at a finite R_c internuclear separation (upper panel) and a schematic diabatic model (lower panel). The Landau-Zener-Stückelberg (LZS) model (Landau, 1932; Zener, 1932; Stückelberg, 1932) has been developed specifically to deal with the case of a strong curve crossing a finite R. Usually, collision systems that involve highly charged ions as projectiles belong to this category because of the very different natures of the initial and final potential curves, e.g., an attractive polarization potential between ions and neutral atoms in the initial ion-atom

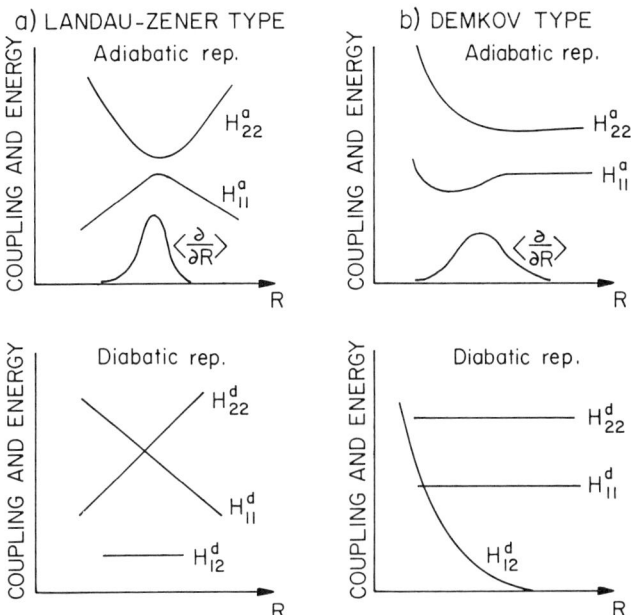

FIG. 9. (a) Landau-Zener type. Adiabatic potentials and corresponding coupling (top). Diabatic potentials and corresponding coupling (bottom). (b) Demkov type. Adiabatic potentials and corresponding coupling (top). Diabatic potentials and corresponding coupling (bottom).

channel and a strong Coulomb repulsive potential between the two ions in the final electron capture channel.

Figure 9b displays a system with a near-degeneracy at $R \to \infty$ but with no strong curve crossing at a finite value of R (upper panel) and a schematic diabatic model (lower panel). The Demkov model (Demkov, 1966; Rosen and Zener, 1932) has been developed particularly to describe this latter case.

Recast in the diabatic representation (see Sec. II,B,4), the two-state MO coupled equations [Eq. (9), without the ETF for simplicity] take the form

$$i\dot{a}_1 = H^d_{12} a_2 \exp\left(i \int^t \Delta\varepsilon \, dt'\right), \quad i\dot{a}_2 = H^{d*}_{21} a_1 \exp\left(-i \int^t \Delta\varepsilon \, dt'\right), \quad (32)$$

where

$$\Delta\varepsilon = H^d_{11} - H^d_{22} \quad (33)$$

and where the initial conditions are given by

$$a_i(-\infty) = \delta_{1i}. \quad (34)$$

1. The Landau-Zener-Stückelberg (LZS) Model

If the diabatic curves cross at R_c, we have approximately

$$\Delta\varepsilon = C_0(R - R_c), \tag{35}$$

and if only the coupling at the crossing point (exact degeneracy) is important, we take

$$H_{12}^d = \text{constant}.$$

Then, the coupled equations [Eq. (32)] can be solved exactly, subject to the initial condition [Eq. (34)]. The LZS transition probability can be written as

$$P_{\text{LZS}} = 4p(1-p)\sin^2 \eta_{\text{LZS}}, \tag{36}$$

where the crossing probability is given by

$$p = \exp(-\pi H_{12}^{d^2}/v\alpha), \tag{37a}$$

where

$$\alpha = \left| \frac{dH_{11}^d}{dR} - \frac{dH_{22}^d}{dR} \right|_{R=R_c}, \tag{37b}$$

and where the phase is

$$\eta_{\text{LZS}} = \int (\Delta\varepsilon^2 + 4H_{12}^{d^2})^{1/2}\, dt. \tag{37c}$$

The $\sin^2 \eta_{\text{LZS}}$ term in Eq. (36) was introduced by Stückelberg to describe the interference effect due to the difference in semiclassical phase angles developed along classical trajectories corresponding to potentials H_{11}^d and H_{22}^d, respectively. An extension of the two-state LZS model to the multichannel case with the inclusion of angular (rotational) coupling has been achieved (see, for example, Janev et al., 1985).

2. The Demkov (or Rosen-Zener) Model

For a collision system where Fig. 9b is appropriate, we set (in the diabatic representation)

$$\Delta\varepsilon = A = \text{constant} \tag{38}$$

and

$$H_{12}^d = C_1 e^{-\gamma R}. \tag{39}$$

With these simplifications, the coupled equations [Eq. (32)] offer another set of exact solutions, again subject to the initial conditions of Eq. (34). The

Demkov transition probability is given by

$$P_D = \text{sech}^2(\pi A/2\gamma)\sin^2 \eta_D, \qquad (40)$$

where

$$\eta_D = \int H_{12}^d \, dt. \qquad (41)$$

Again, the Stückelberg-type oscillatory function $\sin^2 \eta_D$ is present. Interference effects due to the $\sin^2 \eta$ terms in Eqs. (36) and (40) are important in the calculation of differential cross sections, but their effects usually average out in the integrated total cross sections.

F. Relationships of Close-Coupling to Perturbation Methods

Although it is not our intention to review all theoretical methods used to describe atomic collisions, it might be useful to give a brief overview of how the close-coupling methods relate to various perturbative treatments. From Eq. (26), the two-state, two-center AO (TS-TC-AO) expansion can be written in the form

$$\psi(\mathbf{r}, t) = a_i(t)\phi_i^{AO}(\mathbf{r}_T)\exp[-i(\tfrac{1}{2}\mathbf{v}\cdot\mathbf{r} + \tfrac{1}{8}v^2 t + \varepsilon_T t)]$$
$$+ a_j(t)\phi_j^{AO}(\mathbf{r}_P)\exp[-i(-\tfrac{1}{2}\mathbf{v}\cdot\mathbf{r} + \tfrac{1}{8}v^2 t + \varepsilon_P t)], \qquad (42)$$

where T (P) denotes the active electron attached to the target (projectile) nucleus. Then the coupled equation [Eq. (9)] can be recast into a two-state form after a suitable phase transformation of the scattering amplitude from a to b [i.e., $a_i(t) = b_i(t)\exp(-\int^t V_{ii}(t')dt')$] (McDowell and Coleman, 1970) as follows:

$$i\dot{b}_T = b_P\left(\frac{h_{TP} - S_{TP}h_{PP}}{1 - S^2}\right)\exp\left(i\int^t (U_T - U_P)dt\right), \qquad (43a)$$

$$i\dot{b}_P = b_T\left(\frac{h_{PT} - S_{PT}h_{TT}}{1 - S^2}\right)\exp\left(-i\int^t (U_T - U_P)dt\right), \qquad (43b)$$

where

$$h_{AB} = \langle A|h_{el}|B\rangle, \qquad (43c)$$

$$S_{AB} = \langle A|B\rangle, \qquad (43d)$$

with

$$A = \phi_A^{AO}(\mathbf{r}_A)\exp[-i(\tfrac{1}{2}\mathbf{v}\cdot\mathbf{r} + \tfrac{1}{8}v^2 t + \varepsilon_A t)] \qquad (43e)$$

and an analogous form for B, but with $\mathbf{v}$ replaced by $-\mathbf{v}$, and where

$$U_T = \varepsilon_T + \frac{h_{TT} - S_{TP}h_{PT}}{1 - S^2} \tag{43f}$$

$$U_P = \varepsilon_P + \frac{h_{PP} - S_{PT}h_{TP}}{1 - S^2} \tag{43g}$$

represent the distortion potentials of the electrons in the field of the target and projectile, respectively, and h_{AB} is the potential coupling matrix. [Actually, $(h_{AB} - S_{AB}h_{AA})/(1 - S^2)$ represents the full potential coupling.] These distortion potential terms and dynamical coupling terms cause the probability amplitude to oscillate as a function of velocity. However, when $v \simeq v_{el}$, these oscillating terms have the same amplitudes with opposite signs, and the electron capture cross section possesses a peak. However, if the capture probability $|b_P|^2$ is assumed to be small, we can set $b_T = 1$ and integrate Eq. (43b) to obtain

$$b_P(t \to \infty) = -i \int_{-\infty}^{\infty} \left(\frac{h_{PT} - S_{PT}h_{TT}}{1 - S^2}\right) \exp\left(-i \int^t (U_T - U_P) dt'\right). \tag{44}$$

This equation is the so-called "distortion" (distorted wave) approximation for the case where the initial and final states are not orthogonal.

If we further set $U_{T(P)} = \varepsilon_{T(P)}$, ignoring the distortion, then we obtain the impact parameter, first Born approximation in the "post" interaction form,[2]

$$b_P(t \to \infty) = -i \int_{-\infty}^{\infty} h_{PT} \exp[-i(\varepsilon_T - \varepsilon_P)t] dt \tag{45}$$

or

$$b_P(t \to \infty) = -i \int_{-\infty}^{\infty} dt \left\langle F_P \phi_P^{AO} \left| \frac{-Z_T}{r_T} \right| F_T \phi_T^{AO} \right\rangle \exp[-i(\varepsilon_T - \varepsilon_P)t]. \tag{46}$$

Although it is not derived from the TS-TC-AO expansion [Eq. (42)], the eikonal-Glauber approximation for the transition probability can be written in a similar form:

$$b_P^{eikonal}(t \to \infty) = -i \int_{-\infty}^{\infty} dt \left\langle F_P \phi_P^{AO} \left| \frac{-Z_T}{r_T} \right| \exp\left(i \int^t \frac{Z_P}{r_P} dt'\right) F_T \phi_T^{AO} \right\rangle$$

$$\times \exp[-i(\varepsilon_T - \varepsilon_P)t]. \tag{47}$$

---

[2] When the post and prior forms of the interaction are compared, they are known to sometimes give different results. This post–prior paradox arises from approximations made in the initial and final wave functions. If exact eigenstates are used, then results from the post and prior forms of the interaction are the same.

If one removes the eikonal phase factor, Eq. (47) reduces to the first-order Born form (46). The eikonal phase, which describes the Coulomb interaction between the electron and the projectile, is included to improve the simple Born approximation by incorporating some higher order effects. Generally, the eikonal-Glauber approximation predicts better agreement for total electron capture cross sections than does the Born approximation.

For energies approaching the high keV or MeV regime, it is now well known that continuum intermediate states become increasingly important in describing the collision dynamics. In this respect, several forms of the second-order Born and the continuum distorted-wave approximations have been developed and tested extensively for *asymmetric collisions*, particularly in the MeV collision regime (McDowell and Coleman, 1970).

G. Direct Numerical Solution of the Time-Dependent Schrödinger Equation

As an alternative to using an eigenfunction expansion method as has been described, computational technologies for direct numerical integration of Eq. (9) have been developed and applied to several problems of electron capture in heavy-particle collisions (Kulander *et al.*, 1982; Devi and Garcia, 1983). However, because a review of this topic was published recently (Bottcher, 1984), we refer the reader to that work and choose not to discuss the method further here.

H. Observables

1. Differential Cross Sections

Solving the coupled Eqs. (5) or (9), one obtains the asymptotic amplitudes $a_j(t = +\infty)$ (semiclassical representation) or the S matrix (quantum representation). For the semiclassical representation, McCarroll and Salin (1968) proposed a formula for the differential cross section for a transition $i \to j$ within the eikonal approximation,

$$\frac{d\sigma_{ji}(E, \theta)}{d\theta} = 2\pi\mu^2 v^2 \left| \int_0^\infty b \, db \, a_j(b, t = +\infty) J_m\left(2\mu v b \sin\frac{\theta}{2}\right) \right|^2, \quad (48)$$

where μ is the reduced mass of a system, θ is the scattering angle, $J_m(x)$ is the Bessel function of order m, which is the difference between the magnetic quantum numbers in the initial and final states. Note that Eq. (48) was derived under the assumption that the forward scattering is the dominant process; hence, it is valid only for small-angle scattering.

For the quantum representation, the differential cross sections for transition $i \to j$ are obtained in general form (assuming the spherically symmetric case for simplicity) as

$$\frac{d\sigma_{ji}}{d\Omega} = \frac{1}{4k_i^2} \left| \sum_{l=0}^{\infty} (2l+1) S_{ij}^l P_l(\cos\theta) \right|^2, \qquad (49)$$

where k_i is the initial momentum of the projectile relative to the target, S_{ij}^l is an S matrix element, and $P_l(\cos\theta)$ is the Legendre polynomial.

Interference between semiclassical scattering amplitudes associated with different trajectories gives rise to oscillatory structure in the differential cross section, e.g., rainbow scattering or LZS and Demkov oscillations. In symmetric systems oscillatory structure also arises from gerade–ungerade (g–u) interference. In particular, in a semiclassical representation of a symmetric scattering case (where one branch contributes), the differential cross section [Eq. (48)] can be written in the form

$$\frac{d\sigma}{d\theta} \propto \left[\frac{d\sigma_g(\theta)}{d\theta} + \frac{d\sigma_u(\theta)}{d\theta} - 2\left(\frac{d\sigma_g(\theta)}{d\theta} \frac{d\sigma_u(\theta)}{d\theta} \right)^{1/2} \cos(\eta_g - \eta_u) \right], \qquad (50)$$

where $d\sigma_{g(u)}(\theta)/d\sigma$ is the elastic differential cross section for the gerade (ungerade), g (u), state and $\eta_{g(u)}(b)$ represents the phase shift for the specific partial-wave angular momentum (or impact parameter) that corresponds, semiclassically, to the scattering angle θ for the respective g (u) potential. The $\cos(\eta_g - \eta_u)$ term is a source of the oscillation in the differential cross section.

2. Integral Cross Sections

Integral cross sections are derived from partial-wave summation (or from integration over impact parameters) or from the integration of Eqs. (48) and (49) over all scattering angles. Usually, structures that are prominent in the differential cross section are washed out in the integration. However, strong features remain for special cases such as the symmetric alkali systems and the class of Rosenthal oscillations, both of which will be described below.

The total cross section for electron capture in a symmetric collision system may be written as

$$\sigma^{\mathrm{tot}}(E) \propto \sum_l (2l+1)\sin^2[\eta_g(l) - \eta_u(l)], \qquad (51)$$

where $\eta_g(l)$ and $\eta_u(l)$ are the gerade and ungerade phase shifts, respectively. (This simple form is based on the assumption that excitation does not occur in the collision.)

The phase difference $\Delta\eta = [\eta_g(l) - \eta_u(l)]$, which determines the cross section, is closely related to the difference between the gerade and ungerade potentials, $\Delta V = V_g(R) - V_u(R)$. Within the WKB approximation, we can write

$$\Delta\eta \propto \int_b^\infty \frac{\Delta V}{(1 - b^2/R^2)^{1/2}} \, dR. \tag{52}$$

Therefore, it is clear from Eq. (52) that when ΔV has a maximum with respect to R, then $\Delta\eta$ possesses a maximum with respect to b, the latter giving rise to the oscillatory structure in the total cross section (Smith, 1966).

3. Alignment and Orientation Parameters

Recent experimental advances have made it possible to observe at well-defined scattering angles (and corresponding impact parameters) the orientation (O) and alignment (A) of the electron charge clouds of excited states resulting from collisions of ions or atoms with atomic or molecular targets (Hippler, 1988). Such measurements for electron capture or excitation to a specific state give detailed information about the collision dynamics at specific impact parameters and hence can provide sensitive tests of particular theoretical models over a wide range of impact parameters. The MO and AO treatments have been applied to study the O-A parameters for a variety of collision systems and have helped to provide a deeper understanding of the excitation and electron transfer mechanisms. Unfortunately, we do not have the space to include this topic, and we must refer the reader to a recent review (Crowe and Rudge, 1988).

III. Current Status of Theoretical and Experimental Results

As we have attempted to show in the previous section, the concepts of *close collision* and *distant collision* in ion–atom systems are very important for categorizing the collision dynamics. They are valuable for selecting the most suitable theoretical model for a particular application, including extensions of the methods to elastic and inelastic scattering in ion–molecule and ion–surface collisions at low to intermediate energies.

In the first section, we deal with several examples of ion–atom and atom–atom collisions where both theoretical and experimental data are available, regardless of whether the agreement is good. These examples are intended to highlight several important concepts that are fairly well established as well as to point to unresolved problems.

The second section concerns ion–molecule collisions. In comparison with ion–atom systems, rigorous theoretical studies on ion-molecule collisions are extremely rare except for the region of $E \lesssim$ few eV, where chemical reactions become important. The reason for the lack of theoretical activity is the additional complexity involved in obtaining accurate electronic wave functions for the multicenter molecular systems. However, in cases where distant collisions dominate, one can modify the ion–atom collision model for practical application to ion-molecule systems. We will show the applicability of the model as well as its limitations.

The third section concentrates on topics of ion–surface collisions, where experimental activity has increased significantly in recent years, motivated in part by applied research. When atoms or ions collide with a surface, various types of inelastic scattering are possible, including sputtering, channeling, reactions (absorption and adsorbtion), and electron capture and ionization. We will focus on electron capture and ionization and will not touch topics that relate to the condensed phase, e.g., surface plasmon and phonon excitation. For electron capture and ionization processes involving surfaces, a model of ion–atom collisions may be applicable for a certain class of collisions that avoids a large-scale computation of the full multicenter, many-body problems. This fortunate situation is illustrated by way of several specific examples, and the validity and limitations of this approach to ion–surface collisions are discussed.

A. Ion–Atom and Atom–Atom Collisions

It is convenient to discuss the cases of lowly charged ion (or neutral-atom) atom collision (a close collision) and a highly charged ion–atom collision (a distant collision) separately, because these two classes of collision systems show marked distinction with respect to collision dynamics. Thus, the theoretical methods used will be apparent in Sec. II,E.

1. Collision of Neutral Atoms (Lowly Charged Ions) with an Atom (a Close Collision)

A schematic diagram of typical adiabatic potentials associated with a collision system of the class discussed in this section is displayed in Fig. 10. The long-range nature of the potential is essentially determined by the polarizability α of a neutral atom; the potential falls off as $\alpha/2R^4$ at large R. Depending upon the strength of the polarizability and the separation of the asymptotic energy levels of the collision partners, the diabatic potentials show either curve crossing or no curve crossing. (The corresponding adia-

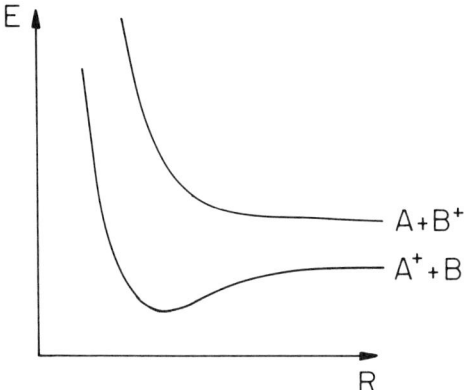

FIG. 10. Schematic adiabatic potentials of lowly charged ion–atom or atom–atom collision systems.

batic potentials will show either no sign of an avoided crossing or weakly avoided crossing.) However, the position of the curve crossing (or weakly avoided curve crossing) generally occurs at relatively small R (compared to what is found in the highly charged ion–atom case). This implies that the MO representation or the MC-AO (or AO+) approach in the AO representation should be employed for the best description of the collision dynamics for this class of collisions. These points are best illustrated with examples:

a. *Symmetric case*: H^+ (p) *Collision with* H. Because a hydrogen atom is most fundamental to an understanding of atomic physics, the one-electron diatomic molecule (OEDM) systems like H_2^+ and HeH^{2+} are also most fundamental to an understanding of both molecular physics and atomic collision physics. The study of such systems exposes rich physics and provides the basic principles that underlie general heavy-particle collisions.

Adiabatic potentials for the H_2^+ system are shown in Fig. 11. This system, along with the He_2^+ system, exhibits the classic oscillatory structure in the resonant electron capture probabilities (Lockwood and Everhart, 1962). This structure is illustrated in Fig. 12, where the probability at a particular angle is plotted against $1/v$. This phenomenon is now well understood as arising from a g-u symmetry interference, i.e., an interference between the two components of the nuclear wave function that correspond to the ground molecular states $^2\Sigma_u$ and $^2\Sigma_g$, which are degenerate at $R \to \infty$, as is seen in Eq. (50) in Sec. II,H,l. Excitation and electron capture to Rydberg states take place by two different mechanisms, as was first discovered by Bates and Sprevak (1970) [and was later extensively studied by Knudson and Thorson (1970)], namely (i) strong angular coupling at the united-atom limit, which connects $2p\sigma_u$

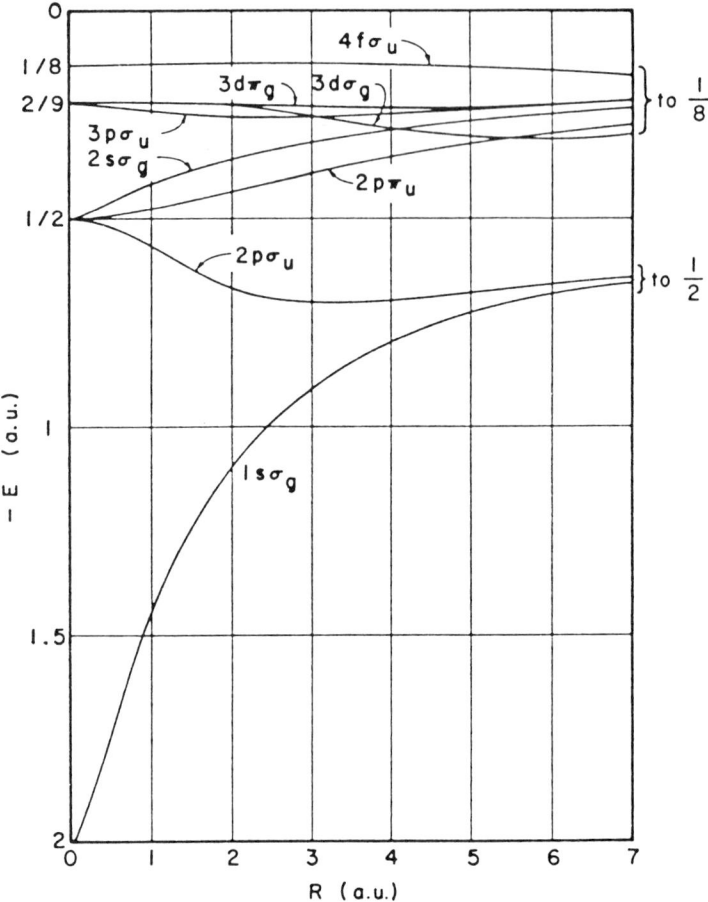

FIG. 11. Adiabatic potentials of H_2^+.

and $2p\pi_u$ states (see Fig. 11), and (ii) radial couplings that mix $2p\sigma_u$ with excited u states and $1s\sigma_g$ with excited g states. Recent experimental measurements of the integral alignment $[A_{20} = (\sigma_{2p_{\pm 1}} - \sigma_{2p_0})/(\sigma_{2p_0} + 2\sigma_{2p_{\pm 1}})]$ of H(2p) in the $H^+ + H$ collision (Hippler et al., 1988) unambiguously demonstrate (see Fig. 13) that the $2p\sigma_u$–$2p\pi_u$ angular coupling is the dominant mechanism for $H(2p_{\pm 1})$ excitation and electron capture in $H^+ + H$ collisions below 10 keV. As the energy is increased, the radial couplings become important, and hence the population of the $H(2p_0)$ state sharply increases. This is seen in Fig. 13, where the integral alignment A_{20} decreases with increasing energy, becoming negative for $E \gtrsim 16$ keV.

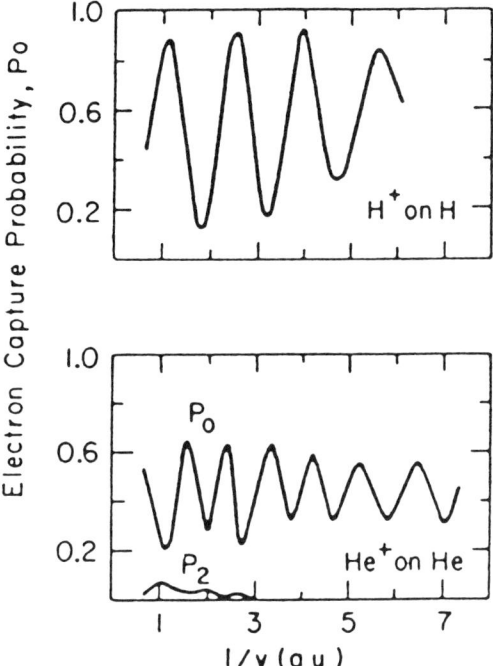

FIG. 12. Electron capture probabilities as a function of reciprocal velocity in $H^+ + H$ and $He^+ + He$ collisions. (From Lockwood and Everhart, 1962.)

This detailed understanding of the relative importance of the different coupling mechanisms was extremely important in gaining some guiding principles of heavy-particle collisions that were extended to understanding a variety of phenomena observed in ion–atom collisions over a wide range of collision energies, including K-shell capture and ionization (Briggs, 1976). Fig. 14 shows cross sections for electron capture to 2s and 2p states; recent AO+ (Fritsch and Lin, 1982) and MC-AO (Cheshire et al., 1970; Rapp and Dinwiddie, 1972; Winter and Lin, 1984) close-coupling results are shown along with measurements (Morgan et al., 1980; Hill et al., 1979). (See also Fig. 1 for earlier theoretical results.) Although general trends in the cross sections are in reasonably good accord among the large-scale close-coupling results, differences in the details of the structure are still sufficient to be of some concern. A difference of 15–25% in the 2s capture and excitation cross sections below $E = 10$ keV is common to all theoretical models including AO-MO matching (Kimura and Lin, 1986) and the MO, AO+, and MC-AO methods. Because this process is dominated by close collisions (the probability times the impact parameter versus the impact parameter peaks at

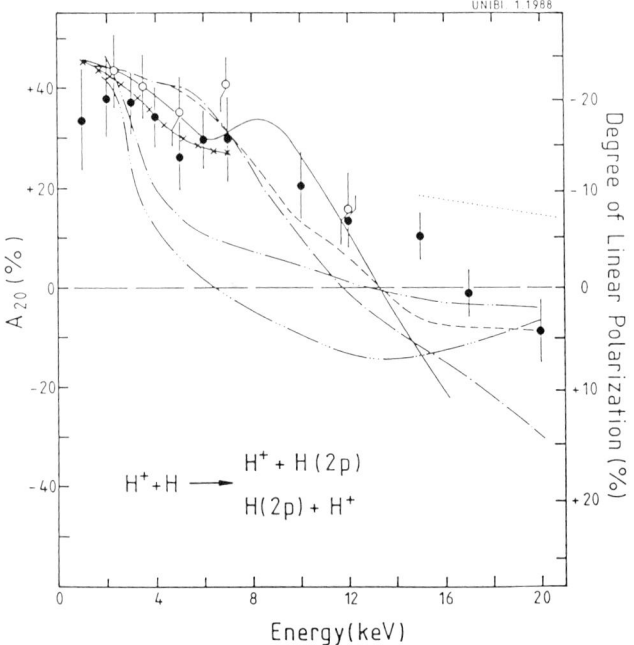

FIG. 13. Experimental alignment parameter in $H^+ + H$ collisions. Theory: $-\cdot-$, Lüdde and Dreizler (1982); ———, Fritsch and Lin (1983); $-\times-$, Kimura and Thorson (1981); $-\cdot\cdot-$, Rapp and Dinwiddie (1972); $--\bigcirc--$, Gaussorgues and Salin (1971); $----$, Shakeshaft (1978). Experiment: ●, ○ (from Hippler et al., 1988).

about $1a_0$, at $E \lesssim 5$ keV), the explicit inclusion of the molecular nature of the collision system is essential in any attempt to determine accurate probabilities and cross sections. In view of this, the AO-MO matching method or the MO method with proper MO-ETF should be more appropriate for a study of this process unless very large basis sets are used in the MC-AO or AO+ methods. The MO result of Crothers and Hughes (1978; "MO '78" in Fig. 1) has a pronounced shoulder at 5 keV, whereas other theories, including other MO results ("MO '81" in Fig. 1), show no pronounced structure. The difference between the two sets of MO results can be attributed to the different treatments of the MO-ETF.

Ionization cross sections obtained by the MO, AO+, and MC-AO methods are compared with recent measurements (Shah et al., 1987) in Fig. 15. The MO calculation included about 150 MO states (20 discrete states and 130 continuum states) in a close-coupling treatment (Kimura and Thorson, 1989), while the AO+ and MC-AO calculations included 25 to 40 AO basis functions including pseudostates, which describe the continuum. Agreement

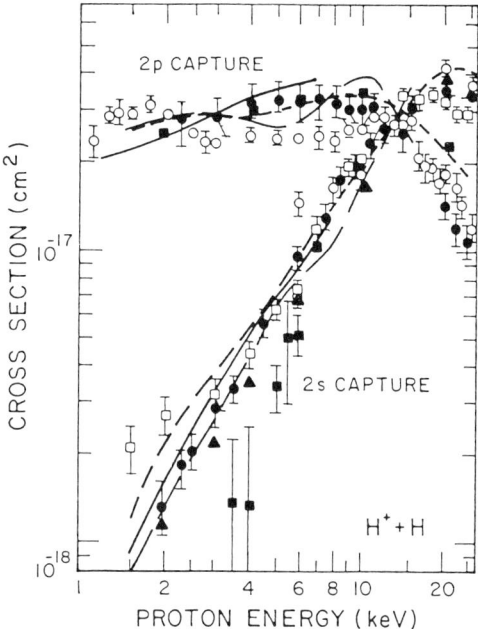

FIG. 14. H(2s) and H(2p) capture cross section in $H^+ + H$ collisions. Theory: AO+, Fritsch and Lin (1982); AO-MO, Kimura and Lin (1985); 3-center, Winter and Lin (1984); MO, Kimura and Thorson (1981). Experiment: ●, Morgan et al. (1973, 1980); ○, Kondow et al. (1974); ■, Bayfield (1969); □, Hill et al. (1979). (From Kimura and Lin, 1985.)

between the theoretical and experimental ionization cross sections is not yet satisfactory. However, these studies have clarified the importance of several physical characteristics of ionization at low energies. These characteristics include (i) electron capture to the continuum, (ii) electron ejection from the equiforce point or saddle point (by the Wannier mechanism of ionization), and (iii) the ladder-climbing process in contrast to direct Coulomb ionization.

b. Antiproton ($\bar{p}$) Collision with H. It is interesting and perhaps informative to compare the collision of an antiproton ($\bar{p}$) with H at low to intermediate energies; this process exhibits characteristics in sharp contrast to those of the p-H collisions described above. For example, recent experiments (Anderson et al., 1986) performed at CERN show that the cross sections for double ionization of rare-gas atoms in collisions with $\bar{p}$ have the interesting feature of a larger double-ionization cross section for $\bar{p}$ than for p at $E \geq 0.5$ MeV. In order to elucidate the CERN finding as well as to find the underlying mechanism, an accurate calculation of the adiabatic potentials of the system

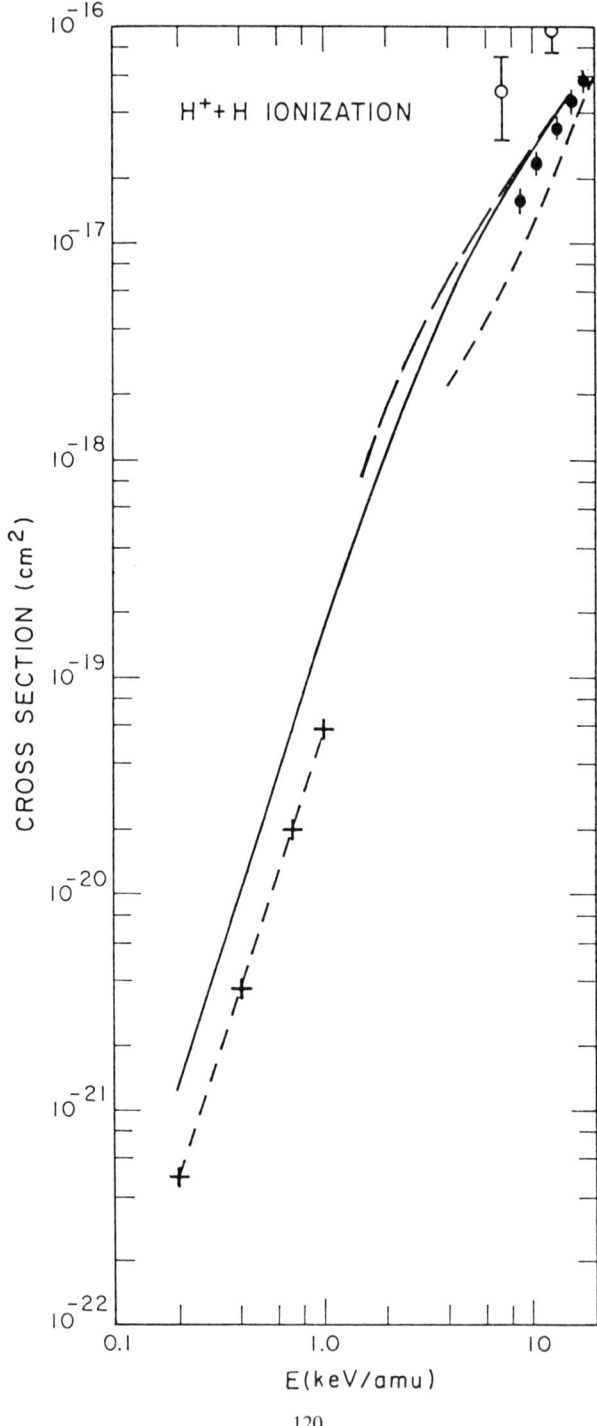

was required. Figures 16a and 16b show adiabatic potentials for the p̄-H and p-H systems, respectively, both including states up to the H(n = 2) manifold (Kimura and Inokuti, 1988). Striking is the fact that the electronic binding energy of every state of the p̄-H system decreases as the internuclear distance decreases and vanishes at a characteristic value of the internuclear separation. This behaviour may be called a curve crossing with the continuum. This phenomenon is now well known in electron–polar-molecule collision systems (see historical background on the topic: Turner, 1977) as the critical dipole moment for bound states of an electron in the field of a stationary permanent electric dipole. Thus, the p̄-H system is in sharp contrast to the p-H system, in which an infinity of electronic states are bound at any internuclear separation. The nature of the p̄-H potential curves suggests that electrons ejected from p̄-H collisions at low energies should have near-zero energies with momenta distributed over a large range of angles. One has a picture of electrons oozing out. Further, the nature of the curves suggests that the ionization cross section for p̄-H collisions at low energies is likely to be much larger than that for the p-H collisions, as is clear from the results displayed in Fig. 17 (Ermolaev, 1987). Although the results in Fig. 17 were obtained by the AO method and are considered to be preliminary, they seem to be qualitatively consistent with the study of adiabatic potentials summarized in Fig. 16.

c. *Symmetric and Near-Symmetric Cases* ($Li^+ + Li$, $Na^+ + Na$, or $Li^+ + Na$). The alkali collision systems have been popular subjects for investigation for some time because of the early experimental discovery of an interesting oscillatory structure in the energy dependence of the total electron capture cross section. A familiar feature of differential cross sections, e.g., for electron capture in $H^+ + H$ or $He^+ + He$ symmetric collisions, is the oscillation (with respect to the scattering angle) that result from various interference effects (Aberth et al., 1965). However, when the differential cross sections are integrated over all scattering angles, the oscillatory structure is smeared out and generally disappears from the integrated total electron capture cross section. The alkali systems are exceptions.

In Fig. 18 experimental total electron capture cross sections are shown for $Li^+ + Li$, $Na^+ + Na$, and $Li^+ + Na$ collisions (Daley and Perel, 1969). Pronounced oscillatory structures are clearly visible in all cross sections. Although the analysis discussed in Section II,H,2 helps to explain the fundamental physical nature of the oscillations, recent theoretical studies

---

FIG. 15. Ionization cross section in $H^+ + H$ collisions. Theory: ———, MO, Kimura and Thorson (1989); - - -, AO+, Fritsch and Lin (1983); — —, 3-center, Winter and Lin (1984); — + —, MO, Sethu Raman et al. (1973). Experiment: ○, Fite et al. (1960); ●, Shah et al. (1987). (From Kimura and Thorson, 1989.)

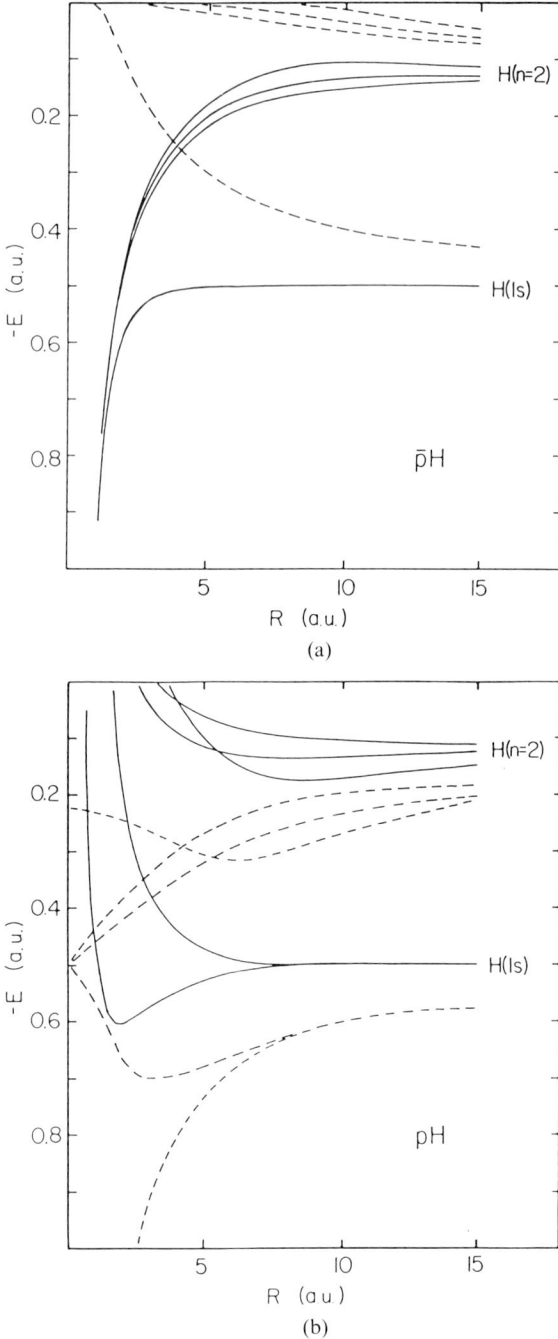

FIG. 16. (a) Adiabatic potentials of the p̄-H system. (From Kimura and Inokuti, 1988.) (b) Adiabatic potentials of the p-H system. (From Kimura and Inokuti, 1988.) The solid line = total energy; the dashed line = electronic energy only.

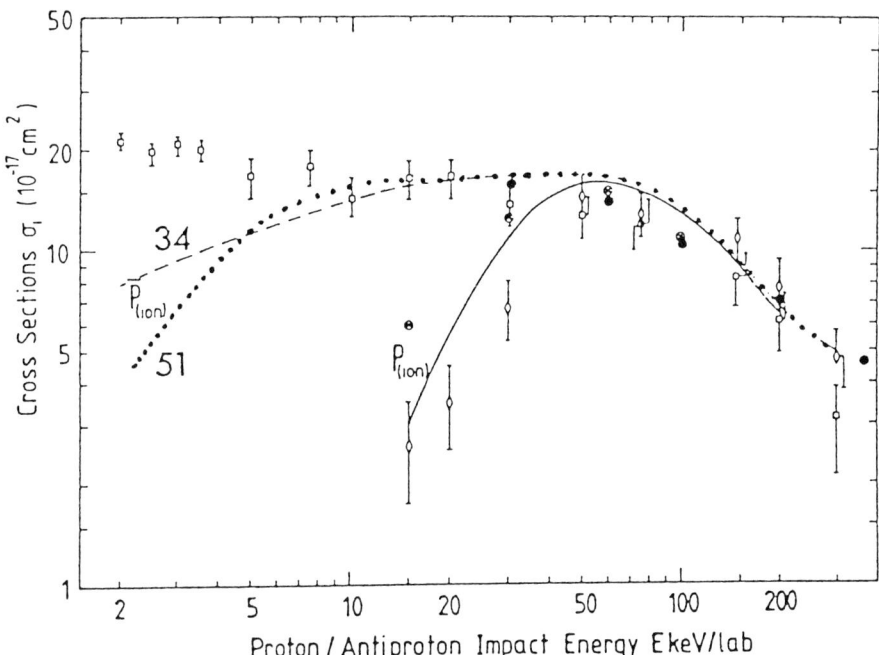

FIG. 17. Ionization cross section in $\bar{p} + H$ and $p + H$ collisions. Proton projectile: —, TC-AO, Shakeshaft (1978); ⊗, one- and one-half center AO, Reading et al. (1981); ◇, classical-trajectory Monte Carlo (CTMC). Antiproton projectile: ○○○, 51TC-AO; ---, 34 TC-AO; □, CTMC, Ermolaev (1987); ●, OC-AO, Martir et al. (1982). (From Ermolaev, 1987.)

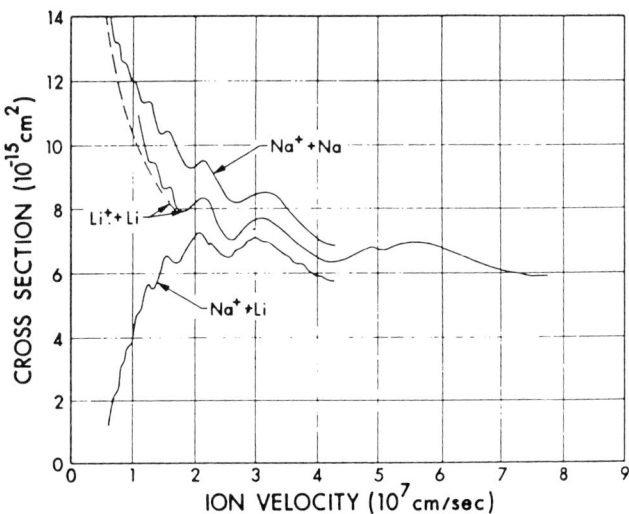

FIG. 18. Experimental resonant electron capture cross sections in $Li^+ + Li$, $Na^+ + Na$, and $Na^+ + Li$ collisions. (From Daley and Parel, 1969.)

(Men et al., 1986; Shingal et al., 1986) using the AO and MO methods suggest that the dominant inelastic channels must also be included in a close-coupling calculation in order to obtain accurate quantitative agreement with the observed phase of the oscillations in the cross section (Fig. 19). Particularly for the alkali systems, it is important to include the Π_u excited state, which crosses the ground Σ_u state at finite R and plays an important role in determining the phase of the oscillation in the cross section because the angular coupling causes flux promotion to the excited state.

   d. Asymmetric Case: Na^+ *Collision with Ne*. In Fig. 20, electron capture and excitation cross sections are shown for Na^+ + Ne collisions (Tolk et al.,

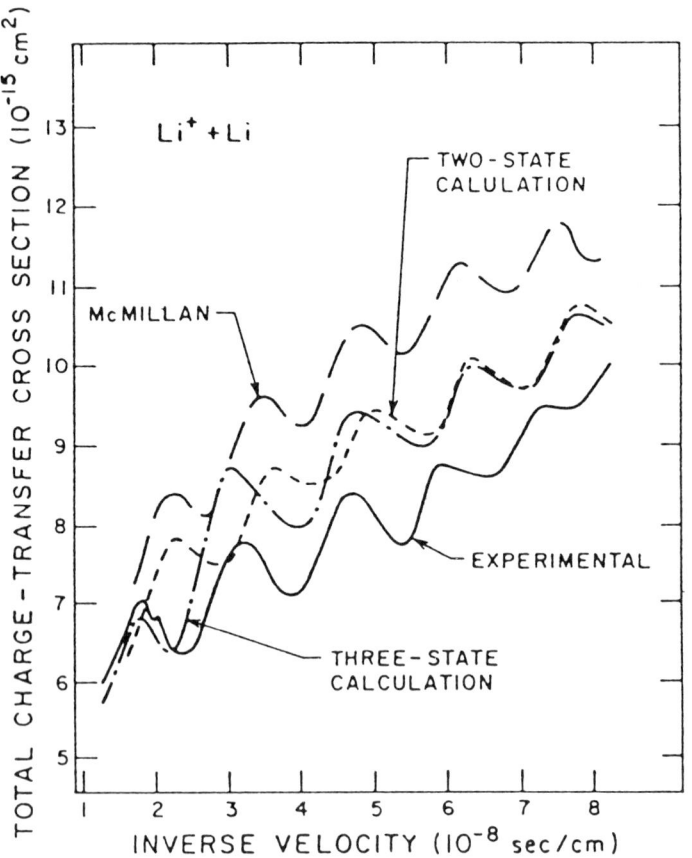

FIG. 19. Theoretical resonant electron capture cross section in Li^+ + Li collisions as a function of inverse velocity. (From Men et al., 1986.)

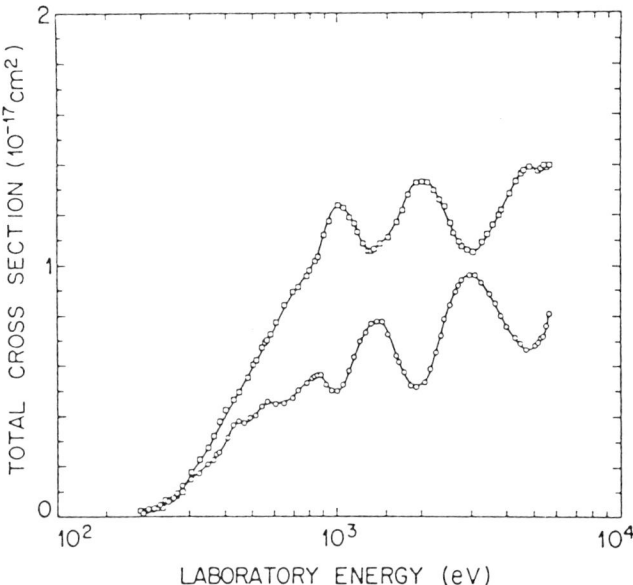

FIG. 20. Experimental electron capture Na(3p) (O: lower curve) and excitation Ne(3p) (□: upper curve) cross sections in Na$^+$ + Ne collisions. Note the out-of-phase (the Rosenthal) oscillations. (From Tolk et al., 1976.)

1975). As in the symmetric and near-symmetric collisions of the Li$^+$ + Li and Li$^+$ + Na systems, respectively, we see an oscillatory structure in the measured cross section. The new and interesting features of these experimental results are the out-of-phase oscillations in the two-electron capture and excitation channels. To explain this phenomena, Rosenthal (1971) proposed a model that is based on weakly avoided curve crossings among three potential curves. A schematic diagram of the potentials in the model is shown in Fig. 21a, and cross sections obtained from the model are shown in Fig. 21b. The essential features of the model are the following: (i) The colliding particles in molecular state O approach the inner avoided crossing regions where the initial state interacts strongly with molecular states 1 and 2. At these inner avoided crossings, the primary excitations take place. (ii) In the outer part of the collision, two excited states 1 and 2 (e.g., one an electron capture and the other an excitation) couple very strongly at a large internuclear separation, i.e., via an outer avoided crossing. The scattering amplitudes associated with these excited states interfere, causing strong flux transfer between states 1 and 2 and, as a result, out-of-phase oscillations in the respective cross sections. These features are called "Rosenthal oscillations" after their pioneer. In recent years, it has become clear that the

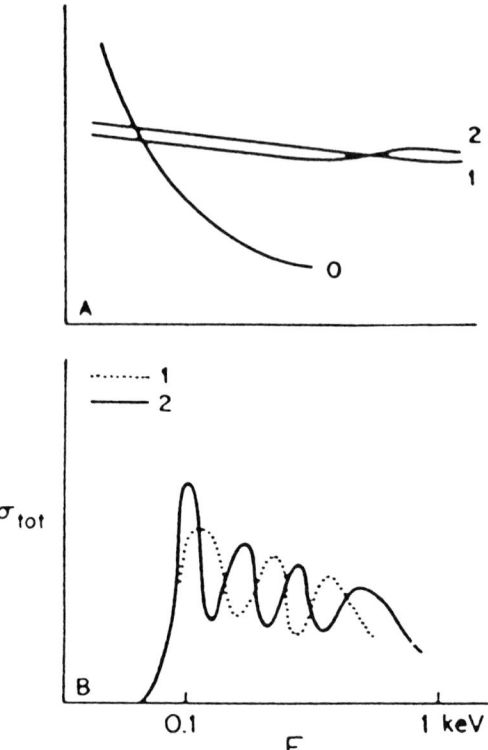

FIG. 21. The Rosenthal model. (a) Schematic diabatic potentials. 0: initial state. 1 and 2: final states. (b) Antiphase oscillation in total cross sections for state 1 and 2. (From Rosenthal, 1971.)

situation just described, viz., small energy separation and strong interaction at large R between excited states and electron capture states, is quite common in ion–atom collisions. An analysis carried out within a semiclassical formalism and similar to that used in the previous section for two potential curves is also useful in describing the case of three potential curves. However, it is important to emphasize that for a multistate curve-crossing problem there is, in general, no reliable rule to tell whether outer curve crossings are adiabatic or diabatic unless the coupling matrix elements are computed and specific close-coupling computations are carried out.

e. Rydberg Atom Collisions: $Na^{**}(n \geq 6)$ Collision with He. Recently, several theoretical attempts to study the quenching of low-lying Rydberg states of atoms in collisions with rare-gas atoms have been initiated (Kumar et al., 1989). The close-coupling MO method is used because it is known that the perturbative treatment is not suitable for low-lying Rydberg atom collisions (Sato and Matsuzawa, 1985). Since a large number of states are usually necessary because of the small energy defects among them, the practical computations based on the method were simply not possible even a few years ago.

For the case of Na** + He(1^1S), Fig. 22 shows representative adiabatic potential curves of 14 Σ states corresponding to initial Na principal quantum numbers ranging from $n = 7$ to $n = 9$ (Kumar et al., 1989). The figure suggests that the dominant process in Na**(9s) + He(1^1S) collisions is the deexcitation of the Na(9s) state to Na**(8p) and Na**(7f) states at avoided crossings at about $\sim 11a_0$ with possible flux transfer to 7g, 7h, and higher-l states that are nearly degenerate with 7f. Calculated cross sections at thermal energies (not shown) indeed show that the e → f (and higher l) process is dominant.

2. Collision of a Highly Charged Ion with an Atom (a Distant Collision)

Figure 23 is a schematic diagram of adiabatic potentials that are representative of highly charged ion–atom collision systems. Because of the presence of an attractive polarization potential in the initial channel and strong Coulomb repulsive potentials in the final excited-state channels, a series of narrow avoided crossings is normally seen. Prominent features of highly charged ion–atom collision studies include the following (Ohtani, 1984; McCullough, 1986):

(i) Electron capture generally dominates other inelastic processes with total cross sections of the order 10^{-14} to 10^{-13} cm^2, and its cross section can be scaled linearly with the projectile charge q.

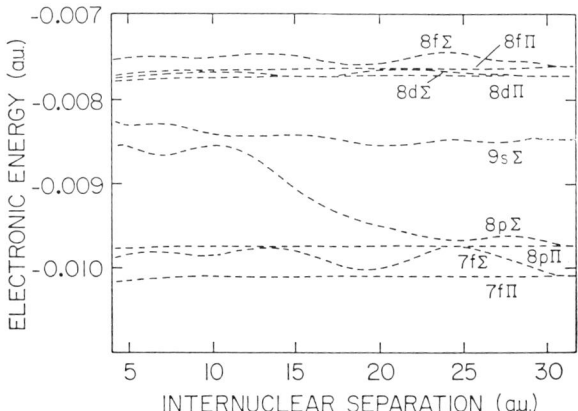

FIG. 22. Adiabatic potentials for [Na**(9s) + He] system. (From Kumar et al., 1989.)

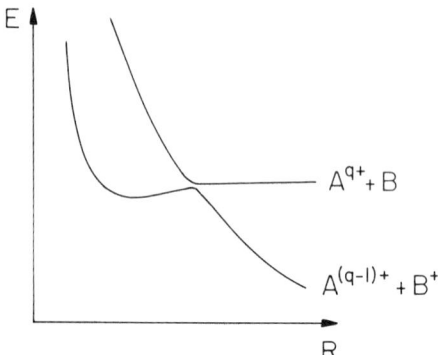

FIG. 23. Schematic adiabatic potentials of the highly charged ion–atom collision system. (Only one representative excited final-state curve is shown.)

(ii) Electron captured states are highly excited. It is obvious from Fig. 23 that as the principal quantum number n of the electron capture state increases, the curve crossing moves farther out (to larger R). Thus, at intermediate collision energies, electron capture to highly excited Rydberg states is a most efficient process.

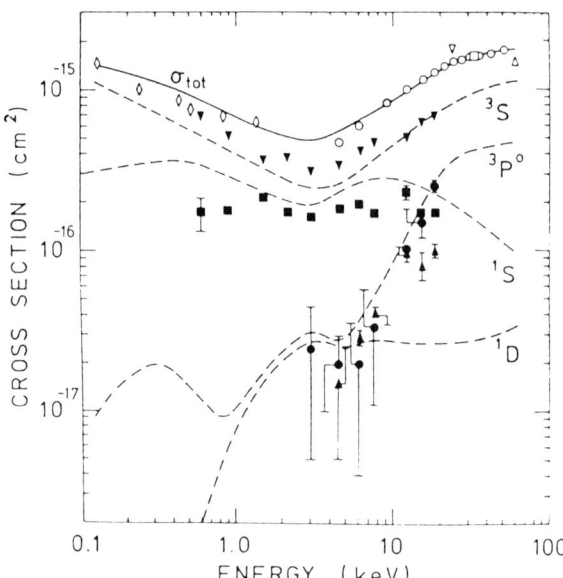

FIG. 24. Experimental and theoretical electron capture cross sections for $C^{3+} + H$ collisions. Theory: ---- MO, Bienstock et al. (1982). Experiment: ●, $C^{2+}(^3P^0)$; ▼, C^{2+} (^3S); ■, C^{2+} (^1S); ▲, C^{2+} (^1D), McCullough et al. (1984); ◇, □, △, Phaneuf et al. (1982). (From McCullough et al., 1985.)

(iii) The total electron capture cross sections often exhibit structure as a function of collision energy because of contributions from a large number of different n, l, and m states, each of which shows a different energy dependence for electron capture (Fig. 24).

(iv) For multiply charged ion–H and –He collisions, a striking oscillatory structure of the total electron capture cross section as a function of projectile charge q at lower fixed collision energies was found; it was interpreted as resulting from a combination of the strength of the coupling at avoided crossings and their locations (reaction windows). At higher energies, a large number of different n states become important to electron capture and a different coupling scheme becomes effective, thus weakening the structure (Fig. 25).

(v) Total electron capture cross sections are in reasonably good agreement with measurements and theories based on MO, MC-AO, and AO-MO methods. However, agreement of the subshell (i.e., l and m states) cross sections is not satisfactory. More precise experimental and theoretical studies are needed.

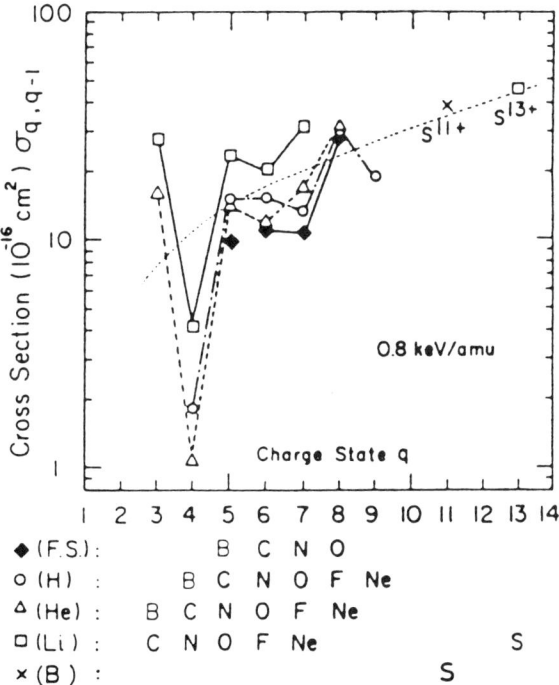

FIG. 25. Measured cross sections as a function of the ionic charge of projectile ions at 0.8 keV/amu. (From Iwai *et al.*, 1982.)

(vi) Multielectron capture processes are a common characteristic feature in multiply charged ion–complex-atom collisions. These features arise from relaxation of target atomic electrons in the strong Coulomb field of the projectile ion, leading to discrete and continuum transition spectra. Obviously, the dynamics of many-electron processes are much more complicated; hence the corresponding theories are less well developed. For example, double electron capture in a multiply charged ion–complex-atom collision often results first in doubly excited (autoionizing) states of the projectile ion, then in Auger ionization or a radiative decay process (McGuire, 1987).

Although many interesting examples of new physics can be identified in these systems, we choose only two for discussion. Several excellent review articles that concentrate specifically on the subject have been recently published (e.g., Janev and Winter, 1985).

a. *Symmetric Case*: He^{2+} *Collision with He*. Because the basic physical phenomena are similar to the $H^+ + H$ case, only a brief description will be given.

Adiabatic potential curves for the singlet gerade and ungerade states are plotted in Fig. 26. Because of the polarization interaction of the neutral He

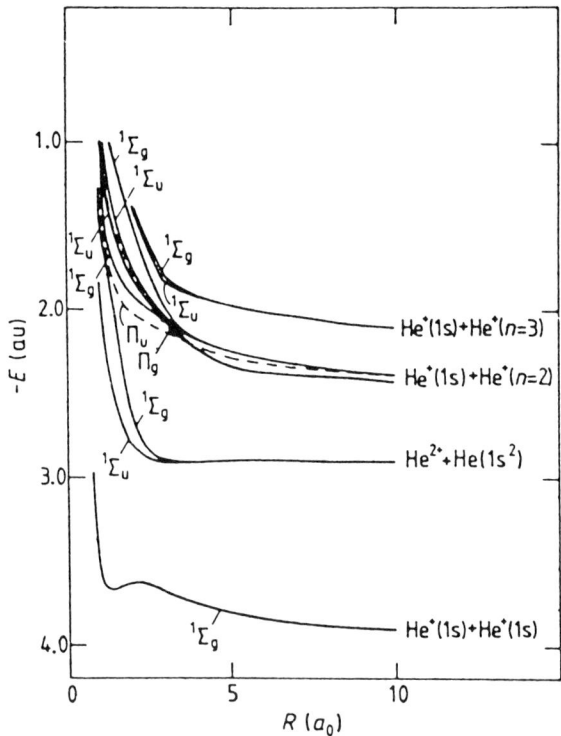

FIG. 26. Adiabatic potentials of $(HeHe)^{2+}$ system. (From Kimura, 1988.)

atom in the initial channel, the potential is rather flat over a wide range of R, except for the region of $R \lesssim 3a_0$. All electron capture channels involve a strong Coulomb repulsion. These potentials possess avoided crossings at small R, typically below $R \lesssim 2a_0$, and no crossings at larger R. Hence, for a symmetric, multiply charged ion–atom collision, a close collision may be the dominant process. In such a collision, single as well as resonant double electron capture processes are possible.

Double electron capture for this system is resonant capture; hence, unlike an example in the next section, the lowest two-electron captured state is the ground He state. As with $H^+ + H$ collisions, g-u oscillations should exist in the differential cross section (Aberth et al., 1965), but for the total electron capture cross section, the oscillations may well be washed out (see Fig. 27). Because the double electron capture process is dominant (a distant collision), reasonable theoretical models can reproduce the measurements well, except for some structure around 30 keV, which seem to be due to excited He states. Electron capture into doubly excited states is very small at these collision

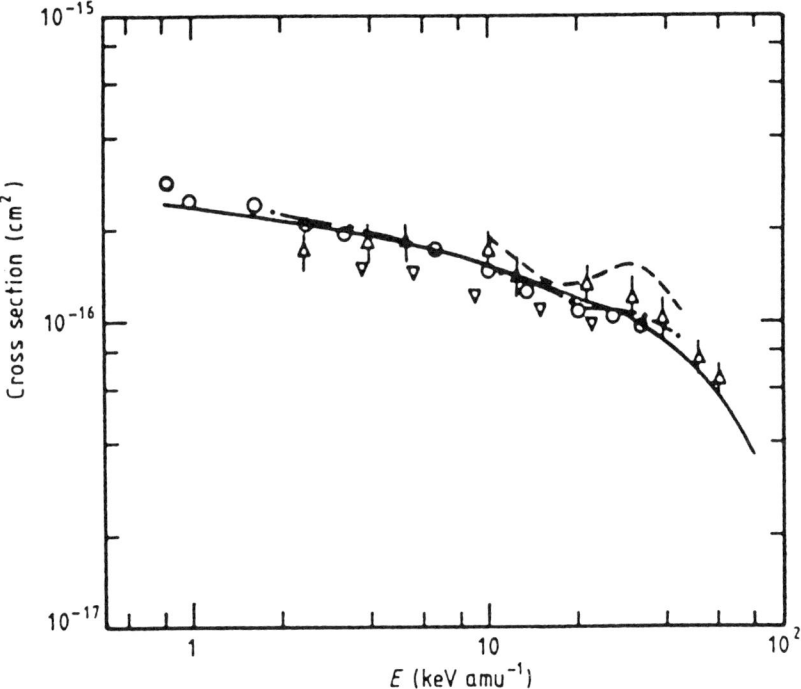

FIG. 27. Double electron capture cross section in $He^{2+} + He$ collisions. Theory: ———, MO, Kimura (1988); ---, numerical solution of Schrödinger equation, Devi and Garcia (1983); -·-, MO, Harel and Salin (1980). Experiment: ○, Afrosimov et al., (1978); △, Berkner et al., (1968); ▽, Bayfield and Khayrallah (1975). (From Kimura, 1988.)

energies and is not included in any theoretical calculations reported to date (see Kimura, 1988). However, the decay of a doubly excited state following an electron capture collision contains a great deal of fundamentally interesting information about the collision physics, and hence, should be studied.

Recent theoretical and experimental single electron capture cross sections are shown in Fig. 28. The dominant processes for the single electron capture involve the $n = 1, 2$, and 3 states of the He^+ ion. All the relevant states can be included rather easily in the close-coupling calculation, and recent large-scale close-coupling results (Harel and Salin, 1980; Devi and Garcia, 1983; Kimura, 1988) are in good accord with one another and with the measurements. Small discrepancies among the theories are due to the use of different-sized bases and different treatments of the ETF (or the neglect of the ETF). Comparisons of the alignment and orientation of the electron clouds of the $He^*(n = 2)$ manifold would be very interesting because these quantities are

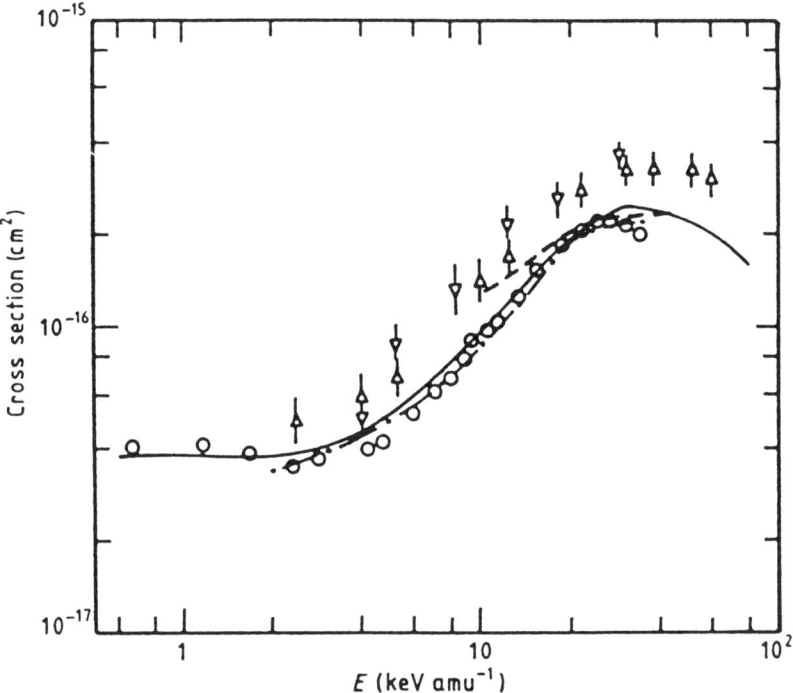

FIG. 28. Single electron capture cross section in He^{2+} + He collisions. Theory: ———, MO, Kimura (1988); - - -, numerical solution of Schrödinger equation, Devi and Garcia (1983); -·-, MO, Harel and Salin (1980). Experiment: ○, Afrosimov et al. (1978); △, Berkner et al. (1968); ▽, Bayfield and Khayrallah (1975). (From Kimura, 1988.)

more sensitive to the theoretical model used. Unfortunately, such studies have not yet been carried out.

b. *Asymmetric Case*: C^{6+} *Collision with H*. A molecular energy diagram of the σ states of the C^{6+} + H system is given in Fig. 29. As was discussed above, the greater the nuclear asymmetry is in the system, the higher is the principal quantum number n involved in the avoided crossings at finite R. Avoided curve crossings between the initial channel (C^{6+} + H) state and electron capture channels for $n > 5$ are considered to be diabatic in the energy range of interest. Hence, all theoretical models to date include channels up to the $n = 5$ C^{5+} shells (Green *et al.*, 1982; Fritsch and Lin, 1984; Kimura and Lin, 1985b). As is clear from Fig. 30a, most of these theories agree well with each other for the dominant capture process of the $n = 4$ C^{5+} shell. However, the situation for capture cross sections for the $n = 5$ C^{5+} shell is quite different, given that the 33-state MO result of Green *et al.* (1982) is larger by an order of magnitude at lower energies than results obtained by the AO+ and the AO-MO methods. Because Green *et al.* (1982) used an additional approximation to evaluate the coupling matrix elements, the

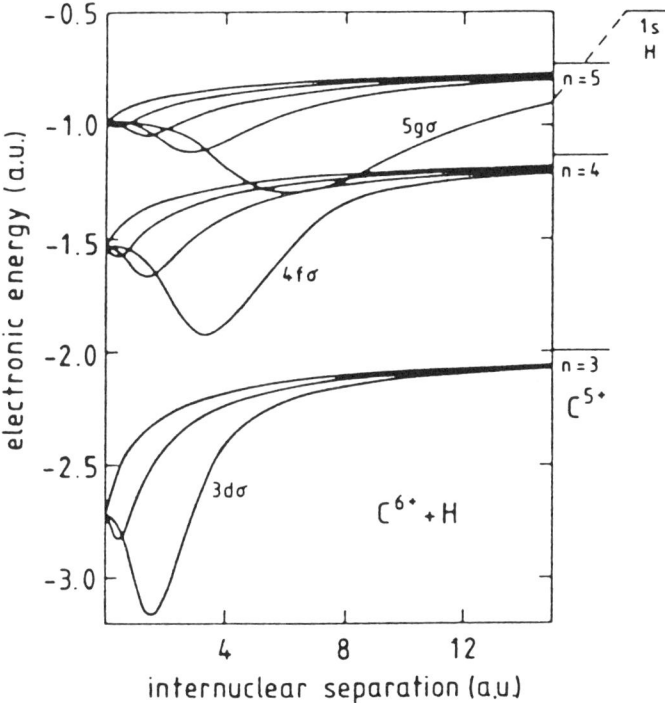

FIG. 29. Adiabatic potentials of CH^{6+} system. (From Fritsch and Lin, 1984.)

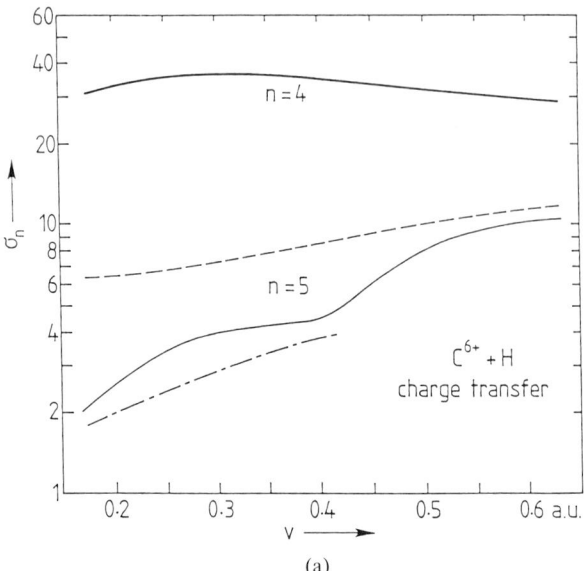

FIG. 30a. Theoretical $C^{5+}(n = 4)$ and $C^{5+}(n = 5)$ electron capture cross section in C^{6+} + H collisions. Theory: - - -, MO, Green et al. (1982); ———, AO+, Fritsch and Lin (1984); -·-, AO-MO, Kimura and Lin (1985). Note that for the $C^{5+}(n = 4)$ capture cross section, all theoretical results are nearly identical.

origin of this discrepancy is not clearly traced to the different treatments of the ETF. Note that recent measurements by Hoekstra et al. (1988) for emission cross sections from the $n = 5$ level show better agreement with the results of Green et al. (1982) (see Fig. 30b). This point needs to be clarified in conjunction with attempting more precise assessments of the applicability of the various methods.

c. *Differential Cross Sections for Asymmetric Case: Selected Ion Collisions with He.* A study of the angular distributions for electron capture to specific states provides more detailed information about the collision dynamics that are most important at various impact parameters. As was discussed earlier, this information along with alignment and orientation parameters provides nearly complete information about the collision dynamics and provides a rigorous test of the various theoretical models over the whole range of contributing impact parameters.

As a first elucidation of the angular distribution of multiply charged ion–atom collisions, we show in Fig. 31a the angular distribution for double electron capture in C^{4+} + He collisions. The relevant molecular state diagram is shown in Fig. 31b. As Fig 31b shows, the dominant channel in the

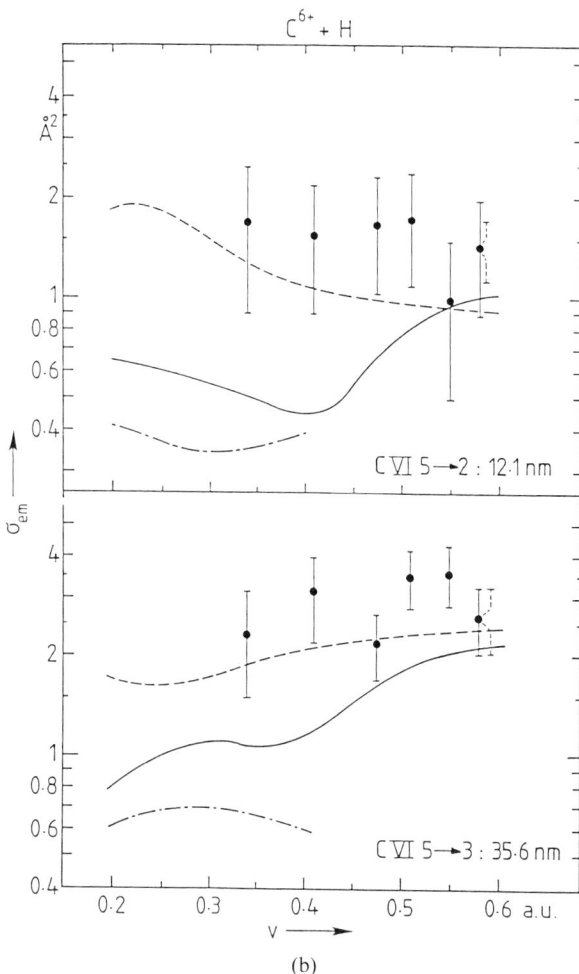

FIG. 30b. Comparison of theoretical and experimental emission cross section. Theory: same as in (a). Experiment: ●, Hoekstra et al. (1988). (From Hoekstra et al., 1988.)

process is double electron capture to the $C^{3+}(1s^22s^2)$ level. Hence, a simple two-channel calculation should represent the collision mechanism quite well. The theoretical results of Tan et al. (1987), based on a two-state quantal calculation, are in excellent agreement with the measurements of Cocke (see Barany et al., 1986). The semiclassical analysis identifies the origin of the oscillation as the interference of two different trajectories of the heavy particles, i.e., the Landau-Zener-Stückelberg oscillations. Experiments have been carried out for other projectile ions including the series C to Ne. Systematic theoretical studies of these systems would be desirable.

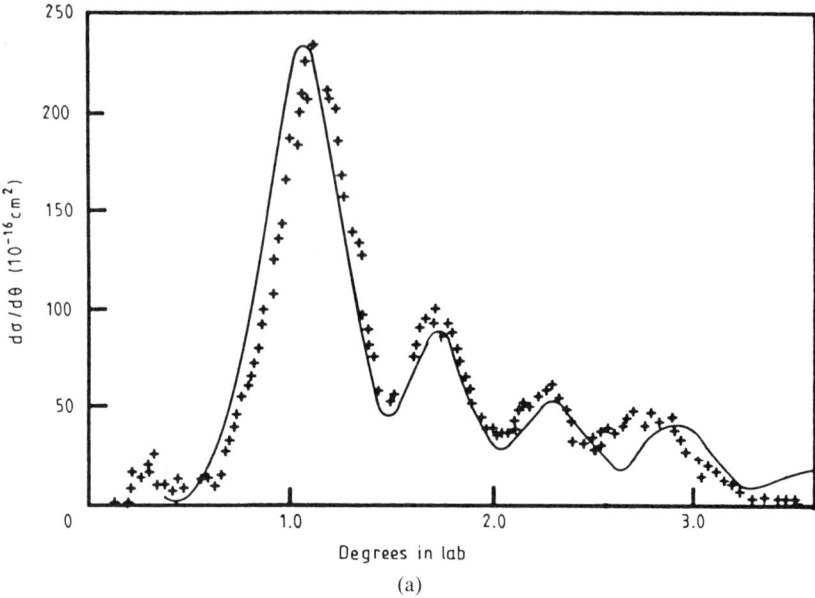

FIG. 31a. Theoretical differential electron capture cross section for C^{4+} collision on He at 1520 eV. Experimental data are from Barany et al., 1986. (From Tan et al., 1987.)

In contrast to the previous example, we next discuss double electron capture in multiply charged ion–He collisions where two distinct capture mechanisms are possible: (i) sequential electron capture, a two-step process, and (ii) direct double electron capture, a one-step process. A schematic diagram of the potential curves for such a system is given in Fig. 32. The collision path A → C → double-electron-capture channel corresponds to mechanism (ii), direct double electron capture (two-electron capture at C, a one-step process), while the path A → B → C → double-electron-capture channel corresponds to mechanism (i), sequential electron capture (one-electron capture at A and a second electron capture at B, a two-step process). Experimentally, Roncin et al. (1986) determined that direct double electron capture favours forward scattering, whereas sequential electron capture occurs at large angles because of the larger deflection associated with particle motion along the path A → B. Their measured differential cross sections for single and double electron capture in O^{8+} + He collisions are shown in Fig. 33. According to their interpretation, mechanism (i) causes large-angle scattering with double electron capture. This also leads to the population of doubly excited (autoionizing) states of the projectile, as was discussed above.

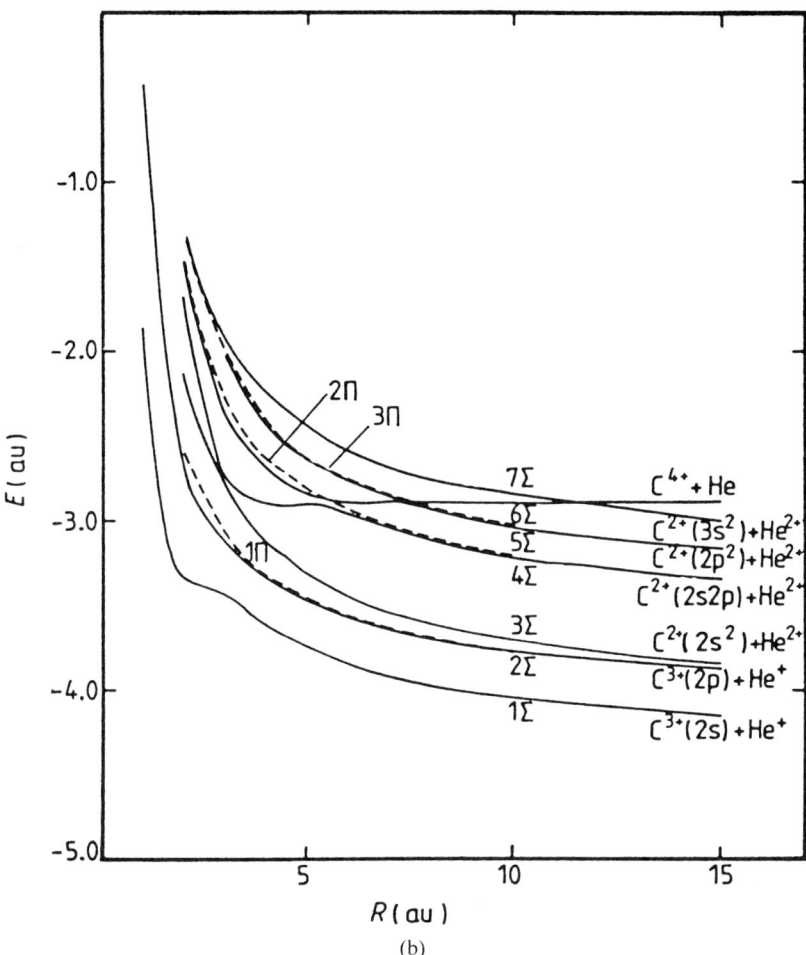

FIG. 31b. Adiabatic potentials of (CHe)$^{4+}$ system. (From Kimura and Olson, 1984.)

Rigorous theoretical studies are likely to be necessary to give a rationale to these observations.

An extremely interesting recent experimental measurement by Kamber *et al.* (1988) is the angular distribution of protons scattered in both singly and doubly ionizing collisions with He atoms at relatively high collision energies. Their measured results show that the ratio of double to single ionization of the He is nearly independent of scattering angle between 0.25 and 0.55 mrad, having a substantially lower value than does the ratio in photoionization. This measurement will provide important additional information regarding the correlation of the two electrons in He, as well as the collision dynamics.

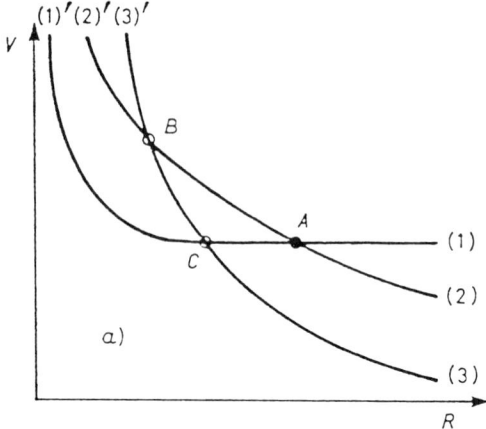

FIG. 32. Schematic adiabatic potentials illustrating the initial O^{8+} + He channel (1), the single electron capture O^{7+} + He^+ channel (2), and the double electron capture O^{6+} + He^{2+} channel (3). (From Roncin et al., 1987.)

B. Ion–Molecule and Atom–Molecule Collisions

We attempted to show in the previous section that our understanding of the collision dynamics of a variety of inelastic events in ion–atom collisions has been extensively broadened in approximately the past decade. In contrast to the flourishing ion–atom collision studies, comprehensive theoretical investigations of ion–molecule collisions are extremely rare (see, for example, Pollack and Hahn, 1985). The difficulties associated with obtaining accurate electronic wave functions for the colliding ion–molecule systems and the complexity of carrying out practical computations of the collision observables are responsible. However, the basic concepts of close and distant collisions, as they have been used to describe the collision mechanisms in ion–atom collisions, should be applicable to the study of ion–molecule collisions as well (Sidis and Dowek, 1984).

For a close collision where the molecular nature of the triatomic collision system plays an important role, explicit representation of the whole ion–molecule system is indispensable to the correct determination of the collision dynamics. Technology developed in quantum chemistry can be adopted to generate reasonably accurate electronic wavefunctions as well as adiabatic potential surfaces for the colliding system.

For a distant collision, where important collision dynamics take place at large internuclear separations and, correspondingly, where the molecular nature of a colliding system does not greatly influence the collision dynamics,

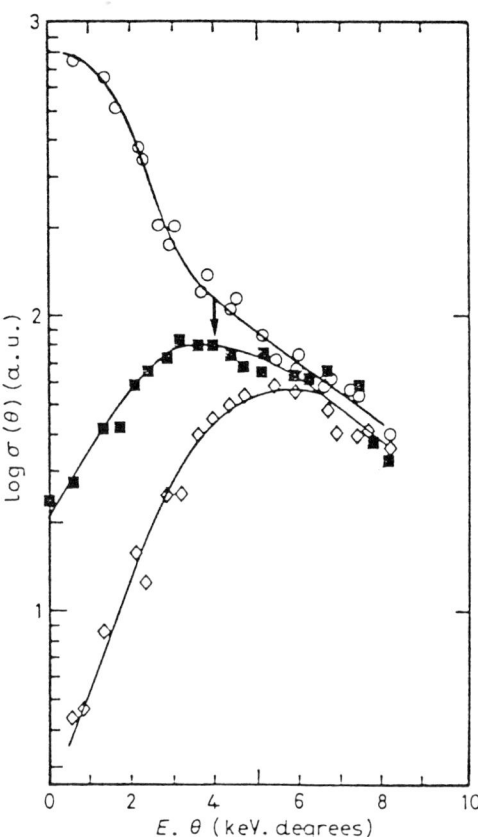

FIG. 33. Differential cross sections for double and single electron capture in O^{8+} + He collisions. ○, single electron capture and ■, ◇, double electron capture. (From Roncin et al., 1987.)

a simple model analogous to that for an ion–atom collision system may describe the ion–molecule collision reasonably well. The latter approach is to approximate the molecule by a model potential and to neglect the strong-coupling region of formation of the ion—molecule collision complex. Generally, the lower ($Z = 1, 2$) charged–ion–molecule collision system belongs to the close-collision category, while the higher ($Z \gtrsim 3$) charged–ion–molecule collision system belongs to the distant-collision category.

For close collisions, both the diatoms-in-molecules (DIM) method (Piacentini and Salin, 1978; Kimura, 1985) and the more sophisticated *ab initio* molecular structure method have been employed to obtain suitably accurate molecular wave functions and potential surfaces as functions of internuclear distances and molecular orientations. For distant collisions, a pseudopotential or model-potential method (Kimura and Lane, 1987) has been employed to represent the molecular-ion core, which allows us to treat a target molecule as an "atom" having an ionization potential of the molecule.

In this section, we will give representative examples of close and distant collisions where a detailed comparison between experimental and theoretical results is possible. Through this comparison, we will try to identify the advantages as well as the difficulties of each model, and some guidelines will be provided for future work.

In the close-coupling method, the scattering wave function of an ion–molecule (diatomic) colliding system is expanded in terms of the molecular adiabatic wave function as

$$\Psi(\mathbf{r}, \rho, t) = \sum_i a_i(t)\phi_i^{MO}(\mathbf{R}, \rho, \mathbf{r})X_{v_i}(\rho)F_i(\mathbf{R}, \mathbf{r})\exp(x), \tag{53}$$

where $F_i(\mathbf{R}, \mathbf{r})$ represents the ETF in Eq. (29), $\phi_i^{MO}(\mathbf{R}, \rho, \mathbf{r})$ denotes the adiabatic electronic system wave function, and $X_{v_i}(\rho)$ denotes the vibrational wave function of the target molecule, where $\mathbf{r}$ and ρ denote the coordinates of electron and vibrating molecule, respectively, and $\mathbf{R}$ is the position of the ion relative to the target molecule. $\exp(x)$ represents the phase factor and is given in Eq. (8). By substituting the scattering wave function [Eq. (53)] into the time-dependent Schrödinger equation [Eq. (7)], one obtains linear first-order coupled equations similar to those seen in ion-atom scattering problems in Eq. (9) except for the presence of matrix elements involving the vibrational wave function of the target molecule.

In the next section, we will discuss representative examples of calculated results along the experiments, to assess our understanding of the collision dynamics of ion-molecule collisions. The coordinates of the colliding system are shown in Fig. 34.

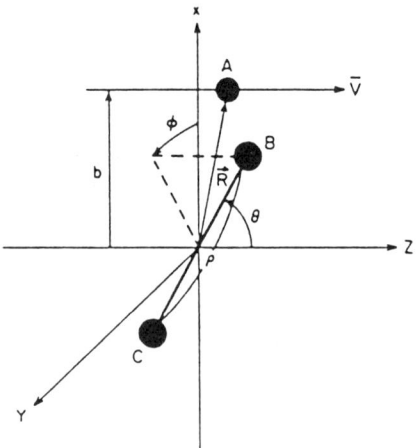

FIG. 34. System of coordinates for ion–molecule system. (From Kimura, 1985.)

1. Examples of Close Collisions

The adiabatic potential curves for the H_3^+ system are shown in Fig. 35 for several values of molecular orientation angle θ with $\phi = 0$ (Kimura, 1985). Note that the molecular orientation effect is not noticeable until the nuclei approach separations of about $R \sim 2$ a.u. A constant energy gap between the initial and electron-capture channels for values of $R \gtrsim 3.5$ a.u. suggests the existence of a peak in the radial coupling for slightly larger R that may induce the Demkov coupling effect. Indeed, the calculated radial coupling matrix elements display a sharp peak at $R \sim 3.5$ a.u., although detailed shapes and magnitudes depend on the molecular orientation for $R < 4$ a.u.

The transition probability times the impact parameters is plotted at $E = 1$ keV for several choices of θ in Fig. 36a. Generally, three main peaks correspond to impact parameters of 2, 3, and 4.5 a.u. for all orientations. This oscillatory structure in the probability is characteristic of the Demkov coupling effect (Yenen et al., 1984). This figure provides clear evidence that a close-collision mechanism is indeed dominant for electron capture in the $H^+ + H_2$ collision. To test the energy dependence of the molecular effect, the probability times the impact parameter versus the impact parameter is plotted in Fig. 36b at $E = 10$ keV for different values of molecular orientation θ. Two main peaks are found at $b \sim 1$ a.u. and 3.5–4.0 a.u. for all θ. The inner

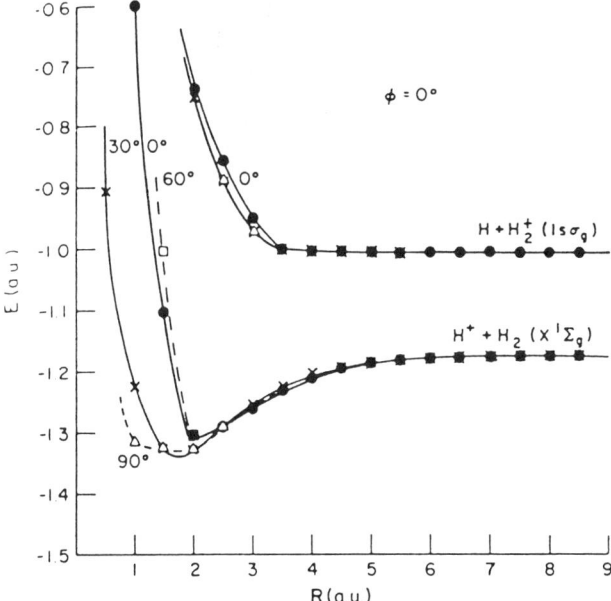

FIG. 35. Adiabatic potential of $(H^+ + H_2)$ system as a function of internuclear distance R and molecular angle θ. (From Kimura, 1985.)

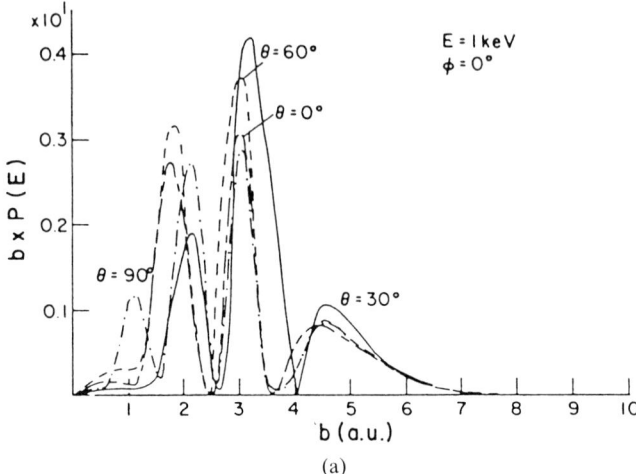

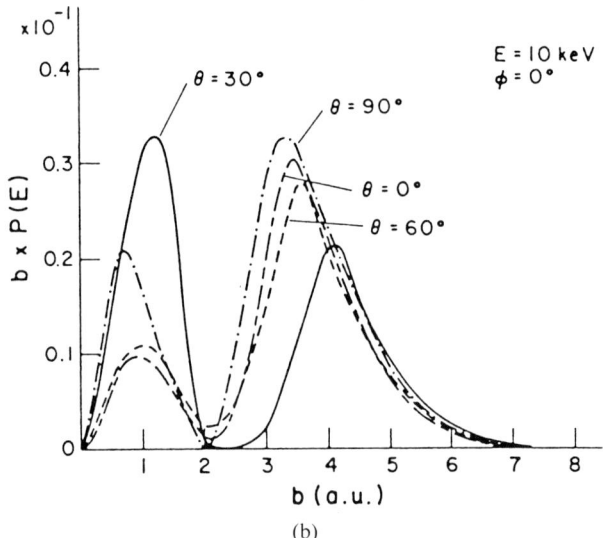

FIG. 36. (a) Transition probability times impact parameter versus impact parameter as a function of angle θ at $E = 1$ keV. (From Kimura, 1985.) (b) Transition probability times impact parameter versus impact parameter as a function of angle θ at $E = 10$ keV. (From Kimura, 1985.)

peak located at $b \sim 1$ a.u. at 1 keV grows dramatically with increasing energy at 10 keV, in contrast to the other two peaks.

The total cross section obtained by summing over all vibrational states of the H_2^+ ($v' = 0\text{-}10$) molecular ion is plotted in Fig. 37 along with relevant experimental measurements. The theoretical results shown are qualitatively in good accord with all measurements in the energy range studied, suggesting that the model used to describe the close collision is physically sound. However, at lower and higher energies, where the calculations were not performed, the present model may not be reliable because of the neglect of other channels, namely, the vibrational excitation or electronically excited states of the atomic or molecular ion, which are considered to be important processes at lower and higher energies, respectively.

Figure 38 includes the recent experimental result for linear polarization of the charge cloud of the excited H(2p) state obtained by Hippler et al. (1988) for $H^+ + H_2$ collisions. The striking difference between the H atom (see Fig. 13) and H_2 molecular targets requires a theoretical rationale.

2. Examples of Distant Collisions

It is well known from previous theoretical investigations of highly-charged-ion–atom collisions that electron capture takes place predominantly by means of avoided crossings located within a narrow range of internuclear

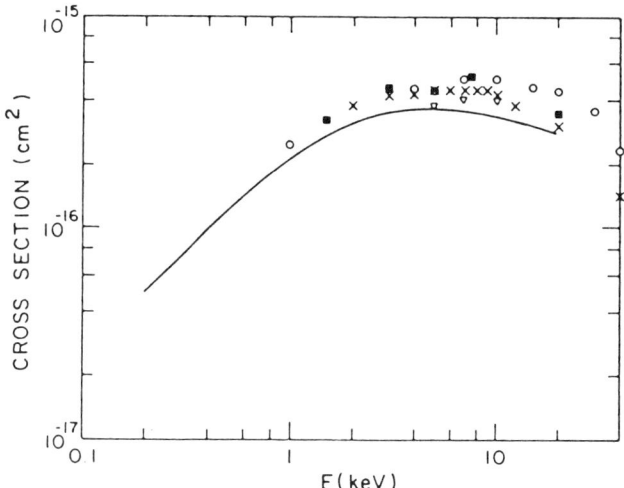

FIG. 37. Electron capture cross section in $H^+ + H_2$ collisions. Theory: ———, MO, Kimura (1985). Experiment: ○, Godeev et al. (1964); ×, Williams and Dunbar (1966); ▽, de Heer et al. (1966); ■, Hollricher (1969). (From Kimura, 1985.)

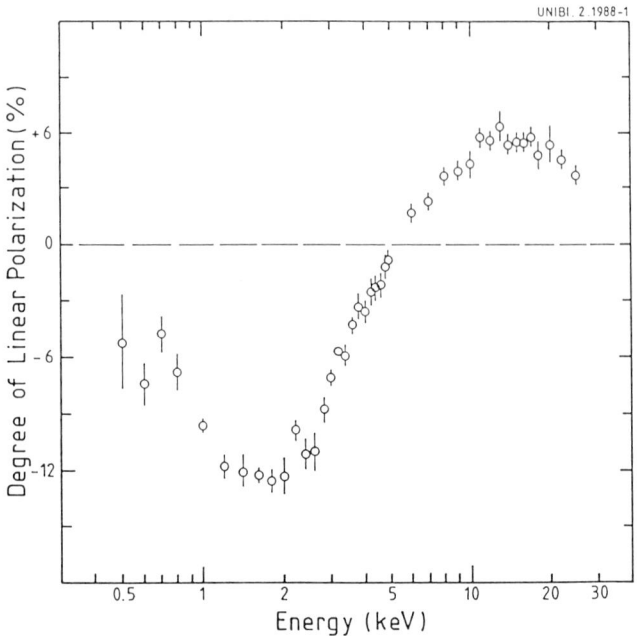

FIG. 38. Experimental alignment parameter for $H^+ + H_2$ collisions. (From Hippler et al., 1988.)

separations at large internuclear distances ($R > 5$ a.u.). This feature of the collision dynamics for atomic targets, H for example, can be expected to hold for the H_2 target as well, because the ionization potentials of H and H_2 are similar. The adiabatic potential curves calculated by means of a pseudopotential are shown in Fig. 39 for the $N^{7+} + H_2$ collision system. The very weakly avoided crossing between the initial and electron capture channels can be observed at $R \sim 15$ a.u. Actually, the avoided crossing at this point is so sharp that the behavior of the system becomes nearly diabatic for energies higher than the range discussed here. The important radial coupling matrix element between the initial and electron capture channels ($n = 5$) has a sharp peak at about $R \sim 15$ a.u. This distance is quite large compared to the size of the H_2 molecule, and therefore the transition is likely to have occurred before the projectile encounters a nonspherical molecule. The corresponding total electron capture probability at $E = 5$ keV/amu is plotted against impact parameters in Fig. 40. The probability grows abruptly at $R \leq 15$ a.u., reflecting the important radial coupling. The contribution to the transition probability from the region where the molecular effect (i.e., nonspherical field) must be influential, i.e., the region near $R < 2$ a.u., may be $\sim 10\%$ of the total transition probability.

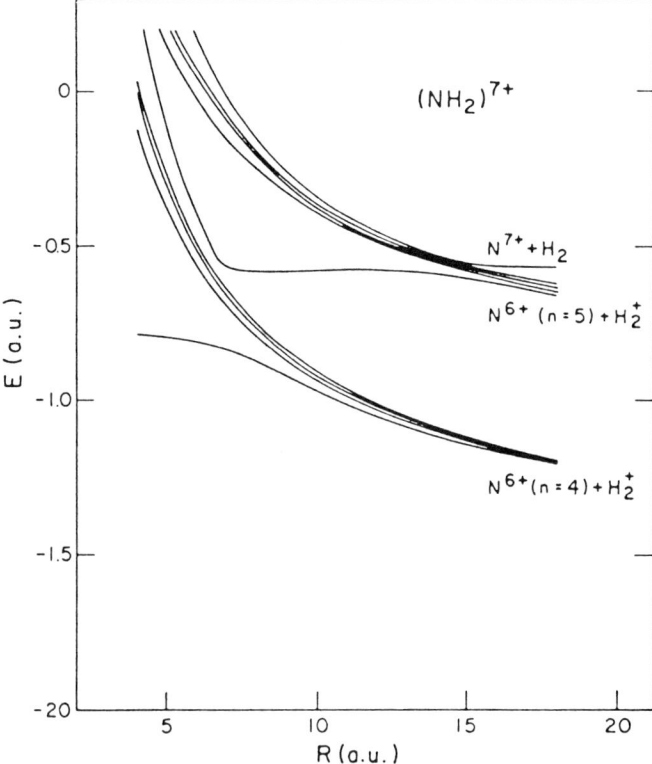

FIG. 39. Adiabatic potentials of $(N^{7+} + H_2)$ system. (From Kimura and Lane, 1987a.)

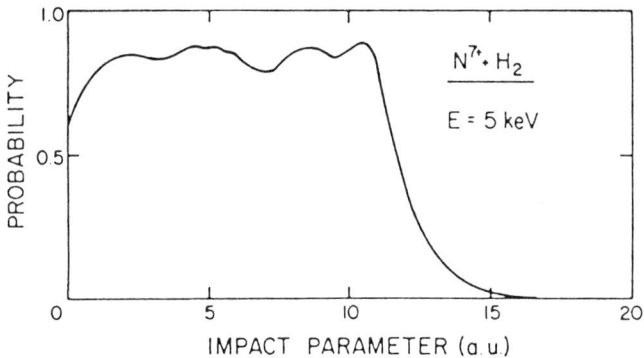

FIG. 40. Electron capture probability versus impact parameter for $N^{7+} + H_2$ collisions at $E = 5$ keV. (From Kimura and Lane, 1987a.)

In Fig. 41, we show the total electron capture section for the $N^{7+} + H_2$ system. Agreement of the theoretical results, which are based on a model of highly-charged-ion–atom collisions using a pseudopotential with recent measurements is very good both in magnitude and shape. Application of this model to other selected systems including C^{6+}, O^{8+}, F^{9+}, and Ne^{10+} projectiles has given consistently good agreement with measurements of total charge transfer cross sections (Kimura and Lane, 1987a), providing encouragement for further application of the ion–atom model to ion–molecule collisions.

C. Ion–Surface Collisions

In certain circumstances that can occur in the scattering of ions at surfaces, the dynamics that describe changes in the quantum states of the incident ion or the constituents of the target surface are very similar to those that describe ion–atom or ion–molecule collisions. In general electron capture, excitation, or ionization of the incident ion or of ions sputtered from the surface are complex phenomena, and their understanding requires a broad knowledge of physics, ranging from atomic collisions to the solid state (Tully and Tolk, 1977). However, for inelastic events that involve the inner shells of either the

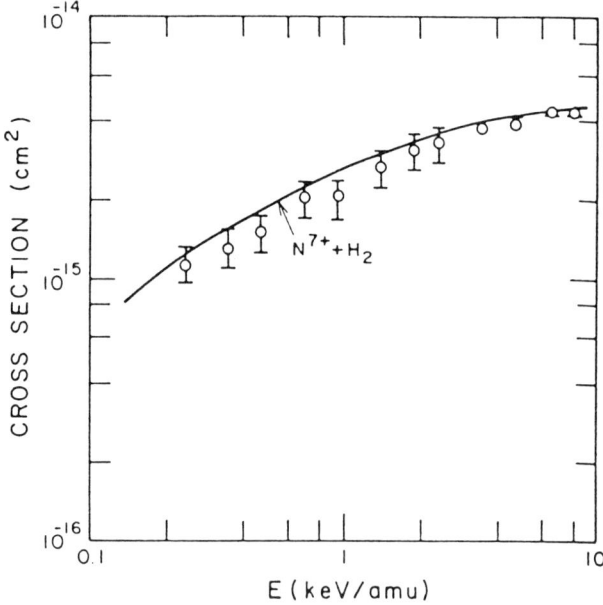

FIG. 41. Electron capture cross section in $N^{7+} + H_2$ collisions. Theory: ———, MO, Kimura and Lane (1987). Experiment: ◯, Meyer et al. (1985). (From Kimura and Lane, 1987a.)

projectile ions or target surface atoms, or for certain single, glancing collisions, the collision dynamics relate less to the condensed-matter nature of the target surface than to properties of the constituent atoms. These are processes where knowledge of ion–atom/molecule collisions is directly applicable. On the other hand, inelastic collisions that involve electrons in the valence band of the solid or multiple collision phenomena do not lend themselves to such treatments. In these cases, the delocalized nature of the valence electrons, the symmetry of the lattice structure, and surface effects must be taken into account in the theory. We will not discuss the latter in this article.

Now, returning to the earlier categorizations, we will classify as a close collision an inner-shell process and as a distant collision a single, glancing collision. The two distinct classes of inelastic mechanisms identified for ion–surface collisions by analogy to ion–atom collisions are (i) an adiabatic collision, where state-changing processes proceed without the exchange of energy between electronic and nuclear motion, and (ii) a nonadiabatic collision, where an exchange of energy is required between electronic and nuclear motion. Processes included in category (i) are resonance electron capture (both valence shell and inner shell), Auger electron capture, and radiative electron capture (see Fig. 42; Hentschke *et al.*, 1986; Imke *et al.*, 1986).

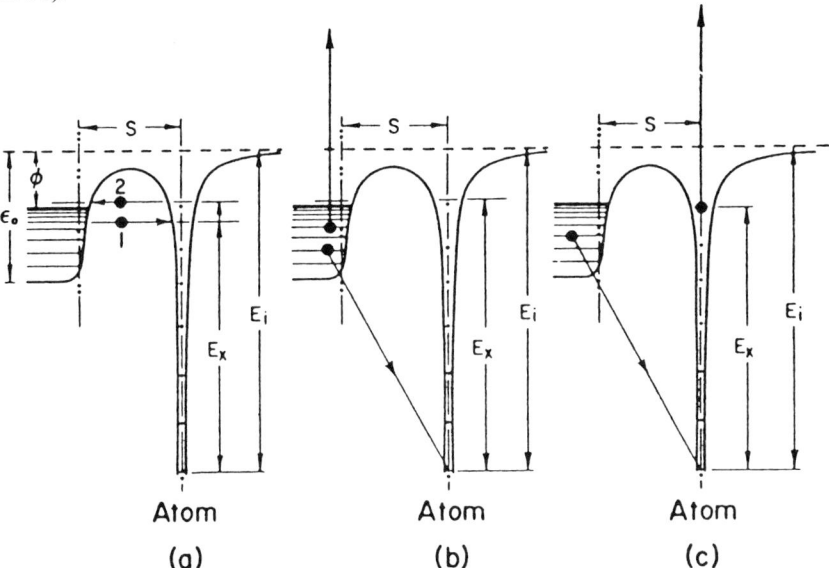

FIG. 42. Potential configurations for an atom at a distance S from the surface. (a) The process of resonance neutralization; (b) Auger neutralization; (c) Auger deexcitation. (From Thomas, 1983.)

The following comparisons of calculated results for He$^+$ + Si atom collisions with those for He$^+$ + Si$_7$ cluster collisions are intended to assess the applicability and make clear the limitations of the ion–atom collision model for treating ion–surface collisions. Of course, it is important to know how accurately one can simulate a surface by including a minimum number of constituent atoms in a practical computation. We give representative examples from each category within the ion–atom collision model.

1. Comparison of He$^+$ + Si Collisions with Si Atoms and Si$_7$ Cluster

a. *Ion-Atom case*: He$^+$ *on Si*. In Fig. 43 the calculated orbital energies of the He$^+$ + Si system (Fujima *et al.*, 1988) are shown. The 1s orbital energy (4σ) of He$^+$ is located between the Si 2p and 3s orbital energies (3σ and 5σ in

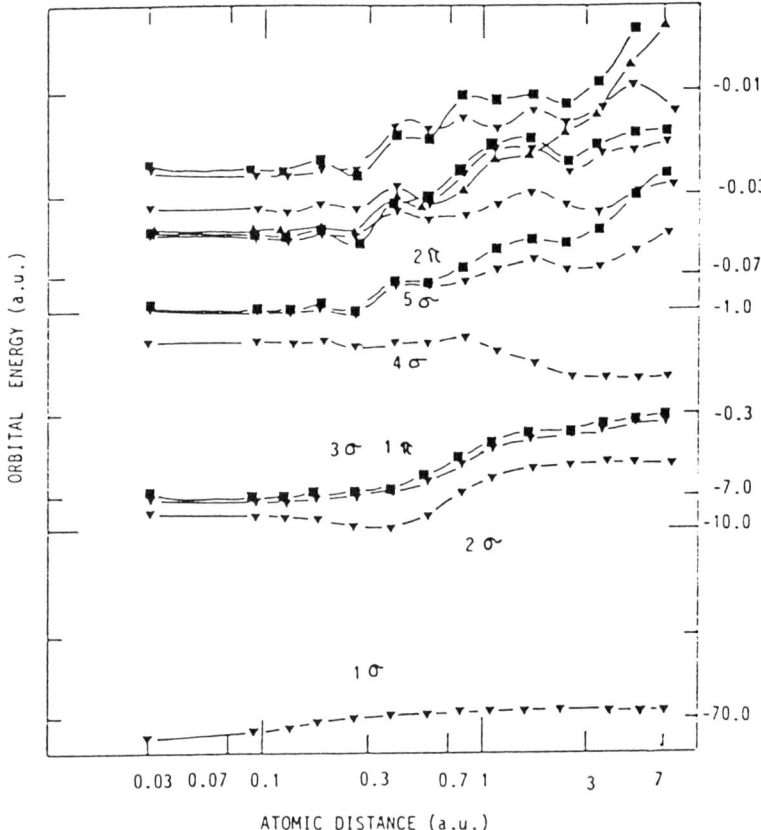

FIG. 43. Energy diagram for He$^+$ + Si system. (From Fujima *et al.*, 1988.)

the molecular designation) at the separated-atom limit. When the internuclear distance between He$^+$ and Si becomes small, the orbital energy of He$^+$(1s) increases (becoming less bound), and finally mixes with the 3s orbital of the Si atom below $R \sim 0.3$ a.u.

b. *Ion–Cluster case: He$^+$ on Si$_7$.* In Fig. 44 the orbital energy diagram of the system of He$^+$ + Si$_7$ (Fujima *et al.*, 1988) is illustrated. In the united-atom limit this system becomes a Si$_6$S$^+$ cluster. Because the 2s and 2p orbitals of the Si atoms are inner-shell orbitals, these orbitals do not mix with each other; hence, they are expected not to participate in the collision dynamics for the entire energy region of interest. The 3s and 3p orbitals give rise to an SP valence band. The 3d orbitals of the Si atom are unoccupied and give rise to an unoccupied band, which is not indicated in the figure. The 1s orbital energy of He$^+$ is located at the middle of the L and M shells of the Si atoms, as is the case for the diatomic system at the separated-atom limit. When a He$^+$ ion approaches the surface at about 1 a.u., the orbital energies of the

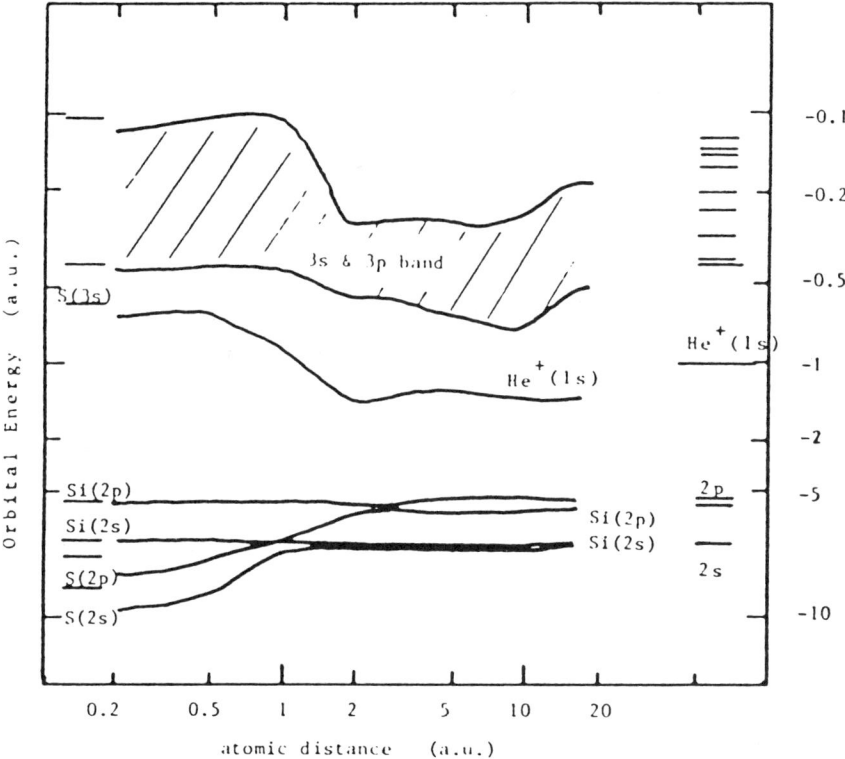

FIG. 44. Energy diagram for He$^+$ + Si$_7$ system. (From Fujima, *et al.*, 1988.)

cluster change little. However, as the He$^+$ approaches more closely to the surface, say less than 1 a.u., the energies of the 2s and 2p orbitals of the central Si atom begin to decrease because of the attractive core potential of the He$^+$ ion. As a result of the lowering of the inner-shell orbital energies, the 3s and 3p orbitals of the central Si mix to form a molecular orbital with the 1s orbital of the He$^+$ ion. The energy difference becomes small between the bottom of the valence band of the Si surface and the He$^+$ 1s energy, and the latter finally merges into the 3s and 3p band at the united-atom limit.

This feature is quite similar to that seen in the He$^+$ + Si system and suggests a large probability of electron capture from the bottom of the (3s 3p) Si band. Because the general aspects of the energy level diagrams are very similar for the two systems He$^+$ + Si$_7$ and He$^+$ + Si, the electron capture cross sections may be expected to be quite similar as well. Furthermore, we expect that within this model of ion–solid-cluster collisions, the analogy with ion-atom collision processes will be helpful in describing the inelastic events and physical mechanisms.

2. Adiabatic and Nonadiabatic Neutralization

In experimental studies of low-energy (0.2–2 keV) He$^+$ ion scattering from a variety of solid surfaces, Erickson and Smith (1975) observed striking oscillatory structures in the intensities of He$^+$ scattered from Pb, Ge, Be, and In surfaces as a function of incident ion energy. Similar observations of oscillations have been made by various experimental groups (Tully and Tolk, 1977; Brongersma and Buck, 1976). The oscillations are known to occur when the ionization energy of the incident atom is within 5 eV of the binding energy of the d electrons in the solid. These phenomena are found to have a close relation to similar oscillatory structure familiar in resonant electron capture in ion-atom collisions, like that shown in Fig. 12 for H$^+$ + H collisions. Tully (1977) developed a theoretical framework that takes advantage of this similarity to describe neutralization of ions colliding with solid surfaces. On the basis of the semiclassical two-state MO close-coupling approach for ion-atom collisions, discussed in Sec. II,A, Tully (1977) arrived at a simple analytical formula for the transition amplitude by using the Moliere potential for both resonant neutralization (the adiabatic case) and nonadiabatic neutralization. This theory gave a physical explanation of the oscillatory structure in He$^+$ + solid surface collisions that is exactly the same as that given to explain the oscillations in the resonant electron capture cross section in gas phase ion-atom collisions discussed above for the H$^+$ + H case. Figure 45 shows experimental data (right panel) of Rusch and Erickson (1975) and theoretical predictions (left panel) by Tully (1977) of the intensity of He$^+$ ions scattered from Cd, In, Sn, and Sb surfaces. For the Cd surface,

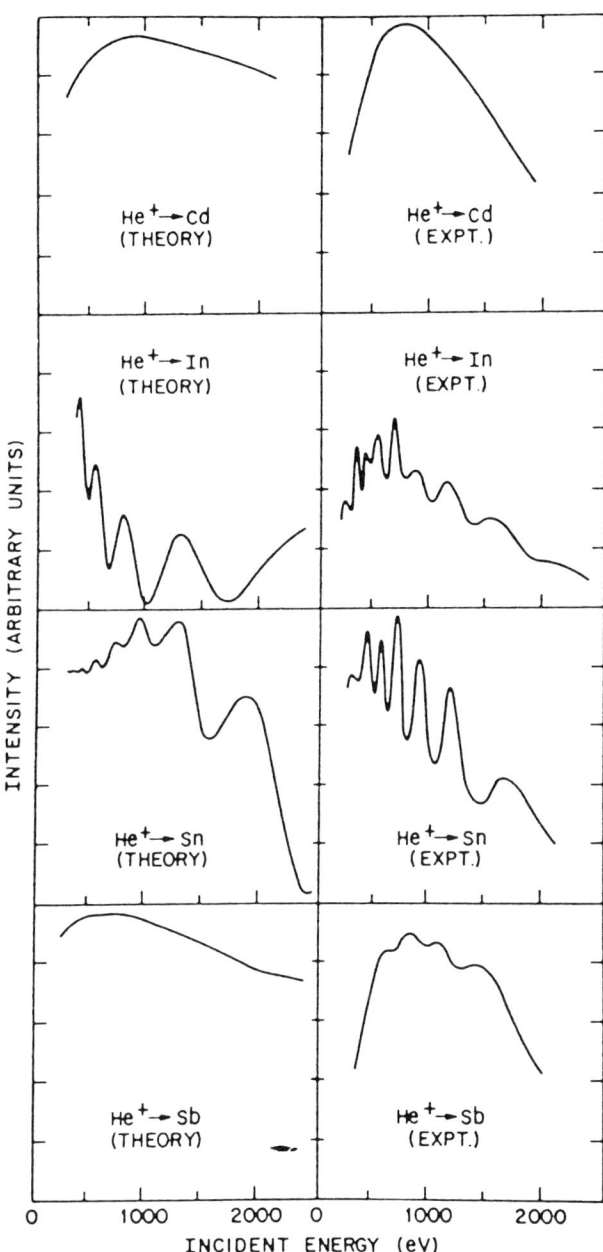

FIG. 45. He$^+$ scattering intensities for Cd, In, Sn, and Sb targets for $\theta = 90°$. Left panel: theoretical results. Tully (1977). Right panel: experimental results. Rusch and Erickson (1976). (From Tully, 1977.)

the neutralization mechanism is resonant adiabatic neutralization; hence, the intensity varies smoothly as a function of incident energy. In contrast, for In, the intensity consists of the sum of a smoothly varying adiabatic part plus an oscillatory nonadiabatic part. Surprisingly, the simple ion–atom collision model can reproduce the qualitative features of the experimental observation quite well for both the adiabatic and nonadiabatic transitions.

3. Re-Ionization Process

Recently, Souda et al. (1985, 1987) performed a series of three-dimensional angle-resolving experiments by ion-scattering spectroscopy using He^+, Ne^+, and Ar^+ ions scattered from a TaC(001) surface in the energy region of the order of 1 keV. Their experimental data (see Fig. 47b) show large differences among the energy spectra of these ions; in particular, the data exhibit two pronounced peaks, denoted A and B, in the energy spectra of He^+ ions on the TaC surface. Peak A corresponds to a pure elastic event, while peak B is displaced to lower energy than peak A by an amount that corresponds to about the He ionization energy. This is especially clear when the incident energy exceeds ~ 300 eV. Because the energy spectra observed in the case of neutral He incident on the same surface appear to match peak B in the He^+ data, peak B is considered to result from a process in which the He^+ ion is neutralized in the incoming channel before being re-ionized on the surface and leaving as an ion in the exit channel. In addition, depending on the nature of the surface constituent elements, peak B can be completely absent. For example, this second peak is missing for Cl, Cu, and other elements, while it is clearly present for Na, Ta, and certain others.

In Fig. 46 a schematic diagram illustrates the relevant potential surfaces. The actual calculated potential surfaces show extremely complex features (Kimura et al., 1987b). The solid line on a single surface, He^+ + Surface (S), in Fig. 46 represents the pure elastic path. The broken line represents the case where neutralization at large internuclear separations is followed by reionization at small internuclear separations, i.e., the mechanism that is considered to cause peak B in the experimental data. Results for the potential surfaces, obtained by using the self-consistent-field–$X\alpha$ method (Johnson, 1973), suggest that d orbitals in the surface atom play an important role in giving rise to peak B. For the Ta case, where peak B is seen, a narrow avoided crossing between He^+ + Ta and He + Ta^+ is found at about $R \simeq 1.8$ a.u. Unoccupied 5d orbitals in Ta and the 1s orbital in He are responsible for this avoided crossing, which eventually leads to the re-ionization process in He^+ + TaC surface collisions. On the other hand, for the case of Ag where peak B is not observed, there is no apparent curve crossing that could cause the re-ionization process for a relatively large range of internuclear separa-

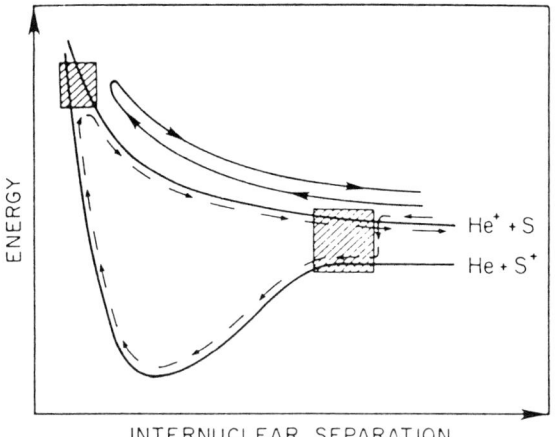

FIG. 46. Schematic diagram of potential surfaces for an ion-surface system. (From Kimura et al., 1987b.)

tions. This observation is related to the fact that all d orbitals in Ag are filled. Similar features of the relevant potential curves have been recently discussed qualitatively by Tsukada et al. (1985) for different surface constituents.

The calculated intensity for He$^+$ + TaC (001) surface collisions for $\theta = 30°$ and $E = 389$ eV is shown in Fig. 47a, and the experimental result of Souda et al. (1987) is in Fig. 47b. The general trend in the theoretical peaks is in good

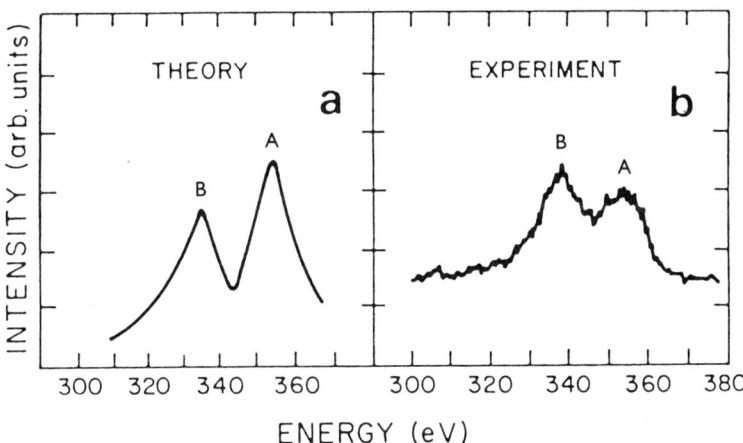

FIG. 47. Typical energy spectra of He$^+$ ion scattered from Ta atoms at the TaC(001) surface for $\theta = 30°$ at $E = 389$ eV. (a) Theoretical results. Kimura et al. (1987b). (b) Experimental results. Souda et al. (1987). (From Kimura et al., 1987b.)

accord with the measurement. In the present collision energy regime, the neutralization and subsequent re-ionization of the He$^+$ ion occur through a strong radial coupling between two channels. This is an alternative representation of the resonance tunneling mechanism. The model based on the ion-atom collisions model provides a reasonable interpretation of the experimental findings.

IV. Conclusions and Perspectives

In approximately the last decade, there has been extraordinary progress in the application of theoretical methods, particularly those based on close-coupling expansions, to various inelastic atomic collision processes at low to intermediate energies. Most applications have involved systems with one active electron. In parallel with these theoretical advances, the techniques for measurement have also been markedly improved. Comparisons of calculated cross sections and impact-parameter-specific observables, such as orientation and alignment, with experimental results have shown impressive agreement.

Thus, our theoretical understanding of the mechanisms that control the collision dynamics is reasonably sound for systems having one active electron. We conclude that for such systems the close-coupling method for elastic scattering, excitation and charge transfer in heavy-particle collisions has grown out of its infancy, having developed into a mature and reliable approach to obtaining quantitatively accurate results. Theoretical progress in collisional ionization has also been impressive. More precise measurements of differential cross sections, subshell distributions, and orientation and alignment parameters will pose more stringent tests of the theoretical methods, leading undoubtedly to further refinements that will make the methods more "user friendly."

The situation for atomic collision processes in which two or more electrons are actively involved is far less satisfying. There have been far fewer applications because of the difficulty in treating such systems. In addition to the usual complications of adequately accounting for electron translation effects, selecting a minimum basis of electronic functions, and solving the coupled differential equations, researchers must determine far more accurately the electronic states of the target and projectile as well as the transient system. In particular, if the collision process to be described involves two simultaneously active and therefore highly correlated electrons, then the electronic states must accurately represent this correlation. Indeed, so little had been accomplished until very recently that even our qualitative understanding of the underlying collision dynamics is unsatisfactory (McGuire, 1987).

Rigorous theoretical treatments of ionization lag behind those of excitation and electron capture because of the additional complications of having to represent the ionization electronic state and dealing with the large number of open channels that often are important. Recent MO calculations of ionization cross sections in the collision of H^+ and H, using 130 discretized continuum and 20 discrete states (Kimura and Thorson, 1989), may provide specific information about the coupling scheme involving the continuum and discrete excited states. From this and similar studies we hope to understand how to select the states that are most important to the process and thereby to simplify applications to other systems.

Collisions of atoms and ions with molecules, clusters, and surfaces present additional challenges to the theory. Semi-rigorous theoretical studies are just beginning to be carried out. With the help of the quantum chemistry community and in response to demands for applications, we expect to see significant progress in the next few years.

The first large-scale MO close-coupling calculation seems to have been the 22-state semi classical study by Winter and Lane (1978) of elastic scattering and electron capture in a collision of He^{2+} with H. This benchmark study stretched the computational capability near its limit at the time. Five years later, 20- to 30-channel calculations had become quite standard, and now 400- to 500-channel calculations are being carried out. Perhaps in a few years the equivalent of several thousand or more channels will be routinely included in such studies. Debates over whether one should use MO-ETF or AO-ETF should be behind us by then. Of course, even though such large calculations might well become routine in some sense, that will not happen without a great deal of research on the methods. We cannot simply increase the number of channels as more computer power becomes available. New algorithms will be needed along with new software and hardware to take advantage of parallel computing environments, including, perhaps, special-purpose chips. As in most other fields of computational science, there is an increasing need for applied mathematicians, computer scientists, and scientists from other disciplines to share their computational knowledge.

As the necessary computational tools become more readily available to those studying atomic collisions, the analytic models and weak-coupling approximations will be used less frequently, except under specific conditions where they are clearly justified. (Even now, students can run few-channel, close-coupling calculations on desk-top computers.) For many applications, the large-scale close-coupling calculation may become the dominant tool within the next few years. An early indication of this trend is found in the recent work of Toshima and Eichler (1988), who have solved the relativistic Dirac close-coupled equations for a basis of 30–40 channels. Larger applications to high-energy collisions are likely in the future.

Computational science is still criticized by some who feel that little can be learned from giant calculations or that true understanding of physical phenomena requires simple explanations. However, the practitioners of the craft know that most of their "quality" time is spent in trying to understand the underlying physics, using the simplest pictures that come to mind, seeking generalizations that can help to explain other phenomena in other systems, and then in attempting to improve the theoretical treatment to obtain more accurate results. With rare exceptions, e.g., some QED calculations, the computational atomic and molecular physicist is attempting not to prove that quantum mechanics is correct, but rather to describe the complicated many-particle dynamics of a physical process in the simplest and most general possible terms. For those who are committed to making advances in the field, there is no escaping the physics.

ACKNOWLEDGEMENTS

This work was supported in part by the U.S. Department of Energy, Office of Health and Environment Research (M.K.); the office of Basic Energy Sciences, Division of Chemical Sciences; and the Robert A. Welch Foundation (N.F.L.). We especially thank Dr. Jean Gallagher of the JILA Atomic Collisions Information Center for her assistance and Mrs. Sandy Gotlund for skillfully typing the manuscript.

REFERENCES

Aberth, W., Lorents, D. C., Marchi, R. P., and Smith, F. T. (1965). *Phys. Rev. Lett.* **14**, 776.
Afrosimov, V. V., Basalaeu, A. A., Leiko, G. A., and Panov, M. N. (1978). *Sov. Phys. JETP (Engl. Transl.)* **47**, 837.
Andersen, L. M., Hvelplund, P., Knudsen, H., Møller, S. P., Elsener, K., Rensfelt, K.-G., and Uggerhøj, E. (1986). *Phys. Rev. Lett.* **57**, 3147.
Anderson, D. G. M., Antal, M. J., and McElroy, M. B. (1974). *J. Phys. B* **7**, L118.
Barany, D., Danard, H., Cederquist, H., Hvelplund, P., Knudsen, H., Pedersen, J. O. K., Cocke, D. L., Tunnel, L. N., Waggoner, W., and Giese, J. P. (1986). *J. Phys. B* **19**, L427.
Barat, M. (1986). *In* "Atomic Processes in Electron-Ion and Ion-Ion Collisions" (F. Brouillard, ed.), p. 271. Plenum Press, New York.
Bates, D. R. (1958). *Proc. R. Soc. London Ser. A* **274**, 294.
Bates, D. R. (1961). *Proc. Phys. Soc.* **77**, 59.
Bates, D. R., and Dalgarno, A. (1953). *Proc. Phys. Soc.* **67**, 972.
Bates, D. R., and McCarroll, R. (1958). *Proc. R. Soc. London, Ser. A* **247**, 158.
Bates, D. R., and Sprevak, D. (1970). *J. Phys. B* **3**, 1483.
Bayfield, J. E. (1969). *Phys. Rev.* **182**, 115.

Bayfield, J. E., and Khayrallah, G. A. (1975). *Phys. Rev. A* **11**, 920.
Berkner, K. H., Pyle, R. V., Stearns, J. W., and Warren, J. C. (1968). *Phys. Rev.* **166**, 44.
Bienstock, S., Heil, T. G., Bottcher, C., and Dalgarno, A. (1982). *Phys. Rev. A* **25**, 2850.
Bottcher, C. (1984). *In* "Electronic and Atomic Collisions" (J. Eichler, I. V. Hertel, and N. Stolterfoht, eds.), p. 187. North-Holland, Amsterdam.
Bransden, B. H., and Noble, C. J. (1981). *J. Phys. B* **14**, 1849.
Bransden, B. H., and Noble, C. J. (1982). *J. Phys. B* **15**, 451.
Briggs, J. S. (1976). *Rep. Prog. Phys.* **39**, 217.
Brongersma, H. H., and Buck, T. M. (1976). *Nucl. Instrum. Meth.* **132**, 559.
Burke, P. G., and Robb, W. D. (1975). *In* "Advances in Atomic and Molecular Physics," Vol. 11, p. 143. Academic Press, New York.
Cheshire, I. M., Gallaher, D. F., and Taylor, A. J. (1970). *J. Phys. B* **3**, 813.
Child, M. S. (1974). "Molecular Collision Theory." Academic Press, New York.
Chong, Y. P., and Fite, W. L. (1977). *Phys. Rev. A* **16**, 933.
Coleman, J. P., and McDowell, M. R. C. (1965). *Proc. Phys. Soc.* **85**, 1097.
Crothers, D. S. F. (1971). *Adv. Phys.* **20**, 405.
Crothers, D. S. F., and Hughes, J. G. (1978). *Proc. R. Soc. London, Ser. A* **359**, 349.
Crowe, A., and Rudge, M. R. H. (1988). "Correlation and Polarization in Electronic and Atomic Collisions." World Scientific, Singapore.
Daley, H. L., and Perel, J. (1969). *VI-ICPEAC Abstract*, p. 1051.
deHeer, F. J., Schutten, J., and Moustafa, H. (1966). *Physics (Utrecht)* **32**, 1768.
Delos, J. B. (1981). *Rev. Mod. Phys.* **53**, 287.
Delos, J. B., and Thorson, W. R. (1979). *J. Chem. Phys.* **70**, 1774.
Demkov, N. Y. (1966). *Dokl. Akad. Nauk. S.S.S.R.* **166**, 1076.
Devi, K. R. S., and Garcia, J. D. (1983). *J. Phys. B* **16**, 2837.
Dose, V., and Semini, C. (1974). *Helv. Phys. Acta* **47**, 609.
Erickson, R. L., and Smith, D. P. (1975). *Phys. Rev. Lett.* **34**, 297.
Ermolaev, A. M. (1987). *In* "Atomic Physics with Positrons" (J. W. Humberston and Z. A. G. Armour, eds.), p. 393. Plenum Press, New York.
Errea, L. F., Mendez, L., and Riera, A. (1982). *J. Phys. B* **15**, 101.
Fano, U., and Lichten, W. (1965). *Phys. Rev. Lett.* **14**, 627.
Fite, W. L., Stebbings, R. F., Hummer, D. G., and Brackmann, R. T. (1960). *Phys. Rev.* **119**, 663.
Flannery, M. R. (1969). *J. Phys. B* **2**, 1044.
Flannery, M. R., and McCann, K. J. (1974). *J. Phys. B* **7**, 1349.
Ford, A. L., Fitchard, E. F., and Reading, J. F. (1977). *Phys. Rev. A* **16**, 133.
Fritsch, W., and Lin, C. D. (1982). *J. Phys. B.* **15**, 1255.
Fritsch, W., and Lin, C. D. (1983). *Phys. Rev. A* **27**, 3361.
Fritsch, W., and Lin, C. D. (1984a). *Phys. Rev. A* **29**, 3039.
Fritsch, W., and Lin, C. D. (1984b). *In* "Electronic and Atomic Collisions" (J. Eichler, I. V. Hertel, and N. Stolterfoht, eds.), p. 331. North-Holland, Amsterdam.
Fujima, K. Adachi, H., and Kimura, M. (1988). *Nucl. Instrum. Meth. B* **33**, 455.
Gallaher, D. E., and Wilets, L. (1968). *Phys. Rev.* **169**, 139.
Gaussorgues, C., and Salin, A. (1971). *J. Phys. B* **4**, 503.
Gordeev, Yu. S., and Panov, M. N. (1964). *Sov. Phys. JETP (Engl. Transl.)* **9**, 656.
Green, T. A., Shipsey, E. J., and Browne, J. C. (1982). *Phys. Rev. A* **25**, 1364.
Harel, C., and Salin, A. (1980). *J. Phys. B* **13**, 785.
Harel, C., and Salin, A. (1988). *In* "Electronic and Atomic Collisions" (H. B. Gilbody, W. R. Newell, F. H. Read, and A. G. H. Smith, eds.), p. 631. North-Holland, Amsterdam.
Heil, T. G., Butler, S. E., and Dalgarno, A. (1981). *Phys. Rev. A* **23**, 1100.
Hentschke, R., Snowdon, K. J., Hertel, P., and Heiland, W. (1986). *Surf. Sci.* **173**, 565.

Hill, J., Geddes, J., and Gilbody, H. B. (1979). *J. Phys. B* **12**, L341.
Hippler, R. (1988), *In* "Electronic and Atomic Collisions" (H. B. Gilbody, W. R. Newell, F. H. Read, and S. C. H. Smith, eds.), p. 241. North-Holland, Amsterdam.
Hippler, R., Madeheim, H., Harbich, W., Kleinpoppen, H., and Lutz, H. O. (1988). *Phys. Rev. A* **38**, 1662.
Hoekstra, R., Ćirić, D., Zinoviev, A. N., Gordeev, Yu. S., de Heer, F. J., and Morgenstern, R. (1988). *Z. Phys. D* **8**, 57.
Hollricher, O. (1969). *Z. Phys.* **187**, 41.
Hund, F. (1927). *Z. Phys.* **40**, 742.
Imanishi, B., and von Oertzen, W. (1987). *Phys. Rep.* **155**, 29.
Imke, U., Snowdon, K. J., and Heiland, W. (1986). *Phys. Rev. B* **34**, 41.
Iwai, T., Kaneko, Y., Kimura, M., Kobayashi, N., Ohtani, S., Okuno, S., Takagi, S., Tawara, H., and Tsurubuchi, S. (1982). *Phys. Rev. A* **26**, 105.
Janev, R. K., and Winter, H. (1985). *Phys. Rep.* **117**, 265.
Janev, R. K., Presnyakov, L. P., and Shevelko, V. P. (1985). "Physics of Highly Charged Ions". Springer-Verlag, Berlin.
Johnson, K. H. (1973). *Adv. Quantum Chem.* **7**, 143.
Kamber, E. Y., Cocke, C. L., Cheng, S., and Varghese, S. L. (1988). *Phys. Rev. Lett.* **60**, 2026.
Kimura, M. (1985). *Phys. Rev. A* **32**, 802.
Kimura, M. (1986). *In* "Electronic and Atomic Collisions" (D. C. Lorentz, W. E. Meyerhof, and J. R. Peterson, eds.) p. 431. North-Holland, Amsterdam.
Kimura, M. (1988). *J. Phys. B* **21**, L19.
Kimura, M., and Inokuti, M. (1988). *Phys. Rev. A* **38**, 3801.
Kimura, M., and Lane, N. F. (1987). *Phys. Rev. A* **35**, 70.
Kimura, M., and Lin, C. D. (1985a). *Phys. Rev. A* **31**, 590.
Kimura, M., and Lin, C. D. (1985b). *Phys. Rev. A* **32**, 1357.
Kimura, M., and Lin, C. D. (1986). *Phys. Rev. A* **34**, 176.
Kimura, M., and Lin, C. D. (1987). *Comm. At. Mol. Phys.* **20**, 35.
Kimura, M., and Olson, R. E. (1984). *J. Phys. B* **17**, L713.
Kimura, M., and Thorson, W. R. (1981). *Phys. Rev.* **24**, 1780.
Kimura, M., and Thorson, W. R. (1989) To be published.
Kimura, M., Lane, N. F., Fujima, K., and Sato, H. (1987). *Nucl. Instrum. Meth. A* **262**, 114.
Kondow, T., Girnius, R. J., Chong, Y. P., and Fite, W. L. (1974). *Phys. Rev. A* **10**, 1167.
Kulander, K. C., Devi, K. R. S., and Koonin, S. E. (1982). *Phys. Rev A* **25**, 2968.
Knudson, S. K., and Thorson, W. R. (1970). *Can. J. Phys.* **48**, 313.
Kumar, A., Lane, N. F., and Kimura, M. (1989). *Phys. Rev.* A**39**, 1020.
Landau, L. D. (1932). *Phys. Z. Sow.* **1**, 46.
Lebeda, C. F., Thorson, W. R., and Levy II, H. (1971). *Phys. Rev. A* **4**, 900.
Levy II, H., and Thorson, W. R. (1969). *Phys. Rev.* **181**, 252.
Lichten, W. (1963). *Phys. Rev.* **131**, 229.
Lichten, W. (1967). *Phys. Rev.* **164**, 131.
Lin, C. D., Soong, S. C., and Tunnell, L. N. (1978). *Phys. Rev. A* **17**, 1646.
Lin, C. D., Winter, T. G., and Fritsch, W. (1982). *Phys. Rev. A* **25**, 2395.
Lockwood, G. J., and Everhart, E. (1962). *Phys. Rev.* **125**, 567.
Lüdde, H. J., and Dreizler, R. M. (1982). *J. Phys. B* **15**, 2703.
Macek, J., and Briggs, J. (1972). *J. Phys. B* **5**, 579.
Macek, J., and Briggs, J. (1973). *J. Phys. B* **6**, 841.
Macek, J., and Briggs, J. (1974). *J. Phys. B* **7**, 1312.
Martir, M. H., Ford, A. L., Reading, J. F., and Becker, R. L. (1982). *J. Phys. B* **15**, 1729.
Massey, H. S. W., and Smith, R. A. (1933). *Proc. R. Soc. London, Ser. A* **142**, 142.
McCarroll, R., and Salin, A. (1966). *Ann. Phys. (Paris)* **1**, 283.

McCarroll, R., and Salin, A. (1968). *J. Phys. B* **1**, 163.
McCullough, R. W. (1986). *In* "Electronic and Atomic Collisions" (D. C. Lorentz, W. E. Meyerhof, and J. R. Peterson, eds.), p. 463. North-Holland, Amsterdam.
McCullough, R. W., Wilkie, F. G., and Gilbody, H. B. (1985). *J. Phys. B* **17**, 1373.
McDowell, M. R. C., and Coleman, J. P. (1970). "Introduction to the Theory of Ion-Atom Collisions." Elsevier, New York.
McGuire, J. H. (1987). *Phys. Rev. A* **36**, 1114.
Men, F. K., Kimura, M., and Olson, R. E. (1986). *Phys. Rev. A* **33**, 3800.
Meyer, F. W., Howald, C. C., Havener, C. C., and Phaneuf, R. A. (1985). *Phys. Rev. A* **32**, 3310.
Mittleman, M. H., and Tai, H. (1973). *Phys. Rev. A* **8**, 1880.
Morgan, T. J., Geddes, J., and Gilbody, H. B. (1973). *J. Phys. B* **6**, 2118.
Morgan, T. J., Stone, J., and Mayo, R. (1980). *Phys. Rev. A* **22**, 1460.
Mulliken, R. S. (1928). *Phys. Rev.* **32**, 186.
Ohtani, S. (1984). *In* "Electronic and Atomic Collisions" (J. Eichler, I. V. Hertel, and N. Stolterfoht, eds.), p. 353. North-Holland, Amsterdam.
Olson, R. E. (1980). *In* "Electronic and Atomic Collisions" (N. Oda and K. Takayangi, eds.), p. 391. North-Holland, Amsterdam.
Phaneuf, R. A., Alvarez, I., Meyer, F. W., and Crandall, D. H. (1982). *Phys. Rev. A* **26**, 1892.
Piacentini, R. D., and Salin, A. (1978). *J. Phys. B* **11**, L323.
Pollack, E. and Hahn, Y. (1986). *In* "Advances in Atomic and Molecular Physics" (D. Bates and B. Bederson, eds.), Vol. 22, p. 243. Academic Press, New York.
Ponce, V. H. (1979). *J. Phys. B.* **12**, 3731.
Presnyakov, L. P., and Ulantsev, A. D. (1975). *Sov. J. Quantum Electron (Engl. Trans.)* **4**, 1320.
Rankin, J., and Thorson, W. R. (1979). *Phys. Rev. A* **18**, 1990.
Rapp, D., and Dinwiddie, D. (1972). *J. Chem. Phys.* **57**, 4919.
Reading, J. F., Ford, A. L., and Fitchard, E. F. (1976). *Phys. Rev. Lett.* **36**, 573.
Reading, J. F., Ford, A. L., and Becker, R. L. (1981). *J. Phys. B* **14**, 1995.
Riley, M. E., and Green, T. A. (1971). *Phys. Rev. A* **4**, 619.
Roncin, P., Barat, M., and Laurent, H. (1986). *Europhys. Lett.* **2**, 371.
Rosen, N., and Zener, C. (1932). *Phys. Rev.* **40**, 502.
Rosenthal, H. (1971). *Phys. Rev. Lett.* **27**, 635.
Rusch, T. W., and Erickson, R. L. (1976). *J. Vac. Sci. Technol.* **13**, 374.
Ryufuku, H. (1982). *Phys. Rev. A* **25**, 720.
Ryufuku, H., and Watanabe, T. (1978). *Phys. Rev. A* **18**, 2005.
Ryufuku, H., and Watanabe, T. (1979a). *Phys. Rev. A* **19**, 1838.
Ryufuku, H., and Watanabe, T. (1979b). *Phys. Rev. A* **20**, 1828.
Salin, A. (1982). *In* "Electronic and Atomic Collisions" (S. Datz, ed.), p. 483. North-Holland, Amsterdam.
Sato, Y., and Matsuzawa, M. (1985). *Phys. Rev. A* **31**, 1366.
Schaefer III, H. F. (1972). "The Electronic Structure in Atoms and Molecules." Addison-Wesley, New York.
Schneiderman, S. B., and Russek, A. (1969). *Phys. Rev. A* **181**, 311.
Sethu Raman, V., Thorson, W. R., and Lebeda, C. F. (1973). *Phys. Rev. A* **8**, 1316.
Shakeshaft, R. (1975). *J. Phys. B* **8**, 1114.
Shakeshaft, R. (1976). *Phys. Rev. A* **14**, 1626.
Shakeshaft, R. (1978). *Phys. Rev. A* **18**, 1930.
Shah, M. B., Elliott, D. S., and Gilbody, H. B. (1987). *J. Phys. B* **20**, 2481.
Shingal, R., and Bransden, B. H. (1987). *J. Phys. B* **20**, 4815.
Shingal, R., Noble, C. J., Bransden, B. H., and Flower, D. R. (1986). *J. Phys. B* **19**, 3951.
Sidis, V., and Dowek, D. (1984). *In* "Electronic and Atomic Collisions" (J. Eichler, I. V. Hertel, and N. Stolterfoht, eds.), p. 403. Elsevier, Amsterdam.

Smith, F. J. (1966). *Phys. Lett.* **20**, 271.
Smith, F. T. (1969). *Phys. Rev.* **179**, 111.
Souda, R., Aono, M., Oshima, C., Otani, S., and Ishizawa, Y. (1985). *Surf. Sci.* **150**, L59.
Souda, R., Aono, M., Oshima, C., Otani, S., and Ishizawa, Y. (1987). *Surf. Sci.* **176**, 657.
Stückelberg, E. C. G. (1932) *Helv. Phys. Acta.* **5**, 369.
Tan, J., Lin, C. D., and Kimura, M. (1987). *J. Phys. B* **20**, L91.
Thormas, E. W. (1983). *In* "Applied Atomic Collision Physics" (S. Datz, ed.), Vol. 4, p. 299. Academic Press, New York.
Toepfer, A., Henne, A., Lüdde, H. J., Horbatsch, M., and Dreizler, R. M. (1987). *Phys. Lett. A* **126**, 11.
Tolk, N. H., Tully, J. C., White, C. W., Krans, J., Monge, A. A., Simms, D. L., Robbins, M. F., Neff, S. H., and Lichten, W. (1976). *Phys. Rev. A* **13**, 969.
Toshima, N., and Eichler, J. (1988). *Phys. Rev.* **38**, 2305.
Tsukada, M., Tsuneyuki, S., and Shima, N. (1985). *Surf. Sci.* **164**, L811.
Tully, J. C. (1977). *Phys. Rev. B* **16**, 4324.
Tully, J. C. and Tolk, N. H. (1977). *In* "Inelastic Ion-Surface Collisions" (N. H. Tolk, J. C. Tully, W. Heiland, and C. W. White, eds.), p. 105. Academic Press, New York.
Turner, J. G. (1977). *Am. J. Phys.* **45**, 758.
Vaaben, J. and Taulbjerg, K. (1981). *J. Phys. B* **14**, 1815.
Vainshtein, L., Presnyakov, L., and Sobel'man, I. (1964). *Sov. Phys. JETP* **18**, 1383.
Williams, J. F., and Dunbar, D. N. (1966). *Phys. Rev.* **149**, 62.
Winter, T. G. (1986). *Phys. Rev. A* **33**, 3842.
Winter, T. G. (1987). *Phys. Rev. A* **35**, 3799.
Winter, T. G. (1988). *Phys. Rev.* **37**, 4656.
Winter, T. G., and Lane, N. F. (1978). *Phys. Rev. A* **17**, 66.
Winter, T. G., and Lane, N. F. (1985). *Phys. Rev.* **31**, 2698.
Winter, T. G., and Lin, C. D. (1984a). *Phys. Rev. A* **29**, 567.
Winter, T. G., and Lin, C. D. (1984b). *Phys. Rev. A* **29**, 3071.
Yenen, O., Jaecks, D. H., and Macek, J. (1984). *Phys. Rev. A* **30**, 597.
Zener, C. (1932). *Proc. R. Soc. London, A* **137**, 696.

VIBRONIC PHENOMENA IN COLLISIONS OF ATOMIC AND MOLECULAR SPECIES

V. SIDIS

Laboratoire des Collisions Atomiques et Moléculaires
Université de Paris-Sud
Orsay Cedex, France

I. Introduction.	161
II. Outline of the Quantum Mechanical Formulation.	164
A. Coordinate Systems.	164
B. Perturbed Stationary States Expansion.	167
C. Basis Sets.	167
III. Quantum Treatment of Vibronic Excitation: The IOS Approximation.	170
IV. Semiclassical Treatment of Vibronic Excitation.	175
A. Background.	175
B. Time-Dependent Description of Relative Motion: The Common Trajectory Treatment of (Ro-) Vibronic Transitions.	176
V. On Franck-Condon-Type Approximations.	181
VI. Studies of Vibronic Transition Processes.	187
A. Vibronic Quantum Beats Accompanying a Process of Ion-Pair Formation.	188
B. Vibronic Effects in Ion-Molecule Charge Transfer.	191
C. Dissociative Charge Exchange.	196
VII. Concluding Remarks.	203
References.	204

I. Introduction

The present contribution concerns the study of phenomena that are associated with simultaneous electronic and vibrational transitions occurring in collisions between (ionic or neutral) atomic and molecular species. The field is immense; it includes various categories of excitation, quenching, charge transfer, bond formation and breakup processes. The scope of this article is intentionally restricted to nonreactive atom-molecule ($A^{(+)} + BC$ or $BC^{(+)} + A$) collisions in the energy range $0.1 \text{ eV/amu} < E \leq 1 \text{ keV/amu}$ and its primary concern is to survey some aspects of the related theoretical endeavor.

From a practical point of view the considered processes play an important role in energy deposition and flow in gases and plasmas and determine both the nature and the state population of the involved species. Direct applications of these properties are found in the study of planetary atmospheres, interstellar media, beam technology, gas discharges, and chemical lasers.

From a fundamental point of view the study of atom-molecule collisions constitutes after that of atom-atom collisions a crucial step in the understanding of the many-body heavy particle collision dynamics. The subject demands the treatment of problems of a high degree of complexity. These arise from the intricate coupling between the electronic, (ro-)vibrational and translational degrees of freedom in a general atom-molecule collision system. It is this formidable complexity that is responsible for the belated development of the theory of vibronic transitions in molecular collisions.

It should be reminded that in the mid 1960s one could barely treat electronic transitions in low and medium energy collisions between the simplest atomic partners (e.g., H and He). The state of the theory was such that it was believed that the detailed study of molecular collision processes would remain for long an exclusive theme of experimental physics. In reality the field started evolving steadily since the early 1970s. The period extending from 1970 up to the present has witnessed tremendous progress in the detailed understanding and actual quantitative description of electronic transition in atom-atom collisions. The related achievements have inspired attempts at extending the so-called theory of nonadiabatic collisions to the more complex atom + molecule systems. The review of Tully (1976), the book of Bernstein. (1979), the report of Kleyn *et al.* (1982) and that of Baer (1983) trace the important stages that have marked the development of molecular collision theory up to the recent past. From the bulk of compilations made available by these reviews one clearly sees that the most widely used method to treat both reactive and nonreactive atom-molecule collisions involving electronic transition has been the (quasiclassical) trajectory surface hopping (TSH) approach of Tully and Preston (1971). As is now well known, the method is based on a classical treatment of all nuclear degrees of freedom and a quantum mechanical treatment of the electronic motion. As originally suggested by Bjerre and Nikitin (1967), an electronic transition may be viewed as the result of a hop between two electronic potential energy surfaces; each time the trajectory of the classical representative point on a given surface encounters a so-called transition seam (characteristic of a pair of states) a prescription determines both the probability of a hop from one surface to the other and the way the momentum of the classical representative point is adjusted to conserve energy. The TSH approach has been reviewed many times and only a brief reminder of its current status will be presented later on in the present review. Continuing our retrospective overview of molecular collision theory, one should mention another important area of research, namely that of *rotational excitation* without any vibrational or electronic transition. Though apparently disconnected from the present concern, the results of successive studies on that topic, initiated in the early 1970s by Pack and co-workers (Tsien and Pack, 1970, 1971; Pack, 1972; Tsien et al., 1973; Pack, 1974), have opened up wide perspectives by introducing the so-called infinite order sudden (IOS) approximation. This approximation permits (under certain conditions) to simplify drastically the treatment of the molecular rotational degree of freedom. It replaces the huge set of coupled equations appearing in the quantum mechanical partial wave treatment of rotational excitation in an atom + rigid rotor collision by a mere radial Schrödinger equation for elastic scattering of an atom by a rotor that is held fixed with respect to the intermolecular axis. This radial equation has to be solved for a series of values of the angle between the rotor axis and the intermolecular axis; the actual scattering matrix is obtained after proper integration over that angle. The IOS approximation is fundamental to the actual undertaking of realistic quantum mechanical calculations on vibronic excitation in molecular collisions. It has first been used by Gerber (1976) for the investigation of vibrational excitation without change of electronic state. The quality of quantal IOS close-coupling calculations for (ro-)vibrational excitation up to collision energies of 10 eV/amu has been established by Schinke et al. (1980), (see also Schinke and McGuire, 1978a, b). Quantal close-coupling calculations on *vibronic* transition processes that made use of the IOS approximation were first reported by McGuire and Bellum (1979). Since then, very few IOS calculations on vibronic processes have appeared in the literature (Becker and Saxon, 1981; Becker, 1982, Baer et al., 1988). Because this approach is thought to have great promise and was little discussed previously it will be considered to some extent in Sec. III. Another quite promising approach is that based on the semiclassical description of the atom-molecule relative motion. Happily, this description applies particularly well in the moderate and high energy range $E \geq 10$ eV/amu where the quantal IOS description is expected to fall into troubles. The relation between the semiclassical description (based on the JWKB approximation) and the so-called time-dependent (common classical path) treatment of the internal motion of a heavy particle collision system (Mott, 1931) has been known for a long time (Bates and Holt, 1966; Berson, 1968; Cross, 1969; Bates and Crothers, 1970, Delos et al., 1972; Gaussorgues et al, 1975; McCann and Flannery, 1978). The use of this method in molecular collision studies was inaugurated by Bates and Reid (1969) in their investigation of charge transfer in $H_2^+ + H_2$ collisions. The method was later extended and developed by Flannery et al. (1973) and Moran et al. (1974, 1975, 1981) on other $AB^+ + AB$ charge transfer systems. Only a single time-dependent study of vibronic transitions in atom-molecule collisions was reported during the same period (Hedrick et al., 1977). The state of the field has rapidly evolved since the early 1980s owing to several contributions (Spalburg and Klomp, 1982; Spalburg et al. 1983, 1985a, b, Spalburg and Gislason, 1985; De Pristo, 1983; Klomp et al. 1984, Lee and De Pristo, 1984; Sidis and De Bruijn, 1984; Kimura and Chapman, 1986; Parlant and Gislason, 1985, 1987, 1988; Cole and De Pristo, 1986; Sidis and Courbin, 1987; Sidis et al., 1988). The new basic knowledge provided by these studies ought to be discussed in perspective; this is done in Secs. IV–VII.

In order to formulate the general problem of (ro-)vibronic transitions in atom-molecule collisions at extrathermal energies we start from a complete quantum-mechanical description. Then, in a somewhat heuristic way, we proceed to make approximations suitable for the low (0.1–10 eV/amu) moderate (1–500 eV/amu) and high (>500 eV/amu) energy ranges. The resulting equations of motion thereby lend themselves to the actual undertaking of practical calculations. Spectacular results of a selected sample of such calculations are presented and discussed. Trends of future theoretical work are indicated in the concluding section.

II. Outline of the Quantum-Mechanical Formulation

A. Coordinate Systems

After removal of the center of mass motion, the (nonrelativistic) Hamiltonian of an atom + molecule $(A + BC)$ collision system containing N electrons (i) writes (in atomic units)

$$\mathcal{H} = -\frac{1}{2}\left(\frac{\Delta_{\mathbf{R}}}{\mu} + \frac{\Delta_{\mathbf{r}}}{m} + \sum_{i=1}^{N}\Delta_{\mathbf{q}_i}\right) + \mathscr{V}(\{\mathbf{q}_i\}, R, r, \gamma)$$

$$-\left(\frac{p}{m_A} - \frac{q}{M}\right)\nabla_{\mathbf{R}} \cdot \sum_{i=1}^{N}\nabla_{\mathbf{q}_i}, \qquad (1)$$

where (Fig. 1)

$$\mathbf{r} = BC; \qquad \mathbf{R} = OA; \qquad \mathbf{q}_i = Ii.$$

O is the center of mass of nuclei B and C, and I is an arbitrary point between O and A (OI = pOA, IA = qOA). The masses μ, m, and M are defined as:

$$M = m_B + m_C, \qquad m = m_B m_C / M, \qquad \mu = m_A M / (m_A + M).$$

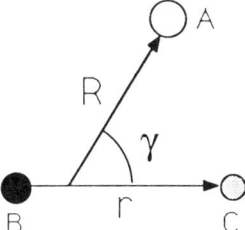

FIG. 1. Set of internal coordinates R, r, γ for an $A + BC$ collisional system.

Clearly, the choice $p/q = m_A/M$ cancels the last term in Eq. (1) and identifies I with the nuclear center of mass (G). Such a choice is not mandatory but will be used throughout for convenience (unless stated otherwise). Equation (1) neglects the small mass polarization terms $(p^2/m_A + q^2/M)(\sum_i \mathbf{V}_{\mathbf{q}_i})^2$.

Traditionally, the electronic motion is described in the so-called body-fixed (BF) reference frame that accompanies the overall rotation of the three nuclei (A, B, C) about the point G (or I). The nuclear motions may be treated in space fixed (SF) or BF frames (see, e.g., Baer 1976; Bellum and McGuire 1983). In view of the forthcoming discussions, the **R** and **r** motions will be treated in the SF frame.

The relevant Euler angles (ϕ, θ, δ) associated with the transformation from the SF frame $\mathscr{S}(X, Y, Z)$ to the BF frame $\mathscr{I}(x, y, z)$ defined as having the **R** axis along **G**z and the **r** axis in the xGz plane are shown in Fig. 2. The relation

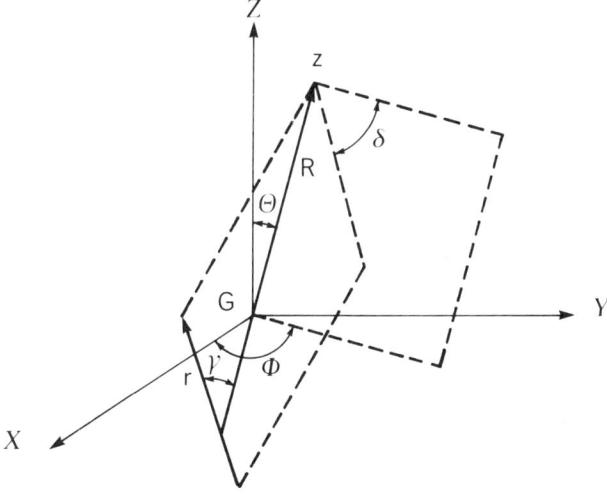

FIG. 2. Space fixed $\mathscr{S}(GXYZ)$ and body fixed $\mathscr{I}(Gxyz)$ coordinate reference frames. The x axis lies in the ABC plane.

between the cartesian coordinates of a vector **u** in the $\mathscr{S}$ and $\mathscr{a}$ frames is given by (Edmonds 1957):

$$u_x = (\cos \delta \cos \theta \cos \phi - \sin \delta \sin \phi)u_X + (\cos \delta \cos \theta \sin \phi + \sin \delta \cos \phi)u_Y$$
$$- \cos \delta \sin \theta \, u_Z, \tag{2a}$$

$$u_y = (-\sin \delta \cos \theta \cos \phi - \cos \delta \sin \phi)u_X + (-\sin \delta \cos \theta \sin \phi$$
$$+ \cos \delta \cos \phi)u_Y + \sin \delta \sin \theta \, u_Z, \tag{2b}$$

$$u_z = \sin \theta \cos \phi \, u_X + \sin \theta \sin \phi \, u_Y + \cos \theta \, u_Z \tag{2c}$$

When applied to the **r** vector this transformation yields

$$\sin \gamma = [\cos \delta \cos \theta \cos(\phi - \phi_r) - \sin \delta \sin(\phi - \phi_r)]$$
$$\times \sin \theta_r - \cos \delta \sin \theta \cos \theta_r, \tag{3a}$$

$$0 = -[\sin \delta \cos \theta \cos(\phi - \phi_r) + \cos \delta \sin(\phi - \phi_r)]$$
$$\times \sin \theta_r + \sin \delta \sin \theta \cos \theta_r, \tag{3b}$$

$$\cos \gamma = \sin \theta \sin \theta_r \cos(\phi - \phi_r) + \cos \theta \cos \theta_r, \tag{3c}$$

thereby establishing all relevant relationships between the angles γ and δ and the polar θ, θ_r as well as azimuthal ϕ, ϕ_r angles of vectors **R** and **r**, respectively.

Owing to Eqs. (2), the BF electronic coordinates noted $\boldsymbol{\rho}_i$ throughout depend implicitly on ϕ, θ and δ. This dependence introduces so-called rotational or Coriolis coupling terms that are related to the following properties:

$$\left.\frac{\partial}{\partial \theta}\right|_{\mathscr{S}} f(\{\boldsymbol{\rho}_i\},\ldots) = \left[\left.\frac{\partial}{\partial \theta}\right|_{\mathscr{a}} - i(\sin \delta \, L_x + \cos \delta \, L_y)\right] f(\{\boldsymbol{\rho}_i\},\ldots), \tag{4a}$$

$$\left.\frac{\partial}{\partial \phi}\right|_{\mathscr{S}} f(\{\boldsymbol{\rho}_i\},\ldots) = \left[\left.\frac{\partial}{\partial \phi}\right|_{\mathscr{a}} - i \cos \theta \, L_x + i \sin \theta(\cos \delta \, L_x - \sin \delta \, L_y)\right]$$
$$f(\{\boldsymbol{\rho}_i\},\ldots), \tag{4b}$$

$$\left.\frac{\partial}{\partial \delta}\right|_{\mathscr{S}} f(\{\boldsymbol{\rho}_i\},\ldots) = \left[\left.\frac{\partial}{\partial \delta}\right|_{\mathscr{a}} - iL_z\right] f(\{\boldsymbol{\rho}_i\},\ldots). \tag{4c}$$

In these equations f is an arbitrary function of the set of BF electronic coordinates (among other dependences noted $\cdots$); $L_{x,y,z}$ are the components of the electronic angular momentum **L**, with respect to G (or I), along the BF axes and the subscripts accompanying the derivative operators specify whether the electronic coordinates are kept fixed in the SF ($\mathscr{S}$) or BF ($\mathscr{a}$) frame when carrying out the differentiations. The final expression of the total hamiltonian H related to the selected mixed set of SF coordinates for the nuclei and BF coordinates for the electrons has been derived by Baer (1976) and Bellum and McGuire (1983).

B. Perturbed Stationary States Expansion

The most general solution of the Schrödinger equation for three nuclei and N electrons;

$$H\Psi(\{\boldsymbol{\rho}_i\}, \mathbf{R}, \mathbf{r}) = E\Psi(\{\boldsymbol{\rho}_i\}, \mathbf{R}, \mathbf{r}), \tag{5}$$

may be sought in the form of an expansion

$$\Psi(\{\boldsymbol{\rho}_i\}, \mathbf{R}, \mathbf{r}) = \sum_n \mathcal{N}_n(\mathbf{R}, \mathbf{r}) \Phi_n(\{\boldsymbol{\rho}_i\}; R, r, \gamma), \tag{6}$$

$$\mathcal{N}_n(\mathbf{R}, \mathbf{r}) = \sum_v F_{nv}(\mathbf{R}) G_{nv}(\mathbf{r}; \ldots). \tag{7}$$

Φ_n and G_{nv} form complete basis sets for the variables $\{\boldsymbol{\rho}_i\}$ and $\mathbf{r}$, respectively. Other indices following a semicolon specify additional (optional) parametric dependences. Both for convenience and in order to allow for transparent interpretations of the various matrix elements of H in the selected bases we assume that

$$\langle \Phi_n | \Phi_m \rangle = \delta_{nm}, \tag{8a}$$

$$\langle G_{nv} | G_{mv'} \rangle = (1 - \delta_{mn}) \mathcal{S}_{nv, mv'} + \delta_{mn} \delta_{vv'}. \tag{8b}$$

The use of Eqs (6, 7) in the Schrödinger Equation (5) provides after projection onto the basis states $G_{nv} \Phi_n$ the familiar set of coupled equations:

$$\langle G_{nv} \Phi_n | H - E | G_{nv} \Phi_n \rangle F_{nv}(\mathbf{R}) = -\sum_{n'v'}{}' \langle G_{nv} \Phi_n | H | G_{n'v'} \Phi_{n'} \rangle F_{n'v'}(\mathbf{R}) \tag{9}$$

Expanded expressions of Eq. (9) may be found elsewhere (Tully 1976; Sidis 1989). Solution of Eq. (9) yields for $R \to \infty$ the scattering amplitudes which determine the state-to-state cross sections.

C. Basis Sets

A great variety of possible choices of Φ_n and G_{nv} basis sets exist.

The electronic basis functions Φ_n are usually determined with the help of quantum chemistry techniques. It should be kept in mind however that the requirements which are placed on the Φ_n are not always the same as those which prevail in spectroscopic studies or when determining equilibrium geometries or structural properties of polyatomic molecules. The sought Φ_n functions and the relevant matrix elements of H in Eq. (9) should be determined for a broad assortment of geometries of the ABC triangle and hopefully for a dense grid of R, r, γ parameters. The number of such functions and their characteristics depend on the collision velocity and are often determined by both the nature of the investigated processes and the range of values of the R, r, γ parameters where they are most likely to occur. It is in this context that the concepts of *diabatic* versus *adiabatic* representations for heavy particle collisions have emerged. The theory and practice of Φ_n representations suitable for the treatment of atom-atom collisions is now flourishing. In order to follow the evolution of this notion one may refer to Hellman and Syrkin (1935), Lichten (1963), Smith (1969), O'Malley (1969, 1971), Browne (1971), Barat and Lichten (1972), Sidis (1976, 1979), Gauyacq (1978), Levy (1981), Garett and Truhlar (1981), Courbin-Gaussorgues et al. (1983), Spiegelman and Malrieu (1984), and Cimiraglia et al. (1985). Much less material is available for atom-molecule collisions involving electronic transitions. Extending the ideas of Smith (1971), Baer (1976) established the formal transformation $\mathbb{C}(R, r, \gamma)$ from the celebrated adiabatic representation (Born and Oppenheimer 1927) defined by

$$H_{el}\Phi_n^a(\{\rho_i\}; R, r, \gamma) = E_n^a(R, r, \gamma)\Phi_n^a(\{\rho_i\}; R, r, \gamma) \tag{10}$$

with

$$H_{el} = -\frac{1}{2}\sum_{i=1}^{N}\Delta_{\rho_i} + \mathscr{V}(\{\rho_i\}; R, r, \gamma) \tag{11}$$

to a diabatic representation (*stricto sensu*) defined by

$$\langle\Phi_n^d(\{\rho_i\}; R, r, \gamma)|\omega|\Phi_{n'}^d(\{\rho_i\}; R, r, \gamma)\rangle \equiv 0 \tag{12}$$

with

$$\omega = (\partial/\partial R, \partial/\partial r, \partial/\partial\gamma)|_{\{\rho_i\}}. \tag{13}$$

The merits of each of these representations have been discussed many times (Hellman and Syrkin 1935; Smith 1969; O'Malley 1971; Baer 1976, 1983; Tully 1976; Sidis 1976, 1989). Therefore it will suffice us to just recall that adiabatic basis functions properly describe all distortions undergone by the electronic cloud of a polyatomic system when its geometry is changed *infinitely slowly*. Its main flaws have to do with the calculation and actual use of matrix elements of the ω operator that arise in the r.h.s. of Eq. (9). These are particularly badly behaved near so-called avoided crossing seams and conical intersections of the $E_n^a(R, r, \gamma)$ potential energy surfaces where electronic transitions are most likely to occur. Equation (12) is the essential property of a diabatic basis. Excepting Coriolis coupling terms [Eqs. (4)] (which exist for both adiabatic and diabatic bases), it implies that all nuclear-electronic couplings in the r.h.s of Eq. (9) are concentrated in matrix elements of the H_{el} operator [Eq. (11)]. These are normally available from quantum chemistry calculations. Implicit in the definition (12) is the notion of *conservation (or slow variation of)* the main characteristics of a diabatic wave functions Φ_n^d with the change of nuclear geometry. Based on this notion several methods have been built that achieve Eq.(12) exactly or approximately or that provide satisfactory compromises between definitions (10) and (12) (Sidis 1989).

Concerning the $G_{nv}(\mathbf{r}; \ldots)$ basis functions these are usually written as

$$G_{nv}(\mathbf{r}; \ldots) = \frac{g_{nv}(r; \ldots)}{r} |j_n, m_{j_n}\rangle \tag{14}$$

thereby making the collective index v explicit. $|j_n, m_{j_n}\rangle$ represents a rotational wave function of the isolated BC molecule. In current practice the functions g_{nv} are also chosen as the vibrational eigenfunctions of the isolated BC molecule in the electronic state which correlates with state n:

$$\left[-\frac{1}{2m}\frac{d^2}{dr^2} + W_n(r)\right]g_{nv}(r) = \mathscr{E}_{nv}g_{nv}(r), \tag{15a}$$

$$W_n(r) = \lim_{R \to \infty} \langle \Phi_n | H_{el} | \Phi_n \rangle. \tag{15b}$$

However, certain studies of vibrational excitation without change of electronic state have advocated the use of *adiabatic vibrational bases* $g_{nv}^a(r; R, \gamma)$ for low-energy collisions (Eno and Balint-Kurti, 1981), i.e.

$$\left[-\frac{1}{2m}\frac{d^2}{dr^2} + \langle \Phi_n | H_{el} | \Phi_n \rangle\right]g_{nv}^a(r; R, \gamma) = \mathscr{E}_{nv}^a(R, \gamma)g_{nv}^a(r; R, \gamma) \tag{16}$$

instead of Eq. (15). Clearly Eqs. (16) and (15) are the analogues of Eqs. (10) and (12), respectively for the vibrational motion. It was rightly pointed out by Baer et al. (1980) that $g_{nv}^a(r; R, \gamma)$ bases will in general have the disadvantage of introducing $\partial/\partial R$, $\partial/\partial \gamma$ coupling terms which are not easily handled and especially so when pseudo crossings of $\mathscr{E}_{nv}^a(R, \gamma)$ energy surfaces may occur. Baer et al. (1980) thereby suggested to perform a unitary transformation $[\mathbb{A}_n(R, \gamma)]$ of the $\{g_{nv}^a\}$ basis that cancels the $\partial/\partial R$ (and $\partial/\partial \gamma$) coupling terms (Smith, 1970; Baer, 1976):

$$g_{nv}^d(r; R, \gamma) = \sum_w g_{nw}^a(r; R, \gamma)\mathbb{A}_{n,wv}(R, \gamma), \tag{17a}$$

$$\left\langle g_{nv}^d \left| \frac{\partial}{\partial R}\left(\text{or } \frac{\partial}{\partial \gamma}\right) \right| g_{n'v'}^d \right\rangle \equiv 0. \tag{17b}$$

Obviously, in the limit of complete basis sets, the resulting $g_{nv}^d(r; R, \gamma)$ functions should coincide with the $g_{nv}(r)$ functions of Eq. (15) (except possibly for a constant unitary transformation). Still $\{g_{nv}^d\}$ representations may be useful since only truncated bases are explicitly used in practice. Methods developed to build diabatic electronic bases may be extended to the construction of the g_{nv}^d functions. In particular, the maximum overlap method of Cimiraglia et al. (1985) (see also Sidis, 1989; Grimbert et al., 1988) may help reach that goal in a straightforward way.

III. Quantum Treatment of Vibronic Excitation: The IOS Approximation

In order to solve the system of coupled equations (9) one usually expresses the relative nuclear wavefunctions $F_{nv}(\mathbf{R})$ as a partial wave expansion

$$F_{nv}(\mathbf{R}) = \sum_{l,m} \frac{F_{nv}^l(R)}{R} Y_{lm}(\theta, \phi) \tag{18}$$

For convenience, it will be assumed throughout that the expansion basis $\Phi_n G_{nv}$ is diabatic (*stricto-sensu*). This is done without loss of generality since unitary transformations of this basis yield the sought results in an arbitrary basis (including of course the adiabatic one). We thence drop the superscript d of Eqs. (12) and (17) to lighten the notation. With these specifications one arrives, at a set of close-coupling equations of the form (Arthurs and Dalgarno, 1960; Secrest, 1975)

$$\left[-\frac{1}{2\mu}\frac{d^2}{dR^2} + \frac{l(l+1)}{2\mu R^2} + \frac{j_n(j_n+1)}{2mr_n^2} + U_{nvj_n}^l(R) - E\right] F_{nvj_n}^l$$
$$= -\sum_{l', n', v', j_{n'}'} \Omega_{nvj_n, n'v'j_{n'}'}^{ll'}(R) F_{n'v'j_{n'}'}^{l'} \tag{19}$$

Where the potential $U_{nvj_n}^l(R)$ and the coupling terms $\Omega_{nvj_n, n'v'j_{n'}'}^{ll'}(R)$ are readily determined with the help of Eqs. (1, 4, 8, 9, 11, 12, 15, and 18). Note that in the left hand side of Eq. (19) the equilibrium bond-length r_n characteristic of state n appears; thus $U_{nvj_n}^l(R)$ contains a term of the form $j_n(j_n + 1)[r^{-2} - r_n^{-2}]/2m$. Actually, the bulk of Eq. (19) is quite far beyond present computational capability. It has been already mentioned in the introduction that the way to overcome this difficulty emerged from the study of rotational transitions in scattering problems (see, e.g., Tsien et al., 1973; Pack, 1974; Secrest, 1975; Schinke and McGuire, 1978; Kouri, 1979). It has long been recognized that when the rotational periods of the molecule are large compared to a characteristic collision time one could use a sudden approximation *vis a vis* the molecular rotation; i.e., the collision occurs for a rotor that is actually held fixed in space fixed (Takayanagi, 1963). Moreover, when the radial relative motion is faster than the angular relative motion a centrifugal sudden approximation may be used; i.e., one may assume that the **R** axis stays fixed in space throughout the collision. In each case the associated angular wavefunction (in $\hat{r}$ or $\hat{R}$ for the former or latter case resp.) does not change during the atom-molecule encounter. The *infinite order sudden approximation* (IOSA) is based on both assumptions. Thence within the IOSA there is only a *trivial distinction between SF and BF reference frames*. Expressed mathematically, the two IOS assumptions amount to replacing all coupled values of l and j compatible with a given value of the conserved total angular momentum by common values $\bar{l}$ and $\bar{j}$, respectively. These replacements are referred to as *centrifugal-sudden* and *energy-sudden* approximations, respectively. The exact quantal close-coupling equations (19) may thenceforth be partially diagonalized in the form (see, e.g., Secrest, 1975; Schinke and McGuire, 1978a, b)

$$\left[\frac{d^2}{dR^2} - \frac{\bar{l}(\bar{l}+1)}{R^2} + k_{nv}^2\right]\mathscr{F}_{nv}^{\bar{l}}(R, \gamma) = 2\mu \sum_{n'v'} \mathscr{H}_{nv,n'v'}(R, \gamma)\mathscr{F}_{n'v'}^{\bar{l}}(R, \gamma), \quad (20)$$

where, in a diabatic vibronic basis (Sec. II,C)

$$\mathscr{H}_{nv,n'v'}(R, \gamma) = \langle g_{nv}|\langle \Phi_n|H_{el}|\Phi_{n'}\rangle - W_n(r)\delta_{nn'}|g_{n'v'}\rangle \quad (21)$$

and

$$k_{nv}^2 = 2\mu\left(E - \mathscr{E}_{nv} - \langle g_{nv}|\frac{\bar{j}(\bar{j}+1)}{2mr^2}|g_{nv}\rangle\right). \quad (22)$$

Consistently with the above IOSA assumptions, the electron-nuclei Coriolis coupling terms arising from the SF-BF transformation [Eqs. (2)–(4)] have been omitted from Eq. (21).

The system of Equations (20) is far easier to solve than the original one [Eq. (19)] and is well within present computational capability. The $\mathscr{F}_{nv}^{\bar{l}}(R, \gamma)$ functions have still to be determined for several values of the *continuous parameter* γ. As usual these functions are subject to the boundary conditions

$$\mathscr{F}_{nv}^{\bar{l}}(R = 0, \gamma) = 0, \quad (23a)$$

$$\mathscr{F}_{nv}^{\bar{l}}(R \to \infty, \gamma) \approx -\frac{1}{2i}\left\{k_{nv}^{-1/2}\exp\left[-i\left(k_{nv}R - \frac{\bar{l}\pi}{2}\right)\right]\delta_{nn'}\delta_{vv'} \right.$$

$$\left. - k_{n'v'}^{-1/2}S_{nv,n'v'}^{\bar{l}}(\gamma)\exp\left[i\left(k_{n'v'}R - \frac{\bar{l}\pi}{2}\right)\right]\right\}. \quad (23b)$$

The actual scattering matrix related to an $nvjl \to n'v'j'l'$ transition for a total angular momentum J is obtained from (Schinke and McGuire, 1978b)

$$S^J_{nvjl,n'v'j'l'} = (-1)^{j+j'} i^{-(l+l'+2\bar{l})} \frac{[(2l'+1)(2l+1)]^{1/2}}{2J+1}$$

$$\times \sum_\lambda C(j'l'J;\lambda 0\lambda)C(jlJ;\lambda 0\lambda) S^{\bar{l}\lambda}_{nvj,n'v'j'}, \quad (24)$$

where the C's are Clebsch-Gordan coefficients and

$$S^{\bar{l}\lambda}_{nvj,n'v'j'} = 2\pi \int_0^\pi d\gamma \, \sin\gamma \, Y_{j'\lambda}(\gamma,0) S^{\bar{l}}_{nv,n'v'}(\gamma) Y_{j\lambda}(\gamma,0). \quad (25)$$

The corresponding degeneracy-averaged differential scattering cross section then obtains from

$$\frac{d\sigma_{nvj,n'v'j'}(\theta)}{d\Omega} = \frac{1}{4(2j+1)k^2_{nvj}}$$

$$\times \sum_{\lambda=-\lambda^*}^{\lambda^*} \left| \sum_{l'} (2l'+1)[\delta_{nn'}\delta_{jj'} - S^{l'\lambda}_{nvj,n'v'j'}] P_{l'}(\cos\theta) \right|^2. \quad (26)$$

In this result the identification $\bar{l} = l'$ has been made according the suggestion of Schinke and McGuire (1978a, b) and $\lambda^* = \min(j, j')$. If one is interested in *vibronic transitions irrespective of the final j' rotational state* then the sought differential cross section is obtained by summing Eq. (26) over j':

$$\frac{d\sigma_{nv,n'v'}(\theta)}{d\Omega} = \frac{1}{8k^2_{nvj}} \int_0^\pi d\gamma \, \sin\gamma \left| \sum_{l'} (2l'+1) P_{l'}(\cos\theta)[\delta_{nn'}\delta_{vv'} - S^{l'}_{nv,n'v'}(\gamma)] \right|^2. \quad (27)$$

This result is extremely simple and expresses the vibronic differential cross section as an average over γ of (fixed γ) differential cross sections $d\sigma_{nv,n'v'}(\gamma,\theta)/d\Omega$. The result (27) depends on the initial j only through k_{nvj} which is defined by Eq. (22) where $\bar{j}$ is now replaced by j.

The two basic assumptions of the IOS approximation have been discussed in many places. The energy sudden approximation requires that the characteristic time of rotation of the BC molecule (typically of the order of $mr_e^2/\bar{j}$) be much longer than a characteristic collision time (of the order of $[\mu/2E]^{1/2}$ say). This condition is likely to be fulfilled for most atom-molecule collision problems at energies above 0.1 eV/amu. The centrifugal sudden (CS) approximation requires that the radial relative motion be much faster than the rotational relative motion (this is also equivalent to require negligible Coriolis coupling in the BF frame). As pointed out by Kouri (1979) the criteria of applicability of the CS approximation are rather difficult to assess quantitatively and most of those in current use stem from numerical calculations concerned with rotational excitation. It turns out that the CS approximation applies best for small values of l, when the scattering is dominated by the short-range repulsive walls of the relevant potentials, far from classical turning points and when the collision energy exceeds the depths of eventual wells in the potential. However, these are only sufficient

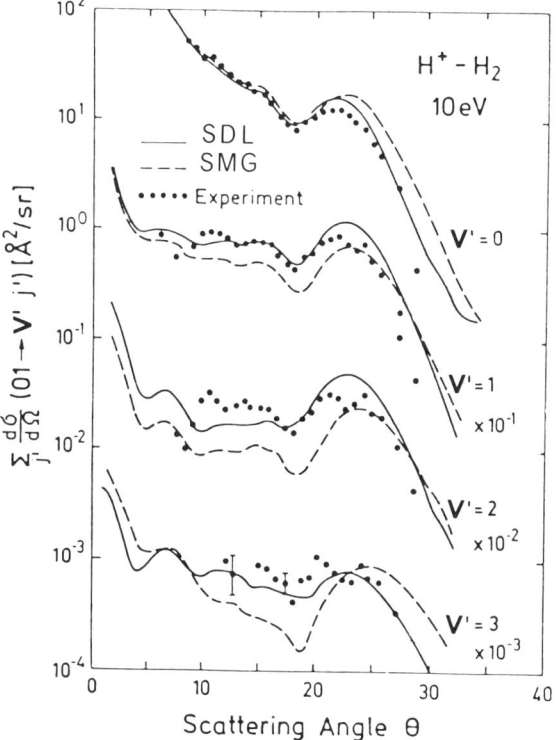

FIG. 3. Rotationally summed differential cross sections for vibrational transitions $v = 0 \rightarrow v' = 0,1,2,3$ in the $H^+ + H_2$ collision at a center of mass energy of 10 eV. Full line (SDL) IOSA calculations of Schinke et al., (1980) involving a fit of an *ab initio* potential energy surface of H_3^+. Dashed line (SMG) IOSA calculations of Schinke and McGuire involving the H_3^+ potential energy surface of Giese and Gentry (1974). Experimental data of Hermann et al. (1978) are normalized at $\theta = 15°$ to the elastic SDL result. [The theoretical calculations neglect non-adiabatic transitions towards $H + H_2^+(v')$ channels lying above $H^+ + H_2(v = 3)$].

conditions. The success of the IOS approximation in the description of both rovibrational excitation (Schinke et al., 1978a, b: Schinke, 1980) and vibronic charge transfer (Baer et al., 1988) in the $H^+ + H_2$ collision in the eV energy range (Figs. 3, 4) are clear indications of the potency of this method [see also Schinke (1984) for the interpretative power of the IOS approach when combined with the semiclassical approximation].

As a final remark concerning quantal treatments of vibronic processes one should mention IOS calculations by McGuire and Bellum (1979) on the $Na + H_s$ collision in perpendicular arrangement ($\gamma = \pi/2$) as well as calculations by Becker and Saxon (1981) and by Becker (1982) on $K + O_2$ and $O_2^+ + O_2$ collisions, respectively. In the latter two calculations the orientation dependences of the relevant interactions were ignored which thereby entailed dramatic simplifications in the theory.

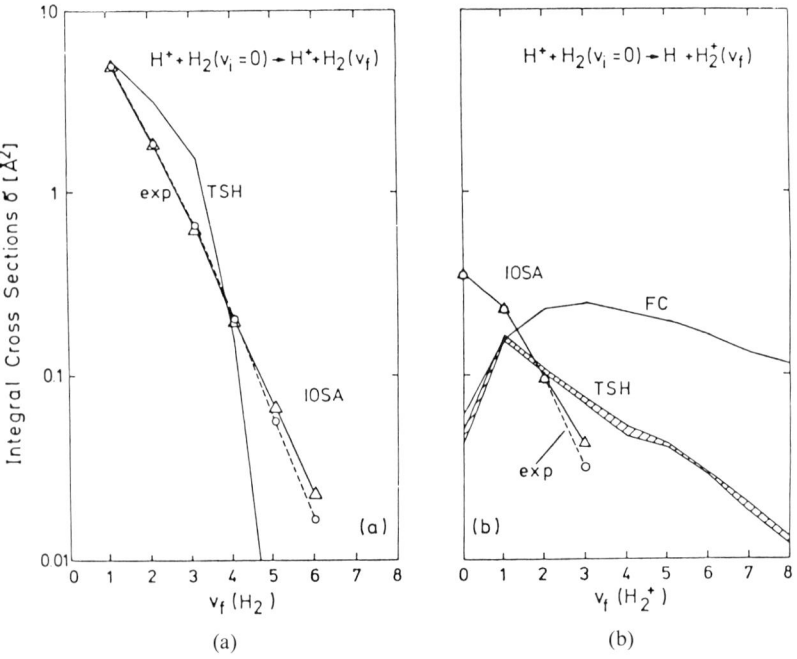

FIG. 4. Rotationally summed integral cross sections for vibrational excitation (a) and vibronic charge transfer (b) in the $H^+ + H_2$ collision at a center of mass energy of 20 eV. —△—, IOSA calculations of Baer et al. (1988); ———, TSH; and —○—, experimental results of Niedner et al. (1987). The curve labelled FC in (b) is the corresponding Franck-Condon prediction. According to the quoted works, charge exchange results from a primary population of states correlated with $H^+ + H_2(v \geq 4)$ followed by population sharings between the latter channels and those associated with $H + H_2^+(v')$.

IV. Semiclassical Treatment of Vibronic Excitation

A. BACKGROUND

A widespread belief in the field of heavy particle collisions is that, owing to their large masses, nuclei are likely to have short associated wavelengths and may thus be treated using classical mechanics (as opposed to electrons which obey quantum mechanics). This is the basis of the so-called (quasi-)classical trajectory method which has been used with some success for a broad class of molecular collisions involving a single electronic state (electronically adiabatic processes). These include rotational excitation (see, e.g., Pattengill, 1979) vibrational excitation (see, e.g., Gentry, 1979), dissociation (see e.g., Schinke, 1989), and reactive scattering (see, e.g., Truhlar and Muckerman, 1979). The method has been described at length in many places; therefore, only its main steps are summarized here. Briefly, the potential energy surface (PES) $E_n^a(R, r, \gamma)$ (see Sec. II,C) that governs the classical nuclear motion is first determined in the framework of the Born-Oppenheimer approximation using *ab initio* (see, e.g., Schaeffer, 1979) or semiempirical (see, e.g., Kuntz, 1979) techniques. Then, the classical equations of motion are solved for a proper sampling of initial conditions. The results of these trajectory calculations are analyzed in terms of the internal energy content of the products and their scattering angle. Finally, the information arising from a large number of trajectories is averaged and properly boxed to represent the cross sections related to the various quantal observables.

Extension of the quasi-classical procedure to vibronic processes which by definition involve *changes of electronic states* (electronically nonadiabatic processes) is not as trivial as might appear from the common use of the so-called trajectory surface hopping (TSH) technique introduced by Tully and Preston (1971)—see also Bjerre and Nikitin (1967). In the TSH approach one proceeds exactly as in the single PES case until the trajectory reaches a so-called transition seam where a hop to a different PES is likely to occur. The probability of this (quantal) event is determined in order to decide whether the system stays on the same PES or actually hops. In the latter case the direction of the momenta are readjusted and the trajectory then evolves on the new surface. There is a rather broad assortment of prescriptions to determine both the characteristics of the transition seam (or hopping locus) and those of the hop itself (Tully and Preston, 1971; Tully 1976; Kuntz *et al.*, 1979, Blais and Truhlar, 1983, Stine and Muckerman, 1976, 1986; Eaker, 1987). There is no general consensus concerning best prescriptions and different choices may produce quite different results (Eaker, 1987). Thus, breakdown of the TSH approach (see, e.g., Fig. 4) may arise either from the inadequacy of the description of the hop or from a failure of classical mechanics to account for tunneling, resonances, subtle interference phenomena, or proper quantization of the initial and final states of the products. These weaknesses of the TSH approach make it desirable to develop approximate quantal treatments as those discussed in Secs. III and IV,B.

B. TIME-DEPENDENT DESCRIPTION OF RELATIVE MOTION: THE COMMON TRAJECTORY TREATMENT OF (RO-)VIBRONIC TRANSITIONS

Contrary to the TSH approach of Sec. IV,A, the so-called common trajectory method treats only the relative degree of freedom classically. The idea for doing so is connected to the fact that *elastic-type* scattering of heavy particles is often quite satisfactorily described within the semiclassical JWKB approximation (see, e.g., Child, 1974; Champion et al., 1970; Mittman et al., 1971; Sidis, 1972; see also Tsien et al., 1973 and Schinke, 1984). Indeed, owing to the large masses of the colliding partners (thousands of atomic units), the De Broglie wavelength of relative motion (for $E \geq 0.1$ eV/amu) is generally much smaller than typical ranges of variation of the intermolecular potential. Relying on this idea one may attempt to describe inelastic scattering of heavy particles using a superposition of elastic-type JWKB wavelets in each channel nv, weighted by proper probability amplitudes. The original problem then reduces to determining the mentioned probability amplitudes. This is the basis of almost all semiclassical theories of inelastic scattering of heavy particles (Bates and Holt, 1966; Berson 1968; Bates and Cross, 1969; Crothers, 1970; Delos et al., 1972; Child, 1974; Gaussorgues et al., 1975; McCann and Flannery, 1978). These theories may however differ in the stage of the derivation at which the JWKB approximation is invoked, i.e., before (McCann and Flannery, 1978) or after (Cross, 1969; Wartell and Cross, 1971) performing the partial wave expansion of Eq.(18). If, in addition to the JWKB conditions, one assumes that the differences between channel wave numbers are small compared to an average wave number K, and that changes of internal (ro-vibronic) angular momentum are negligible compared to the relative angular momentum, then the *quantum mechanical problem of* Sec. II,B [Eq. (9)] *becomes nearly equivalent* to that of solving the time-dependent(-like) Schrödinger equation:

$$\frac{i\partial \chi}{\partial t}(\{\mathbf{\rho}_i\}, \mathbf{R}(t), \mathbf{r}) = H_{\text{int}} \chi(\{\mathbf{\rho}_i\}, \mathbf{R}(t), \mathbf{r}), \tag{28}$$

where H_{int} is the *ro-vibronic* Hamiltonian:

$$H_{\text{int}} = H_{\text{el}} - \frac{1}{2m}\Delta_{\mathbf{r}} \qquad (29)$$

(see definitions and labelling in Secs. II,A and II,B). The parameter t defines the common classical trajectory according to

$$\frac{\mu\, d\mathbf{R}}{dt} = \mathbf{K} \qquad (30)$$

and χ describes the (internal) ro-vibronic state of the system

$$\chi(\{\mathbf{\rho}_i\}, \mathbf{R}(t), \mathbf{r}) = \sum_{nv} A_{nv}(\mathbf{R}(t))G_{nv}(\mathbf{r}; \ldots)\Phi_n(\{\mathbf{\rho}_i\}; R, r, \gamma). \qquad (31)$$

Implicit in the derivation of Eqs. (28) and (31) is the assumption that there exists one trajectory [Eq. (30)] for each impact parameter $\mathbf{b}$ [$b \approx (l + 1/2)/K$, φ_b] and that a common relative wavefunction describing the scattering may be factored out (Flannery and McCann, 1973). In the described context the scattering amplitude for the $nv \to n'v'$ transition may be obtained in the form (Cross 1969; Wartell and Cross, 1971; Flannery and McCann, 1973)

$$f_{nv,\,n'v'}(\theta, \phi) = -\frac{ik_{nv}}{2\pi}\int_0^\infty b\, db \int_0^{2\pi} d\varphi_b\, \exp[-2ik_{nv} b \sin\frac{\theta}{2}\cos(\phi - \varphi_b)]$$
$$\times [S_{nv,\,n'v'}(b, \varphi_b) - \delta_{nn'}\delta_{vv'}], \qquad (32)$$

where

$$S_{nv,\,n'v'}(b, \varphi_b) = A_{n'v'}(\mathbf{R}(b, \varphi_b, t \to \infty)) \qquad (33a)$$

with the condition that

$$A_{n'v'}(\mathbf{R}(b, \varphi_b, t \to -\infty)) = \delta_{nn'}\delta_{vv'}. \qquad (33b)$$

The differential cross section is obtained as usual from the square modulus of Eq. (32).

There are many ways of choosing the trajectory implied by Eq. (30). The trajectory may be as simple as a straight line or may derive from a judicious spherical potential (see, e.g., Tully, 1976). Following McCann and Flannery (1978), De Pristo (1983) has advocated the use of a trajectory that conserves the average total energy of the system (in the sense of Ehrenfest's theorem) and suggested to combine Eq. (30) with

$$\frac{d\mathbf{K}}{dt} = \sum_{nv,\,n'v'} A^*_{nv}(\nabla_R \langle \Phi_n G_{nv}|H_{\text{int}}|\Phi_{n'} G_{n'v'}\rangle)A_{n'v'} \qquad (34)$$

This prescription actually achieves some coupling of internal and relative degrees of freedom. It was shown to provide *excellent* agreement between semi-classical and quantum-mechanical calculations of total cross sections for state-to-state vibronic charge transfer in the $O_2^+ + O_2$ collision in the eV energy range (De Pristo, 1983).

Upon closing this subsection two points should be stressed:

(i) Despite of the terminology used to qualify Eq. (28) this equation stems from a *stationary* state treatment of the collision; the dummy variable t having the dimension of a time describes the variation of **R** along the common trajectory.

(ii) The common trajectory itself is merely a device to extract the probability amplitudes A_{nv}. As Eq. (32) shows, the actual properties of scattering in channel $n'v'$ come from all regions of space and are actually governed by the phases of the $A_{n'v'}$ amplitudes. Each of these phases is mainly determined by the classical action integral along the trajectory.

In certain extreme cases, like those involving the formation of a long-lived complex in one or a few channels and not in the others, the concept of a single trajectory breaks down. These cases often occur at much lower collision energies than those for which the common trajectory approach is expected to apply and therefore they should better be tackled by pure quantum-mechanical means (as described in Sec. III).

1. Sudden Approximation for Rotation

The set of coupled equation that may be obtained by inserting Eq. (31) in (28) and projecting the result onto the $G_{nv}\Phi_n$ basis is computationally lighter than the exact quantum mechanical set given by Eq. (19). Firstly, only the needed row or column of the scattering matrix is determined. Secondly, owing to the large orbital angular momentum assumption, decoupling of internal and relative angular momenta is approximately achieved.[1] Still, the number of coupled equations to be handled remains enormous. Of course, this is due to the rapid increase with collision energy of the number of rotational states associated with the open vibronic channels (of the order of Nj_{max}^2, where j_{max} is the highest value of j to be considered and N the total number of vibronic states). Yet it should be kept in mind that such a large number of rotational states arises because the corresponding energy differ-

---

[1] It should be noted, however, that Wartell and Cross (1971) have formulated a semiclassical theory of inelastic collisions where the coupling between rotational and orbital angular momenta are kept coupled. The sudden approximation for rotation—to be discussed shortly—is then applied in this context. Decoupling of the orbital angular momentum is performed later on in that theory by the use of a classical action-angle transformation.

ences are tiny as compared to the collision energy. But this is just the condition required to apply the sudden approximation *vis a vis* the molecular rotation (i.e., the energy sudden approximation of Sec. III). This approximation has been worked out in different ways (see, e.g., Wartell and Cross, 1971; Sidis and de Bruijn, 1984; Sidis 1989). It amounts to assuming that the collision time is short compared to typical molecular rotation periods so that neither the BC axis rotates nor the initial rotational wave-packet spreads during the *A-BC* encounter. In such conditions the initial rotation states $|j_I, m_{j_I}\rangle$ may be factored out from Eq. (31):

$$\chi(\{\boldsymbol{\rho}_i\}, \mathbf{R}(t), \mathbf{r}) = X(\{\boldsymbol{\rho}_i\}, \mathbf{R}(t), r; \hat{\mathbf{r}})|j_I m_{j_I}\rangle, \tag{35}$$

$$X(\{\boldsymbol{\rho}_i\}, \mathbf{R}(t), r; \hat{\mathbf{r}}) = \sum_{n'v'} B_{n'v'}(\mathbf{R}(t); \hat{\mathbf{r}}) \frac{g_{n'v'}(r)}{r} \Phi_{n'}(\{\boldsymbol{\rho}_i\}; R, r, \gamma), \tag{36}$$

and the *vibronic amplitudes* are determined from

$$i\frac{\partial X}{\partial t}(\{\boldsymbol{\rho}_i\}, \mathbf{R}(t), r; \hat{\mathbf{r}}) = H_{\text{vibronic}} X(\{\boldsymbol{\rho}_i\}, \mathbf{R}(t), r; \hat{\mathbf{r}}), \tag{37}$$

$$H_{\text{vibronic}} = H_{\text{el}} - \frac{1}{2m}\frac{d^2}{dr^2}. \tag{38}$$

Equation (37) is to be solved for a two-dimensional grid of values associated with the pair of angular coordinates $\hat{\mathbf{r}} = (\theta_r, \phi_r)$ that define the orientation of the *BC* axis. Solution of Eq. (37) subject to the condition

$$B_{n'v'}(\mathbf{R}(t \to -\infty); \hat{\mathbf{r}}) = \delta_{nn'}\delta_{vv'} \tag{39}$$

provides with Eqs. (31), (35) and (36) the sudden ro-vibronic probability amplitudes

$$A_{n'v'j'mj'}(\mathbf{R}(t)) = \langle j'm_{j'}|B_{n'v'}(\mathbf{R}(t); \hat{\mathbf{r}})|j_I m_{j_I}\rangle, \tag{40}$$

which may thence be used in Eqs. (32) and (33a). If one is only interested in *vibronic cross sections summed over all final rotational states as well as initial m_{j_I} degeneracies*, then the result is obtained from (Wartell and Cross 1971, Krüger and Schinke 1977):

$$\frac{d\sigma_{nv,n'v'}(\theta, \phi)}{d\Omega} = \frac{1}{4\pi}\int_0^\pi \sin\theta_r\, d\theta_r \int_0^{2\pi} d\phi_r |f_{nv,n'v'}(\theta, \phi; \hat{\mathbf{r}})|^2, \tag{41}$$

where the *vibronic scattering amplitude* $f_{nv,n'v'}(\theta, \phi; \hat{\mathbf{r}})$ is defined similarly to Eq. (32) save that $\nu = (vjm_j)$ is now replaced by v and the relevant vibronic scattering matrix is [with Eq. (39)]

$$S_{nv,n'v'}(b, \phi_b; \hat{\mathbf{r}}) = B_{n'v'}(\mathbf{R}(b, \varphi_b, t \to \infty); \hat{\mathbf{r}}). \tag{42}$$

The foregoing approximation has been the basis of a few theoretical studies of vibrational (Schinke, 1977) and vibronic processes (Hedrick et al., 1977; Lee and De Pristo 1983; Sidis and Courbin-Gaussorgues, 1987; Gislason et al., 1987). Other calculations by Bates and Reid (1969), Flannery et al., (1973); Moran et al., (1974, 1975, 1981), Spalburg et al., (1983, 1985a, b) and De Pristo (1983a, b) have also been carried out in the just described context but with the simplifying approximation of *isotropic interactions* [i.e., by dropping the parametric $\hat{\mathbf{r}}$ dependence of Eqs.(36)–(42)].

The above treatment only makes use of the (vibrational) energy-sudden approximation (Sec.III) and is thus of wider applicability than the IOSA at energies that are high enough to warrant the use of the semiclassical approach of Sec.IV,B. However when both the energy-sudden and centrifugal-sudden approximations are compatible with the JWKB and common trajectory assumptions a semiclassical IOS treatment may be worked out. Starting from Eq.(20) and proceeding as indicated by Cross (1969) and others (see the introductory part of Sec.IV,B) one easily arrives at a vibronic time dependent Schrödinger equation similar to Eq.(37) except that the fixed parameter $\hat{\mathbf{r}} = (\theta_r, \phi_r)$ is replaced by $\gamma = (\mathbf{r}, \mathbf{R})$. The semiclassical IOS scattering matrix $S^l_{nv,n'v'}(\gamma)$ is readily seen to be given by an equation similar to Eq.(42) with the appropriate replacement of $\hat{\mathbf{r}}$ by γ. Calculations based on the semiclassical IOSA have been reported by Parlant and Gislason (1986, 1987) (see Sec.VI,B). Gislason and Sachs (1975) have also used this approximation in conjunction with the Bauer-Fischer-Gilmore (1969) model (see also Desfrançois et al., 1988). As is well known that model assumes that the vibronic energy terms $\mathscr{E}_{nv} + \mathscr{H}_{nv,nv}$ and $\mathscr{E}_{n'v'} + \mathscr{H}_{n'v',n'v'}$ [Eq. (21)], for $n \neq n'$, form a grid of crossings where Landau-Zener (1932) type transitions may occur. This model has often been presented in the literature; therefore we just indicate here some references where its conditions of applicability are considered (Child, 1974; Child and Baer, 1981; Kleyn et al., 1982; Klomp et al., 1984; Spalburg et al., 1986).

2. Sudden Approximation for Both Vibration and Rotation

As the collision energy increases and exceeds several hundreds of eV/amu the time $T_{\text{coll}} \approx L(\mu/2E)^{1/2}$ taken by the system to travel a distance L (of a few atomic units say) may become much shorter than typical vibration times $T_{\text{vib}} \approx \Delta E_{\text{vib}}^{-1}$ [$= |\mathscr{E}_{nv'} - \mathscr{E}_{nv}|^{-1}$, Eq. (15)] and *a fortiori* of typical rotation times (Secs. III, IV,B,1). One may therefore invoke a sudden approximation *vis a vis* the molecular (ro-)vibration motion. In this approximation the $\mathbf{r} = (r, \theta_r, \phi_r)$ vector *remains fixed* and the initial ro-vibration wavepacket does not evolve during the collision, i.e. (Wartell and Cross, 1971; Sidis and de Bruijn, 1984; Dhuicq et al., 1985; Sidis, 1989);

$$\chi(\{\mathbf{\rho}_i\}, \mathbf{R}(t), \mathbf{r}) = Y(\{\mathbf{\rho}_i\}, \mathbf{R}(t); \mathbf{r}) \frac{g_{nv}}{r} |j_I m_{j_I}\rangle, \tag{43}$$

$$Y(\{\mathbf{\rho}_i\}, \mathbf{R}(t); \mathbf{r}) = \sum_{n'} C_{n'}(\{\mathbf{\rho}_i\}, \mathbf{R}(t); \mathbf{r}) \Phi_{n'}(\{\mathbf{\rho}_i\}; R, r, \gamma). \tag{44}$$

The sudden *electronic* probability amplitudes are thence determined (as in the atom + atom case) from

$$i \frac{\partial Y}{\partial t}(\{\mathbf{\rho}_i\}, \mathbf{R}(t); \mathbf{r}) = H_{el} Y(\{\mathbf{\rho}_i\}, \mathbf{R}(t); \mathbf{r}) \tag{45}$$

with $C_{n'}(\mathbf{R}(t \to -\infty); \mathbf{r}) = \delta_{nn'}$, where n is the initial electronic state of the system. Comparison of Eqs. (43) and (44) with Eqs. (35) and (36) provides the *sudden vibronic probability amplitudes*

$$B_{n'v'}(\mathbf{R}(t); \mathbf{r}) = \int_0^\infty dr\, g_{n'v'}(r) C_{n'}(\mathbf{R}(t); \mathbf{r}) g_{nv}(r). \tag{46}$$

The ro-vibronic amplitudes and relevant cross sections are thenceforth determined as discussed in Sec. IV,B,1 (Eqs. 40–42).

V. On Franck-Condon-type Approximations

It is often thought that at high collision energies the population of vibrational states of the $BC^{(*)}$ molecule subsequent to electronic transition processes obey the Franck-Condon (FC) principle. In other words, the $nv \to n'v'$ transition probabilities $P_{n'v'}$ and cross sections (total cross sections say) $Q_{n'v'}$ would be written

$$P_{n'v'} = \frac{1}{4\pi} \int_0^\pi \sin\theta_r\, d\theta_r \int_0^{2\pi} d\varphi_r |B_{n'v'}(\mathbf{R}(t \to \infty); \hat{\mathbf{r}})|^2$$

$$\approx P_{n'} |\langle g_{n'v'}|g_{nv}\rangle|^2, \tag{47}$$

$$Q_{n'v'} \approx Q_{n'} |\langle g_{n'v'}|g_{nv}\rangle|^2. \tag{48}$$

The discussion in Sec. IV,B,2 clearly shows that "high collision energy" means *high enough for the vibrational-sudden approximation to hold*. Equations (47, 48) imply another simultaneous condition namely: that $C_{n'}(\mathbf{R}(t); \mathbf{r})$ should have a weak dependence upon the molecule bond distance r. The later condition is of course trivially satisfied when both the relevant electronic energies $\langle \Phi_n | H_{el} | \Phi_n \rangle$ and coupling terms $\langle \Phi_n | H_{el} | \Phi_{n'} \rangle$ have little or no dependence upon r. Dhuicq and Sidis (1986, 1987) and Sidis (1989) discussed the case when $C_{n'}(\mathbf{R}(t \to \infty); r)$ in an $n \to n'$ charge transfer problem would be given by a formula inspired by results of the Demkov (1964) model:

$$C_{n'}(\mathbf{R}(t \to \infty); r) \approx \frac{1}{\sqrt{2}} \operatorname{sech} \frac{\pi \Delta W(r)}{2\Lambda u}, \qquad (49)$$

where u is the collision velocity and

$$\Delta W(r) = |\langle \Phi_n | H_{el} | \Phi_n \rangle - \langle \Phi_{n'} | H_{el} | \Phi_{n'} \rangle| \approx |W_n(r) - W_{n'}(r)|, \qquad (50)$$

$$\langle \Phi_n | H_{el} | \Phi_{n'} \rangle \approx A \exp(-\Lambda R). \qquad (51)$$

In this case the only r-dependence of the problem lies in the electronic resonance energy defect $\Delta W(r)$ [Eq. (50)]. Equation (49) shows that values of r where $\pi \Delta W(r)/(2\Lambda u)$ is less than about 0.3 will provide, within a few percent, equal $C_{n'}(\mathbf{R}(t \to \infty); r)$ values. Thence, a (*sine qua non*) condition to obtain FC-type distributions in this case is that the main contributions to Eq. (46) actually come from such values of r. The extension of the corresponding r range is easily seen to narrow as Λu when the collision velocity decreases. The observation by Parlant and Gislason (1987) that the FC prediction fails to reproduce the results of theoretical calculations on the

$$\text{Ar}^+ + \text{N}_2(X^1\Sigma_g^+, v = 0) \to \text{Ar} + \text{N}_2^+(A^2\Pi, v')$$

charge transfer collision at keV energies is to be cast in the just discussed context. This was indeed understood by Gislason and Parlant (1987) in similar terms by making reference to a weak coupling approximation (see below). Another example related to the same discussion is provided by the results of both experiment and theory (De Bruijn et al., 1984; Sidis and de Bruijn, 1984) on the state selected dissociative charge exchange process:

$$\text{H}_2^+(X^1\Sigma_g^+; v = 0, 1, 2 + 3) + \text{Mg} \to (\text{H} + \text{H})_\varepsilon + \text{Mg}^+$$

at keV collision energies. In this example the final vibration function in Eq. (46) is actually a continuum wavefunction describing the H-H dissociation at energy ε (see, e.g., Sidis and De Bruijn, 1984). Figure 5 shows clear departures from the FC predictions for $v = 0$ and 1 initial states of H_2^+ and illustrates the role of the r-dependence of the $C_{n'}$ probability amplitudes in Eq. (46).

Actually, Eq. (49) specifies only the r-dependence of $|C_{n'}|$ since, for simplicity, all phase factors have been ignored there. In fact, the r-dependence of the *phase* of the $C_{n'}$ amplitude in Eq. (46) is also quite important. This is simply illustrated by the process of vibrational excitation for *pure electronically elastic scattering* (i.e., $P_{n'} = |C_{n'}|^2 = \delta_{nn'}$). In this case the sudden vibrational excitation amplitude Eq. (46) is written (see, e.g., Collins and

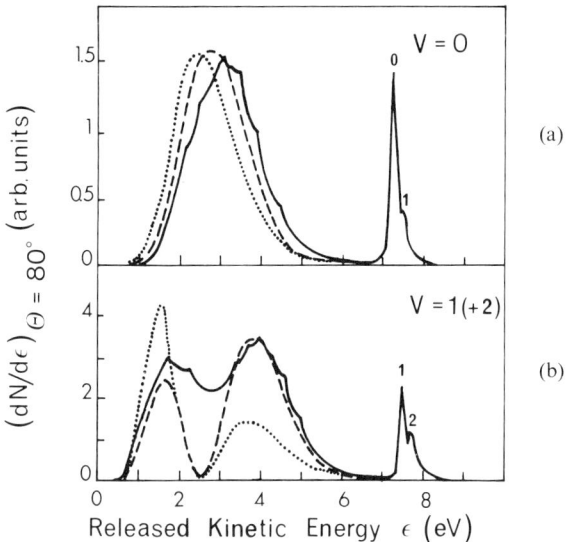

FIG. 5. Energy spectra of H atoms produced in dissociative charge collisions of H_2^+ with Mg at a laboratory energy of 5 keV. Spectra (a) and (b) correspond respectively to $H_2^+(v=0)$ and $H_2^+(v=1)$ incident ions. The broad structure appearing for kinetic energies of the H fragments below 7 eV corresponds to both direct dissociation of $H_2(b^3\Sigma_g^+)$ and radiative dissociative dissociation of $H_2(a^3\Sigma_g^+)$; peaks appearing at high energies correspond to predissociation of $H_2(c^3\Pi_u)$—see Sec. VI,A. The solid curves are experimental data of De Bruijn et al.; dashed curves are theoretical results of Sidis and De Bruijn (1984); dotted curves show predictions based on the Franck-Condon principle. The dip appearing in the $v=1$ data for $2 < \varepsilon < 3$ eV reflects the nodal structure of the $H_2^+(v=1)$ vibrational wavefunction.

Cross, 1976; Dhuicq and Sidis, 1987)

$$B_{nv'}(\mathbf{R}(t \to \infty); \hat{\mathbf{r}}) = \langle g_{nv'}|\exp[2i\eta_n(\mathbf{r})]|g_{nv}\rangle, \qquad (52)$$

where $\eta_n(\mathbf{r})$ is the JWKB phase shift (or an approximation thereof) for elastic scattering in the potential $\langle \Phi_n|H_{el}|\Phi_n\rangle$ with the $\mathbf{r}$ vector held fixed. Again, only when $\eta_n(\mathbf{r})$ is independent of r does Eq. (52) match the FC result of Eq. (47) which in this case predicts no vibrational excitation at all. On the other hand, when $\eta_n(\mathbf{r})$ varies slowly about the equilibrium distance r_e of the molecule BC (in state n) it may be expanded in Taylor series around that point; this enables one to obtain the result of Eq. (52) in closed analytical form (Collins and Cross, 1976) for Morse or harmonic oscillator wavefunctions (when linear terms or both linear and quadratic terms respectively are retained in the expansion). Expressions of electronic probability amplitudes for various two-state models, including proper phases, may be found in the review of Crothers (1981).

Dhuicq et al. (1985) and Sidis (1989) discussed the case when, due to long-range interactions (e.g., charge-permanent dipole, charge-quadrupole, and charge-induced dipole) the length L entering in the definition of the collision time in Sec. IV,B,2 may be too large, thereby thwarting the use of the sudden approximation (and *a fortiori* that of the FC approximation). For $n \to n'$ electronic excitation processes occurring at interparticle distances $R \leq L^*/2 \approx$ few a_0, these authors proposed a piecewise approach:

(i) For $R \in [L^*/2, \infty[$ and $-t \in [T, \infty[$, treat only *vibrational transitions* $nv \leftrightarrow nv'$ within the manifold belonging to the *entrance* electronic channel n [i.e., using Eqs. (36)–(38) with $B_{n'v'} = B_{nv}(\mathbf{R}; \hat{\mathbf{r}})\delta_{nn'}$].

(ii) For $R \in [L^*/2, b]$ and $t \in [-T, +T]$ treat *sudden electronic transitions* according to Eq. (45).

(iii) At $t = +T(R = L^*/2)$, extract $B_{n'v'}(\mathbf{R}(+T); \hat{\mathbf{r}})$ for all relevant v' states associated with the considered electronic exit channel n' using Eq. (46) in which the modified "initial" vibrational wave packet $\sum_{v'} B_{nv}(\mathbf{R}(-T); \hat{\mathbf{r}})g_{nv'}(r)$ now replaces $g_{nv}(r)$.

(iv) Repeat step (i) with $t \in [T, \infty]$, for the vibrational manifold associated with the *exit* electronic channel n' subject to the starting conditions steming from step (iii).

When the results of step (ii) have little or no dependence upon bond distance r the aforementioned procedure introduces the notion of *perturbed FC approach*. This notion is the basis of the so-called Lipeles (1969) model where one tentatively assumes that in step (i) the vibrational motion of the system is *strictly adiabatic*. In that case the modified "initial" vibrational wave packet of step (iii) is one of the adiabatic vibrational wavefunctions $g_{nv}^a(r; R, \gamma)$ defined in Eq. (16). The original formulation of the Lipeles model assumes in addition that step (iv) is sudden, thereby implying that the mentioned long-range interactions do not come into play in the exit channel n'. In that case the FC factor $|\langle g_{n'v'}|g_{nv}\rangle|^2$ appearing in Eqs. (47) and (48) is replaced by the perturbed FC factor: $|\langle g_{n'v'}|g_{nv}^a(R^*, \gamma^*)\rangle|^2$. This result is seen to depend on the point R^*, γ^* along the considered trajectory (impact parameter and geometry of the ABC triangle) where the sudden electronic transition is likely to take place. Seemingly, the consequences of this remark have not been fully appreciated by Lipeles (1969) (see Dhuicq et al., 1985).

The above discussion shows how stringent the conditions of applicability of the FC principle are, even in the keV energy range where it is often invoked (Dhuicq and Sidis, 1987).

At energies below a few hundred eV/amu the FC predictions are expected to fail (Fig. 4) owing to the concomitant breakdown of the vibrational-sudden approximation of Sec. IV,B,2 (see, e.g. Hedrick et al., 1977; Kleyn et al., 1982; Los and Spalburg, 1984; Gislason and Parlant, 1987; Parlant and

Gislason, 1988; Dhuicq et al., 1985; Dhuicq and Sidis, 1986; Noll and Toennies, 1987; Baer et al., 1988). The proper description of vibronic phenomena at low energies thus implies the treatment of the *correlated* electronic and vibrational motions (Los and Spalburg 1984) as done in Secs. III and IV,B,1. Notwithstanding such conclusions, comparisons of actual vibronic distributions with the FC predictions were still made until quite recently in some discussions of experimental results of ion (atom) + molecule collisions at low energies (see, e.g., Guyer et al., 1983; Noll and Toennies, 1986). Moreover, in some attempts at rationalizing experimental data, use was made of model two-state probabilities and cross sections multiplied by FC factors (Campbell et al., 1980; Kato et al., 1982a, b; Kusunoki and Ishikawa, 1985). In order to examine the validity of FC-type approximations at low energies it should first be remarked that FC factors appear in the theory when the vibronic coupling matrix element $\mathcal{H}_{nv,n'v'}$ of Eq. (21) may be written in the form

$$\mathcal{H}_{nv,n'v'}(R,\gamma) \cong h_{nn'}(R,\gamma)\langle g_{nv}|g_{n'v'}\rangle \tag{53}$$

Of course, this approximation is legitimate when $\langle \Phi_n|H_{el}|\Phi_{n'}\rangle - W_n(r)\delta_{nn'}$ [Eq. (21)] is nearly independent of r. Notwithstanding, it has often been assumed in theoretical models to palliate the lack of knowledge of the relevant r dependences. Assuming this approximation to hold it is easily seen that in perturbation situations where *all* vibronic transition probabilities are weak (i.e., $|B_{n'v'}| \ll 1 \ \forall n' \neq n, v' \neq v$, and $|B_{nv}| \approx 1$), the FC overlap factors out (Child, 1973):

$$B_{n'v'}(\mathbf{R}(t);\hat{\mathbf{r}}) \approx i\langle g_{n'v'}|g_{nv}\rangle \exp\left(-i\int_{-\infty}^{t}(h_{n'n'} + \mathcal{E}_{n'v'})dt'\right)$$
$$\times \int_{-\infty}^{t} h_{n'n} \exp\left(i\int_{-\infty}^{\tau} D_{nv,n'v'}\,dt''\right)d\tau, \tag{54}$$

where

$$D_{nv,n'v'}(R,\gamma) = h_{n'n'}(R,\gamma) - h_{nn}(R,\gamma) + \mathcal{E}_{n'v'} - \mathcal{E}_{nv}. \tag{55}$$

In cases when the *Landau-Zener* (1932) model assumptions hold, namely (i) $h_{nn'}$ is a slowly varying function of $R(t)$ and $\gamma(t)$, and (ii) a crossing seam exists along which $D_{nv,n'v'}$ vanishes linearly, Eq. (54) may be treated by the stationary phase method (Child, 1974; Nikitin and Umanskii, 1984) which yields

$$B_{n'v'}(\mathbf{R}(t \to \infty);\hat{\mathbf{r}}) \approx -i\langle g_{n'v'}|g_{nv}\rangle \exp\left[-i\int_{-\infty}^{+\infty}(h_{n'n'} + \mathcal{E}_{n'v'})dt'\right]$$
$$\times \sum_j \left[\frac{2\pi h_{nn'}^2}{idD_{nv,n'v'}/dt}\right]_{t_j}^{1/2} \exp\left(i\int_{-\infty}^{t_j} D_{nv,n'v'}\,dt'\right), \tag{56}$$

where t_j are the points of stationary phase in Eq. (54). Usually two points of stationary phase exist: one in the incoming path of the trajectory ($t < 0$) and one in the outgoing path ($t \geq 0$). Consequently, interferences result from the summation in Eq. (56). When the points of stationary phase coalesce the discussed integral [Eq. (54)] may be solved by making reference to the properties of Airy functions (Child, 1974; Nikitin and Umanskii, 1984). If instead of obeying the Landau-Zener (1932) model assumptions, $h_{nn'}$ and $D_{nv,n'v'}$ behave as supposed in the Demkov-model (1964), i.e., (i) $h_{nn'}$ given by Eq. (51) and (ii) $D_{nv,n'v'} = \mathscr{E}_{nv} - \mathscr{E}_{n'v'}$, then (Spalburg et al., 1985a)

$$B_{n'v'}(\mathbf{R}(t \to \infty)) \approx i\langle g_{n'v'}|g_{nv}\rangle A \frac{2\Lambda b^2 K_1(\xi)}{u\xi}, \tag{57}$$

where A and Λ are defined in Eq. (51), K_1 is the modified Bessel function and b is the impact parameter for a straight line trajectory ($\mathbf{R} = \mathbf{b} + \mathbf{u}t$) with constant velocity $\mathbf{u}$ and

$$\xi = \Lambda b[(1 + D_{nv,n'v'}^2)/\Lambda^2 u^2]^{1/2} \tag{58}$$

[see Sidis and Courbin-Gaussorgues (1987) and Sidis (1989) for an extension of these results to some anisotropic interactions and to cases when the vibronic set $n'v'$ ($n' \neq n$) is dense enough to convey a description where the initial state would be embedded in a sort of quasicontinuum]. Equations (54), (56) and (57) bear the form assumed by Campbell et al. (1980) and Kato et al. (1982a,b) in that they do involve FC overlaps multiplying two-state probability amplitudes; a major difference however is that the latter are *weak coupling* amplitudes as opposed to the expressions assumed by the mentioned authors. It has been pointed out by Spalburg et al. (1985a) that not only are such expressions generally incorrect but that their use for vibronic near resonant charge transfer problems is unjustified owing to the presence of more than two states.

Child (1974) has investigated the conditions in which the vibronic level crossing network of Bauer et al. (1969)—Sec. IV,B,2—would give rise to "some sort of FC distribution." Though his procedure is quite instructive, attention is called here to some limitations thereof. It should be stressed first that the reasoning of Child (1974) applies to a single passage across the $n \to n'$ electronic transition zone R_c. From the typical width Δx of that zone, within the Landau-Zener (1932) model, Child (1974) determined the velocity conditions for which the time spent by the system to travel the distance Δx is much smaller than the typical time over which the initial vibrational wave packet g_{nv} may spread. Anticipating that this velocity would be rather important, Child (1974) used the so-called Landau-Zener *dynamical* width:

$$\Delta x_d = (u/\Delta F_{nn'})^{1/2}, \tag{59}$$

$$\Delta F_{nn'} = \left|\frac{d}{dR}(\langle\Phi_n|H_{el}|\Phi_n\rangle - \langle\Phi_{n'}|H_{el}|\Phi_{n'}\rangle)\right|_{R_c}. \tag{60}$$

Actually, at low velocities (u) one should use an expression involving both *static* and *dynamic* widths (Spalburg et al., 1986):

$$\Delta x \approx \left| \frac{2\langle \Phi_n | H_{el} | \Phi_{n'}\rangle}{\Delta F_{nn'}} \right| \left\{ \left[1 + \left(\frac{u \Delta F_{nn'}}{2\langle \Phi_n | H_{el} | \Phi_{n'}\rangle^2} \right)^2 \right]^{1/2} + 1 \right\}^{1/2}. \quad (61)$$

Moreover, in discussing the validity of the FC predictions one is usually interested in the vibrational distributions after the system has traversed the transition zone twice (which lasts typically for a time $T \approx 2[(R_c + \Delta x)^2 - b^2]^{1/2}/u$); this is generally different from what one would get at the immediate exit of one of these zones. Finally, it should always be kept in mind that the Bauer-Fincher-Gilmore (1969) model makes explicit use of the simplifying assumption of Eq. (53). Hence, it may not describe direct vibrational excitation in each electronic channel and is therefore bound to miss some causes of breakdown of the FC approximation.

VI. Studies of Vibronic Transition Processes

In the preceding description of the theoretical treatments of vibronic transitions in atom + molecule collisions, reference has been continuously made to actual studies of these processes. Here, some examples are presented to illustrate the type of new information that may be extracted from such studies. The considered samples have been deliberately selected among a set of theoretical works where particular attention is paid to the *quantum-mechanical description of the internal motion* of the $A^{(+)} + BC$ collisional system (Secs. III and IV,B).

A common characteristics of all the cases considered here is that they involve an *electron transfer* from one collision partner to the other, e.g.,

(i) $\quad A + BC(v_0) \rightarrow A^+ + BC^-(v'_-)$
$\qquad\qquad A^{(*)} + BC(v'_{0(*)})$,

(ii) $\quad A^+ + BC(v_0) \rightarrow A + BC^+(v'_+)$
$\qquad\qquad A^+ + BC(v'_0)$,

(iii) $\quad BC^+(v_+) + A \rightarrow (B + C)_\varepsilon + A^+$
$\qquad\qquad BC^+(v'_+) + A.$

Moreover, all of the considered reactions are characterized by rather *small* (<5 eV) *energy losses or gains*. Owing to these characteristics the considered processes often occur at relatively large interparticle distances ($R \geq 5a_0$). Therefore, the relevant diabatic states of the problem (Sec. II,C) are essentially described by wavefunctions of the separated collision partners. The mentioned features are of great help for modelling purposes and simplify considerably the task of determining the relevant potential energy surfaces and coupling terms ($\langle\Phi_n|H_{el}|\Phi_{n'}\rangle$ and $\mathcal{H}_{nv,n'v'}$ in Sec. II). Actually, the necessary ingredients of vibronic close-coupling calculations may be obtained by many different ways ranging from more or less empirical models to various degrees of *ab initio* sophistication [see, e.g., Janev, 1977; Kleyn et al., 1982; Truhlar et al., 1982; Yarkony, 1986; and Gadéa et al., 1986 for discussions and references related to processes of type (i) and Olson et al., 1971; Smirnov, 1973; Nikitin and Smirnov, 1978; Hedrick et al., 1977; Lee and De Pristo, 1983; Evseev et al., 1980, 1982, Sidis and de Bruijn, 1984; Tully and Preston 1971; Kuntz et al., 1979; Chapman, 1985; Archirel and Levy, 1986; Gislason et al., 1987; Grimbert et al., 1988; and Staemmler and Gianturco, 1985 for an assortment of methods related to processes (i) and (ii)]. It turns out, however, that much physical insight into the dynamics of vibronic phenomena in processes (i)–(iii) has actually emerged from calculations based on rather modest descriptions of the potential energy surfaces and interactions as will be seen in the forthcoming discussions.

A. VIBRONIC QUANTUM BEATS ACCOMPANYING A PROCESS OF ION-PAIR FORMATION

The experimental study of the collision of K or Cs with O_2 ($X^3\Sigma_g^-$, $v = 0$) as a function of scattering angle in the 25–100 eV energy range (Kleyn et al., 1980a, b) revealed interesting new structure in addition to the rainbow-type features which normally appear in atom-atom collisions involving the interaction between covalent and ionic states (process i). The first interpretation of these structures has been proposed within the classical TSH approach (Sec. IV,A) by Kleyn et al., (1980a, b, 1982). This interpretation is based on the O–O vibration when the system undergoes a transition from the covalent state (C): Cs, K + O_2 to the ionic state (I): Cs^+, $K^+ + O_2^-$ and then back from I to C. To see that, let for simplicity the potential energy surfaces and coupling be described by

$$\langle C|H_{el}|C\rangle = W_C(r), \tag{62a}$$

$$\langle I|H_{el}|I\rangle = W_I(r) + \mathcal{I}_{Alk} - 1/R, \tag{62b}$$

$$\langle I|H_{el}|C\rangle = \langle C|H_{el}|I\rangle = A\exp(-\Lambda R), \tag{62c}$$

where anisotropy (γ-dependence) is disregarded; $W_C(r)$ is the $O_2(X^3\Sigma_g^-)$ potential energy curve, $W_I(r)$ the $O_2^-(X^2\Pi_g)$ curve placed relative to $W_C(r)$ and $\mathcal{I}_{Alk}$ the ionization potential of K or Cs. Clearly, the *crossing seam* is given by

$$R_x(r) = [W_I(r) + \mathcal{I}_{Alk} - W_C(r)]^{-1}. \tag{63}$$

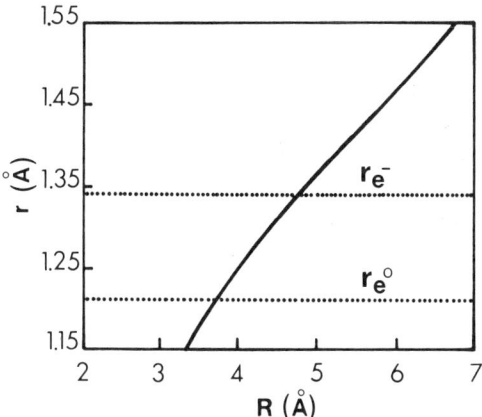

FIG. 6. Location of the crossing seam between the covalent Cs + $O_2(X^3\Sigma_g^+)$ and ionic Cs$^+$ + $O_2^-(X^2\Pi_g)$ potential energy surfaces used in the TSH calculations of Kleyn et al. (1982). Dotted lines indicate the equilibrium bond distance in O_2 and O_2^-.

For the considered collisional systems R_x is an increasing function of bond distance r and the equilibrium distance of O_2^- ($r_e^- \cong 2.53a_0$) is larger than that of O_2 ($r_e^0 \cong 2.28a_0$) (see, e.g., Fig. 6). As the collision partners approach and reach the crossing seam $R_x(r_e^0)$ a hop from the covalent to the ionic energy surface may take place. When this occurs, the representative point of the system finds itself at W_I (r_e^0), i.e., on the repulsive wall of the O_2^- potential; the molecule thence stretches. As the collision partners recede they reach the crossing seam at $R'_x(r > r_e^0) > R_x(r_e^0)$ (Fig. 6). Owing to the exponential behavior of the I-C coupling [Eq. (62c), Janev, 1971] the transition probability at $R'_x(r)$ is much weaker than its value at $R_x(r_e^0)$. Hence, hopping trajectories for which the time lag (t_{coll}) between the passages at $R_x(r_e^0)$ and $R'_x(r)$ differs from an integer number of O_2^- vibration periods τ_{vib} hardly undergo the I-C transition at $R'_x(r)$ thereby entailing a leakage from the neutral K, Cs + O_2 channel. Varying the impact parameter (which is related to the scattering angle) causes a variation of the mentioned time lag and provides a way of tuning through the condition for reneutralization. This was shown to be the classical origin of the mentioned new structures in the differential cross sections (Kleyn et al., 1980a, b, 1982).

Using the approach described in Sec. IV,B,1, Spalburg et al. (1983) have investigated what would be the quantal analogue of the aforementioned classical phenomenon. Their calculation is based on electronic potential energy surfaces and coupling terms of the form given in Eqs. (62) except at small distances R where an exponential function describes the relevant repulsive walls of the potentials. For didactic purposes the calculation involved a vibronic basis made only of: $|C, v = 0\rangle$ and $|I, 0 \leq v' \leq 3\rangle$. The transition probabilities $C, v = 0 \to I, v' \leq 2$ as functions of impact parameter displayed a system of high and low frequency oscillations (for the state I, $v' = 3$, the low frequency oscillation was absent). The low frequency oscillation persists in the convoluted sum of these transition probabilities as shown in Fig. 7. The quantal analogue of the classical vibration of the molecule in the electronic ionic state has been identified by tracing the origin of this beating effect (Spalburg et al., 1983). As shown schematically in Fig. 8, essentially three waves contribute to the observed phenomenon. The large

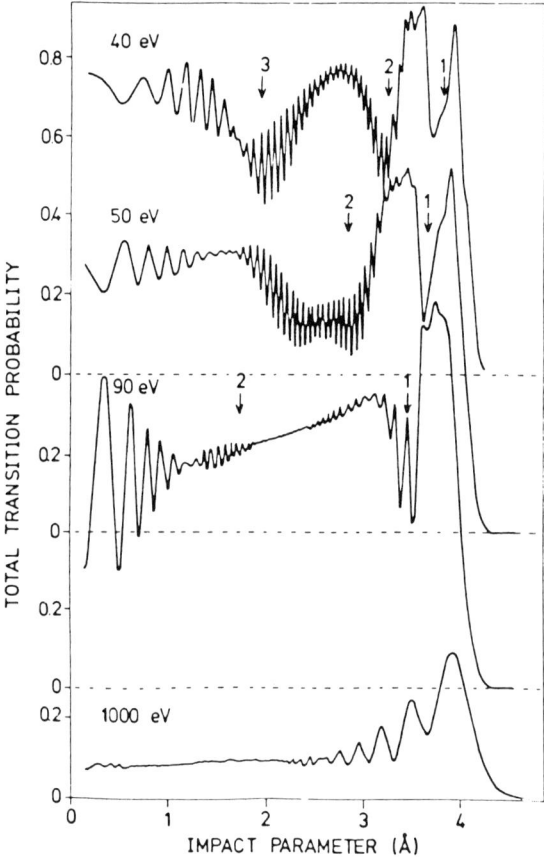

FIG. 7. Impact parameter dependence of the convoluted sum (over final vibrational states) of transition probabilities for ion-pair formation in a prototype collisional system simulating $Cs + O_2$ (Spalburg et al., 1983). The corresponding calculations involve one covalent ($C, v = 0$) and four ionic ($I, v' = 0,1,2,3$) vibronic states. The arrows indicate the impact parameters for which the O_2^- ion can perform complete vibrations during the collision. (As discussed in Sec.IV,B,2, at the highest energy the O_2^- ion can barely vibrate during the collision.)

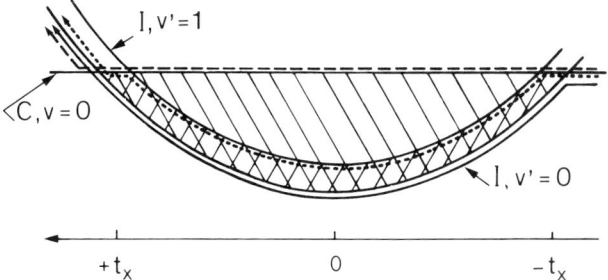

FIG. 8. Scheme of the three paths relevant to he discussion of the interference phenomena resulting in the quantum beats seen in Fig. 7 (from Spalburg et al., 1983).

phase difference between the covalent path and either of the two ionic paths gives rise to a Stueckelberg-type interference pattern (Olson and Smith, 1971; Nikitin and Umanskii, 1984) which manifests itself as the mentioned high frequency oscillations. These rapid oscillations are more or less smeared out in Fig. 7. The smaller phase difference that develops between two adjacent ionic states v', $v' + 1$ (Fig. 8) causes a *modulation* of the Stueckelberg oscillation pattern. It is immediately seen (Spalburg et al., 1983) that the latter phase difference is given by

$$\omega t_{coll} = 2\pi t_{coll}/\tau_{vib}, \tag{64}$$

where t_{coll} is the time lag between the entrance and exit crossing points (Fig. 8) and ω is the energy separation between v' and $v' + 1$ levels of the O_2^- molecule. Equation (64) thus entails the same tuning condition as that of the classical description of Kleyn et al. (1980a, b, 1982), i.e., $t_{coll}/\tau_{vib} = N$ integer; this is clearly seen in Fig. 7. Detailed comparisons between experiment and theoretical calculations involving larger vibronic basis sets have been reported by Spalburg et al. (1985b) (see also Becker and Saxon, 1981).

B. Vibronic Effects in Ion-Molecule Charge Transfer

The ion-molecule charge exchange process

[1] $N_2^+(X^2\Sigma_g^+, v) + Ar(^1S_0) \leftrightarrow$ [2] $N_2(X^1\Sigma_g^+, v') + Ar^+(^2P_{3/2, 1/2})$

at low energy (1–100 eV) has been a subject of intensive experimental and theoretical activity (see, e.g., the review of Guyon et al., 1986 and later work by Shao et al., 1987 as well as the review of Parlant and Gislason, 1987). The particular interest in this and similar ion-molecule collision processes lies in a few pending questions as to the role of near resonance, the existence or not of curve crossings and the importance of FC factors.

The most detailed calculations of integral cross sections for the above process have been performed by Parlant and Gislason (1986, 1987). These calculations are based on an IOSA-type semiclassical close-coupling approach (Sec. IV,B,1) and make use of *ab initio* potential energy curves and interactions determined by Archirel and Levy (1986). Sample results of these calculations are shown in Fig. 9. The first salient feature of these results is that the N_2^+ $(X, v = 0)$ state does not react though it lies only 0.18 eV lower than its closest charge exchange neighbor $Ar^+(^2P_{3/2}) + N_2(X, v' = 0)$ and despite of a large FC factor ($|\langle g_{10}|g_{20}\rangle|^2 \approx 0.92$). This low reactivity of $N_2^+(X, v = 0)$ contrasts with the relatively large and comparable cross sections obtained with $N_2^+(X, v \geq 1)$ ions ($Q \approx 25 \pm 5$ Å^2). In addition to these features theory predicts a propensity for $v = v' + 1$ transitions (Spalburg and Gislason 1985). The understanding of these phenomena has progressively emerged in successive contributions. Spalburg et al. (1985a) provided a first clue to these riddles by considering an empirical *isotropic* model of the $(N_2Ar)^+$ vibronic (diabatic) interaction matrix

$$\left\langle g_{nv} \left| \left[-\frac{1}{2m}\frac{d^2}{dr^2} + \langle \Phi_n|H_{el}|\Phi_{n'}\rangle \right] \right| g_{n'v'} \right\rangle = \mathcal{E}_{nv}\delta_{nn'}\delta_{vv'} + \mathcal{H}_{nv,n'v'}(R), \quad (65a)$$

$$\mathcal{H}_{nv,n'v'}(R) = (1 - \delta_{nn'})h_{nn'}(R)\langle g_{nv}|g_{n'v'}\rangle, \quad (65b)$$

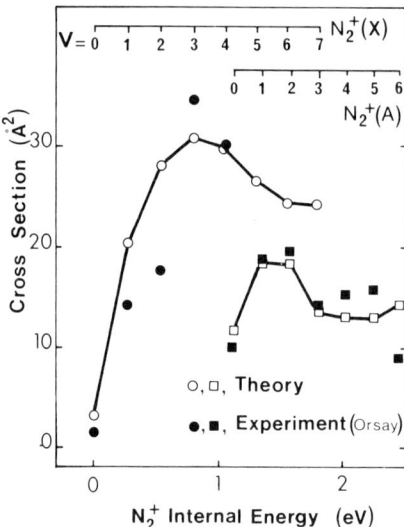

FIG. 9. Dependence on the initial vibronic state of the integral cross section for charge transfer in the $N_2^+(v) + Ar$ collision at a center of mass collision energy of 20 eV: solid symbols are experimental data points of Govers et al. (1984). The solid line joining open symbols corresponds to theoretical results of Parlant and Gislason (1986).

where $h_{nn'}(R)$ assumes the exponential form given in Eq. (51). They pointed out that *diagonalization of this matrix reveals $1v - 2v'$, $v \neq v'$ avoided level crossings* whose characteristics have to do with the just-mentioned features. Actually, Eq. (65) lacks the anisotropic charge quadrupole and charge-induced dipole interactions which were invoked in a Bauer-Fischer-Gilmore-type interpretation of experimental data in a former work by Govers et al. (1984). That work provided another clue to the understanding of the N_2^+ + Ar charge transfer dynamics. Indeed, it showed that, notwithstanding the existence of a crossing in the colinear arrangement ($\gamma = 0$) at $R_x \approx 6a_0$ between the $\mathscr{E}_{10} + \mathscr{H}_{10,10}$ and $\mathscr{E}_{20} + \mathscr{H}_{20,20}$ levels the corresponding Landau-Zener (1932) transition probability in a double passage though R_x is negligible. *Similar results were obtained for all $1v \to 2v'$ transitions with $v \equiv v'$.* Using their own *ab initio* results as inputs to the Landau-Zener formula Archirel and Levy (1986) essentially confirmed the latter conclusions and further examined a few cases of Demkov (1964) type transitions between noncrossing vibronic levels. Again, transitions with $v \equiv v'$ (corresponding to asymptotic resonance energy defects of $\cong 0.2$ eV) were found to have negligible average transition probabilities (a few percent say) at $E_{cm} \leq 20$ eV (see, e.g., Guyon et al., 1986; Sidis, 1989). These small probabilities thereby indicate that at the considered low energies the system behaves in a near *adiabatic* fashion as concerns $1v \to 2v'$ transitions with $v = v'$. Thence, the appropriate vibronic energy levels corresponding to such an adiabatic behavior should be those which result from diagonalizations of 2×2 matrices: $\mathscr{E}_{nv}\delta_{nn'}\delta_{vv'} + \mathscr{H}_{nv,n'v'}(v \equiv v')$. Since the $N_2^+(X^2\Sigma_g^+)$ and $N_2(X^1\Sigma_g^+)$ equilibrium distances and vibrational frequencies are but slightly different these 2×2 matrices may be *approximately* diagonalized by a common transformation, e.g., that diagonalizing the electronic $\langle \Phi_n | H_{el} | \Phi_{n'} \rangle$ interaction matrix (Nikitin et al., 1987; Sidis, 1989). The new vibronic basis may thus be built as

$$\xi_{\overline{1v}} = c(R, \gamma)\Phi_1 g_{1v} + s(R, \gamma)\Phi_2 g_{2v}, \tag{66a}$$

$$\xi_{\overline{2v'}} = -s(R, \gamma)\Phi_1 g_{1v'} + c(R, \gamma)\Phi_2 g_{2v'}, \tag{66b}$$

$$c(R \to \infty, \forall \gamma) \to 1 \quad \text{and} \quad s(R \to \infty, \forall \gamma) \to 0.$$

The resulting vibronic energy levels $\overline{1v}$, $\overline{2v'}$ depicted schematically in Fig. 10 provide the last clue to the understanding of the N_2^+ + Ar charge transfer process at low energy. Indeed, this figure shows crossings between the $\overline{1v}$ and $\overline{2v'}$ energy levels having $v > v'$. The corresponding couplings $\mathscr{H}_{\overline{1v},\overline{2v'}}$, are given by (Nikitin et al., 1987; Sidis, 1989)

$$\mathscr{H}_{\overline{1v},\overline{2v'}} \cong (c^2 \langle g_{1v} | g_{2v'} \rangle - s^2 \langle g_{2v} | g_{1v'} \rangle)h_{12}, \quad v > v'. \tag{67}$$

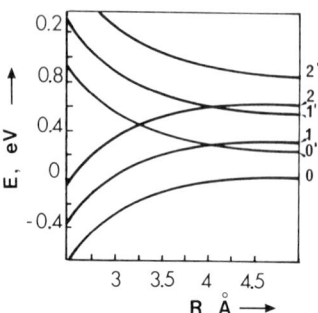

FIG. 10. Schematic diagram illustrating the behaviour of the two sets of vibronic levels (1v, 2v') obtained after diagonalization of the electronic $\langle \Phi_n | H_{el} | \Phi_{n'} \rangle$ matrix in a system like $(Ar-N_2)^+$. $v + 1 \to v'$ transitions are seen to be possible at curve crossings of vibronic states. The low reactivity of $N_2^+(v = 0) + Ar$ at low energy is immediately understood from such a diagram (the presented sketch is inspired from the results of Spalburg et al., 1985a, and Nikitin et al., 1987).

With the $\overline{1v}, \overline{2v'}$ vibronic sets at hand, the propensity for $v = v' + 1$ transitions stems naturally from Eq.(67) since among $v \neq v'$ states the $\langle g_{1v} | g_{2v'} \rangle$FC overlaps are largest for $|v - v'| = 1$. In addition since the characteristics of the $\overline{1v}, \overline{2v'}$ ($v = v' + 1$) pairs of levels at their crossings and the locations thereof are but slightly dependent on v' one easily understands why $N_2^+(X^2\Sigma_g^+, 1 \leq v \leq 3) + Ar$ collisions yield comparable charge transfer cross sections at low energies. More detailed discussions concerning the role of spin-orbit effects and the importance of the $N_2^+(A^2\Pi) + Ar$ channels may be found in the works of Spalburg et al., (1985a), Spalburg and Gislason (1985), Parlant and Gislason (1986, 1987), and Nikitin et al., (1987).

Similar calculations have been reported by Parlant et al., (1987, 1988) on the following processes:

$$Ar^+(^2P_{3/2}) + CO(X^1\Sigma_g^+, v = 0) \to Ar + CO^+(X, A, v')$$
$$\to Ar^+(^2P_{1/2}) + CO(X, v')$$
$$\to Ar^+(^2P_{3/2}) + CO(X, v').$$

The most interesting feature of this collisional system is that the largest FC overlaps [involved in Eq.(53)] for the $CO(X, v = 0) \to CO^+(X, v')$ transition correspond to the lowest two v' states (0.9635, 0.0365 for $v' = 0,1$, respectively) whereas the smallest resonance energy defects for charge transfer occur for $CO^+(X, v' = 7)$ ($\Delta E = 0.046$ eV). Experiments by Marx et al., (1983) at thermal energies suggested that only $CO^+(X, v' = 4)$ ions are produced by charge transfer whereas Lin et al., (1985) and Hamilton et al., (1985) measured a distribution of CO^+ ions with $0 \leq v' \leq 6$ peaking at $v' = 5$. The

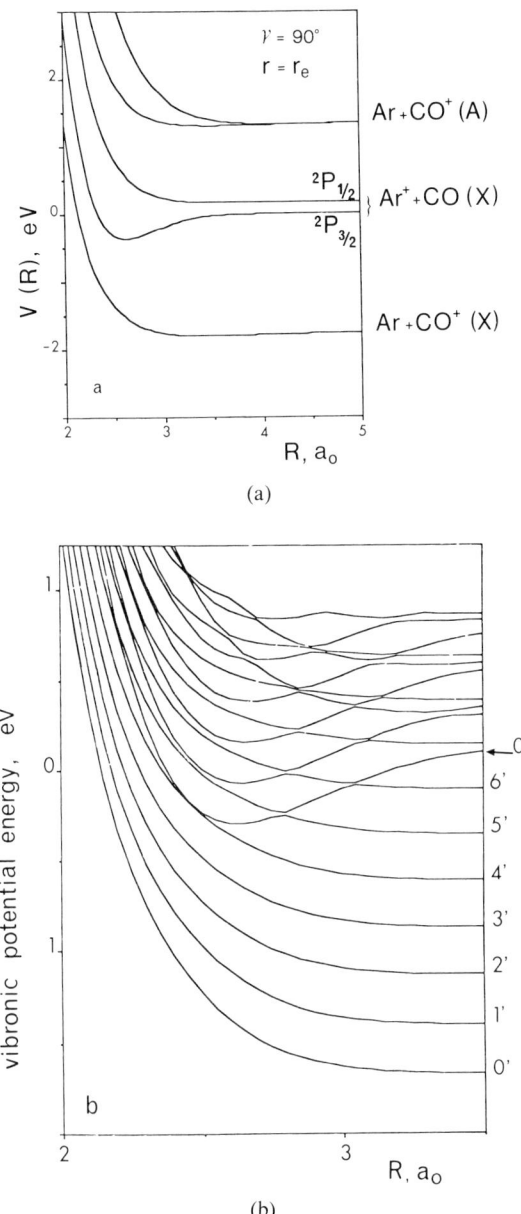

FIG. 11. Sample cuts of the potential energy hypersurfaces used by Gislason et al. (1987) in their study of charge transfer in the Ar$^+$CO collision at low energy. (a) Adiabatic electronic energy curves for $\gamma = 90°$ and $r = r_e$(CO) $= 2.132a_0$. (b) Expanded view of the low-lying adiabatic vibronic levels for $\gamma = 90°$.

total charge transfer cross section was found to be much smaller than that of $Ar^+ + N_2$ in the eV energy range (Dotan and Lindinger, 1982). The calculations by Gislason et al., (1987) were performed at a relative energy of 2 eV using the formalism described in Sec. IV,B,1. The basis sets involved the four electronic states implied in the above reactions each associated with fifteen vibrational states. The relevant shapes of the diabatic electronic potential energy curves $\langle \Phi_n | H_{el} | \Phi_n \rangle$ were determined in an *ad hoc* manner whereas the electronic coupling matrix elements $\langle \Phi_n | H_{el} | \Phi_{n'} \rangle$ have been obtained by similar *ab initio* means as those used by Archirel and Levy (1986). Though the comparison between that theory and experiment is still demanding, its conclusions are quite instructive. Indeed, it stresses both the crucial importance of the electronic state correlating with $Ar + CO^+(A^2\Pi)$ and the role of the anisotropy of the relevant interactions. One of the related effects is illustrated in Fig. 11 which shows some *adiabatic vibronic levels* obtained by diagonalizing the $\mathscr{E}_{nv}\delta_{nn'}\delta_{vv'} + \mathscr{H}_{nv,n'v'}$ matrix in the perpendicular approach. Close to $\gamma = 100°$ the coupling between the $Ar^+(^2P_J) + CO(X)$ and $Ar + CO^+(A)$ electronic diabatic states is largest. This coupling is seen in Fig. 11 to give rise to adiabatic vibronic energy levels displaying potential wells and undergoing a series of avoided crossings. This effect thereby qualitatively suggests a plausible Bauer-Fischer-Gilmore type of mechanism for the population of $Ar + CO^+(X; 4 \leq v' \leq 6)$ charge transfer states at low energy. Another effect of the admixture of an $Ar + CO^+(A)$ component into the *adiabatic* state correlating with $Ar^+ + CO(X)$ is a dynamic change of the CO vibration characteristics owing to the different equilibrium distances and vibrational frequencies of the $CO^{(+)}(X)$ and $CO^+(A)$ states. Such a change entails vibrational excitation and of course modifies the FC overlap [Eq.(53)] according to the R, γ geometry of the $(ArCO)^+$ molecule along the trajectory followed by the system. Detailed comparison of the discussed calculations with experiments still awaits a study of the role of the *r* dependence of the interactions. Moreover the possibility of extending the theory at energies of fractions of an eV where formation of a long lived $(ArCO)^+$ complex has been invoked (Gislason et al., 1987) will certainly require consideration of a quantum mechanical approach as that discussed in Sec. III.

C. Dissociative Charge Exchange

An important vibronic phenomenon in collisions of atomic and molecular species is that involving a change of the initial electronic state of the system and the simultaneous *breakup of the molecular bond*. A prototype of such a phenomenon is provided by the dissociative charge exchange (DCE) process specified by reaction (iii) in the introductory part of Sec. VI. One may distinguish three types of DCE processes. In direct-DCE the electronic potential energy curve of the product molecule is repulsive and leads to *spontaneous dissociation*. On the the other hand, for indirect processes the potential energy curve of the product molecule is attractive. In such cases

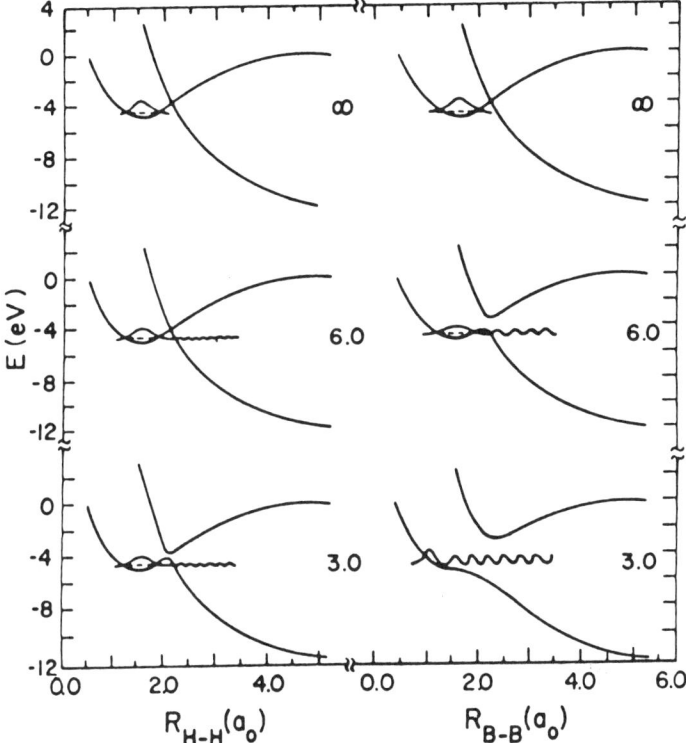

FIG. 12. Cuts of adiabatic electronic potential energy hypersurfaces involved in direct dissociative charge exchange (DCE) processes or more generally in so-called collision induced predissociation (CIP) phenomena. The figure (borrowed from Preston and Tully, 1978) views such processes as resulting from the tunneling of the system through the barrier formed at finite distances of approach on the lowest adiabatic energy surface near an avoided crossing seam. The left side of the figure corresponds to the actual case of the $He^+ + H_2$ collisional system in collinear geometry. (The numbers indicated to the right of each curve correspond to the distance of the He center to the nearest H center of the molecule). The right side of the figure illustrates a strong coupling case for a hypothetical $A^{(+)} + B_2$ system. Alternatively, DCE or CIP may be viewed as the decay at finite distances of approach of the initially bound vibrational state (associated with the underlying attractive diabatic curve) in the dissociation continuum (associated with the underlying repulsive diabatic curve)—see, e.g., Sidis and Courbin-Gaussorgues (1987).

dissociation occurs either via a *radiative transition to a repulsive electronic state* or via *predissociation*. All three DCE processes have been studied both theoretically and experimentally in collisions of H_2^+ with Mg and keV energies (Sidis and De Bruijn, 1984; De Bruijn et al., 1984). For such high energies the theoretical approach presented in Sec. IV,B,2 was actually applied (see, e.g., Fig. 5). The relevant exchange and polarization matrix elements $\langle \Phi_n | H_{el} | \Phi_{n'} \rangle$ [involved in Eqs.(44)–(45)] were described using formulae appropriate to distant collisions (Sidis and De Bruijn, 1984; see also Evseev et al., 1980). As already pointed out in Sec.IV,B,2 the simplifying feature of vibronic transitions at high energies is that they may be viewed as occurring in two steps. An electronic transition takes place first; then after a comparatively long time, the molecule dissociates according to one of the above mentioned breakup processes. On the contrary, at low energies (≤ 100eV) the molecule may vibrate and thence dissociate *during* the very collision event. Up to the early eighties the only general method of treatment of such phenomena at low energies was the quasiclassical TSH approach (see Kuntz and Whitton, 1976 and Kuntz et al., 1979). Yet Preston et al., (1978) proposed a perturbation model (based on a IOS type approximation) where DCE is viewed as a tunneling process through the barrier appearing on the lowest member of an avoided crossing between two potential energy surfaces (Fig. 12). This model stands somewhere in between the TSH approach (Sec. IV,A) and the close-coupling vibronic method of Sec. IV,B,1. Actually, the whole procedure presented in the latter section may also be applied to dissociative vibronic processes save for one important difference: the expansion in Eq.(36) now contains both discrete and continuum vibrational states. For instance, considering two electronic states, Eq.(36) becomes

$$X(\{\boldsymbol{\rho}_i\}, \mathbf{R}(t), r; \hat{\mathbf{r}}) = \Phi_1(\{\boldsymbol{\rho}_i\}; R, r, \gamma) \sum_{v'} B_{1v'}(\mathbf{R}(t); \hat{\mathbf{r}}) \frac{g_{1v'}(r)}{r}$$
$$+ \Phi_2(\{\boldsymbol{\rho}_i\}; R, r, \gamma) \int d\varepsilon \, B_2(\varepsilon, \mathbf{R}(t); \hat{\mathbf{r}}) \frac{g_2(\varepsilon, r)}{r}. \quad (68)$$

FIG. 13. Theoretical prediction of the dependence of direct dissociative charge exchange spectra upon orientation θ_r of the diatom axis with respect to the incident direction (Sidis and Courbin-Gaussorgues, 1987). The spectra shown are averaged over the azimuthal angle ϕ_r. They are obtained for a collision energy of 10 eV (laboratory) and a model case having the same characteristics as the H_2^+ + Mg system at $R \to \infty$. The corresponding calculations assume that the discrete vibronic energy is constant and that the bound-continuum interaction $\mathcal{H}_{1v,2}(\varepsilon; R, \gamma)$ behaves exponentially with R and involves an anisotrophy factor behaving as $\cos \gamma$. (p is the impact parameter and ε the kinetic energy of the fragments after charge transfer.) Note the dip occurring at the resonance energy $\varepsilon_r = \mathscr{E}_{1v}$ in the rising front of the $\theta_r = 0$ spectrum.

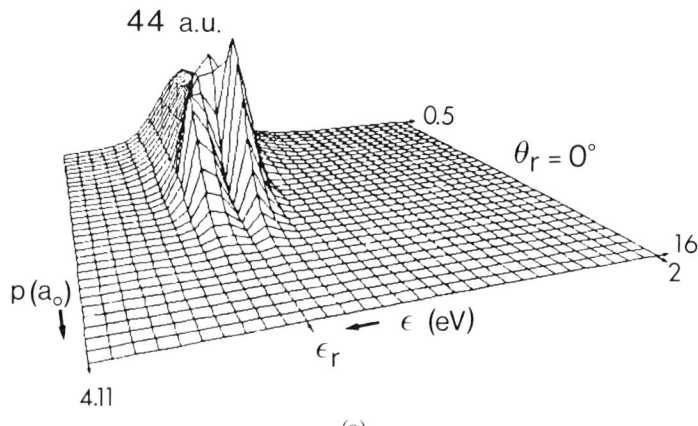

(a)

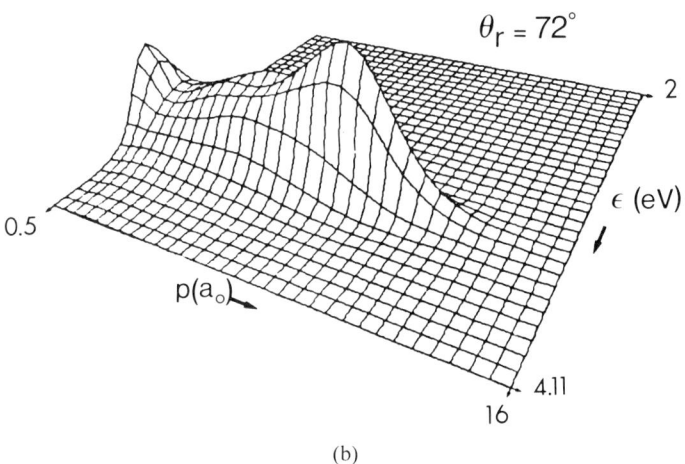

(b)

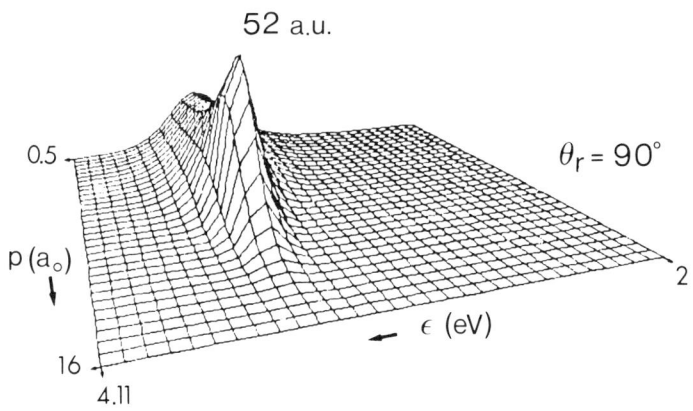

(c)

As a result one is faced with the difficulty of determining the infinite continuous set of coefficients $B_2(\varepsilon, \mathbf{R}; \hat{\mathbf{r}})$. Sidis and Courbin-Gaussorgues (1987) and Sidis et al., (1988a) have tackled this problem for the direct dissociative near resonant charge exchange process in model systems having similar characteristics as those of the $H_2^+ + Mg$ collisional system. It was pointed out by these authors that great similarities exist between DCE processes and Penning ionization (see, e.g., Pesnelle and Runge, 1983 and references therein) thereby suggesting the use of a similar method of treatment based on the so-called *complex local potential approach*. As recalled by Sidis et al., (1988a) the method first consists of replacing the actual $B_2(\varepsilon, \mathbf{R}(t); \hat{\mathbf{r}})$ amplitudes by zero-order estimates:

$$B_2^0(\varepsilon, \mathbf{R}(t); \hat{\mathbf{r}}) \approx -i\pi \exp\left(-i \int^t \varepsilon \, dt'\right) \sum_{v'} B_{1v'}(\mathbf{R}(t); \hat{\mathbf{r}}) \delta(\varepsilon - \mathscr{E}_{1v'}) \quad (69)$$

which when inserted in the relevant coupled equations for the discrete amplitudes ($B_{1v'}$) give rise to *complex* vibronic energies ($\mathscr{E}_{1v} + \mathscr{H}_{1v,1v}$) and coupling terms ($\mathscr{H}_{1v,1v'}$; $v \neq v'$). Emergence of imaginary parts in the interactions expresses both the decay of the discrete states and their coupling via the continuum (i.e., they represent both the breakup of the molecule and the partial recapture of the corresponding fragments into bound states, see Sidis et al., 1988a, b). The mentioned procedure then enables one to determine approximate probability amplitudes for the discrete states (B_{1v}) which then serve to extract first order estimates of the sought $B_2(\varepsilon, \mathbf{R}(t); \hat{\mathbf{r}})$ coefficients. The energy and angular distributions of the dissociation products are then obtained from the probability densities $|B_2(\varepsilon, \mathbf{R}(t \to \infty); \hat{\mathbf{r}}) \mathscr{R}(\hat{\mathbf{r}})|^2$, where $\mathscr{R}(\hat{\mathbf{r}})$ represents the initial rotation state of the molecule. A detailed study of the effects of these spectra of the discrete-continuum energy differences ($\mathscr{E}_{1v} - \varepsilon$) and coupling strengths as well as collision energy, impact parameter, molecular orientation and initial vibration state has been reported by Sidis and Courbin-Gaussorgues (1987) (see also Sidis et al., 1988a). Samples of their results are shown in Figs. 13–15. Particularly spectacular are the effects of the molecule orientation (and thus that of ejection of the fragments) with respect to the direction of the incident velocity vector when the interaction $\mathscr{H}_{1v,2}(\varepsilon; R, \gamma)$ is an odd function of $\cos\gamma$ (Figs. 13–15). The dip appearing in Fig. 13 in the onset range of the DCE spectrum near the energy of the entrance vibronic channel ($\mathscr{E}_{1v}$) for $\theta_r = 0$ is characteristic of such interactions; it was shown to result from cancelations between the contributions of the incoming and outgoing paths of the trajectory (Sidis and Courbin-Gaussorgues 1987, Sidis 1989). Another remarkable vibronic phenomenon appears in the impact parameter dependence of the DCE spectra when the discrete energy level $\mathscr{E}_{1v} + \mathscr{H}_{1v,1v}$ is a function of the relative

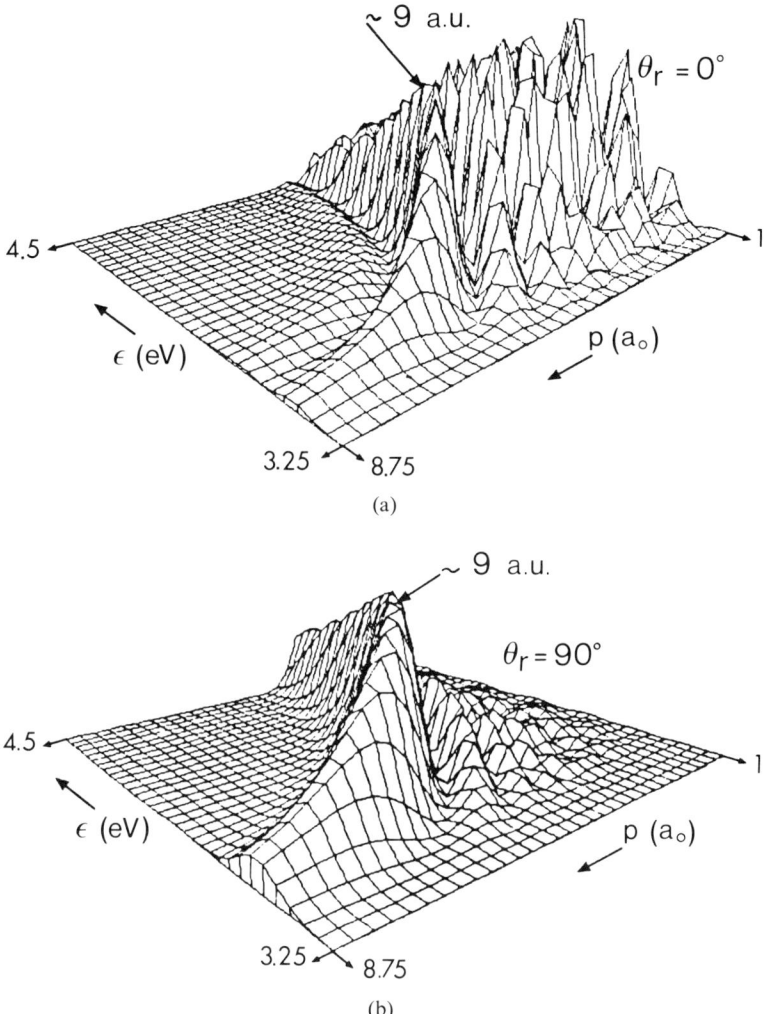

FIG. 14. Same as Fig. 13 except that the discrete energy level $(\mathcal{E}_{1v} + \mathcal{H}_{1v,1v})$ behaves as $\mathcal{E}_{1v} + 4\exp(-0.5R)/3R$ and that the bound continuum interaction is weaker by a factor 10 (from Sidis and Courbin-Gaussorgues, 1987).

distance R. The DCE spectra (Fig. 14) are found to reflect this R dependence and, when the $\mathcal{H}_{1v,2}(\varepsilon; R, \gamma)$ coupling is weak enough structures appear which result from interferences between waves associated with the breakup of the molecule either when the collision partners approach or when they recede. As shown in Figs. 14, 15 the general appearance of these structures dramatically depends on the angle of ejection of the DCE fragments.

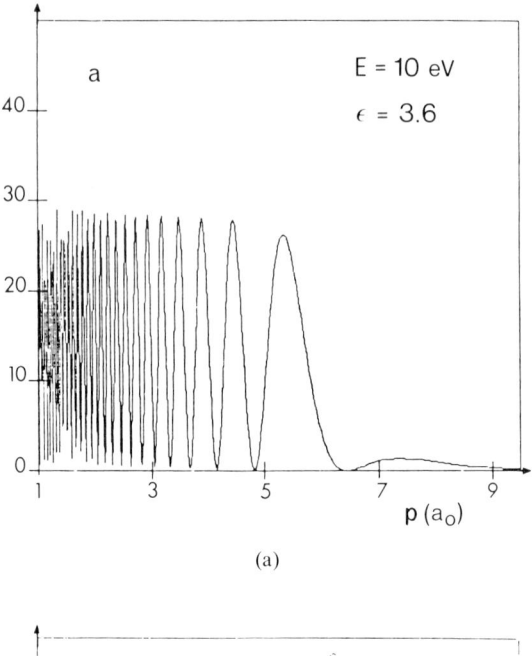

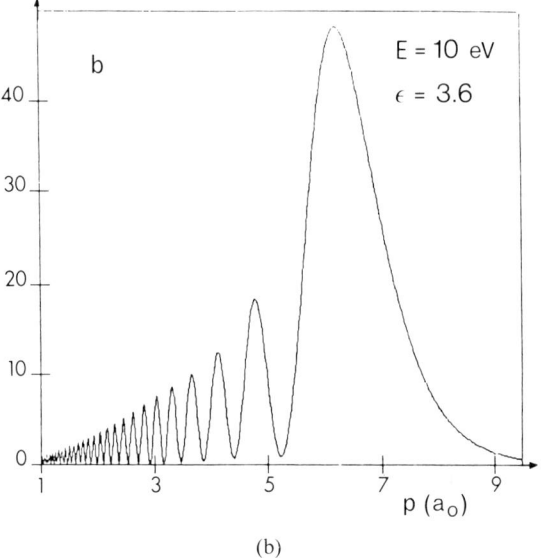

FIG. 15. Slices of dissociative charge exchange spectra obtaned for $\varepsilon = 3.6$ eV in the same conditions as those of Fig. 14: (a) $\theta_r = 0$, $\forall \phi_r$, (b) $\theta_r = 90°$, $\phi_r = 0$. The figures show details of the impact parameter dependence of the probability densities $|B_2(\varepsilon, \mathbf{R}(t \to \infty); \hat{\mathbf{r}})\mathcal{R}(\hat{\mathbf{r}})|^2$. The various features of the presented plots have been discussed by Sidis and Courbin-Gaussorgues (1987) using a weak coupling approach similar to that entailing Eq. (56) in text.

VII. Concluding Remarks

The field of atom-molecule collisions has lately witnessed persevering efforts to design manageable quantal and semiclassical theories of vibronic transition processes. Undeniably, this endeavor has been stimulated by the concomitant emergence of a new generation of experiments concerned with the acquisition of state specific dynamical information and angular distributions of the scattered products (see, e.g., Koyano *et al.*, 1986; Guyon *et al.*, 1986; Noll and Toennies, 1986: Futrell 1987; Linder, 1988; De Bruijn *et al.*, 1984; Dhuicq and Sidis, 1987). These combined theoretical and experimental advances are progressively remodelling our conceptual views of atom-molecule vibronic phenomena.

Many constraining difficulties of the quasiclassical TSH approach (Sec. IV,A) simply do not exist in the quantal and semiclassical formulations of the theory. The celebrated level-crossing model is still a quite vivid notion in our qualitative understanding of transitions between vibronic states; yet it is advantageously complemented with the as famous notion of near resonance. Despite their high interpretative value, explicit reference to these notions is not a prerequisite of close-coupling calculations (except perhaps for the purpose of enforcing the physico-chemical content of certain expansion basis sets with a view to achieve a reduction of computational effort).

Quantal phases play an eminent role in different stages of the discussed theory: (i) they determine the conditions under which different degrees of freedom may be kept frozen or play an active role during the collision; (ii) via interference and beating phenomena they reveal the underlying classical paths available to the quantal system; (iii) they contain the proper information concerning the actual scattering (thereby overriding the constraint imposed by the choice of a common trajectory in the semiclassical approach); and (iv) in certain sudden collision conditions, they induce ro-vibrational transitions.

Investigation of the various conditions under which the predictions of the Franck-Condon principle apply to vibrational distributions of the product molecule lead to the conclusion that the necessary requirements are quite stringent and are therefore seldom met in actual systems. These FC predictions should only be used as an asymptotic reference.

Further progress in the considered field depends of course upon the availability of the relevant $\langle \Phi_n | H_{el} | \Phi_{n'} \rangle$ interactions for both broad and dense grids of the variables R, r, γ; so far, too few quantal and semiclassical calculations have taken full consideration of the latter two. Moreover, standardization of such calculations (as is presently the case of the so familiar TSH approach) require that the repetitive (i.e., 10^3–10^4 times) resolution of several dozen coupled differential equations becomes a routine matter. Rapid progress of modern computer technology is quite promising in this respect.

Satisfactory results have been extracted from quantal and semi-classical calculations of integral cross sections for vibronic transitions. Complementary work is still needed to obtain theoretical differential cross sections which, as everyone knows, are sensitive to finer details of the atom-molecule interactions. There are good grounds to expect the semi-classical vibronic treatment to be a valid approach for collision velocities $u \geq 5.10^{-3}$ a.u. (i.e., $E \geq 1$ eV/amu). As discussed above, for $E < 100$ eV/amu molecular vibration actually occurs during the collision event; it is this very feature which makes vibronic processes particularly interesting. Still, in the 0.1 eV/amu energy range and below a quantal approach appears to be necessary. This is particularly obvious in cases when the formation of a long lived complex is invoked. It is felt that the quantum mechanical IOSA approach is likely to provide new quantitative information and stimulating ideas in the discussion of "warm" gas-phase physico-chemistry.

Note added in proof: Attention is also called to a new method (Gauyacq and Sidis, 1989) based on a coupled wavepacket description of vibronic transitions that is particularly suited to the study of dissociative collision processes.

REFERENCES

Archirel, P., and Levy, B. (1986). *Chem. Phys.* **106**, 51.
Arthurs, A. M., and Dalgarno (1960). *Proc. R. Soc. A* **256**, 540.
Baer, M. (1976). *Chem. Phys.* **15**, 49.
Baer, M. (1983). In "*Topics in Current Physics*, Vol. 3: Molecular Collision Dynamics" (J.M. Bowman, ed.), p. 117. Springer-Verlag, Berlin.
Baer, M., Drolshagen, G., and Toennies, J.P. (1980). *J. Chem. Phys.* **73**, 1690.
Baer, M., Niedner, G., and Toennies J.P. (1988). *J. Chem. Phys.* **88**, 1461.
Barat, M., and Lichten, W., (1972). *Phys. Rev. A* **6**, 211.
Bates, D. R., and Holt, A. R. (1966). *Proc. R. Soc. London, Ser. A* **292**, 168.
Bates, D. R., and Crothers, D. S. F. (1970). *Proc. R. Soc. London, Ser, A* **315**, 465.
Bates, D. R., and Reid, R. H. G. (1969). *Proc. R. Soc. London, Ser. A* **310**, 1.
Bauer, E., Fischer, E. R., and Gilmore, F. R. (1969). *J. Chem. Phys.* **51**, 4173.
Becker, C. H. (1982). *J. Chem. Phys.* **76**, 5928.
Becker, C. H., and Saxon, R. P. (1981). *J. Chem. Phys.* **75**, 4899.
Bellum, J. C., and McGuire, P. (1983). *J. Chem. Phys.* **75**, 765.
Bernstein, R. B. (1979). "Atom-Molecule Collision Theory". Plenum, New York.
Berson, I. Ja. (1968). *Latv. PSR Zinat. Akad. Vestis. Fiz. Teh. Zinat. Ser. N. 4*, 47.
Bjerre, A., and Nikitin, E. E. (1967). *Chem. Phys. Lett.* **1**, 179.
Blais, N., and Truhlar, D. G. (1983). *J. Chem. Phys.* **79**, 1334.

Born, M., and Oppenheimer, J. R. (1927). *Ann. Phys.* **84**, 457.
Browne, J. C., (1971). *Adv. At. Mol. Phys.* **7**, 47.
Campbell, F. M., Browning, R., and Latimer, C. J. (1980). *J. Phys. B* **13**, 4257.
Champion, R. L., Doverspike, L. D., Rich, W. G., and Bobbio S. M., (1970). *Phys. Rev. A* **2**, 2327.
Chapman, S. (1985). *J. Chem. Phys.* **82**, 4033.
Child, M. S., (1973). *Farad. Disc. Chem. Soc.* **55**, 30.
Child, M. S., (1974). "*Molecular Collision Theory*". Academic Press, London.
Child, M. S., and Baer, M. (1981). *J. Chem. Phys.* **74**, 2832.
Cimiraglia, R., Malrieu, J. P., Persico, M., and Spiegelmann, F. (1985). *J. Phys. B* **18**, 3073.
Cole, S. K., and De Pristo, A. E. (1986). *J. Chem. Phys.* **85**, 1389.
Collins, F. S., and Cross, R. J. (1976). *J. Chem. Phys.* **65**, 644.
Courbin-Gaussorgues, C., Sidis, V., and Vaaben, J. (1983). *J. Phys. B* **16**, 2817.
Cross, R. J. (1967). *J. Chem. Phys.* **47**, 3724.
Cross, R. J. (1969). *J. Chem. Phys.* **51**, 5163.
Crothers, D. S. F. (1981). *Adv. At. Mol. Phys.* **17**, 55.
De Bruijn, D. P., Neuteboom, J., Sidis, V., and Los, J. (1984). *Chem. Phys.* **85**, 215.
Delos, J. B., Thorson, W. R., and Knudson, S. (1972). *Phys. Rev. A* **6**, 709.
Demkov, Yu. N. (1964). *Sov. Phys. JETP* **18**, 138.
De Pristo, A. E. (1983a). *J. Chem. Phys.* **78**, 1237.
De Pristo, A. E. (1983b). *J. Chem. Phys.* **79**, 1741.
Desfrançois, G., Astruc, J. P., Barbé, R., and Schermann, J. P. (1988). *J. Chem. Phys.* **88**, 3037.
Dhuicq, D., Brenot, J. C., and Sidis, V. (1985). *J. Phys. B.* **18**, 1395.
Dhuicq, D., and Sidis, V. (1986). *J. Phys. B* **19**, 199.
Dhuicq, D., and Sidis, V. (1987). *J. Phys. B* **20**, 5089.
Dotan, I., and Lindinger, W., (1982). *J. Chem. Phys.* **76**, 4972.
Eaker, C. W. (1987). *J. Chem. Phys.* **87**, 4532.
Edmonds, A. R. (1957). "Angular Momentum in Quantum Mechanics" Princeton University, Princeton.
Eno, L., and Balint-Kurti, G. G. (1981). *J. Chem. Phys.* **75**, 690.
Evseev, A. V., Radtsig, A. A., and Smirnov, B.M. (1980). *Sov. Phys. JETP* **50**, 283.
Evseev, A. V., Radtsig, A. A., and Smirnov, B. M. (1982). *J. Phys. B* **15**, 4437.
Flannery, M. R., Cosby, P. C., and Moran, T. F. (1973). *J. Chem. Phys.* **59**, 5494.
Flannery, M. R., and McCann, K. J. (1973). *Phys. Rev. A* **8**, 2915.
Futrell, J. H. (1987). *In* "Structure Reactivity and Thermochemistry of Ions" (P. Ausloss and S. G. Lias, eds.), p. 57. NATO ASI, Reidel, Dordrecht.
Gadéa, F. X., Spiegelmann, F., Pelissier, M., and Malrieu, J. P. (1986). *J. Chem. Phys.* **84**, 4872.
Garett, B. C., and Truhlar, D. G. (1981). *In* "Theoretical Chemistry: Advances and Perspectives," Vol. 64, p. 215. Academic Press, New York.
Gaussorgues, C., Le Sech, C., Masnou-Seeuws, F., McCarroll, R., and Riera, A. (1975). *J. Phys. B* **8**, 239.
Gauyacq, J. P. (1978). *In* "Electronic and Atomic Collisions" (G. Watel, ed.), p. 431. North-Holland), Amsterdam.
Gauyacq, J. P., and Sidis, V. (1989). XVIth Int. Conf. on the Phys. of Electronic and Atomic Collisions (Book of Abstracts), New York.
Gentry, W. R. (1979). *In* "Atom-Molecule Collision Theory" (R.B. Bernstein, ed.), p. 391. Plenum, New York.
Gerber, R. B. (1976). *Chem. Phys.* **16**, 19.
Giese, C. F., and Gentry, W. R. (1974). *Phys. Rev. A* **10**, 2156.
Gislason, E. A., and Sachs, J. G. (1975). *J. Chem. Phys.* **62**, 2678.
Gislason, E. A., and Parlant, G. (1987). *Comments At. Mol. Phys.* **19**, 157.

Gislason, E. A., Parlant, G., Archirel, P., and Sizun, M. (1987). *Faraday Disc. Chem. Soc.* **84**, 1.
Govers, T. R., Guyon, P. M., Baer, T., Cole, K., Fröhlich, H., and Lavollée, M., (1984). *Chem. Phys.* **87**, 373.
Grimbert, D., Lassier-Govers, B., and Sidis, V. (1988). *Chem. Phys.* **124**, 187.
Guyer, D. R., Hüwel, L., and Leone, S. R. (1983). *J. Chem. Phys.* **79**, 1259.
Guyon, P. M., Govers, T. R., and Baer, T. (1986). *Z. Phys. D* **4**, 89.
Hamilton, C. E., Bierbaum, V. M., and Leone, S. R. (1985). *J. Chem. Phys.* **83**, 601.
Hedrick, A. F., Moran, T. F., McCann, K. J., and Flannery, M. R. (1977). *J. Chem. Phys.* **66**, 24.
Hellmann, H., and Syrkin, J. K. (1935). *Acta Phys. Chem. USSR* **2**, 433.
Herman, V., Schmidt, H., and Linder, F. (1978). *J. Phys. B* **11**, 433.
Janev, R. K. (1976). *Adv. At. Mol. Phys.* **12**, 1.
Kato, T., Tanaka, K., and Koyano, I. (1982a). *J. Chem. Phys.* **77**, 334.
Kato, T., Tanaka, K., and Koyano, I. (1982b). *J. Chem. Phys.* **77**, 834.
Kleyn, A. W., Los, J., and Gislason, E. A. (1982). *Phys. Rep.* **90**, 1.
Klomp, U. C., Spalburg, M. R., and Los, J. (1984). *Chem. Phys.* **83**, 33.
Kouri, D. J. (1979). In "Atom-Molecule Collision Theory" (R. B. Bernstein, ed.), p. 301. Plenum, New York.
Koyano, I., Tanaka, K., and Kato, T. (1986). In "Electronic and Atomic Collisions" (D. C. Lorents, W. E. Meyerhof, and J. R. Peterson, eds.), p. 524. North Holland, Amsterdam.
Krüger, H., and Schinke, R. (1977). *J. Chem. Phys.* **66**, 5087.
Kuntz, P. J. (1979). In "Atom-Molecule Collision Theory" (R. B. Bernstein, ed.), p. 79. Plenum, New York.
Kuntz, P. J., and Whitton, W. N. (1976). *Chem. Phys.* **16**, 301.
Kuntz, P. J., Kendrick, J., and Whitton, W. N. (1979). *Chem. Phys.* **38**, 147.
Kusunoki, I., and Ishikawa, T. (1985). *J. Chem. Phys.* **82**, 4991.
Landau, L. D. (1932). *Phys. Z. Sowjet Union* **2**, 46.
Lee, C. Y., and De Pristo, A. E. (1984). *J. Chem. Phys.* **80**, 1116.
Levy, B. (1981). In "Spectral Line Shapes" (B. Wende ed.), p. 615, W. de Gruyter. Berlin.
Lichten, W. (1963). *Phys. Rev.* **131**, 229.
Lin, G. H., Maier, J., and Leone, S. R. (1985). *J. Chem. Phys.* **82**, 5527.
Linder, F. (1988). In "Electronic and Atomic Collisions" (H.B. Gilbody, W. R. Newell, F. H. Read, and A. C. H. Smith, eds.), p. 287. North Holland, Amsterdam.
Lipeles, M. (1969). *J. Chem. Phys.* **51**, 1252.
Los, J., and Spalburg, M. R. (1984). In "Electronic and Atomic Collisions" (J. Eichler, I. V. Hertel, and N. Stolterfoht, eds.), p. 393. North Holland, Amsterdam.
Marx, R., Mauclaire, G., and Derai, R. (1983). *Int. J. Mass. Spectrom. Ion. Phys.* **47**, 155.
McCann, K. J., and Flannery, M. R. (1978). *J. Chem. Phys.* **69**, 5275.
Mittmann, H. U., Weise, H. P., Ding, A., and Heinglein, A. (1971). *Z. Naturf.* **26a**, 1112.
Moran, T. F., Flannery, M. R., and Cosby, P. C. (1974). *J. Chem. Phys.* **61**, 1261.
Moran, T. F., McCann, K. J., and Flannery, M. R. (1975). *J. Chem. Phys.* **63**, 3857.
Moran, T. F., McCann, K. J., Cobb, M., Borkman, R. F., and Flannery, M. R. (1981). *J. Chem. Phys.* **74**, 2325.
Mott, N. F. (1931). *Proc. Cambridge Phil. Soc. (London) A* **143**, 142.
Niedner, G., Noll, M. Toennies, J. P., and Schlier, C. (1987). *J. Chem. Phys.* **87**, 2686.
Nikitin, E. E., and Smirnov, B. M. (1978). *Sov. Phys. Usp.* **21**, 95.
Nikitin, E. E., and Umanskii, S. Ya. (1984). "Theory of Slow Atomic Collisions". Springer, Berlin.
Nikitin, E. E., Ovchinnikova, M. Ya, and Shalashilin, D. V. (1987). *Chem. Phys.* **111**, 313.
Noll, M., and Toennies, J. P. (1986). *J. Chem. Phys.* **85**, 3313.
O'Malley, T. F. (1969). *J. Chem. Phys.* **51**, 322.

O'Malley, T. F. (1971). *Adv. At. Mol. Phys.* **7**, 223.
Olson, R. E., and Smith, F. T. (1971). *Phys. Rev. A* **13**, 1607.
Olson, R. E., Smith, F. T., and Bauer, E. (1971). *Appl. Opt.* **10**, 1848.
Pack, R. T. (1972). *Chem. Phys. Lett.* **14**, 393.
Pack, R. T. (1974). *J. Chem. Phys.* **60**, 633.
Parlant, G., and Gislason, E. A. (1986). *Chem. Phys.* **101**, 227.
Parlant, G., and Gislason, E. A. (1987). *J. Chem. Phys.* **86**, 6183.
Parlant, G., and Gislason, E. A. (1988). *In* "Electronic and Atomic Collisions" (H. B. Gilbody, W. R. Newell, F. H. Read, and A. C. H. Smith, eds.), p. 357. North Holland, Amsterdam.
Pattengill, M. D. (1979). *In* "Atom-Molecule Collision Theory" (R. B. Bernstein, ed.), p. 359. Plenum, New York.
Pesnelle, A., and Runge, S. (1984). *In* "Electronic and Atomic Collisions" (J. Eichler, I. V. Hertel, and N. Stolterfoht, eds.), p. 559. Elsevier, CITY.
Preston, R. K., Thompson, D. L., and McLaughlin, D. R. (1978). *J. Chem. Phys.* **68**, 13.
Schaeffer III, H. F. (1979). *In* "Atom-Molecule Collision Theory" (R. B. Bernstein, ed.), p. 45. Plenum, New York.
Schinke, R. (1977). *Chem. Phys.* **24**, 379.
Schinke, R. (1984). *In* "Electronic and Atomic Collisions" (J. Eichler, I.V. Hertel, and N. Stolterfoht, eds.), p. 429. North Holland, Amsterdam.
Schinke, R. (1989). *In* "Collision Theory for Atoms and Molecules" (F.A. Gianturco, ed.) NATO ASI, Plenum, New York.
Schinke, R., and McGuire, (1978a). *Chem. Phys.* **28**, 129.
Schinke, R., and McGuire, (1978b). *Chem. Phys.* **31**, 391.
Schinke, R., Dupuis M., and Lester, Jr. W.A. (1980). *J. Chem. Phys.* **72**, 3909.
Secrest, D. (1975). *J. Chem. Phys.* **62**, 710.
Shao, J. D., Li, Y. G., and Ng, C. Y. (1987). *J. Chem. Phys.* **86**, 170.
Sidis, V. (1972). *J. Phys. B* **5**, 1517.
Sidis, V. (1976). *In* "The Physics of Electronic and Atomic Collisions" (J. D. Risley and R. G. Geballe, eds.), p. 295. University of Washington Press, Seattle.
Sidis, V. (1979). *In* "Quantum Theory of Chemical Reaction" (R. Daudel, A. Pullman, L. Salem, and A. Veillard, eds.), Vol. 1, p. 1. Reidel, Dordrecht.
Sidis, V. (1989). *In* "Collision Theory for Atoms and Molecules" (F. A. Gianturco, ed.). NATO ASI, Plenum, New York.
Sidis, V., and De Bruijn, D. P. (1984). *Chem. Phys.* **85**, 201.
Sidis, V., and Courbin-Gaussorgues, C. (1987). *Chem. Phys.* **111**, 285.
Sidis, V., Grimbert, D., and Courbin-Gaussorgues, C. (1988a). *In* "Electronic and Atomic Collisions" (H. B. Gilbody, W. R. Newell, F. H. Read, and A. C. H. Smith, eds.), p. 485. North Holland, Amsterdam.
Sidis, V., Grimbert, D., and Courbin-Gaussorgues, C. (1988b). *J. Phys. B.*
Smirnov, B. M. (1973). "Asymptotic Methods in the Theory of Atomic Collisions". Atomizdat, Moscow [in Russian].
Smith, F. T. (1969). *Phys. Rev.* **179**, 111.
Spalburg, M. R., and Klomp, U. C. (1982). *Comp. Phys. Comm.* (1982).
Spalburg, M. R,, Sidis, V., and Los, J. (1983). *Chem. Phys. Lett.* **96**, 14.
Spalburg, M. R., Los, J., and Gislason, E. A. (1985a). *Chem. Phys.* **94**, 327.
Spalburg, M. R., and Gislason, E. A. (1985). *Chem. Phys.* **94**, 339.
Spalburg, M. R., Vervaat, M. G. A., Kleyn, W., and Los, J. (1985b). *Chem. Phys.* **99**, 1.
Spalburg, M. R., Los, J., and Devdariani, A. Z. (1986). *Chem. Phys.* **103**, 253.
Spiegelmann, F., and Malrieu, J. P. (1984). *J. Phys. B* **17**, 1259.
Staemmler, V., and Gianturco, F. A. (1985). *Int. J. Quant. Chem.* **XXVIII**, 553.

Stine, J. R., and Muckerman, J. T. (1976). *J. Chem. Phys.* **65**, 3975.
Stine, J. R., and Muckerman, J. T. (1987). *J. Phys. Chem.* **91**, 459.
Takayanagi, K. (1963). *Prog. Theor. Phys. Suppl.* **25**, 1.
Tsien, T .P., and Pack, R. T. (1970). *Chem. Phys. Lett.* **6**, 54.
Tsien, T. P., and Pack, R. T. (1971). *Chem. Phys. Lett.* **8**, 579.
Tsien, T. P., Parker, G. A., and Pack, R. T. (1973). *J. Chem. Phys.* **59**, 5373.
Tully, J. C., and Preston, R. K. (1971). *J. Chem. Phys.* **55**, 562.
Tully, J. C. (1976). *In* "Dynamics of Molecular Collisions" (W.H. Miller, ed.), p. 217. Plenum, New York.
Truhlar, D. G., and Muckerman, J. T. (1979). *In* "Atom-Molecule Collision Theory" (R. B. Bernstein, ed.), p. 505. Plenum, New York.
Truhlar, D. G., Duff, J. W., Blais, N. C., Tully, J. C., and Garrett, B. C. (1982). *J. Chem. Phys.* **77**, 764.
Wartell, M. A., and Cross, R. J. (1971). *J. Chem. Phys.* **55**, 4983.
Zener, C. (1932). *Proc. Roy. Soc. A* **137**, 696.

ASSOCIATIVE IONIZATION: EXPERIMENTS, POTENTIALS, AND DYNAMICS

John Weiner

Department of Chemistry and Biochemistry
University of Maryland
College Park, Maryland

Françoise Masnou-Seeuws

Laboratoire des Collisions Atomiques et Moléculaires
Université Paris-Sud
Orsay, France

and

Annick Giusti-Suzor

Laboratoire de Photophysique Moléculaire
Université Paris-Sud
Orsay, France

I. Introduction	210
II. Experiments	211
A. Cross Sections and Rate Coefficients	212
B. Velocity and Polarization Dependence of the AI Cross Section	220
C. Associative Ionization at Ultracold Temperatures	235
III. The Problem of Molecular Potentials	240
A. Potential Curves for the H_2 System. Molecular Quantum Defect	240
B. Calculation of the Potential Curves for the Alkali Dimers and Their Cations	244
IV. Dynamics of Associative Ionization	261
A. Introduction	261
B. Semiclassical Treatments	264
C. Quantum Mechanical Treatments	268
V. Summary, Conclusions, and Perspectives	289
Acknowledgements	291
References	292

I. Introduction

Associative ionization (AI) is an elementary process,

$$A + B \to AB^+ + e, \tag{1}$$

demonstrating two important features of inelastic collisions: the production of charged particles and the formation of a molecular bond. Plasma ignition at low temperatures in alkali vapors (Lucatorto and McIlrath, 1980; Koch et al., 1982) and the nature of gas-phase chemistry in flames (Fontijn, 1985) illustrate natural phenomena in which the charged particle production of AI plays an important (sometimes a crucial) role. The formation of molecules in interstellar space must proceed either on the surface of grains or by two-body associative processes, and the bond-forming feature of AI may be an important first link in a complex chain of ion–molecule reactions leading to the astonishing variety of molecules identified in the interstellar gas (Dalgarno and Black, 1976).

At a more fundamental level associative ionization provides a simple, tractable inelastic process for investigating general questions of chemical reactivity: How does the probability of reaction vary with collision energy? What is the effect of orientation and alignment of the partners prior to collision? What is the distribution of internal states after the collision and what can we learn about the orbital motion prior to collision from the final disposition of angular momentum? In what ways can reactivity be altered by the presence of optical fields?

This article presents an account of research into these fundamental questions, spanning roughly the past decade, and providing continuity from earlier reviews of collisional ionization (Muschlitz, 1966; Berry, 1970). The closely related process of Penning ionization (PI),

$$A + B \to A + B^+ + e, \tag{2}$$

will not be treated since earlier authors have already dealt thoroughly with this subject (Niehaus, 1981; Smirnov, 1981). We will further restrict this review to associative ionization in which the asymptotic energy of $A + B$ lies below the dissociation limit of AB^+ such that the electronic states of the entrance channels either cross directly into the ground electronic state of AB^+ (Fig. 1a) or pass below the ion curve minimum (Fig. 1b). Neutral-state coupling to the continuum in these two cases generally arises from physically distinct processes requiring a somewhat deeper analysis than the simpler case of collisional ionization in which the entrance channel energy lies far above the AB^+ dissociation limit (Fig. 1c). We discuss progress along three principal lines: experimental investigation of the fundamental questions listed above, the calculation of molecular potential curves required for the interpre-

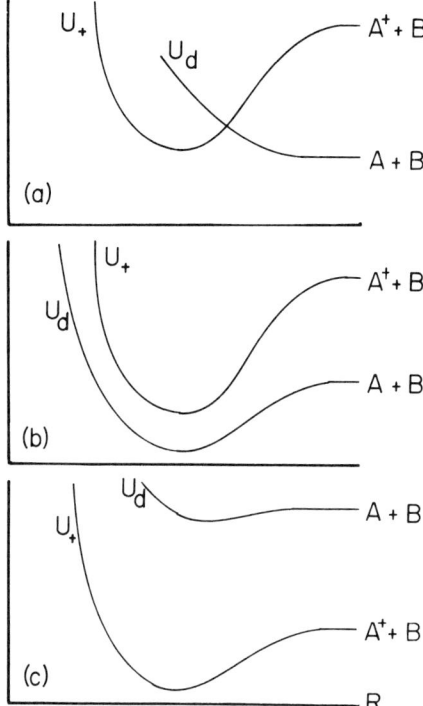

FIG. 1. Relative position of neutral and ion potential curves giving rise to three generically different forms of collisional ionization: (a) neutral curve penetrates the ion curve below the dissociation limit, electronic coupling; (b) neutral curve stays below but closely parallels ion curves, rovibrational coupling; (c) neutral curve remains well above ion curve, Penning ionization with energetic electron release, *local* electronic coupling.

tation of experiments, and a review of theoretical approaches developed to grapple with the physics of electron bound–free coupling in heavy particle associative and dissociative processes.

II. Experiments

We divide the discussion of experiments into four major categories: measurement of total cross sections and rate coefficients of AI; the effect of velocity, orientation, and alignment on the reaction probability; product distributions of internal states and angular momentum; and the behavior of AI under "ultracold" conditions where the kinetic energy of colliding partners corresponds to temperatures in the millikelvin range.

A. CROSS SECTIONS AND RATE COEFFICIENTS

1. Introduction

Because of the importance of AI to plasma, flame, atmospheric, and interstellar physics and chemistry, models of these phenomena depend heavily on accurate determination of absolute cross sections or rate coefficients. At first glance these measurements might appear to be a simple matter since the ion product can be detected with almost unit efficiency. However, nature has generously provided several pitfalls for the unwary. First, the determination of the rate coefficient, which is defined as the product of cross section and velocity averaged over an appropriate velocity distribution function,

$$k = \int_0^\infty \sigma(v) f(v) v \, dv, \qquad (3)$$

depends on the experimental arrangement since the relative velocity distribution for cells, single beams, and crossed beams are all different. Second, for homonuclear AI between resonantly excited atoms, $2A^* \to A_2^+ + e$, the volume of excited vapor is subject to radiation trapping and depends sensitively on vapor pressure, detailed geometry of the experimental apparatus, and the elapsed time after initial excitation. The interaction volume determines excited state density which enters the rate coefficient expression as the square,

$$k = \frac{\frac{d}{dt}[A_2^+]}{[A^*]^2}. \qquad (4)$$

Therefore an accurate measurement of the rate coefficient requires great care in the determination of the excited atom volume. Third, AI between ground-state atoms and those initially populated in high Rydberg states is subject to an ambiguity in the initial state definition due to population diffusion into neighboring Rydberg states by stimulated absorption and emission of background black-body radiation. We discuss several specific cases illustrative of these difficulties and describe how different investigators have grappled with and overcome them.

2. The rate coefficient in cells, single beams and crossed beams

It is important to remember that the rate coefficient for a two-body collision is a velocity-averaged quantity and that velocity distribution functions differ in cells, single beams, and crossed beams. All of the early work on AI was carried out in some sort of cell (Mohler and Boeckner, 1930; Hornbeck and Molnar, 1951). In more recent times a series of experiments on alkali AI has been carried out by Klyucharev and his collaborators in a cell apparatus using resonance lamp excitation (Klyucharev and Ryazanov, 1972; Dobrolezh *et al.*, 1975; Borodin *et al.*, 1975; Klyucharev *et al.*, 1977). Cheret and coworkers at Saclay (Cheret *et al.*, 1981; Djerad *et al.*, 1985; Barbier and Cheret, 1987; Djerad *et al.*, 1987) measured rate coefficients by exciting rubidium and potassium in a cell with laser excitation. Leventhal and coworkers (Kushawaha and Leventhal, 1980, 1982) measured AI rate coefficients in a cell-like oven, while Huennekens and Gallagher (1983b) carried out a very careful measurement of the rate coefficient for AI between resonantly excited sodium atoms at cell pressures (5×10^{14} cm^{-3}), high enough to collisionally mix the two ^2P fine structure states. In contrast, Zagrebin and Samson (1985a, b) measured AI rate coefficients in single beams of sodium and lithium. The Utrecht atomic physics group, in the course of their initial studies on the polarization dependence of AI, also reported a Na rate coefficient determination in a single beam (de Jong and van der Valk, 1979). Finally, Weiner and Boulmer (1986) at Orsay performed experiments on Rydberg states of Na in a single beam, extending to higher principal quantum number the crossed-beam work of Weiner and coworkers begun at Maryland (Boulmer *et al.*, 1983).

As discussed by Bezuglov *et al.*, (1987) and by Wang and Weiner (1987), the velocity distribution functions for cells, single beams (sb), and crossed beams (cb) are, respectively,

$$f_{\text{cell}}(v) = \frac{4}{\sqrt{\pi}} v^2 \alpha^{-3} \exp(-v^2/\alpha^2), \tag{5a}$$

$$f_{\text{sb}}(v) = \frac{1}{\alpha^6 \pi^3} \exp\left(-\frac{v^2}{2\alpha^2}\right) \left[\exp\left(-\frac{v^2}{2\alpha^2}\right) \left(\frac{2}{32} v\alpha^4 - \frac{v^3}{32} \alpha^2\right) \right.$$
$$\left. + \left(\frac{3}{16}\alpha^4 - \frac{v^2}{8}\alpha^2 + \frac{v^4}{16}\right) \int_{v/2}^{\infty} \exp\left(-\frac{2s^2}{\alpha^2}\right) ds \right], \tag{5b}$$

$$f_{\text{cb}}(v) = \frac{v^5}{\alpha^6} \exp(-v^2/\alpha^2), \tag{5c}$$

where $\alpha = (2k_B T/M)^{1/2}$, with k_B the Boltzman constant and M the reduced mass, and v is the relative velocity. Figure 2 shows that while crossed-beam and cell distributions look quite similar, the single-beam distribution markedly favors lower velocities. Useful ratios of average velocities for the three cases are (Bezugulov *et al.*, 1987)

$$\frac{\langle v \rangle_{\text{sb}}}{\langle v \rangle_{\text{cell}}} = 0.336, \qquad \frac{\langle v \rangle_{\text{cb}}}{\langle v \rangle_{\text{cell}}} = 1.041. \tag{6}$$

It is interesting to note that the "most probable" relative velocity in a

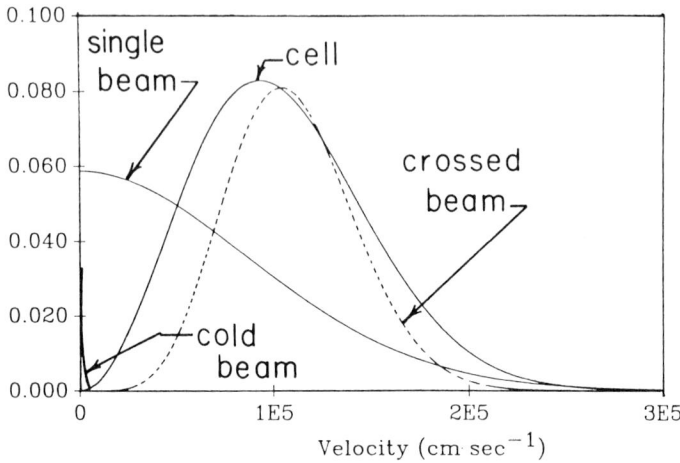

FIG. 2. Velocity (speed) distributions for various experimental arrangements. These specific curves are calculated for sodium atoms at 450 K although the relative positions and widths are universal. The "cold beam" distribution refers to collisions between velocity-selected, excited sodium atoms in a single beam (see Sec. II,C).

single-beam experiment is zero! In order to compare correctly rate coefficients determined from these different experimental arrangements, Eq. (3) shows that the velocity dependence of the cross section must also be known. As shown later in this review (see Sec. II,C,5), the velocity dependence of the cross section varies rather markedly with the polarization state of the colliding partners. Therefore, as a practical matter, results from cell, crossed-beam, and single-beam experiments cannot be directly compared.

3. Associative ionization between Na(3p) atoms

Early efforts to model laser-initiated, alkali vapor plasma ignition underscored the necessity of "seed electron" producing mechanisms (Lucatorto and McIlrath, 1980) and consequently sparked a lively interest in the absolute cross section of the AI process,

$$\text{Na}(3p) + \text{Na}(3p) \rightarrow \text{Na}_2^+ + e. \tag{7}$$

The first reported measurements varied over several orders of magnitude probably because of poorly measured and controlled densities of the excited states. The principal problem appears to be spatial diffusion of the excited state population due to radiation trapping in the alkali vapor. Two separate determinations of the rate coefficient for process (7), one in a high-density cell (Huennekens and Gallagher, 1983b); the other in a low-density crossed-beam environment (Bonanno et al., 1983, 1985), have treated carefully the radiation trapping problem and have resulted in good agreement, even after taking account of the somewhat different relative velocity distributions in the two environments. In the cell apparatus used by Heunnekens and Gallagher (1983b), a single-frequency dye laser tuned near the Na resonance transition traverses the cell in the z-direction, and ions are collected on an electrode mounted along the x-axis and x-z plane. The spatial intensity profile of the excited state density in the y direction (resonance fluorescence) is observed directly with a slit scanning parallel to the y axis. Application of slab-geometry radiation diffusion theory (Holstein, 1947, 1951) provides the excited-state density along the x-direction. Spatial variation along z is measured directly by absorption of the exciting laser beam. With these three "form factors," $\phi(x)$, $Y(y)$, $Z(z)$, in hand the excited state density is expressed as

$$[\text{Na}^*(x, y, z)] = [\text{Na}^*(0, 0, 0)]\phi(x)Y(y)Z(z) \tag{8}$$

and the task reduces to determining an absolute value for the excited state density at the center of the cell, $[\text{Na}^*(0, 0, 0)]$. Since this is the critical number on which the entire measurement depends, Heunnekens and Gallagher determined $[\text{Na}^*(0, 0, 0)]$ in three independent ways: by laser power absorption, absolute resonance fluorescence intensity, and measurement of the fluorescence intensity from the Na(5s) → Na(3p) transition (proportional to the product of $[\text{Na}^*(0, 0, 0)]^2$ and the rate coefficient for the energy transfer collision),

$$\text{Na}(3p) + \text{Na}(3p) \rightarrow \text{Na}(5s) + \text{Na}(3s) \tag{9}$$

which was determined in previous work (Huennekens and Gallagher, 1983a). The final results of various measurements for process (7) are entered in Table I.

In contrast to the cell arrangement, Bonanno et al. (1983, 1985) performed a crossed-beam experiment in which two thermal atomic beams intersect a broad-band pulsed laser beam tuned to the $^2S_{1/2} \rightarrow {}^2P_{3/2}$ fine-structure Na transition. Product ions were mass-analyzed by a time-of-flight tube mounted vertically above the collision plane. The rate coefficient for AI is expressed in terms of the ratio of molecular ion to atomic ion density in which the Na_2^+ arises from AI, and Na^+ results from selective photoionization of Na*(3p) by a uv pulse tuned just above the Na*(3p) photoionization threshold. The expression for the AI rate coefficient is

$$k = \frac{I_{\text{Na}_2^+}}{I_{\text{Na}^+}} = \frac{\sigma_i \phi}{[\text{Na}(3p)]\delta t}, \tag{10}$$

TABLE I
Selected Rate Coefficients and Cross Sections for $Na(3p\ ^2P_{3/2}) + Na(3p\ ^2P_{3/2})$ Associative Ionization

Rate coefficient (10^{-12} cm^3 sec^{-1})	Cross section (Å)^2	Reference	Remark
1.3	0.13	Kushawaha and Leventhal (1982)	Cross section estimated by taking $\langle v \rangle = 10^5$ cm sec^{-1}
38 ± 0.4		Klyucharev et al. (1977)	Cell experiment
5.6	0.51	Huennekens and Gallagher (1983b)	Fine-structure levels $^2P_{3/2}$, $^2P_{1/2}$ not distinguished; cell experiment
8.1 ± 0.3	0.73 ± 0.27	Huennekens and Gallagher (1983b)	Estimate $^2P_{3/2}$ component of entry 3
		Bonanno et al. (1983), (1985)	
11 ± 0.4	0.89 ± 0.4	Bonanno et al. (1983)	Crossed beam
		Bonanno et al. (1985)	
6.3 ± $^{1.8}_{1.1}$		Thorsheim et al. (1989)	Crossed beam
11 ± $^{1.3}_{0.5}$	860	Gould et al. (1988)	Measurement in optical trap at 0.75 mK

where σ_i is the photoionization cross section of Na(3p), ϕ is the uv photon radiation density (photons cm^{-2}) flux, and δt is the duration of the resonant excitation pulse. Since almost all apparatus-dependent variables cancel in the ion signal ratio ($I_{Na_2^+}/I_{Na^+}$) the task reduces to a determination of the cross-beam interaction volume defining the excited-state density. Increase in the apparent lifetime of the excited atoms, due to radiation trapping in the Na vapor, can be related to the excited-state density by applying the Milne theory of radiation diffusion (Milne, 1926; Garver et al., 1982). Care must be taken, however, to include the absorption of both ground-state hyperfine levels when using a broad-band excitation source (Bonanno et al., 1985). Note that the crossed-beam experiment measures the rate coefficient for collision only between Na(3p $^2P_{3/2}$) atoms, while the cell experiment of Heunnekens and Gallagher populates both fine-structure levels. Comparison requires the plausible assumption that the rate coefficients are statistically related so that the contribution from Na(3p $^2P_{3/2}$) can be extracted from the cell result. Note that the cross sections entered in Table I are calculated from the measured rate coefficients by normalizing to the average velocity, which implies the assumption of weak velocity dependence in the cross sections.

4. Associative ionization between ground-state and Rydberg atoms

In 1979 Janev and Mihajlov began a series of articles (Janev and Mihajlov, 1979, 1980; Mihajlov and Janev, 1981) proposing a general theory for excitation and ionization in slow collisions between Rydberg atoms and ground-state atoms. Their theory (hereafter called the JM theory) predicted the AI rate constant for collisions.

$$A + B^*(n) \to AB^+ + e \qquad (11)$$

as function of the principal quantum number n. An early version treated only homonuclear collisions ($A = B$) (Janev and Mihajlov, 1979), but a later version encompassed heteronuclear collisions as well (Janev and Mihajlov, 1980). Lack of space prohibits a detailed examination of this theory, but we introduce the main features here simply to motivate discussion of a number of experiments designed to test its predictions. The essential elements of the theory are (1) nuclear motion is adiabatic (slow collisions), (2) the Rydberg electron remains outside the core, and adiabatic potential curves on which the collision takes place are defined by the ion–atom core interaction. For homonuclear and "quasiresonant" collisions (i.e., between partners whose ionization potentials are much greater than the difference between them), the long range ion–atom interaction is modeled as a two-state system: one state attractive, the other repulsive. Only collisions along the repulsive curve lead to AI, and (3) relaxation of the core electronic potential transfers energy to the Rydberg electron via dipole coupling. Janev and Mihjalov term this process the "exchange" channel. The basic mechanism is schematized in Fig. 3.

Early experimental work on homonuclear systems by Kluycharev's group; (Devdariani et al., 1978; Dobrolezh et al., 1975; Borodin et al., 1975) was carried out in cells with Cs(np) ($9 \leq n \leq 15$), and Rb(np) ($7 \leq n \leq 14$) levels in which total (atomic + molecular) ionization rates were determined. Barbier and Cheret (1987), using cw laser photoexcitation and mass analysis, made AI measurements on a few low-lying s and d levels of Rb. Zagrebin and Samson (1985a) carried out single-beam measurements on Na(np), ($5 \leq n \leq 21$) and on Li(np) (Zagrebin and Samson, 1985b) over about the same n range. Boulmer et al. (1983) in a crossed beam arrangement used pulsed laser excitation to populate np levels of Na and time-of-flight mass analysis to distinguish AI in collisions between Na atoms:

$$Na^*(np) + Na(3s) \rightarrow Na_2^+ + e. \qquad (12)$$

The time resolution of this experiment eliminates contribution to the ion production rate from lower excited states by cascade. Results of the crossed

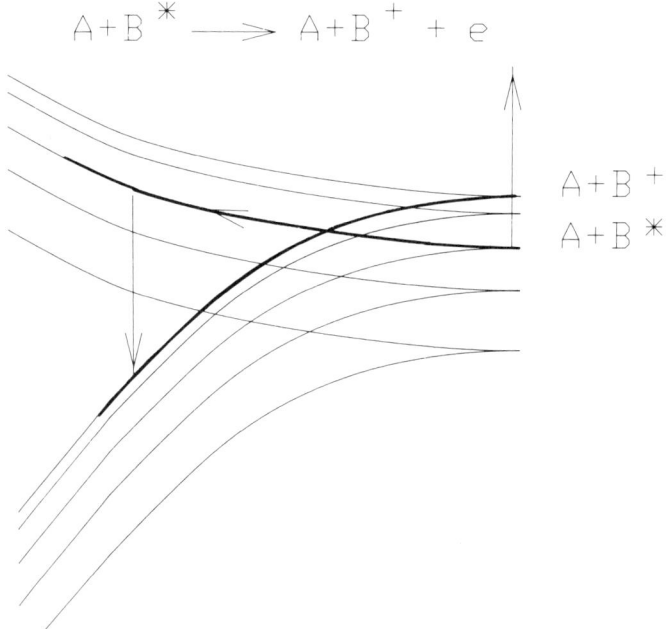

FIG. 3. The Janev model of associative ionization. Entrance channel is a repulsive Rydberg state. At the classical turning point, the core relaxes (vertical downward arrow) and dipole couples the released energy to the ionizing Rydberg electron (vertical upward arrow).

beam studies in Na all show three common features: (1) rapid rise of the rate coefficient with n_{eff}, the "effective" principal quantum number, ($n_{eff} = n - \delta$ where δ is the quantum defect) to a maximum around $n_{eff} = 8$–10 followed by a slow decline, (2) large rate coefficients ($\approx 10^{-9}$ cm^3 sec^{-1}) near the maximum, and (3) only fair agreement with JM theory both in absolute magnitude of rate coefficient and functional dependence on the principal quantum number. In a subsequent paper Weiner and Boulmer (1986) extended the study of process (12) to $n = 27$ and, by applying Stark-field switching, populated s and d levels as well as p levels from the s-level atomic ground state. In contrast to the Maryland experiments in crossed beams (Boulmer et al., 1983), the Orsay experiments (Weiner and Boulmer, 1986) were carried out in a single beam, and therefore the rate coefficients, determined with very different velocity distribution functions, cannot be concatenated. However, the relative behavior of the rate coefficient as a function of n_{eff} can still be compared to JM prediction over the entire range of principal quantum number covered in the two experiments. The results clearly show that JM theory produces more satisfactory agreement at high principal quantum number. This finding is not surprising since JM theory is predicated on the assumption of core impenetrability by the Rydberg electron. However, experimentally measured ordering of rate coefficients as a function of l quantum number (fixed n) does not accord with JM prediction, (determined by the relative ordering of s, p, d photoionization cross sections at fixed n). The experimental ordering does, however, correlate with core penetrability: results show that s levels have the highest rate coefficients and d levels the lowest.

In the case of quasiresonant heteronuclear associative ionization, JM makes two predictions verifiable by experiment (Mihajlov and Janev, 1981): (1) a pronounced enhancement of the rate coefficient compared to the homonuclear value from either partner. For example, the rate coefficient for RbK$^+$ production is predicted to be 4.5 times the value for either Rb$_2^+$ or K$_2^+$, and (2) a strong preference for the "exchange" channel over the "direct" channel in which the system follows the attractive rather than the repulsive incoming potential curve. The Saclay group (Djerad et al., 1987) reported measurements for the RbK$^+$ which do not show the predicted enhancement factor over their earlier Rb$_2^+$ results. The Maryland group (Johnson et al., 1988) carried out a series of crossed-beam experiments on the NaLi$^+$ system by measuring the heteronuclear rate coefficient *interbeam* and the homonuclear rates *intrabeam* all within the same experiment. Their results show that not only is the expected enhancement factor of 8 not found (in fact the ratio of heteronuclear to homonuclear rate coefficients is less than unity) but also the so-called "direct" channel,

$$Na^* + Li \to NaLi^+ + e \qquad (13)$$

has larger rate coefficients than the "exchange" channel,

$$Na + Li^* \rightarrow NaLi^+ + e \qquad (14)$$

in direct contradiction to JM predictions. It appears clear that JM theory is inadequate to explain the experimental results and that a new theoretical point of departure, taking into account the interaction of the Rydberg electron with the core, will probably be necessary. A promising development in that direction is the application of multichannel quantum defect theory (MQDT). Ross and Jungen (1987) have shown how molecular MQDT can take account of vibronic coupling in a whole series of Rydberg states without explicit calculation on a state-by-state basis. In a recent paper Urbain *et al.* (1986) have used an MQDT approach to calculate AI cross sections for the H^+, H^- system and have compared their calculations to the experimental results of Poulaert *et al.*, 1978, (see Sec. IV,C). The agreement appears to be quite good over about four orders of magnitude in collision energy, although it remains to be seen if MQDT can accurately predict cross section behavior as a function of Rydberg level.

B. Velocity and Polarization Dependence of the AI Cross Section

1. Associative ionization between He* and He(1 ^{1}S)

Starting in the mid 1970s, Pesnelle and coworkers began a series of experiments (e.g. Fort *et al.*, 1976; 1978a, b; Runge *et al.* 1985) in which they measured the velocity dependence of associative ionization between He in various metastable states and target gases such as He, H, D, H_2 and D_2 in their ground states. Space does not permit a comprehensive discussion of this important family of experiments, but we choose a recent representative example (Runge *et al.*, 1985) for two reasons: first, to illustrate the experimental features common to the whole series; and second, to compare the experimental results with the predictions of the multistate curve-crossing model (MSCC) of Cohen (1976). The specific AI process is

$$He^*(5\ ^3P) + He(1\ ^1S) \rightarrow He_2^+ + e, \qquad (15)$$

the relevant potential curves of which are shown in Fig. 4. The experimental setup is a crossed-beam arrangement in which metastable He beam source emits a mixture of $He^*(2\ ^1S)$ and $He^*(2\ ^3S)$. The beam is velocity selected by a rotating-wheel chopper and enters an interaction zone where a uv laser excites the He^* ($2\ ^3S$) to $He^*(5\ ^3P)$. The optically excited beam subsequently interacts with a crossed, effusive beam of ground-state He, producing He_2^+ ions from process (15). A quadrupole mass analyser and particle multiplier detector count the number of product ions produced as a function of arrival

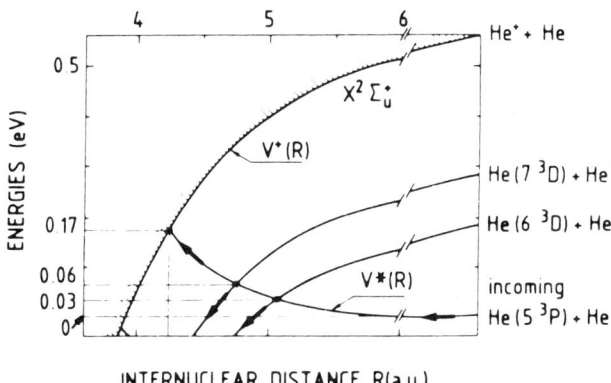

FIG. 4. Relevant diabatic curves for application of MSCC to process (15).

time at the detector. After normalization to the He* density the resulting time-of-flight spectrum transforms to a cross section for process (15) as a function of relative collision energy; see Fig. 5. Runge *et al.* (1985) then compare the experimental results to those predicted by the MSCC model. This model is described more fully in Sec. IV,B. We simply note here that the model envokes a semiclassical curve crossing mechanism in which the entrance diabatic curve is crossed by asymptotically higher-lying Rydberg states. At each crossing a Landau-Zener expression (Cohen, 1976) evaluates the probability of jumping to the new curve, and finally the overall AI probability is determined by summing the probabilities over all possible pathways to the AI region. The theory depends on (1) accurate diabatic potential curves and (2) the assumption that localized interactions between the diabatic curves, confined to the crossing region, adequately describe the AI probability. Comparison with experimental results can only be regarded as fair. Runge *et al.* (1985) conclude that, in the case of $n = 5$, the assumption of localized coupling at the curve crossings is dubious, and they introduce an "effective" interaction parameter to try to compensate for a region of nonlocalized interaction at long range where the two diabatic curves are almost parallel. This *ad hoc* procedure yields a better fit to experiment in the middle energy range between 0.05 eV and 0.10 eV but gives poor agreement near reaction threshold (Fig. 5c). Another adjustment, a uniform shift to all the crossing energies of the incoming channel, improves the threshold behavior but overestimates the cross section at higher energies (Fig. 5b). The breakdown of assumption (2), interaction localized to a curve crossing, runs counter to the fundamental idea of the MSCC model and indicates that this semiclassical treatment, while appropriate for many situations, cannot be considered a universal nostrum for all the headaches of AI dynamics.

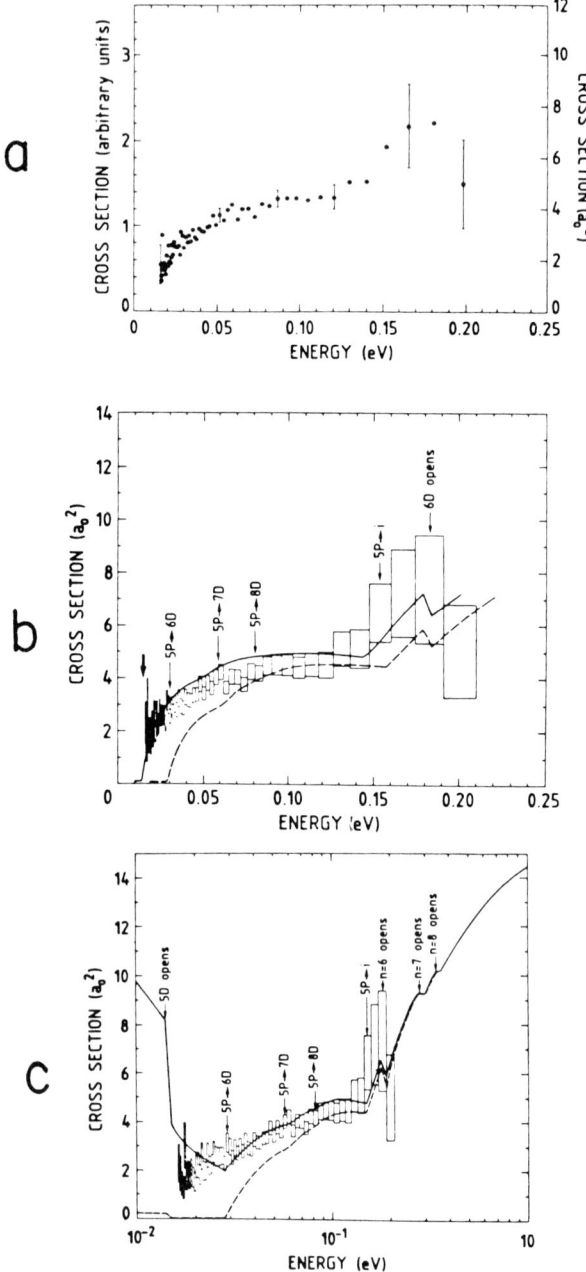

2. Optical velocity selection and polarization

a. Introduction. The total energy contained in two Na atoms, both of which are excited to the first resonance level, is sufficient to open the AI channel [see Fig. (6)]. Most experiments have concentrated on collisions between atoms populated in the higher fine-structure level,

$$\text{Na}(3p\ ^2P_{3/2}) + \text{Na}(3p\ ^2P_{3/2}) \rightarrow \text{Na}_2^+ + e. \tag{16}$$

Therefore a single-mode laser, selectively exciting narrow velocity groups by Doppler tuning within a broad Maxwell-Boltzmann distribution in a single beam or in crossed beams, acts as a precise, optical velocity selector for both partners of the collision. By tuning the laser over the thermal velocity distribution and at the same time judiciously preparing the laser polarization, one measures the velocity and polarization dependence of the reaction probability. The results bear a direct relationship to the symmetries of the molecular states responsible for the AI event, and they show that the reaction dynamic is not only quite sensitive to these symmetries but also that their relative importance changes over the range of velocities available. The experiments discussed in this section probe the reaction mechanism at a more detailed level than the rate coefficient and begin to uncover fascinating features usually obscured by velocity and polarization averaging.

b. Early work on the polarization effect in associative ionization. Investigators in the Utrecht atomic physics group (Kircz et al., 1982) were the first to report a polarization dependence of the AI probability in process (16). They crossed an effusive Na beam with a single-mode laser at right angles, tuned to the resonance line. Rotating the laser polarization axis from parallel to perpendicular with respect to the atomic beam axis produced a modulation of the AI rate by a factor of 1.7. One of the authors proposed a density matrix theory to account for the polarization behavior as a function of the angle between polarization and atomic beam axes (Nienhuis, 1982). This theory interprets the polarization effect in terms of initial LM_L or JM_J populations of the colliding partners and a laboratory spatial axis of quantization along the Na beam. In their initial report the authors concluded that collisions between atoms in the $^2P_{3/2}$, $M_J = \frac{1}{2}$ states were the most important, that the rate coefficient for collisions between atoms, one of which is in the $M_J = \frac{3}{2}$ and the other in the $M_J = \frac{1}{2}$ state, is very small; and that the theory only yielded

---

FIG. 5. Cross section vs. collision energy for process (15): (a) experimental results of Runge et al. (1985); (b) application of MSCC theory, dashed line; MSCC theory with 14 meV shift to all crossing energies, solid line. (c) MSCC theory, dashed line; MSCC theory modified to include correction for long-range couplings, solid line. In (b) and (c) the experimental data are represented by rectangles to reflect 4 μ sec channel width in time-of-flight measurement.

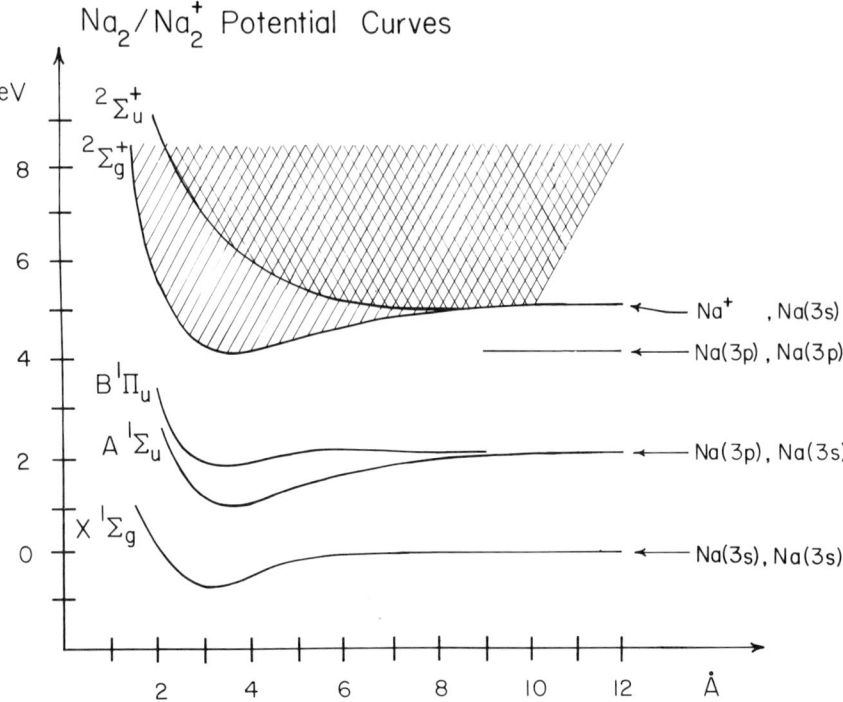

FIG. 6. Relative position of some asymptotic energies of neutral states and the ion ground state in sodium. Note that the asymptotic energy of Na(3p) + Na(3p) lies above the minimum of the Na_2^+ potential well.

physically reasonable results (non-negative rate coefficients) if JM_J quantum labels were used rather than the LM_L labels of the electronic orbital angular momentum. From this finding it was concluded that spin-orbit coupling cannot be ignored. Although the theory makes no explicit reference to a *molecular* quantization axis, if one assumes that the AI probability is determined at far internuclear separation where the angle between laboratory and molecular axes is small, then the preference for angular momentum states with minimum projection on the quantization axis suggests that molecular Σ states play a dominant role.

c. Angular momentum of product Na_2^+. Motivated by the early results of the Utrecht group, Wang, et al. (1986a) performed a photofragmentation analysis of the dimer ions produced by Na(3p) AI. The experiment consists of two parts: (1) measurement of the polarization modulation in the AI cross section, and (2) determination of spatial anisotropy in the distribution of rotational angular momentum J vectors of the product Na_2^+. Figure 7 shows a schematic of the single-atomic-beam, two-laser-beam apparatus. In the first

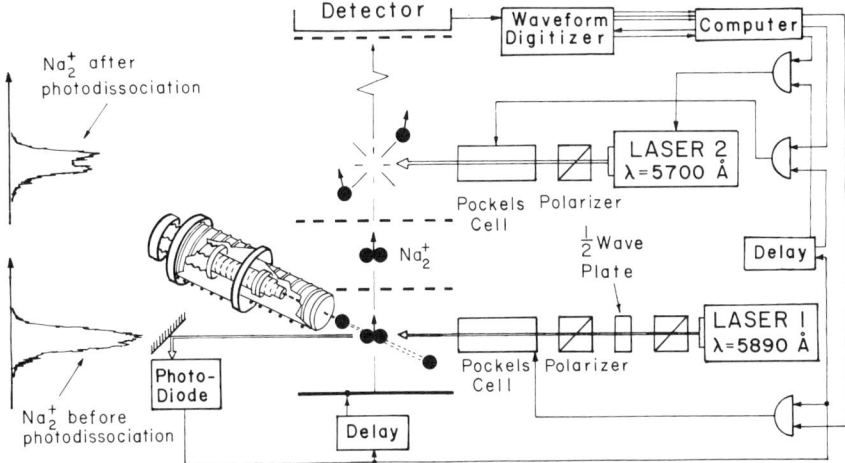

FIG. 7. Schematic of the polarization/photofragmentation apparatus used to determine angular momentum anisotropy in Na_2^+ arising from process (16).

part of the experiment laser 1 populates the $^2P_{3/2}$ level of Na(3p), switching polarization alternately parallel and perpendicular to the atomic beam axis. The relative yield of Na_2^+ product from AI, as a function of the polarization axis, confirms the preference for alignment along the atomic beam axis. In the second part of the experiment laser 2 photodissociates the ensemble of Na_2^+ ions produced in part 1. The amplitude and kinetic energy distribution of the dissociated species is analyzed by time-of-flight (TOF) mass spectroscopy as a function of laser-2 polarization axis. By toggling the polarization axis of laser 2 synchronously with laser 1 (but at twice the frequency), the relative yield of photofragment ions with laser 2 aligned parallel or perpendicular to the atomic beam axis reflects the spatial distribution of J-vectors in the X-Y plane. Figure 8 diagrams the relation between laser-2 polarization, the distribution of rotational angular momentum vectors, and Σ (longitudinal) or Π (transverse) approach of the two electron charge distributions prior to collision. The result of a rather involved analysis (M.-X. Wang et al., 1986b) is that spatial anisotropy of the J-vector distributions implies a preference for Σ over Π approach by about a factor of 6.

In addition to the intensity of the TOF dissociation spectrum, the line shape contains information on the internal state distribution of the TOF product fragments. Keller et al. (1986) carried out measurements of the line shape and, after examining several different combinations of rotational and vibrational distributions, concluded that (1) only a relatively small amount of rotational energy is deposited in the product Na_2^+, and (2) vibrational energy linearly increasing with quantum level up to the maximum allowed by energy

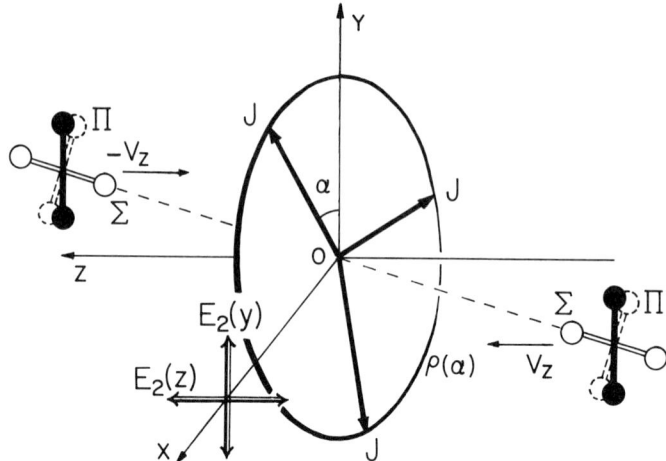

FIG. 8. Diagrams showing relations between polarization in the laboratory frame (blackened dumbbells correspond to E_2 polarization) and projection onto the relative collision axis. The J disc in the x-y plane shows uniform distribution of J vectors for Σ approach but shows anisotropy (greater density of J vectors along the x axis) for Π approach.

conversation was the only distribution consistent with the TOF line shape. Borrowing some ideas from the multichannel quantum defect approach to dissociative recombination (Giusti, 1980) and applying the principle of micro-reversibility, Wang and Weiner (1989) have concluded that the vibrational energy distribution arises from incoming Na_2 potential curves attractive in the range of internuclear separations where AI can take place.

An alternative technique to photofragmentation is energy analysis of the ionizing electron. Carré et al. (1984, 1986) were the first to measure the electron energy distribution from Na(3p) associative ionization, using a cylindrical energy analyzer together with both cw and pulsed laser excitation. Although these pioneering studies identified the ensemble of low-energy electrons associated with AI, the resolution of the energy analyser near threshold and ambiguities in the local electric field environment (contact potentials and ambipolar diffusion) precluded a reliable measure of the electron energy distribution within the ensemble. Preliminary results from another electron-energy experiment (Müller et al., 1989), in which associative ionization is observed in intrabeam collisions, indicate that electrons emitted in process (16) have kinetic energy very close to zero and a distribution barely exceeding the spectrometer resolution. More extensive results which can be compared to conclusions drawn from the photo-fragmentation data should be forthcoming from this experiment in the near future.

d. Velocity selection and polarization in atomic beams. In the spring of 1985 Weiner and coworkers carried out the first in a series of experiments measuring directly the AI cross section for process (16) as a function of collision velocity over the thermal distribution from two crossed beams (Wang et al., 1985). In the parlance of chemical kinetics, the behavior of a reactive cross section as velocity increases is termed the "excitation function," and, though rarely measured, figures importantly in the theory of chemical reaction rates. In the initial experimental arrangement a monomode dye laser (bandwidth $\approx$ 1–2 MHz) excited a narrow velocity group within the Maxwell-Boltzmann distribution (width $\approx$ 1.5 GHz) emanating from each of two crossed beams. The selected velocity v is related to the laser frequency v by,

$$v = c\left(\frac{v - v_0}{v_0}\right)\cos\theta, \tag{17}$$

where v_0 is the resonance frequency with the atom at rest, v the Doppler-shifted resonance frequency, and θ the angle between atomic and laser beams. Tuning the laser frequency effectively tunes the relative collision velocity of the AI process. This technique is particularly well suited to inelastic collisions between two beams of excited-state particles because the laser selects a narrow velocity group from each beam, permitting determination of the excitation function with high resolution. Ions are detected by a particle multiplier mounted above the collision plane at the end of a TOF drift tube. A counterpropagating uv beam ($\lambda = 351$ nm) from an Ar$^+$-ion laser source selectively ionizes only Na(3p) atoms. The resultant ion signal is then directly proportional to the excited-state density and can be used to normalize the AI yield. The cross section is related to the experimentally measured parameters by,

$$\sigma_{AI} = [I(Na_2^+)/I^2(Na^+)][(\Phi_{ph}\sigma_{ph})^2/v], \tag{18}$$

where $I(Na_2^+)$, $I(Na^+)$ are the AI and photoionization signals, and Φ_{ph}, σ_{ph} the uv photon flux and atomic photoionization cross section, respectively, and v is defined by Eq. (17). The results of this early experiment showed a trough in the cross section at about 1.6×10^3 msec^{-1} with rather steeply rising slopes on both sides. A preliminary interpretation indicated that more than one potential curve must be invoked to explain an increasing cross section on both sides of the minimum. Although this initial experiment demonstrated interesting possibilities of the technique, the importance of atom–light–field interactions such as optical pumping, power broadening, radiation trapping, and the effect of stray magnetic fields on polarization were not fully appreciated or controlled. Two new experiments, therefore, were carried out under more carefully controlled conditions to systematically exploit not only the velocity selectivity, but also optical polarization so as to align or orient the collision partners prior to collision. The first of these experiments studied the effect of linear polarization on the collision as a

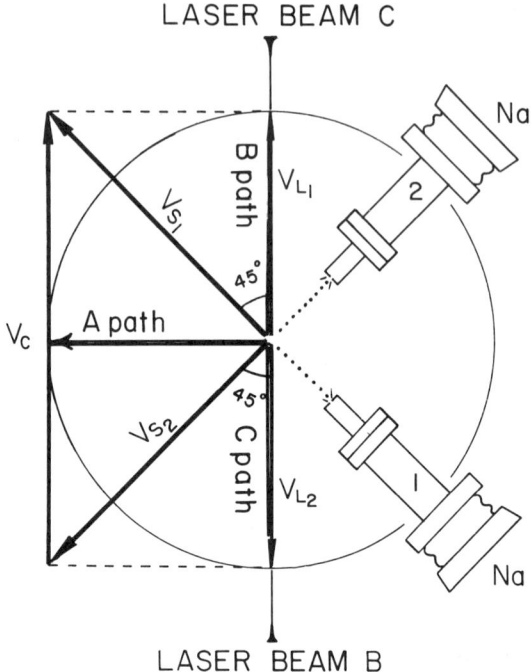

FIG. 9. Diagram of crossed-beam, velocity and polarization selected experiment. The paths A, B, C correspond to various axes of propagation on which experiments were carried out. The vector V_c shows the relative collision velocity vector selected by narrowband optical excitation.

function of velocity (Wang *et al.*, 1986). The orientation between the laser propagation axis and the particle beam axes was chosen either to bisect the right angle between the two atomic beams or to cross them at an angle of 45°; see Fig. 9. In the latter case the laser frequency excites a velocity group in only one of the atomic beams (e.g., beam 1) while a counterpropagating laser excites the same velocity group in beam 2. This arrangement permits separate polarization of the velocity groups in each atomic beam prior to collision. The excitation function for several different choices of relative polarization are shown in Fig. 10. The velocity dependence of the AI cross section is striking in the case of "axial-axial" approach in which both velocity groups are polarized linearly with the axis of polarization perpendicular to the "*A* path" of Fig. 9 and in the collision plane. A consistency check with the thermal beam polarization effect initially reported by Kircz *et al.* (1982) and later confirmed by Meijer *et al.* (1986) was carried out by averaging the cross section results for polarization parallel and perpendicular to the collision plane. The thermally averaged ratio of 1.7 ± 0.3 is in good agreement with

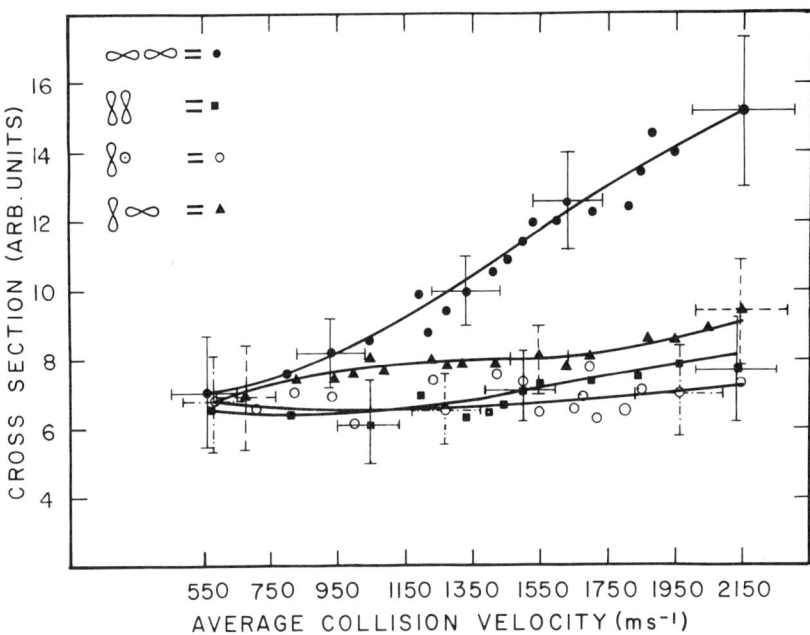

FIG. 10. Excitation functions from linear polarization experiments carried out in the arrangement of Fig. 9. The figure-eights indicate the relative polarization of beams (1) and (2) in Fig. 9 with respect to V_c. Note strong velocity dependence when p-orbital charge distribution is aligned along the collision axis.

the ratio of 1.8 reported by the Utrecht group and shows that in-plane, axial-axial collisions are increasingly favored at higher velocity. In a subsequent article Wang and Weiner (1987b) analyzed the results in terms of a "locking radius" model based on the notion of a localized point along the reaction coordinate at which the quantization axis transforms from a laboratory-fixed to a molecular-fixed reference frame. Populations of atomic substates produced by polarized excitation go over (via a frame transformation) into populations of molecular orbitals at the locking radius. The rank-ordering of the molecular states built up from these molecular orbitals reveal their relative importance as a function of collision velocity. From this analysis Wang and Weiner concluded that the locking radius could be identified with an internuclear distance greater than 25 Å and that $^3\Sigma_u^+$ and $^1\Sigma_g^+$ states dominate the AI process at collision velocities greater than 1000 msec^{-1}. Using the same experimental setup, Weiner and coworkers also measured the excitation function for circular polarization (M.-X. Wang et al., 1987). By tuning the laser onto the $F = 2$, $M_F = 2 \rightarrow F = 3$, $M_F = 3$ hyperfine transition in sodium, the atomic population is transferred entirely to these two

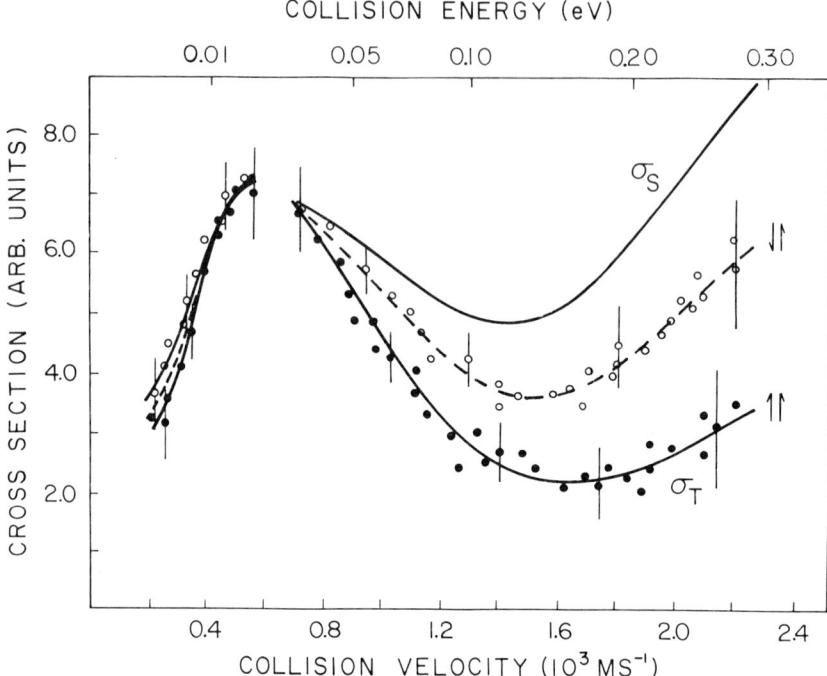

FIG. 11. Excitation functions from circular polarization experiments carried out in the arrangements of Fig. 9. States contributing to the spins-parallel curve (↑ ↑) are triplets. States contributing to the spins antiparallel curve (↑ ↓) are singlet-triplet mixtures. The curve labeled σ_S is a pure singlet curve calculated from the spin antiparallel curve with the assumption that one singlet and one triplet contribute equally. Low-velocity data are on the left, after the break, obtained with experimental arrangement shown in Fig. 12.

states after several Rabi cycles. Associative ionization collisions therefore take place between excited-state velocity groups (one in each beam) which are maximally oriented with their total angular momentum parallel or antiparallel (depending on the sense of the circular polarization vector) to the laser propagation axis. The spin-selected excitation functions are shown in Fig. 11. The "spins-parallel" curve consists only of triplet states, but the "spins-antiparallel" result contains contributions from both singlet and triplet states. Assuming that the locking radius determined from the linear polarization studies is still valid, the "spins-parallel" excitation function must consist principally of the $^3\Delta_u$ state and the "spins-antiparallel" a linear combination of $^3\Sigma_u^-$, $^1\Sigma_u^-$, $^3\Sigma_u^+$, $^1\Sigma_g^+$ states. The dip in the cross section reported earlier (Wang *et al.*, 1985) still appears at a velocity of about 1.6×10^3 msec^{-1}, but the slope on the high-velocity side is less abrupt. The

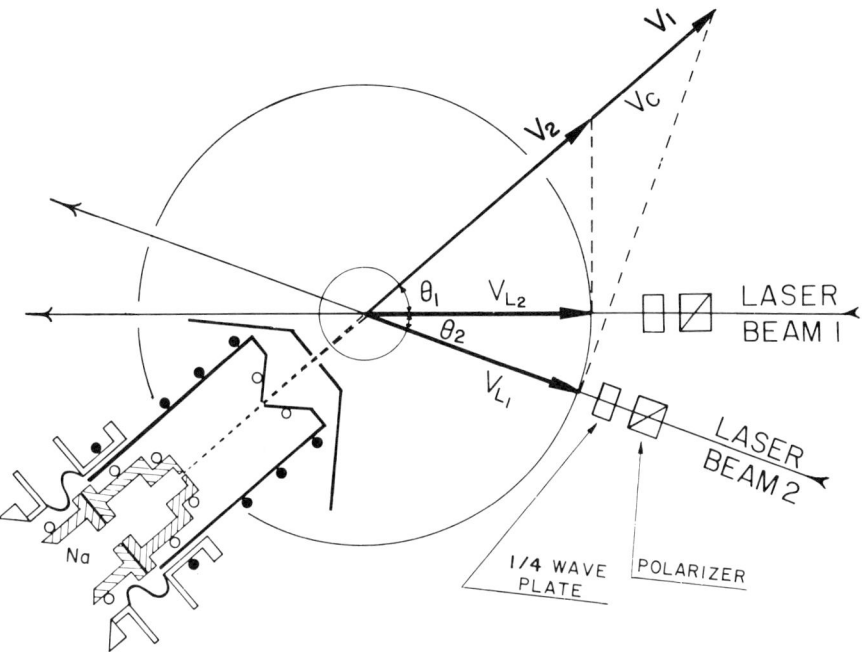

FIG. 12. Experimental arrangement used to obtain low-velocity data in Fig. 11. Laser beams 1 and 2 separated by angle θ_2 traverse single atomic beam. Collision velocity along the atomic beam axis, V_c, is the difference between V_1 and V_2, the velocity groups selected by lasers 1 and 2, respectively. Relation between angles and V_c is given by Eq. 19.

difference may be due to an improved signal normalization procedure in the later experiment and to the fact that the different relative orientations of laser and atomic beams between the two experiments lead to populations of different molecular orbitals or molecular states.

In an effort to extend the excitation function to lower velocities, the Maryland group devised a technique to observe collisions between the two different velocity groups in a single atomic beam. Fig. 12 shows the arrangement and Eq. (19) expresses the relative velocity v_c as a function of the angles, θ_1, θ_2, and velocity groups v_{L1}, v_{L2} between the two laser beams and the atomic beam:

$$|v_c| = \frac{|v_{L1}|}{\cos(\theta_1 + \theta_2)} - \frac{|v_{L2}|}{\cos \theta_1}. \tag{19}$$

The results in Fig. 11 show that the cross section drops off rapidly with decreasing velocity. This drop-off is not unexpected and is probably due to the long-range interaction between two excited Na(3p) states. The leading term in the multipole expansion of the electrostatic interaction between two P states is the quadrupole-quadrupole term, C_5/R^5, which is attractive or repulsive depending on the relative orientation of the two atomic quadrupole distributions. In fact attractive orientation leads only to Π states, and as the kinetic energy of the collision decreases, all other repulsive entrance Σ, Δ channels begin to close. We will come back to this point when we discuss associative ionization in ultracold collisions.

In contrast to the experiments at Maryland, the Utrecht atomic physics group investigated the velocity and polarization dependence of process (16) using coaxial opposing atomic beams crossed at 87° by a single-mode excitation laser. Using linear polarization, they measured the AI rate $R(\theta)$ as a function of angle θ between the axis of polarization and the atomic beam axis. A Fourier expansion with the appropriate angular symmetry,

$$R(\theta) = R_0 + R_1 \cos(2\theta) + R_2 \cos(4\theta), \tag{20}$$

is fit to the experimental curve and the resulting experimentally determined expansion coefficients are associated with a density matrix representation of the ion production rate,

$$R = \text{Tr}_A \text{Tr}_B G_{\rho_A \rho_B}. \tag{21}$$

In Eq. (21) G is the "detection operator" acting on the density matrices ρ_A, ρ_B representing the population distributions of atoms A and B. The atomic density matrices reflect states of polarization produced by the laser excitation. Taking into account the necessary angular transformation of the atomic state polarization from the optical axis to the atomic beam axis, the right side of Eq. (21) is expressed as a multipole expansion of the detection operator and the atomic density matrices with the atomic beam axis as the axis of quantization,

$$R(\theta) = \sum_{kq} \sum_{k'q'} g(kq, k'q') c_k c_{k'} d_q^k(\theta) d_{q'}^{k'}(\theta). \tag{22}$$

In Eq. (22) $g(kq; k'q')$ are the coefficients of the multipole expansion of the detection operator, $c_k c_{k'}$ are coefficients in the expansion of the atomic density matrices which reflect the experimentally prepared atomic populations, and $d_q^k(\theta) d_{q'}^{k'}(\theta)$ are coefficients of the frame transformation taking the atomic population from the laser to the atomic beam axis of quantization. Association of the Fourier coefficients [Eq. (20)] with those from the multipole expansion leads to a set of linearly independent equations from which the contributions of the various irreducible tensors of the detection operator can be determined. These components themselves can then be expressed as linear combinations of the detection operator matrix elements in the magnetic substate representation, $\langle M'_A M'_B | G | M_A M_B \rangle$, which is the point at which the experimental quantity $R(\theta)$ can be interpreted in terms of the initial atomic states. The expression of the detection matrix in terms of irreducible tensor operators specifies the maximum number of elements that completely describe the AI rate and thus provides a criterion for judging the maximum amount of information that can be extracted from a set of experiments. By employing various combinations of linear and circular polarization and a symmetry-breaking small magnetic field, the Utrecht group has determined the four on-diagonal (cross sections) and four off-diagonal (coherences) irreducible matrix elements of G in the LM_L atomic basis (Meijer et al., 1988, 1989a, b). However, measurement of the off-diagonal matrix elements (Meijer et al., 1987) shows that they are of comparable size to the on-diagonal "cross section," suggesting that the atomic state representation, although formally not incorrect, may not be the most physically revealing choice to represent the collision channels. Furthermore Meijer et al. (1989a) determine a cross section corresponding to collisions between two Na atoms populating $M_L = 1$, with negative values over most of the velocity range covered in the experiment. This unphysical result leads Meijer et al. (1989b) to conclude that the LM_L basis is an inappropriate representation because spin-orbit coupling takes place during the collision event. An obvious alternative is to recast G, ρ_A, ρ_B in the JM_J atomic basis. However, the irreducible tensor elements of G increase from eight in LM_L representation to sixteen in JM_J and the symmetry-breaking magnetic field experiment no longer determines a single tensor element but a linear combination of five: four off-diagonal and one on-diagonal term. A unique five-parameter fit to the data appears to be out of reach and Meijer et al. conclude that G in the JM_J basis cannot be completely determined from their battery of experiments. The final picture that emerges from the G matrix analysis is the relative contribution of various *atomic* state pairs (LM_L or JM_J) to the AI rate. The detection matrix analysis provides a succinct, compact formalism delineating the information required for a complete description of the collision event in terms of a chosen set of initial atomic states. Furthermore, multipole expansion of the detection operator and atomic density matrices provides invaluable insight into the proper choice of experimental conditions needed to obtain maximal information. However, the *dynamics* of bond-formation through which the initial atomic states transform to final molecular states lies outside the scope of the G-matrix analysis. The Utrecht results do show that the principal contributions to AI come from atomic states with minimum angular momentum projection along the atomic beam axis. With the assumption that the beam axis and collision axis roughly coincide (i.e., that quantization axis recoupling from laboratory to collision coordinates takes place at far internuclear separation) one can conclude from their results that the most important molecular state symmetries contributing to AI are $^{1,3}\Sigma$ states. This conclusion is in accord with the Maryland results. The excitation functions obtained by the two groups differ substantially, however, with the Utrecht group reporting cross sections with less sensitivity to collision velocity (see, for example, Figs. 11 and 14 in Meijer et al., 1989a). The origin of the disagreement is not evident and is still under investigation. The differing geometries of atomic beam and laser beam crossing mean that the initial atomic state populations are not the same, and that AI does not take place along the same reaction paths even under identical conditions of polarization and excitation intensity. Another important consideration is the relative widths of the velocity classes excited in the two experiments. The velocity distribution of excited atoms is a Lorentzian folded with a thermal Boltzmann distribution of ground state atoms emerging from the effusive atomic beam. Because the Lorentzian width varies inversely as the cosine of the angle between the atomic and laser beams, 87 degree geometry produces rather wide velocity distributions—especially for velocity groups corresponding to energies displaced either side of the Boltzmann maximum by more than kT. The increased width will tend to smear out structure in the excitation function. However, velocity resolution alone probably does not account fully for the disparity between the two sets of data, and further work will be required to bring measurements into accord.

In a very recent paper Brencher et al. (1988) report the polarization dependence of AI, PI, and ion pair production in crossed-beam collisions between K(4p $^2P_{3/2}$) atoms. They analyze their results in terms of the Nienhuis G matrix theory and find again that states with minimum angular momentum projection on the quantization axis are the leading contributors. These conclusions cannot be accepted, however, without reserve since the authors analysed the results assuming that off-diagonal coherence terms are negligible. Since the Utrecht measurements showed that coherence terms are comparable to the on-diagonal "cross sections" in Na AI, a compelling case has to be demonstrated that their neglect in $K(^2P_{3/2}) + K(^2P_{3/2})$ AI is legitimate.

As an alternative to the Utrecht atomic state basis approach, Jones and Dahler (1987) have proposed a theory in which the density matrix of the laser-prepared atomic states is represented in a molecular state basis. Initial atomic fine-structure state populations are re-expressed as populations of Born-Oppenheimer quasimolecular states of the two collision partners. Jones and Dahler then write the basic equation relating the total AI cross section to the partial cross sections for each molecular state as

$$\sigma(\epsilon, \theta) = \sum_p \sigma_{pp}(\epsilon)\rho_{pp}(\theta) \qquad (23)$$

in which $\rho_{pp}(\theta)$ is the on-diagonal density matrix element of quasimolecular state p; $\sigma_{pp}(\epsilon)$ is the energy-dependent partial AI cross section for state p; and θ is the angle between the laser linear polarization axis and the collision axis [see Eq. (21)]. Off-diagonal terms $\sigma_{pp'}$, $\rho_{pp'}$ coupling different molecular states do not appear in the sum because, according to Jones and Dahler, they are negligible [at least for the specific case of AI between Na(3p) atoms]. In Eq. (23) the left side, $\sigma(\epsilon, \theta)$, results directly from the experimental measurement, $\rho_{pp}(\theta)$ is determined by the initial state preparation; and $\sigma_{pp}(\epsilon)$, the calculation of which is the goal of the theory, reflects the importance of each specific quasimolecular channel to the AI process. After a rather involved analysis Jones and Dahler reach the conclusion that only four states, $^3\Sigma_u^+(\sigma_g\sigma_u)$, $^1\Sigma_g^+(\pi_g^1\pi_g^{-1})$, $^1\Pi_u(\sigma_g\pi_u^{\pm 1})$, $^3\Pi_u(\sigma_g\pi_u^{\pm 1})$ can contribute significantly to $\sigma(\epsilon, \theta)$ observed in the experiments of the Utrecht group; and of these only $^1\Sigma_g^+$ and $^3\Sigma_u^+$, in the ratio of about 2:1, are required to obtain $\sigma(\epsilon, \theta)$ in accord with experimental results. This overall conclusion, that $^1\Sigma_u^+$ and $^3\Sigma_u^+$ are the principal contributors to AI produced by *linear* polarization, agrees with Wang and Weiner, 1987b. Apparently the Utrecht group (Meijer *et al.*, 1989b) has reached similar conclusions. Nevertheless, Jones and Dahler make several simplifying assumptions the validity of which have yet to be examined: (1) Δ states are ignored although the spins-aligned triplet state prepared in the cross-beam experiment (Wang *et al.*, 1987) is certainly in large measure $^3\Delta_u$ ($^1\pi_g\ ^1\pi_u$). (2) Spin-orbit coupling is ignored, a dubious assumption in light of the inability of the Utrecht group to characterize the G matrix in the LM_L basis set. (3) Atomic–molecular angular momentum recoupling is assumed at infinite internuclear separation; and (4) an *adiabatic* assumption is made in which the molecular states constructed from the Na(3p) + Na(3p) asymptotic atomic basis do not mix significantly with other molecular states constructed from neighboring asymptotic atomic levels [e.g., Na(3s) + Na(4d)] during the course of the collision. This adiabatic assumption is quite likely to be incorrect and the extent to which neighboring configurations mix with the initial asymptotic basis during the course of the collision is revealed by potential curve calculations—the subject of Sec. II.

C. Associative Ionization at Ultracold Temperatures

1. Trap experiments

Under favorable circumstances laser light can be used to cool an ensemble of atoms to temperatures well below 1 mK and hold them in an optical trap (Prodan *et al.*, 1985; Gould *et al.*, 1987). Heavy-particle collisions in this temperature regime exhibit some novel characteristics for the following reasons (Thorsheim et al., 1987): (1) particle deBroglie wavelengths become much greater than the typical range of the chemical bond. In Na_2, for example, the bond length is about 5.8 au while the deBroglie wavelength of Na at 1 mK is 450 au. (2) The angular momentum of the collision reduces to only a few partial waves. Therefore quantum resonance effects normally obscured by averaging over many partial waves become manifest in the total cross section. (3) Long-range interactions control collision dynamics. At 1 mK kinetic energy, variations in the interaction potential of a few cm^{-1} can open or close inelastic entrance channels, or even trap the molecule in bound states (Stwalley et al., 1978). (4) For the special case of collisions between excited states the peculiar condition exists that the spontaneous emission lifetime becomes short compared to the characteristic collision time. Therefore the relative strength of atom–field coupling versus atom–atom coupling enters directly into the probability of the inelastic event (Julienne, 1988).

The investigation of ultracold-temperature collisions is just beginning, and at this writing only a few theoretical (Thorsheim et al., 1987; Julienne, 1988) and experimental (Gould et al., 1988; Prentiss et al., 1988) papers have appeared. For the special case of associative ionization, process (16) has been observed by Gould et al. (1988) and its cross section measured at 0.75 mK using the apparatus shown in Fig. 13. A cooling laser first slows a beam of

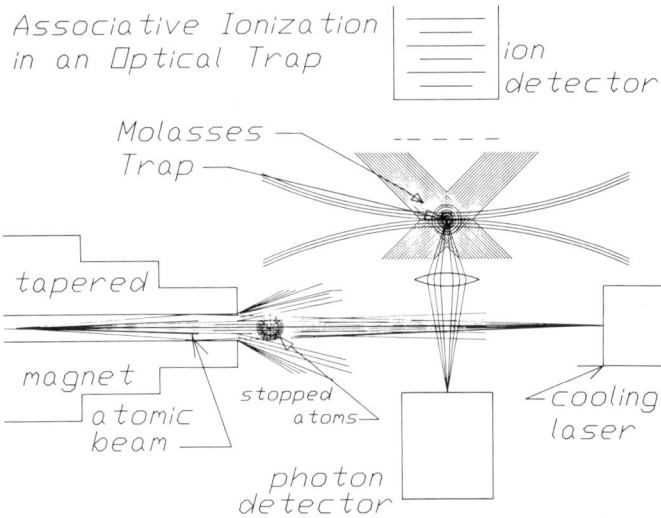

FIG. 13. Schematic diagram of apparatus used to measure cross section of AI collisions between excited sodium atoms at 0.75 mK. Ions and photons are collected simultaneously from dipole trap at the center of the laser focus. Round ball at the end of the tapered magnetic field represents stopped atoms, some of which are held in the molasses region (intersection of six counterpropagating orthogonal laser beams) from which the dipole trap is loaded.

atoms along the axis of a decreasing solenoidal magnetic field (needed to ensure that the decelerating atoms do not shift out of resonance with the fixed frequency laser). The cold atoms congregate in a small spatial region at the intersection point of six orthogonal laser beams tuned to maximize the intensity of this "optical molasses" (Chu et al., 1985). The atomic density in the molasses ($\approx 1 \times 10^7$ cm^{-3}) is insufficient to observe collisional processes; but an optical trap, based on the attractive force between the electric field gradient of the focused laser and the atomic transition dipole, concentrates the density to $\approx 1 \times 10^{10}$ cm^{-3}. A time-of-flight mass sampling of the trap confirms the presence of Na$_2^+$, the intensity of which grows quadratically with atom density. Careful measurement of trap density and ion number permits a determination of the rate constant for AI at 0.75 mK. The result, 8.6×10^{-14} cm^2, is about three orders of magnitude larger than the cross section measured earlier under conventional conditions (Huennekens and Gallagher, 1983b; Bonanno et al. 1983, 1985). This surprising result can be rationalized in terms of the increased deBroglie wavelength at very low temperatures. The quantal cross section expression for a single entrance channel is given by

$$\sigma_{12} = \frac{\pi}{\kappa^2} \sum_{J=0}^{\infty} (2J+1)|S_{12}(\epsilon, \rho)|^2, \qquad (24)$$

where S_{12} is the S-matrix element connecting the initial and final state, J is the collisional angular momentum, and κ the wave vector, equal to 2π times the inverse of the deBroglie wavelength. An approximate expression in terms of the maximum partial wave contribution to the scattering is

$$\sigma_{12} \simeq \pi \lambda^2 (J_{\max} + 1)^2 P_{12}, \qquad (25)$$

from which we see that the measured result is consistent with a reaction probability P_{12} of about 0.1 if only s-wave scattering contributes. In fact, even at 0.75 mK, a few higher partial waves contribute so that the probability is even less than 0.1. Thus the large cross section is not due to an increased reactive probability (opacity function) but comes from the increased size of the collision partners as measured by the deBroglie wavelength. In a recent article Julienne has pointed out (Julienne, 1988) that, in the limit of weak optical field, the two excited partners decouple from the laser excitation and relax to the ground state by spontaneous emission before they reach the maximum internuclear separation permitting AI. Thus the weak-field probability drops to near zero. Under strong-field coupling conditions the two approaching atoms maintain their excited-state populations for a longer fraction of the collision time, effectively increasing the opacity function for AI. The principal conclusion of Julienne's analysis is that the associative ionization cross section is negligible in optical molasses (weak-field condition) and the observed signal comes almost entirely from the strong-field region of the optical trap.

Because the atoms are randomly distributed throughout the trap volume, polarization information and the effect of different approach orientations on the AI cross section cannot be detected easily. Atomic-beam experiments, on the other hand, are well-adapted to this purpose, and the Maryland group has recently measured the AI cross section for collisions in a single beam at average energies equivalent to "temperatures" ($E = \frac{3}{2} kT$) between 60 mK and 1 K (Thorsheim et al., 1989). The experimental arrangement is shown in Fig. 14. The laser beam intersects a single Na atomic beam and excites a velocity group the minimum which is governed by the natural Lorentz width of the resonance transition. Collisions within the velocity group lead to associative ionization, and the relative velocity distribution along the atomic-beam axis is given by

$$P(V) = \int_{-\infty}^{+\infty} dv_2 \, P_1(V + v_2) P_2(v_2), \tag{26}$$

where $V = v_1 - v_2$ is the relative collision velocity, $v_{1,2}$ are the laboratory velocities of particles 1 and 2 within the selected velocity group, and $P_{1,2}$ are the probabilities of finding particles 1,2 in the excited state with velocities $v_{1,2}$

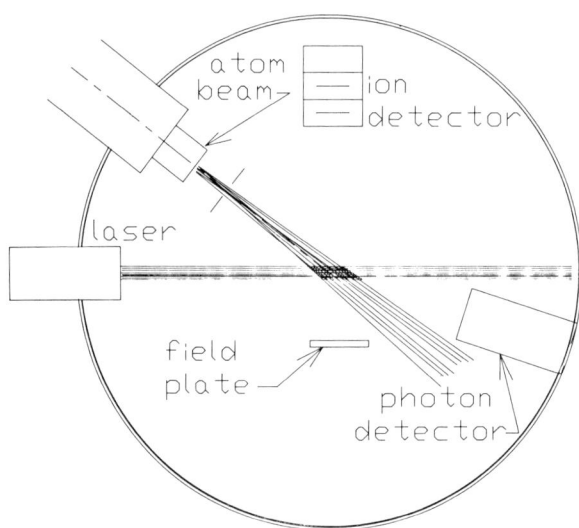

FIG. 14. Single-atomic beam apparatus used to measure collisions within an excited velocity group selected by narrow-band laser. Ions produced at interaction region are accelerated by field plate onto the particle multiplier. Photodiode detector measures fluorescence at intersection region which permits normalization of ion count to excited state population.

(Lorentz distribution). The velocity distribution for a "cold" beam is plotted together with more commonly encountered distributions in Fig. 2. Evaluation of the average relative velocity from this distribution together with contributions from the residual transverse velocity components leads to a collision temperature of 60 mK. A particle multiplier mounted above the interaction plane counts ions, and the Na(3p) density is determined both by absorption and fluorescence. The result of the measurement is a cross section of 4.9×10^{-16} cm^2 for circular polarization. Linear polarization yields a cross section greater by a factor of 2.

Figure 15 summarizes the cross section data from about 600 K to 0.75 mK and compares it to an approximate theory recently proposed by Geltman (1988). The decline in the cross section from 700 K to 70 K is due the closing of incoming channels. As Geltman points out, the leading long-range multipole term between two atoms in a p configuration is the quadrupole-quadrupole interaction, C_5/R^5. This term is attractive only for Π states;

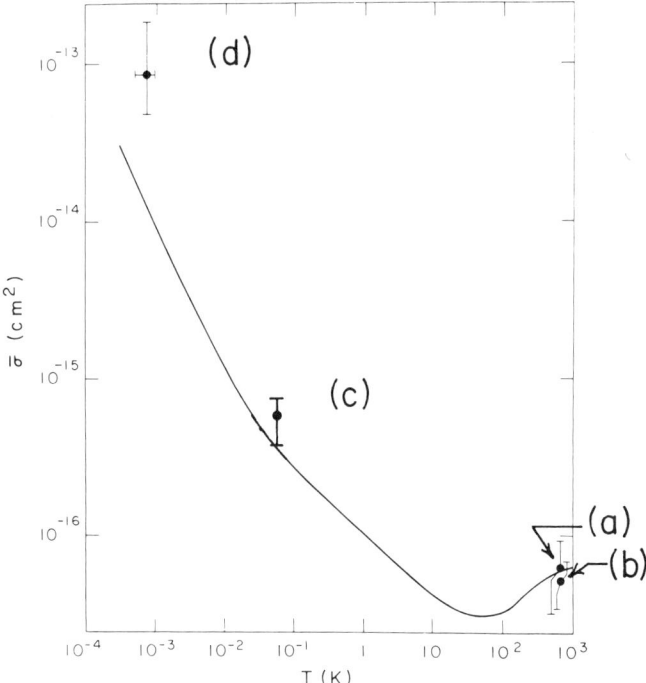

FIG. 15. Theoretical excitation function for process (16) over 6 orders of magnitude in temperature, calculated by Geltman (1989). Points show measured cross section values: (a) Bonanno et al. (1985), (b) Huennekens and Gallagher (1983b), (c) Thorsheim et al. (1989), (d) Gould et al. (1988).

long-range repulsion eventually excludes all other incoming channels from associative ionization. As the temperature continues to decrease, the cross section again begins to rise due to the effect of the increasing deBroglie wavelength [Eq. (23)]. A more detailed interpretation of the difference between circular and linear polarization should soon be forthcoming as well as the effect of power broadening. Investigation of the temperature region around the predicted minimum is an obvious goal that will be pursued in the near future.

III. The Problem of Molecular Potentials

The treatment of associative ionization necessitates knowledge of accurate potential curves and dynamical couplings characterizing the quasimolecule formed during the collision. Two regions of internuclear distances must be considered, which should not necessarily be treated by the same methods:

(i) An inner region, where a chemical bond is formed, and where the main physical effect one has to describe is the interaction between a bound electronic state and a continuum. The calculations should then be capable of predicting Rydberg states of the quasimolecule, and we shall see that the concept of molecular quantum defect is very helpful.

(ii) An outer region, where the system is well described by considering two separate atoms interacting with long range forces, or alternatively a positive and a negative atomic ion. In most cases, the dissociation channels are embedded in Rydberg series of one of the two atoms.

In both regions, the energy spacing between the various channels is small, so that high accuracy is required. It is important to describe correctly the border region, in order to understand how the population of the system of the two atoms is shared between the various molecular states.

In the present review, we shall limit ourselves to the discussion of collisional systems involving either two hydrogen atoms or two alkali atoms; these should be considered as prototype systems, for which many accurate data have already been obtained.

In the present section, atomic units will be used except when otherwise stated.

A. POTENTIAL CURVES FOR THE H_2 SYSTEM. MOLECULAR QUANTUM DEFECT

The hydrogen molecule being a two-electron system, sophisticated methods have been developed long ago for the ground and first excited states

(Kolos and Wolniewicz, 1965, 1969; Kolos, 1978; Wolniewicz and Dressler, 1977, 1985; Dressler et al., 1979, and references therein). A variational method is used, the wave function being given as a linear combination of properly symmetrized two-electron basis functions in elliptic coordinates:

$$\Phi_j(1, 2) = \xi_1^{p_j}\eta_1^{q_j}\xi_2^{r_j}\eta_2^{s_j} \exp(-\alpha_1\eta_1 - \alpha_2\eta_2)$$
$$\times [\exp(\beta_1\eta_1 + \beta_2\eta_2) \pm \exp(-\beta_1\eta_1 - \beta_2\eta_2)]r_{12}^{n_j} \quad (27)$$

with

$$\xi_i = (r_{Ai} + r_{Bi})/R, \quad \eta_i = (r_{Ai} - r_{Bi})/R$$

In Fig. 16a, the coordinates of the two electrons are indicated by (1) and (2), $r_{A1(2)}$ and $r_{B1(2)}$ are the distances of the electron 1 (2) to the two nuclei A and B; p, q, r, s, n are integer numbers and $\alpha_1, \alpha_2, \beta_1, \beta_2$, variational parameters.

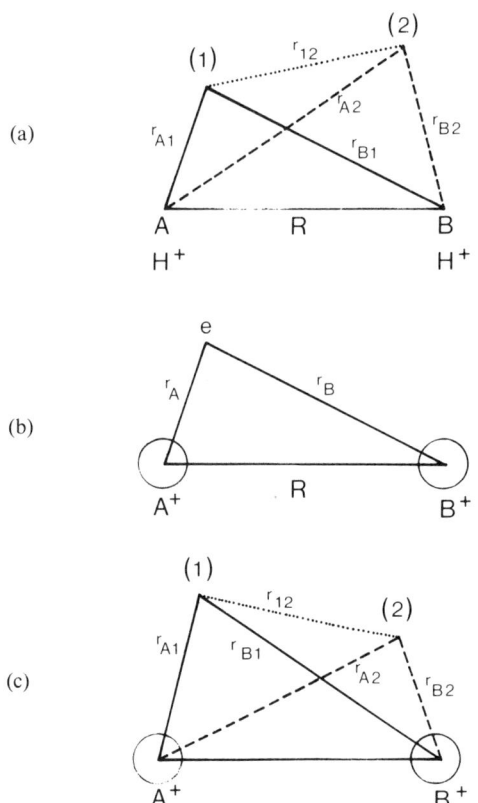

FIG. 16. (a) Coordinate system for H_2, (b) coordinate system for AB^+, (c) coordinate system for AB. Numbers in parentheses (1) and (2) refer to electron 1 and electron 2.

The presence of the exponential terms in Eq. (27) ensures the correct behavior of the wave function when one electron goes to infinity, r_{12} is the interelectronic distance and the explicit dependence of the basis functions on a power of r_{12} ($n = 0, 1, 2$) accelerates the convergence of the expansion.

Very accurate results have been obtained by such a method; the calculated well depth of the ground state agrees with the experimental data within a fraction of a wavenumber, the excitation energies of some excited states being reproduced within one or a few wavenumbers.

The lowest excited $^1\Sigma_g^+$ states in H_2 constitute a well known example of double minimum potential curves. This shape can be interpreted by considering the two electron wavefunction as a mixture of configurations constituted from two H_2^+ orbitals. In the $^1\Sigma_g^+$ symmetry, there is strong mixing between singly excited configurations (1s σ_g, $nl\ \sigma_g$, $l = 0, 2$) and the lowest doubly excited configuration (2p σ_u^2). In the first case, one electron stays on the ground-state bonding H_2^+ orbital, the other one being in an excited state. In the second case, both electrons are in the first excited antibonding H_2^+ orbital. This configuration mixing is the dominant mechanism for dynamical processes in H_2, such as dissociative recombination (Giusti-Suzor et al., 1983; Nakashima et al., 1987) and associative ionization (Urbain et al., 1986; Takagi and Nakamura, 1987) as well as for striking non-Franck-Condon distributions in photoelectron spectra (Pratt et al., 1983, 1986).

The adiabatic potential curves computed by Wolniewicz and Dressler (1977) are displayed on Fig. 17 for several $^1\Sigma_g^+$ excited states of H_2. The

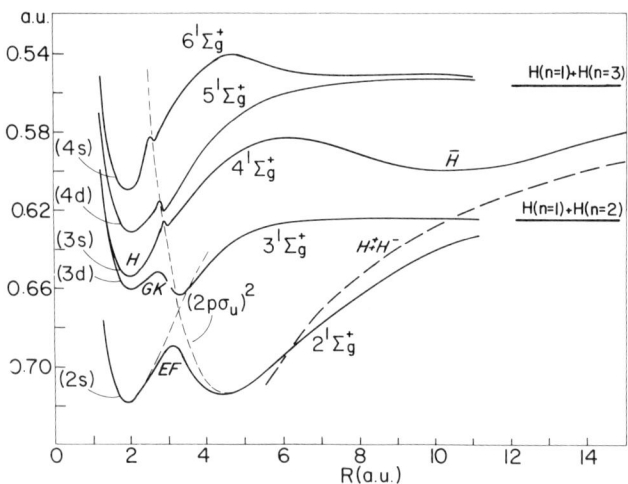

FIG. 17. Adiabatic potential curves for H_2 ($^1\Sigma_g^+$ states) calculated by Wolniewicz and Dressler (1977). Rising dashed curves on the left shows diabatic states.

curves vary greatly from one state to the next, displaying many avoided crossings.

A possible treatment of the associative ionization reaction between two hydrogen atoms would consist in computing first all the adiabatic potential curves of a given symmetry, including the ($H_2^+ + e$) continuum states, and then the dynamical coupling matrix elements between the various states, in particular the often tedious radial coupling d/dR matrix elements. The collision problem would then be treated by solving a large number of coupled equations. It is often more convenient to replace this adiabatic representation by a diabatic one, in which the matrix element of the d/dR operator is negligible (Smith, 1969). The diabatic states are no longer eigenstates of the total electronic hamiltonian and crossings may occur between diabatic curves of the same symmetry. The dynamical treatment will now introduce the electronic coupling between the various diabatic states. This representation is widely used for the treatment of Rydberg-valence mixing in molecules (Lefebvre-Brion and Field 1986) and forms the basis of most of the AI treatments described in Sec. IV.

An elegant way of defining diabatic states has been proposed by Ross and Jungen (1987): in the MQDT framework of the unified treatment of bound and continuum states, they have interpreted the double minimum structure present in the $^1\Sigma_g^+$ states of H_2 as an inelastic "collisional" process taking place at negative energy. Using only the information contained in the first three excited $^1\Sigma_g^+$ adiabatic curves, they have determined a three-channel reaction matrix describing the short-range scattering of an external electron by an H_2^+ ionic core in the 1s σ_g and 2p σ_u state. In other words, the diabatic curves correspond to three Rydberg series, with energies determined by the molecular Rydberg formula which depends upon the internuclear distance R:

$$E_n^\alpha(R) - E_+^i(R) = \frac{-\text{Ryd}}{[n - \mu_\alpha(R)]^2}. \tag{28}$$

In (28) $E_n^\alpha(R)$ is the electronic energy of the molecular Rydberg state α with principal quantum number n, $E_+^i(R)$ is the energy of the ionization limit (H_2^+ ion in the 1s σ_g or 2p σ_u state), $\mu_\alpha(R)$ is the quantum defect *which is a constant along a Rydberg series*, and Ryd is the Rydberg constant. In such a picture, a whole set of bound and continuum diabatic potential curves is characterized by a unique function $\mu_\alpha(R)$. The three diabatic Rydberg series in the $^1\Sigma_g^+$ symmetry can be interpreted, when the Rydberg electron is outside the core, as corresponding to the following physical situations:

(i) One electron with an orbital angular $l = 0$ in the field of a ground state H_2^+ core. The inner electron is then in a bonding orbital, the ionization energy in (28) is $E^0_+(R)$, and the Rydberg series converging to the ground state of H_2^+ is characterized by a quantum defect $\mu_0(R)$.

(ii) Similarly, one may consider the bound and continuum states for the motion of one electron with an orbital angular momentum with $l = 2$ in the field of a ground-state H_2^+ core, defining a Rydberg series characterized by a quantum defect $\mu_2(R)$.

(iii) In contrast, the third Rydberg series, characterized by a quantum defect $\mu_1(R)$, corresponds to the motion of a Rydberg electron in the field of an excited H_2^+ core, the inner electron being in the 2p σ_u antibonding orbital. The ionization energy in (2) is then $E^1_+(R)$.

The three-channel R-dependent reaction matrix K, includes the three quantum defects (diagonal terms) together with the interactions between the three channels (off-diagonal terms). It is almost energy independent so that it contains all the information required to account for the electronic and nuclear dynamics in the $^1\Sigma_g^+$ symmetry, not only for the lowest states but also for the entire set of Rydberg and continuum states. We shall see in Sec. IV how this information has been used (Urbain, 1988) for the calculation of the H + H*(2s) associative ionization cross section.

The determination of the reaction matrix from *ab initio* calculations was possible because of the availability of highly accurate data. We shall see in the following paragraph that, owing to an intense effort during the last fifteen years, a similar accuracy will soon be available for the alkali dimer systems.

B. CALCULATION OF THE POTENTIAL CURVES FOR THE ALKALI DIMERS AND THEIR CATIONS

1. Introduction

The alkali dimers and their cations appear as a prototype for molecular systems with one or two active electrons, and therefore they have been the subject of an intense theoretical effort, with the aim of elucidating the similarities and differences with the H_2 problem, due to the presence of the core electrons. The existence of accurate experimental data make these systems a benchmark for various theoretical methods. We have chosen not to give an exhaustive account of the different approaches but rather to focus on the calculations for the ground and first excited states using two main methods which have developed simultaneously over the past decade and a half:

(i) Model potential (Dalgarno *et al.*, 1970; Dalgarno, 1975) or pseudopotential (Bardsley, 1974) methods, in which only the one or two valence electrons are explicitly included in the calculations, the interaction with the core electrons being represented by effective potentials, which should in particular take account of core-polarization effects.

(ii) *Ab initio* calculations, in the framework of the quantum chemistry methods, in which all the electrons are explicitly included (Konowalow *et al.*, 1977, 1979).

From 1980, several groups (Jeung *et al.*, 1982; Konowalow and Fish, 1984; Müller and Meyer, 1984) have realized that the core-polarization effects should also be included in the *ab initio* treatment. Their inclusion made it possible to obtain excellent agreement with the experimental spectroscopic data for the ground and first excited states. However, more highly excited states, and in particular molecular states correlated to two excited atomic states [Na(3p) + Na(3p), K(4p) + K(4p)] have not yet been correctly predicted in the *ab initio* calculations; it is apparently difficult with such methods to find a prescription for choosing the highly excited diffuse orbitals required in the basis set.

During the same period the model potential approach has been generalized to the highly excited states of the dimer cations by Henriet and Masnou-Seeuws (1983). They have developed a method for the treatment of the dimer problem which has succeeded in predicting the potential curves correlated to the Na(3p) + Na(3p) dissociation limit.

2. *Model potential and pseudopotential treatment for the ground and first excited states: first results (before 1980)*

The valence electron of an alkali atom is loosely bound, and the idea of an adiabatic separation between its slow motion and the rapid motion of the core electrons has been developed by many authors (Veselov and Bersuker, 1958; Veselov and Schottf, 1967; Caves and Dalgarno, 1972). As in the Born-Oppenheimer separation for the molecular problem, the energy $W(\mathbf{r}) - 1/r$ of the core may be determined first for the position $\mathbf{r}$ of the valence electron relative to the nucleus. Then the motion of the valence electron is treated in turn by considering $W(\mathbf{r})$ as a mean potential in the one-electron Schrödinger equation:

$$(-\tfrac{1}{2}\nabla_r^2 + W(\mathbf{r}) - 1/r - E)\phi(\mathbf{r}) = 0. \tag{29}$$

Let us remark that the potential $W(\mathbf{r})$ is the same for the various energy levels of the valence electron.[1]

---

[1] In the case of the lithium atom, Veselov and Schtoff (1967) have performed a variational calculation both of $W(\mathbf{r})$ and of the nonadiabatic correction due to the coupling between the motions of the core electrons and of the valence electron: the latter has been found both small and confined to a region close to the nucleus.

The idea of a model potential very naturally comes from the empirical search of a potential which should be equivalent to $W(\mathbf{r})$. The next step consists of realizing that many properties of an alkali atom are governed by that part of the valence–electron wave function which is outside the core region. Since the long-range part of the potential is coulombic, the wave functions are characterized by their quantum defects in the asymptotic region, and *any short-range potential capable of reproducing the experimental quantum defects will be considered as satisfactory*.

However, the physics of an alkali atom cannot be restricted to the motion of the valence electron in a static potential: the core is not a point charge as in the hydrogen atom, and there exist an infinity of core states. The valence electron creates an electric field which modifies the core wave function, and the equations must take account of core-polarization effects. A perturbation treatment (Hameed *et al.*, 1968; Caves and Dalgarno, 1972) shows that this problem can be handled by modifying the one electron operators acting on the valence electron. For large values of the distance $\mathbf{r}$, the electric dipole operator $e \cdot \mathbf{r}$ is thus modified into:

$$e\mathbf{r}(1 - \alpha_d/r^3), \qquad r \to \infty \tag{30}$$

where e is the electron charge and α_d the dipole polarizability of the core; similarly, the model potential should contain a polarization term:

$$V_{\text{pol}} \to -\frac{\alpha_d}{2r^4} \qquad \text{as} \quad r \to \infty. \tag{31}$$

The divergence of the operators when $r \to 0$ is corrected by introducing a smooth cutoff function, so that V_{pol} is written

$$V_{\text{pol}}(r) = -\frac{\alpha_d}{2r^4} f(\rho), \qquad f(\rho) = \begin{cases} 1 & \text{for} \quad \rho \gg 1, \\ 0 & \text{for} \quad \rho \ll 1, \end{cases} \qquad \rho = \frac{r}{r_0}. \tag{32}$$

The cutoff radius r_0 is typical of the core dimensions. The cutoff function $f(\rho)$ is equal to zero close to the nucleus and 1 outside the core region.

Turning now to the alkali dimer ion, it is treated by considering the motion of one electron in the field of two alkali cores A^+ and B^+ at a distance R [see Fig. (16b)], and the effective Hamiltonian should be

$$h = -\tfrac{1}{2}(\nabla)^2 - \frac{1}{r_A} + V_A(r_A) - \frac{1}{r_B} + V_B(r_B) + V_{cc}(R) + V_3(R, r_A, r_B) \tag{33}$$

r_A and r_B being respectively the distances between the electron and the two cores A and B, the two potentials $V_A(r_A) - 1/r_A$ and $V_B(r_B) - 1/r_B$ represent the interaction potential of the electron with each one of the two cores separately. $V_{cc}(R)$ is the core-core interaction:

$$V_{cc}(R) = \frac{1}{R} + V_{cpol}(R) \tag{34}$$

with

$$V_{cpol}(R) = -\frac{\alpha_d}{2R^4}.$$

Finally, the tensorial term V_3 is a cross polarization term depending upon the angles $(\mathbf{r}_A, \mathbf{R})$ and $(\mathbf{r}_B, \mathbf{R})$ and accounting for the fact that when one core is polarized simultaneously by the electron and by the other core, the electric fields should be added rather than the potentials. This term should be written

$$V_3(R, r_A, r_B) = (\alpha_d/R^2 r_A^2)\cos(\mathbf{R}, \mathbf{r}_A)[f(r_A/r_0)]^{1/2}$$
$$+ (\alpha_d/R^2 r_b^2)\cos(\mathbf{R}, \mathbf{r}_B)[f(r_B/r_0)]^{1/2}. \tag{35}$$

If V_3 is not included, the polarization potentials of one core due to the electron and to the other core are added, as if the corresponding electric fields were parallel. This is not justified, unless the electron and the other core stay in the same direction: *the neglect of the tensorial term V_3 will therefore cause an overestimation of the core-polarization effects.*

The fact that the square root of the cutoff function f used in Eq. (32) should be considered in Eq. (39) has been discussed by Peach (1978) while the choice of the cutoff function itself is discussed by Reeh (1960). We have omitted on purpose the higher-order terms, such as the quadrupole polarizability, which have been considered in the early work (Bottcher and Dalgarno, 1974) and shown later on to be unnecessary (Müller and Meyer, 1984; Henriet and Masnou-Seeuws, 1983).

The *two*-electron effective Hamiltonian for the alkali dimers is (see Fig. 16c)

$$H(1, 2) = h(1) + h(2) + V(1, 2) - V_{cc}(R) \tag{36}$$

In Eq. (36), $h(1)$ and $h(2)$ are the effective Hamiltonians, [see Eq. (33)], for the motion of the electrons (1) and (2), treated independently, in the field of the two cores A^+ and B^+. The core–core interaction $V_{cc}(R)$ [see Eq. (34)], being counted twice, has to be subtracted. Finally, the interaction between the two electrons is represented by the effective operator,

$$V(1, 2) = 1/r_{12} + V_{diel}$$

with

$$V_{\text{diel}} = -\frac{\alpha_d^A \cos(\mathbf{r}_{A1}, \mathbf{r}_{A2})}{r_{A1}^2 \cdot r_{A2}^2} \left[f\left(\frac{r_{A1}}{r_0}\right) f\left(\frac{r_{A1}}{r_0}\right) \right]^{1/2}$$
$$- \frac{\alpha_d^B \cos(\mathbf{r}_{B1}, \mathbf{r}_{B2})}{r_{B1}^2 \cdot r_{B2}^2} \left[f\left(\frac{r_{B1}}{r_0}\right) f\left(\frac{r_{B1}}{r_0}\right) \right]^{1/2}. \quad (37)$$

Like all the other operators, the two-electron interaction $1/r_{12}$ has to be corrected to take account of core polarization effects. The dielectric correction V_{diel} in Eq. (37), first introduced by Chisholm and Opik (1964), comes from the addition of the two electric dipoles created by the electrons (1) and (2) on one polarizable core which are not generally oriented in the same direction. *Its omission also results in an overestimation of the core polarization effects*: for instance, the electron affinity of the alkali negative ion is overestimated (Norcross, 1973).

We shall turn now to the applications of such a formalism. The potentials $V(r)$ in Eqs. (33) and (36) are parametric potentials adjusted so as to reproduce the experimental spectrum of the corresponding alkali atom, by solving the one-electron Schrödinger equation

$$[-\tfrac{1}{2}(\nabla)^2 + V(r) - 1/r]\phi_i(\mathbf{r}) = \epsilon_i \phi_i(\mathbf{r}) \quad (38)$$

and minimizing the difference between the computed ϵ_i and the experimental ones. There exist two main classes of potentials:

(i) model potentials, attractive in the core region, associated with wavefunctions $\phi_i(\mathbf{r})$ which have the correct number of nodes. Such a potential has been proposed by Klapisch (1969, 1971) and subsequently used to compute various properties of the alkali atoms (Aymar, 1978; Aymar, et al., 1976). It is a three parameter potential $V_K(r)$.

$$V_K(r) = -\frac{Z-1}{r} [\exp(-\alpha_1 r) + \alpha_2 r \exp(-\alpha_3 r)] \quad (39)$$

In Eq. (39) Z is the core charge and $\alpha_1, \alpha_2, \alpha_3$ are adjustable parameters. In the Harvard group, a more complicated model potential has been introduced (Weisheit and Dalgarno, 1971; Bottcher et al., 1973; Cerjan et al., 1976; Laughlin, 1978):

$$V_H(r) - \frac{1}{r} = U_{\text{HF}} - \frac{\alpha_d}{2r^4} f_6(r/r_1) - \frac{\alpha_q}{2r^6} f_8(r/r_1)$$
$$+ (a_0 + a_1 r + a_2 r^2)\exp(-r/r_2) \quad (40)$$

In (40) U_{HF} is the core Hartree-Fock potential, α_d and α_q are the dipole and quadrupole polarizabilities of the core, f_6 and f_8 are cutoff functions, $r_1, r_2, a_0,$

a_1, a_2 are adjustable parameters. Although in principle the potential V_H containing the correct polarization contribution V_{pol} [see Eq. (31)] should give better results for the alkali energy levels, the two kinds of potentials lead to the same accuracy (for the sodium atom, mean error 8 cm^{-1}, maximum error 20 cm^{-1}). Henriet (1985) has shown that in V_H the r^{-4} term is dominated by the $r \exp(-r/r_2)$ term, which simulates exchange effects. Therefore neither the cutoff function f nor the cutoff radius r_0, present in (32), can be unambiguously fit to the atomic levels: this fact has important consequences for the molecular problem.

(ii) alternatively, pseudopotentials that are repulsive in the core region have been proposed by Bardsley (1974) and later on by Valance and Nguyen Tuan (1978, 1980). The "valence" solutions are then the only solutions of the Eq. (39), and the atomic ground-state wave function has no nodes in the core region. In the absence of the core orbitals, the Pauli exclusion principle is simulated by an l dependence of the potential; Bardsley (1974) has proposed the following potential:

$$V_B(r) = \sum_l A_l r^p \exp(-B_l r^q) P_l - \frac{\alpha_d}{2(r^2 + d^2)^2} - \frac{\alpha_q}{2(r^2 + d^2)^3}. \quad (41)$$

In (41) P_l is the projection operator on the subspace of atomic valence wavefunctions with an orbital angular momentum l. The parameters A_l, B_l correspond to atomic states with orbital angular momentum $l = 0$, 1 and $l \geq 2$; they are fit, together with the cutoff parameter d, to the experimental spectrum. Both pseudopotentials and model potentials have been capable of providing accurate results for the atomic properties.

Turning now to the treatment of the alkali dimer ion, by solution of the one-electron, two-cores Eq. (33), we must note that the accuracy of the results markedly depends upon the inclusion of the V_3 correction, which we have defined in the Eq. (35). We present in Table II the spectroscopic constants for the ground state of the Li$_2^+$, Na$_2^+$, K$_2^+$ molecular ion obtained in various calculations. *When the cross polarization correction V_3 is neglected, the induced dipole moments are taken parallel, and the polarization forces are overestimated: it is clear from Table II that the calculations then overestimate the dissociation energy of the corresponding molecular ion, and underestimate the bond length.*

In the perturbation theory that we have described above, the cross-polarization correction V_3 is determined asymptotically and depends on a cutoff radius r_0 typical of the core dimensions (see Eq. (32)). We reproduce in Fig. 18 the ground state potential curves of Na$_2^+$ obtained by Henriet and Masnou-Seeuws (1983) for a given cutoff function and for various choices of the cutoff radius r_0. It is clear that the accuracy of the results does depend

TABLE II

Spectroscopic Constants for the Ground States of Li_2^+, Na_2^+, K_2^+.[a]

	D_e (eV)	R_e (Å)	ω_e (cm^{-1})	Method	Ref.
				Neglecting V_3	
Li_2^+	1.280	3.127		ab initio SCF	Konowalow Rosenkrantz (1988)
	1.281	3.096		ab initio SCF	Müller and Meyer (1984)
				Including V_3	
	1.284	3.14		model potential	Cerjan (1975)
	1.28	3.09	283	model potential	Henriet et al. (1983)
	1.30	3.09	268	pseudo-potential	Fuentealba et al. (1982)
	1.295	3.096	265.5	SCF + core polarization	Müller and Meyer (1984)
				Experiment	
	1.298	3.11	262.2 ± 1.5		Bernheim et al. (1983)
	1.300				Eisel et al. (1983)
				Neglecting V_3	
Na_2^+	1.02	3.54		pseudopotential	Bardsley et al. (1976)
	1.02	3.28		pseudopotential	Valance (1978)
	1.02	3.55		model potential	Henriet et al. (1983)
	0.92	3.74	114	ab initio pseudopotential	Jeung et al. (1982)
	0.97	3.718		ab initio SCF	Müller and Meyer (1984)
				Including V_3	
	0.98	3.3		model potential	Cerjan et al. (1976)
	0.986	3.60	123	model potential	Henriet et al. (1983)
	0.993	3.59	119	pseudopotential	Fuentealba et al. (1982)
	0.95	3.52	119	ab initio pseudopotential	Jeung et al. (1982)
	0.987	3.598	120.7	SCF + core polarization	Müller and Meyer (1984)
				Experiment	
	0.984	3.60 ± 0.04	118		Carlson et al. (1980)
	0.986	3.60 + 0.05	120.8 ± 0.8		Martin et al. (1982)
				Neglecting V_3	
K_2^+	0.76	4.18		pseudopotential	Valance (1988)
	0.904			model potential	Henriet et al. (1983)
	0.69	4.83	61	ab initio pseudopotential	Jeung et al. (1982)
	0.804	4.796	66.7	ab initio SCF	Müller and Meyer (1984)
				Including V_3	
	0.823	4.49	72	pseudopotential	Valance (1978)
	0.801	4.44	75	model potential	Henriet et al. (1983)
	0.82	4.50	74	ab initio pseudopotential	Jeung et al. (1982)
	0.823	4.49	72	SCF + core polarization	Müller and Meyer (1984)
				Experiment	
	0.794a	4.4	72.5		Leutwyler et al. (1980)
	0.789a	4.4	73.4		Broyer et al. (1983)

[a] In the case of K_2^+ the experimental determination of D_e depends on an estimation of the K_2 well depth, which might be in error by a few meV (see Broyer et al., 1983).

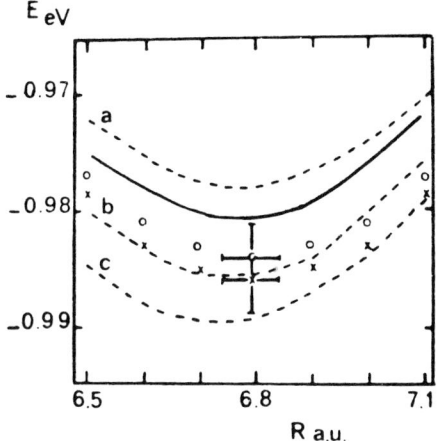

FIG. 18. Potential curves for the ground state of Na_2^+ in the minimum region for various choices of the cutoff radius r_0. Dashed lines: calculations using (a) $r_c = 2$, (b) $r_c = 3$, (c) $r_c = 3.5$. Full line: calculation using the value fitted by Weisheit (1972) on photoionization cross sections. Crosses: experimental results of Martin et al. (1982). Circles: experimental results of Carlson et al. (1980).

upon this choice, which appears delicate and will be discussed in the following paragraphs. This is probably the reason why the results of the Harvard group, although very accurate for Li_2^+ (Cerjan, 1975), underestimate by 10% the bond length for the Na_2^+ ground state (Cerjan et al., 1976). In the pseudopotential calculations of Bardsley et al. (1976) for Na_2^+ and of Valance (1978) for Na_2^+ and K_2^+, the three-body term V_3 has been omitted, so that the core polarization forces are overestimated.

A few applications of the model potential and pseudopotential methods to the alkali dimer problem have been carried out before 1980. After solving the one-electron Schrödinger equation [Eq. (33)] for the alkali dimer ion

$$h_j \chi_j = \epsilon_d \chi_j \qquad (42)$$

the eigenfunctions of the two-electron Hamiltonian [Eq. (36)] are determined by a configuration interaction technique. The calculations of the Harvard group for the molecular properties of Li_2 (Watson et al., 1977; Watson, 1977; Uzer et al., 1978), of Bardsley et al. (1976) for Na_2 and of Valance and Nguyen Tuan (1982) for Na_2 and K_2 do not include the core polarization terms V_3 and V_{diel}: some results for the ground state are compared with more recent results in the paper of Müller and Meyer (1984), and some results have been obtained for the first excited states. In spite of good accuracy, the method has not been extended to more excited states, the calculations being generally performed with a limited basis set of Slater orbitals on each center.

As we shall see below technical progress in the art of computation made the *ab initio* methods more competitive for a while.

3. Ab initio *treatment of the excited states of the alkali dimers*

In 1976, Bertoncini and Wahl proposed an *ab initio* calculation of the ground and first excited states of Na_2 (see Stevens *et al.*, 1977), using the multiconfiguration self-consistent field method of Das and Wahl (1972). In such an approach the 22 electrons are explicitly included. The orbitals for the 20 core electrons are determined first; for the valence electron the interaction with the core is no longer represented by a potential as in the preceding paragraph, but by a Fock operator including the direct and exchange two-electron integrals with the core electrons. The core orbitals are assumed to be unmodified by the presence of the valence electron (frozen core approximation), so that the core polarization effects are neglected. This approach has been developed by Konowalow and his group, who have optimized large basis sets of Slater type atomic orbitals in order to compute the eight lowest potential curves of Li_2 and Na_2 with *different* symmetry, together with dipole moment matrix elements (see Olson and Konowalow, 1977; Konowalow and Olson, 1977, 1979; Konowalow *et al.*, 1980, 1983; Konowalow and Julienne, 1980, and the review by Konowalow and Rosenkrantz, 1982). The accuracy in the well depth of the various curves is about 5%. The results show a tendency to *underestimate* the potential well depth and to *overestimate* the bond length: for instance the Li_2 ground state is too shallow by 0.03 eV (3%), with the potential minimum displaced 0.02 Å toward greater internuclear separation.

Next the availability of a configuration interaction technique, limited to valence excitations, and using portions of the ALCHEMY program of Bagus, *et al.*, made possible the calculation of various states of the *same* symmetry. Konowalow and Fish (1983, 1984) have introduced into the quantum chemistry codes the effective core potentials of Bardsley, (1974) [see Eq. (41)]. The treatment is then similar to the pseudopotential treatment described in Sec. III,B,2, and potential curves have thus been obtained for the 26 lowest lying states of Li_2 (up to the 2p + 2p dissociation limit) over a wide range of internuclear distances (from 3 to 35 a.u.). Such calculations neglect the cross polarization terms V_3 and V_{diel} defined in the preceding paragraph so that the core polarization effects are *overestimated*. Konowalow and his coworkers then realized that in the two kinds of calculations for Li_2 (frozen core multiconfiguration SCF or pseudopotential calculations neglecting the V_3 correction) the errors are of the same magnitude but opposite in sign: Konowalow and Fish (1984) proposed an *ad hoc* way of introducing core polarization effects by considering curves lying between the two previous determinations. Although they could obtain an excellent agreement with the experimental spectroscopic constants for the ground state, it was clear that a more rigorous treatment of core polarization effects was necessary.

One possibility was to get rid of the frozen core approximation by performing *ab initio* calculations in which excited core configurations are explicitly included: this approach, used by Rosmus and Meyer (1976) for the ground state of NaLi and by Partridge *et al.* (1983) for the ground state of Li$_2$ provide substantial improvement of the spectroscopic constants. However, such a procedure reveals a very slow convergence entailing burdensome computations, so that specific methods have been developed in order to introduce the core-polarization effects in a more elegant way.

4. Inclusion of core–valence correlation effects and of pseudopotentials in ab initio calculations

The Toulouse group (Maynau and Daudey, 1981; Jeung *et al.*, 1982, 1983; Jeung, 1983) has made an important contribution to the development of *ab initio* pseudopotential calculations for the alkalis and their dimers. Maynau and Daudey (1981) proposed an *ab initio* pseudopotential for Li, Na, and K:

$$V_{\rm T}(r) = \sum_l \left(\sum_i c_l^i r^{n_l^i} \exp(-\alpha_l r^2) \right) P_l. \qquad (43)$$

In Eq. (43), α_l, c_l^i and n_l^i are adjustable parameters, and the projection operator P_l has been defined in Eq. (41). The aim of the fitting procedure is no longer to reproduce a large number of experimental energy levels of the alkali atom but to obtain for its ground state a wave function $\phi(r)$ as close as possible to the Hartree-Fock wave function. Next the alkali dimer is treated as a two-electron problem by using a large basis set of Gaussian orbitals on each center,[2] fitted on previously determined Slater-type orbitals (Stevens *et al.*, 1977), and by performing an extensive configuration interaction. Because neglect of core-valence correlation effects leads to an overestimation of the bond length by 6%, Jeung *et al.* (1982, 1983) proposed a *perturbative* treatment to include such effects. Their method is similar in principle to the Bottcher and Dalgarno treatment described in Sec. III,B,2, but without the assumption that the valence excitation energies are negligible compared to the core excitation energies. The introduction of an effective operator with a somewhat arbitrary cutoff is therefore avoided: instead the method explicitly considers the perturbation of the valence wave functions, which are corrected up to the second order for core-valence correlation effects. However, as

---

[2] A modified version of the HONDO package is used for the calculations of the two-electron integrals.

discussed by Müller and Meyer (1984) the arbitrary choice of the cutoff radius r_0 is replaced by arbitrariness in the choice of the basis set, which limits the molecular orbital space available to describe this perturbation. The spectroscopic constants computed for the ground states of Li_2^+, Na_s^+, and K_2^+ are reported in Table II.

The main success of the method has been obtained for the excited states of Na_2 and K_2. Jeung (1983, 1987) has computed 38 electronic states of Na_2, with an accuracy never reached before. Comparison with spectroscopic data shows that the excitation energies, which vary from 19×10^3 to $35 \times 10^3 \text{ cm}^{-1}$, are predicted within 200 cm^{-1}, and the position of the minima within 0.1 Å. In contrast, the curves computed for the states correlated to the Na(3p) + Na(3p) dissociation limit (Jeung and Ross, 1988) do not agree with the experiment, the discrepancy being most likely due to the limitation of the basis set. Very good results have also been obtained (Jeung and Ross, 1988) for the 23 lowest states of K_2, up to the (4s + 3d) dissociation limit.

A somewhat different approach has been developed in Kaiserslautern by W. Meyer and his group. In 1984, Müller et al. introduced effective core polarization potentials into the SCF and configuration interaction treatments. The effective core polarization potentials act in the valence orbital space only, and are identical to the operators V_{pol}, V_3, V_{diel}, and V_{cpol} defined in Sec. III,B,2.[3] Those operators depend upon a cutoff function and a cutoff radius which are empirically adjusted to give the best fit to atomic properties (ionization potential of the atomic ground state, dipole moment matrix element). A large Gaussian basis set is used. The high accuracy results for the ground-state spectroscopic constants of Li_2^+, Na_2^+, and K_2^+ obtained by Müller and Meyer (1984) are given in Table II. We have also displayed the spectroscopic constants obtained by the same authors when the effective operators are not included: we note that in this approach the core-polarization corrections are smaller than in the calculations of Jeung et al. (1982).

The spectroscopic constants for the ground state of the alkali dimers computed by Müller and Meyer (1984) reveal a remarkable accuracy: the maximum deviation from the experimental data are 1% or 0.03 Å for the equilibrium distance R_e, 2% or 100 cm^1 for the well depth D_e, 0.5% or 1 cm^{-1} for the vibrational constant ω_e. An even higher accuracy may be found in the calculations of Schmidt-Mink et al. (1985) for eighteen electronic states of Li_2, the largest discrepancies with the available experimental data being 0.006 Å for R_e, 81 cm^{-1} for D_e, that is, nearly one order of magnitude smaller than in the calculations quoted above (Watson et al.,

---

[3] Valence configuration interaction calculations are performed by including V_{pol}, V_3, and V_{cpol} into the Fock operator, and the two-electron integrals are transformed to include V_{diel}.

1977; Konowalow and Olson, 1979; Konowalow and Fish, 1984). We should note however that the Li$_2$ calculations of Schmidt-Mink *et al.* are not extended up to the Li(2p) + Li(2p) dissociation limit, the Gaussian basis set being apparently less convenient than the Slater basis used by Konowalow and Fish.

It will be very interesting to know whether the MQDT analysis developed by Ross and Jungen (1987) for H$_2$ can also be applied to Li$_2$ in order to predict the more highly excited curves, or whether the curves need to be explicitly computed.

5. *Model potential calculations for the intermediate Rydberg states of* Na$_2^+$, K$_2^+$, *and of* Na$_2$

From the preceding paragraph we may conclude that *ab initio* calculations are at present capable of providing the most accurate results for the ground states of the alkali dimers and their cations. Pseudopotential calculations of Fuentealba *et al.* (1982), using the effective core polarization potentials of Müller and Meyer (1984) with a preliminary version of the cutoff, provide molecular constants for the ground states of Li$_2^+$, Na$_2^+$ and K$_2^+$ molecular constants that are slightly less accurate than the all-electron calculations (see Table II). However, this conclusion does not necessarily apply to the excited states.

The Orsay group has developed a two-electron model potential method to compute the excited curves of Na$_2$ up to the 3p + 3p dissociation limit. The Klapisch potential defined in Eq. (39) seems well adapted to represent the ground and excited levels of the alkali atoms: the sodium energy levels are compared in Table III with experiment and with the *ab initio* calculations of Müller *et al.* (1984). The two main problems in the *ab initio* molecular treatment are *the representation of the diffuse orbitals and the convergence of the configuration interaction procedure*.

For the treatment of the excited states of the molecular ion, the one-electron Schrödinger equation (44),

$$\left(-\tfrac{1}{2}\nabla^2 + V_K(r_A) - \frac{1}{r_A} + V_K(r_B) - \frac{1}{r_B} + V_3(R, r_A, r_B) + V_{cc}\right)\chi(r) = \epsilon\chi(r), \quad (44)$$

is solved. V_K, V_3 and V_{cc} have been defined in the Eqs. (39), (35), and (34), respectively. The cutoff radius in V_3 can be varied so as to achieve best agreement with the atomic dipole moment or the photoionization cross sections (Henriet and Masnou-Seeuws, 1983). However, in the framework of a treatment of the molecular states of the Na$_2$ dimer, it seems justified to vary r_0 to get the best agreement with the experimental well depth of the Na$_2^+$ ground state.

TABLE III

IONIZATION ENERGIES FOR THE SODIUM ATOM LEVELS (a.u.)

Atomic level	E_1 (experiment)[a]	$E_2{}^b$	$(E_2 - E_1) \times 10^5$	$E_3{}^c$	$(E_3 - E_1) \times 10^5$
3s	0.188863	0.188831	3.2	0.18886	0
4s	0.071581	0.071666	−7.5		
5s	0.037586	0.037626	−4.0		
6s	0.023133	0.023143	−1		
3p	0.111551	0.111510	4.1	0.11175	24
4p	0.050937	0.051020	−8.3		
5p	0.029196	0.029238	−4.2		
3d	0.055938	0.055977	−3.9	0.05594	0
4d	0.031443	0.031499	−5.6		
4f	0.031261	0.031247	1.4		

[a] Moore tables.
[b] Computed with the Klapisch potential by Henriet (1985).
[c] Computed with the SCF + core polarization potentials by Müller et al. (1984).

Equation (44) is solved by expanding the wave function $\chi(r)$ on a large basis set of generalized Slater orbitals in elliptic coordinates (ξ, η, ϕ) where ξ and η have already been defined in Eq. (27) and ϕ is the azimuthal angle. The basis vectors are (Valiron, 1976):

$$e_{ij} = \xi^{p_j}\eta^{q_j} \exp\left(-\alpha_i \frac{R}{2}(\xi \pm \eta)\right), \qquad 0 \leq p_j, q_j \leq n_i - 1 \qquad (45)$$

and should be compared with Eq. (27). The large basis set contains the spherical Slater orbitals on each center with exponent α_i and orbital momentum $l_i \leq n_i - 1$. Diffuse orbitals with large values of l are thus generated.[4]

The excited and Rydberg states of the Li_2^+, Na_2^+ and K_2^+ molecular ions have been computed (Henriet and Masnou-Seeuws, 1983; Henriet, 1985) up to the states dissociating into a Rydberg atom with a principal quantum number $n = 6$. The results can be interpreted in terms of molecular quantum defects, so that in a given Rydberg series more excited wave functions could be computed as Coulombic functions.

For the treatment of the Na_2 problem, the two-electron Hamiltonian $H(1, 2)$ defined in Eq. (36) is computed within a space of configurations. Two

---

[4] In the actual calculations for Na_2^+, the basis set contains eight s, eight p, six d, and two f basis functions on each center together with elliptic orbitals. The most diffuse orbital has a maximum at r = 33 a.u. of each center.

different expansions have been considered for the two-electron wave function:

$$F(1, 2) = \sum \alpha_{ab}[\chi_a(1)\chi_b(2) \pm \chi_a(2)\chi_b(1)]\Omega(r_{12}), \qquad (46)$$

$$F(1, 2) = \sum \beta_{ab}[\chi'_c(1)\chi'_d(2) \pm \chi'_c(2)\chi'_d(1)]. \qquad (47)$$

In Eqs. (46) and (47) the coordinates of the two electrons are indicated by (1) and (2). In Eq. (46), χ is an orbital of the molecular ion, calculated as a solution to Eq. (44). The term $\Omega(r_{12})$ is a correlation function first introduced by Pluvinage (1950, 1951) and ensuring a correct behaviour for $r_{12} \to 0$. Results for the 55 first excited curves of Na_2 have been obtained by Henriet and Masnou-Seeuws (1987) using a limited (≈ 20) number of such configurations. The accuracy of the spectroscopic constants is not so good as in Jeung's results (see Sec. III,B,4). Although the position of the curve minima is correctly predicted, the excitation energies are overestimated by ≈ 400 cm^{-1}. An analysis of the results shows that the introduction of a correlation function dependent on r_{12} modifies the asymptotic part of the wavefunction for the excited states: it appears that it is in fact very important to describe correctly the Coulombic wave function when a Rydberg electron is outside the Na_2^+ core. The Orsay group therefore has proposed the expansion (47) which uses orbitals χ', eigenfunctions of a modified model Hamiltonian obtained by adding a screened Coulomb potential on each center:

$$h'(i) = h(i) + \left(\frac{0.5}{r_{Ai}} + \frac{0.5}{r_{Bi}}\right)\{1 - \exp[-\gamma(\xi_i - 1)]\}. \qquad (48)$$

In (48) γ is a screening parameter and ξ_i the elliptic coordinate defined above. The lowest χ' orbitals are identical to the solution of Eq. (44) while the Rydberg orbitals asymptotically correspond to the real Na_2 situation where the Coulomb field of each Na^+ ion is partially screened by the innermost of the two electrons. The two-electron interaction $V(1, 2)$ is therefore corrected for this "average" interaction introduced in (48).

A numerical evaluation of the two-electron integrals has been developed, which at present reaches an accuracy of 10^{-5} a.u.—far less than in standard *ab initio* codes. Nevertheless, the quality of the spectroscopic constants is as good as in Jeung's results (see Table IV). Moreover, the method is capable of predicting excited states correlated to the $Na(3p) + Na(3p)$ dissociation limit, for instance the $7\ ^1\Sigma_g^+$ and $8\ ^1\Sigma_g^+$ states indicated in the table.

The conclusion therefore is that when the core polarization effects are correctly included, both *ab initio* and model potential treatments are capable of providing accurate adiabatic potential curves. Obtaining the excited curves depends mainly upon "savoir-faire" in the choice of the basis sets. An

TABLE IV

Spectroscopic Constants for the $^1\Sigma_g^+$ Intermediate Rydberg States of Na_2.[a]

	$4\,^1\Sigma_g^+$	$5\,^1\Sigma_g^+$	$6\,^1\Sigma_g^+$	$7\,^1\Sigma_g^+$	$8\,^1\Sigma_g^+$
T_e (10^3 cm^{-1})					
Orsay work	28.61	31.9	32.7	34.78	35.04
Experiment	28.32561[a]	31.77243[b]	32.56216[a]	34.93967[a]	35.09568[a]
				34.701	34.858
Jeung (1987)	28.5	31.9	32.8		
B_e (cm^{-1})					
Orsay work	0.088	0.110	0.110	0.113	0.113
Experiment	0.08994	0.11363	0.10586	0.10835	0.11087
Jeung (1987)	0.0838	0.107	0.101		
ω_e (cm^{-1})					
Orsay work	105	113–115[b]	123	119	119
Experiment	108.738	109.412	123.6736	114.774	118.612
Jeung (1987)	107	110	119		

[a] The excitation energies T_e, rotational constant B_e and vibrational constant ω_e are compared with the theoretical determination of Jeung (1987) and with the experimental determination of (a) Taylor et al. (1981) and (b) Yan et al. (1987). The underlined excitation energies have been obtained by modifying by 2 units the vibrational numbering (a).

[b] The strong anharmonicity of the $5\,^1\Sigma_g^+$ potential curve makes the determination of ω_e ambiguous.

important factor is apparently the representation of the asymptotic Coulombic part of the Rydberg wave function.

6. Molecular quantum defects. Diabatic potential curves

The adiabatic potential curves for the various symmetries of the Na_2 dimer have been interpreted in terms of molecular quantum defects by Henriet and Masnou-Seeuws (1988, 1989). Perturbations in the Rydberg series do appear, which can be explained (see Sec. III,A) by an interaction between different channels corresponding to one Rydberg electron in the field of a Na_2^+ ion which is either in the ground state or in an excited state.

Quasidiabatic curves can therefore be computed by dividing, at a given internuclear distance R, the configuration space into two subspaces **P** and **Q**: the subspace **P** contains the singly excited configurations with one electron in the bonding σ_g 3s orbital, while the subspace **Q** consists of doubly excited configurations.[5] The electronic Hamiltonian is diagonalized separately with-

---

[5] In the case of the $^1\Sigma_g^+$ symmetry, the two subspaces have to be modified so that P includes the ground state with a correct dissociation limit at large R.

in the subspaces **P** and **Q**: the corresponding quasidiabatic curves are displayed in Fig. 19 for the $^3\Sigma_u^+$ symmetry. One may see that the unperturbed Rydberg series (in **P**) converging to the Na_2^+ ground state are crossed by doubly excited curves (in **Q**) converging to the excited $A\ ^2\Sigma_u^+$ state of Na_2^+. The matrix element of V_{nl} between one state of Q and the state of **P** belonging to the same Rydberg series is shown to verify a scaling law [see Eq. (52) in Sec. IV,A] and therefore can be extrapolated to the electronic continuum. In constrast, the matrix element of the radial coupling operator d/dR may be shown to be negligible. Accurate calculation of these couplings are necessary for the treatment of the dynamical problem (see Sec. IV).

Finally, a look at Fig. 19 together with the consideration of diabatic curves for other symmetries (Henriet and Masnou-Seeuws, 1989) shows that the $^3\Sigma_u^+$ symmetry is probably the best candidate for the associative ionization between two Na(3p) atoms: indeed the diabatic curve with the 3p + 3p dissociation limit then crosses the Na_2^+ ground-state potential curve. This

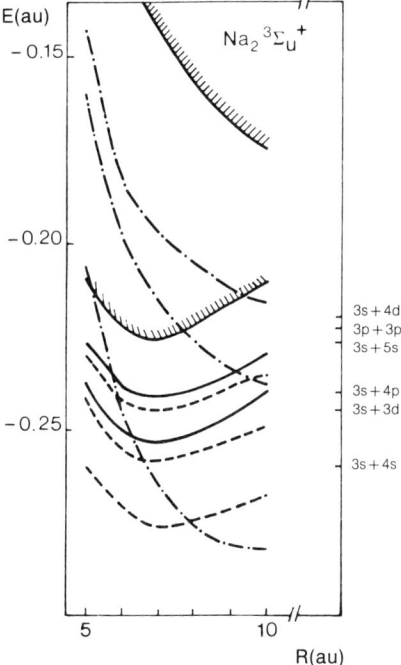

FIG. 19. Quasidiabatic potential curves of $^3\Sigma_u^+$ symmetry for the Na_2 molecule: dashed lines, "p" Rydberg series; solid line "f" Rydberg series; dash-dotted line, doubly excited states. The shadowed curves correspond to the ground and first excited states of the Na_2^+ ion.

selectivity should be confirmed by other calculations, and verified for other systems (K + K). We note that the diabatic curves are at present known only for internuclear distances $R \leq 10$ a.u., and that the long-range part of the potential curves and couplings has still to be determined in order to understand how the initially populated atomic states transform to molecular states prior to the ionizing event.

7. Long-range forces

Some of the *ab initio* calculations for the excited states of the sodium dimers described in Sec. III,B,3 have been interpreted in terms of long-range forces. Konowalow and Fish (1983) have analyzed the energies of the 26 lowest lying states of Li_2, including the states dissociating into Li(2p) + Li(2p) in order to determine the leading terms of the first order electrostatic energies and the second order dispersion energies at large internuclear distances. For several states, an important perturbation by the ionic curve dissociating into $Li^+ + Li^-$ has been found. Similar work has been performed by Konowalow and Rosenkrantz (1982a) for the Na_2 potential curves, while Ross and Jeung (1986) have discussed how a valence bond treatment could reproduce the asymptotic part ($R > 12$ a.u.) of the RKR experimental curve for the ground and first excited state of NaK.

Bussery and Aubert-Frecon (1985a) have used a perturbation treatment to compute the long-range interaction between two identical alkali atoms, at distances such that the overlap between the electronic clouds is negligible. They consider the molecular states correlating either to two ground-state atoms or to one ground-state atom and one atom excited to the first resonance level. Calculations neglecting the spin-orbit effects yield multipole expansion coefficients which are in good agreement with the *ab initio* determination or with previous calculations (Dalgarno, 1967) for Li_2 and Na_2. Some discrepancies appear for the heavier systems (Cs_2, Rb_2). Calculations including the spin-orbit effects show that extrema are present in the long-range part of the potential curves correlated to the $ns + np$ dissociation limit. Similar calculations have been performed by Bussery and Aubert-Frecon (1985b) for the Na(s) + K(4p) system.

Motivated by the experimental results of energy pooling collisions, Kowalczyk (1979, 1980, 1984) has developed calculations where the interaction of the two excited sodium atoms is dominated by the exchange interaction between the asymptotic part of the two atomic wavefunctions. In the case of two interacting Na(3p) atoms, this interaction is considered (Kowalczyk, 1984) together with the dispersion energy. The effect of the spin-orbit coupling is included, and two avoided crossings are found, near 30 and 20 atomic units, with potential curves correlated to the 3s + 4d asymptote which is above the 3p + 3p asymptote: such calculations demonstrate that it is certainly not possible, in the asymptotic region, to limit the expansion to products of atomic states restricted to the Na(3p) manifold, but that a larger space including the 3s + 4d and possibly 3s + 5s dissociation limits should be employed. The importance of exchange effects compared to dispersion forces must be clarified in the future. The connection between this asymptotic region, where spin-orbit effects have to be included, and the region of shorter internuclear distances where associative ionization occurs has to be established.

IV. Dynamics of Associative Ionization

A. INTRODUCTION

This section reviews and classifies the theoretical approaches applied to associative ionization dynamics. We assume here that the basic molecular data, the potential curves and couplings responsible for the ionization process, are either known from first principles or have been parametrically fit to experimental results. Initially we consider only one entrance channel at a time, leaving the problem of several coupled associative channels for the final discussion.

The main contributions to the theory of AI dynamics are summarized in Table V. We focus attention on cases where the asymptotic energy of the entrance potential $U_d(R)$ lies below the ion curve $U_+(r)$ at large R, and stays below (Fig. 1b) or crosses into it at shorter internuclear distance (Fig. 1a).[6] These two cases often (but not always) correspond to two different ionization mechanisms. The neutral curve in Fig. 1b, nearly parallel to the ion potential curve at short distance, corresponds to an ion ground-state configuration on which is superposed an electron in a Rydberg orbital. In this case no electronic interaction couples the neutral state to the ionization continuum, and ionization may only occur via nonadiabatic coupling (radial or Coriolis interaction) with energy exchange between electronic and nuclear motions. In contrast Fig. 1a corresponds to the case of diabatic state electronically coupled to the ionization continuum. We mention nevertheless that even when the neutral curve stays below the ion curve, it *may* correspond to an

---

[6] Metastable helium, He*(2 ^{3}S), colliding with atomic hydrogen does not belong in this category but is nevertheless referred to in Table V because the corresponding papers by Nakamura (1971) and Bieniek (1978) contain important formal developments valid for the present cases.

TABLE V

CALCULATIONS OF AI CROSS SECTIONS

Approach	References	Applications
Semi-classical	Miller (1970)	
	Cohen (1976)	He + He*(n³S, P, D), n = 3, 4
	Nielsen and Dahler (1979)	N*(⁴S, ᴰ) + O(³P)
	Mihajlov and Janev (1981)	Alkali systems
	Runge et al. (1985)	He*(S³P) + He
Quantum mechanical		
1. Vibrational ionization mechanism		
	Nielsen and Berry (1971)	H*(n = 3) + H
	Nakamura (1972)	H*(n = 3) + H
Quantum mechanical:		
2. Electronic ionization mechanism		
a. Configuration mixing or Fesbach operator formalism	Bardsley (1968a, b)[a]	
local complex potential	⎧ Nakamura (1971)	He(2³S) + H
	⎨ Bieniek (1978)	He(2³S) + H
	⎩ Jones and Dahler (1988)	
	⎧ Bieniek (1980)	H + H⁻
nonlocal treatments	⎨ Lam and George	
	⎩ (1984, 1985)	model calculations
b. MQDT	Giusti (1980)[a]	
	Urbain et al. (1986)	H + H*(2s)
	Takagi et al. (1988)	H + H*(2s)
	Golubkov et al. (1988)	N*(²D) + O(³P)
	Urbain (1989)	H + H*(2s)
	Henriet et al. (1988)	Na*(3p) + Na*

[a] Note: formalism developed for dissociative recombination.

excited configuration *electronically* coupled to the ionization continuum (Giusti-Suzor and Lefebvre-Brion 1977; Henriet and Masnou, 1989).

Besides the ion and neutral curves, the third essential quantity required for the analysis of AI dynamics is the bound-continuum coupling $V(\epsilon, R)$. It results from an integration over electronic coordinates of the relevant Born-Oppenheimer states, performed at fixed R, for a given kinetic energy $\epsilon = (\epsilon, \hat{\epsilon})$ of the ejected electron. Whatever the physical nature of $V(\epsilon, R)$, electronic or rovibronic, it is generally calculated separately for each partial wave l of the outgoing electron, thus generating a set of partial couplings

$V_l(\epsilon, R)$. One defines also the total "electronic width", Γ, which is related to the partial couplings by the golden-rule-like expression,

$$\Gamma(\epsilon, R) = 2\pi \sum_l |V_l(\epsilon, R)|^2. \qquad (49)$$

It is worthwhile commenting on the dimensions in Eq. (49). Here and in the rest of this section the continuum wave functions are assumed to be *energy normalized*,

$$\int \phi_l^*(\epsilon, r)\phi_l(\epsilon', r)dr = \delta(\epsilon - \epsilon'),$$

where ϕ_l is the l-partial wave component of the departing electron radial wave function. The dimension of ϕ_l is thus (length × energy)$^{-1/2}$ and the corresponding coupling terms, resulting from integration over electronic coordinates,

$$V_l = \langle \phi_l(\epsilon, r)\Phi_+ | H_{el} | \Phi_d \rangle \qquad (50)$$

have the dimension (energy)$^{1/2}$. In (50) Φ_+ and ϕ_d denote the bound electronic wave functions of the ion ground state and neutral state, respectively. The resulting width Γ given by Eq. (49) appears therefore as an energy, as expected.

It is to be noted that in the "crossing case" of Fig. 1a, the repulsive neutral state crosses not only the ion potential curve but also those of the infinite number of Rydberg states lying below. These attractive Rydberg states mimic the shape of the ion ground state and are termed "singly excited" because they consist of a ground state core and one electron in a Rydberg orbital. The repulsive neutral state belongs to a family of Rydberg levels converging on the first excited state of the molecular ion. The repulsive states are termed "doubly excited" because they are constructed from an excited core plus one Rydberg electron. The singly and doubly excited Rydberg states are coupled by an electronic interaction essentially similar to that responsible for ionization. These additional couplings obviously complicate the dynamics of the AI process and have been included in only a few treatments (Cohen, 1976; Urbain *et al.*, 1986; Takagi and Nakamura, 1988). The effect of these couplings will be described in some detail since they are intrinsic to the cases considered in this review.

We represent the singly excited Rydberg state as the product of a Rydberg electron orbital ϕ_{nl} and the ion core Φ_+. The interaction with the doubly excited state Φ_d is written,

$$V_{nl} = \langle \phi_{nl}\Phi_+ | H_{el} | \Phi_d \rangle. \qquad (51)$$

The density of states in the vicinity of the nth Rydberg state with effective quantum number n^* is given by

$$\rho_n = \frac{dn}{dE} = \frac{n^{*3}}{2\,\text{Ry}}$$

obtained by differentiation of the Rydberg formula $E_n = -\text{Ry}/(n^*)^2$ for the binding energy of a Rydberg electron. Thus the "energy normalized" form of the Rydberg coupling is

$$\sqrt{\rho_n}V_{nl} = \frac{n^{*3/2}}{\sqrt{2\,\text{Ry}}}V_{nl}, \tag{52}$$

which is known to be nearly constant along a Rydberg series (Bethe, 1974) and extrapolates smoothly into the continuum value V_l of Eq. (50).

We describe now the main aspects of each treatment classified in Table V. The aim is to obtain partial cross sections $\sigma_{v^+ J^+}$ for the formation of a bound level of the molecular ion with vibrational and rotational quantum numbers (v^+, J^+), as well as their sum, the total cross section σ_{tot}. For a given total energy E, which also represents the initial kinetic energy of the incoming atoms, each cross section is first calculated for given angular momentum $\hbar J$ or semiclassical impact parameter b. One then performs a summation,

$$\sigma = \sum_{J=0}^{J_{\max}} (2J+1)\sigma_J \tag{53}$$

or the equivalent integration with respect to the impact parameter. We denote $J_{\max}$ as the angular momentum quantum number above which the centrifugal potential barrier prevents penetration into the associative ionization region. In the special case of identical reactants with nuclear spin I, symmetry considerations and nuclear statistics lead to the more elaborate expressions (Cohen, 1976)

$$\sigma = \sum_{J_{\text{even}}} (2J+1)[\lambda\sigma_J^g + (1-\lambda)\sigma_J^u] + \sum_{J_{\text{odd}}} (2J+1)[\lambda\sigma_J^u + (1-\lambda)\sigma_J^g], \tag{54}$$

where σ^g and σ^u are the respective contributions of gerade and ungerade molecular potential curves correlated to the atomic initial states, and

$$\lambda = \begin{cases} (I+1)/(2I+1), & I \text{ integral} \\ I/(2I+1), & I \text{ half integral} \end{cases}.$$

B. Semiclassical Treatments

The classical description of the nuclear motion in a heavy particle collision is most likely to be valid when the deBroglie wavelength of the relative particle is small compared to the range of the interaction potential and when the contribution from classically forbidden regions is negligible. Therefore it has been used mostly for Penning ionization processes with unbound nuclear motion in both the entrance and final channels and still more specifically for cases where the entrance neutral state lies far above the ion potential curve (Fig. 1c). These cases are excluded from our discussion since they have been extensively reviewed elsewhere (Niehaus, 1981). Here we just give a brief outline of the semiclassical formalism, following the basic paper by Miller (1970), and present two applications to diabatic curve-crossing situations which have been explored by Cohen (1976) and by Nielsen and Dahler (1979).

Along a well-defined classical path the number of ionization events per unit length, for orbital angular momentum $\hbar J$ and energy E, is

$$A_J(R) = \frac{\Gamma(R)}{\hbar v_J(R)}, \tag{55}$$

where $\Gamma(R)$ is the electronic width of Eq. (49) and $v_J(R)$ the radial velocity of the nuclear classical motion in the entrance channel,

$$v_J(R) = \left[\frac{2}{M}\left(E - V_d(R) - \frac{\hbar^2(J + \tfrac{1}{2})^2}{2MR^2}\right)\right]^{1/2}. \tag{56}$$

The ionization probability P_J during the collision is obtained as the integral

$$\int_{R_c}^{\infty} [A_J^{\text{in}}(R) + A_J^{\text{out}}(R)]dR,$$

where $A_J^{\text{in (out)}}$ are the ionization probability densities within the interval $[R, R + dR]$ during the inward (outward) part of the collision, and R_c is the classical turning point. Solving the integral expressions for the instantaneous probabilities, one easily gets

$$P_J(E) = 1 - \exp\left(-2\int_{R_c}^{\infty} A_J(R)dR\right) \tag{57}$$

and the total cross section for this inelastic collision is expressed as

$$\sigma_{\text{tot}}(E) = \frac{\pi}{\kappa^2}\sum_J (2J + 1)P_J(E), \tag{58}$$

where κ is the wave vector of nuclear motion, related to kinetic energy $E = \hbar^2\kappa^2/2M$. Note that the total cross section depends only on $\Gamma(R)$ and $V_d(R)$, but not on the ion potential $V_+(R)$, which nevertheless plays a role in the electron energy distribution. The independent conservation of electronic and nuclear energies imposes

$$\epsilon(R) = V_d(R) - V_+(R)$$

for the kinetic energy of the electron ejected at the internuclear distance R. This relation implies also a "local" (or vertical) ionization process. These two characteristics of the classical formulation (insensitivity to the ion potential curve and local ionization) will be shown to be poor approximations for the specific AI cases depicted in Figs. 1a and 1b.

A refinement of the classical approach is provided by the JWKB approximation, in which one calculates probability amplitudes instead of probabilities and introduces the *phase* of the nuclear motion along each of the potential curves $V_d(R)$ and $V_+(R)$. When the different amplitudes are superposed, interference effects appear which are neglected in the classical treatment. We do not give more details on these approaches, thoroughly discussed by Miller (1970) and Niehaus (1981), since the cases reviewed here require a quantal treatment. Interesting connections between the quantal formulas and the semiclassical ones are derived by Miller using the JWKB representation of bound and continuum nuclear wave functions, and stationary phase arguments for the evaluation of radial integrals.

Cohen (1976) developed a semiclassical multistate curve crossing (MSCC) model to analyze associative ionization between metastable and ground-state helium,

$$\text{He*}(n\ ^3\text{S},\ ^3\text{P},\ ^3\text{D}) + \text{He} \rightarrow \text{He}_2^+ + e, \qquad n = 3, 4.$$

A diabatic representation of the relevant manifolds of Rydberg states is depicted in Fig. 20. For each electronic symmetry the diabatic representation yields two series of Rydberg states: (1) an attractive manifold with the ion ground state as the core configuration, and (2) repulsive manifold with an excited state of the ion as the core configuration. The populated entrance channel may belong to either of these manifolds; and, as is evident from Fig. 20, will undergo a series of curve-crossings with members of the opposite Rydberg series. At each curve intersection the probability of staying on the same diabatic curve is evaluated with the help of the Landau-Zener formula,

$$p = e^{-2\pi\gamma_l}, \qquad \gamma_l = \frac{|V_{12}(R_x)|^2}{v_R(R_x)|d/dR(V_{11} - V_{22})|_{R_x}} \tag{59}$$

where $V_{12}(R_x)$ is the radial (or angular) coupling between the curves evaluated at the crossing R_x; v_R is the radial velocity, and $|d/dR(V_{11} - V_{22})|_{R_x}$ the difference of the slopes of the interacting curves at the crossing. To apply the model one needs to (1) calculate accurate diabatic curves that the relevant entrance channels are likely to encounter; (2) at each crossing calculate with

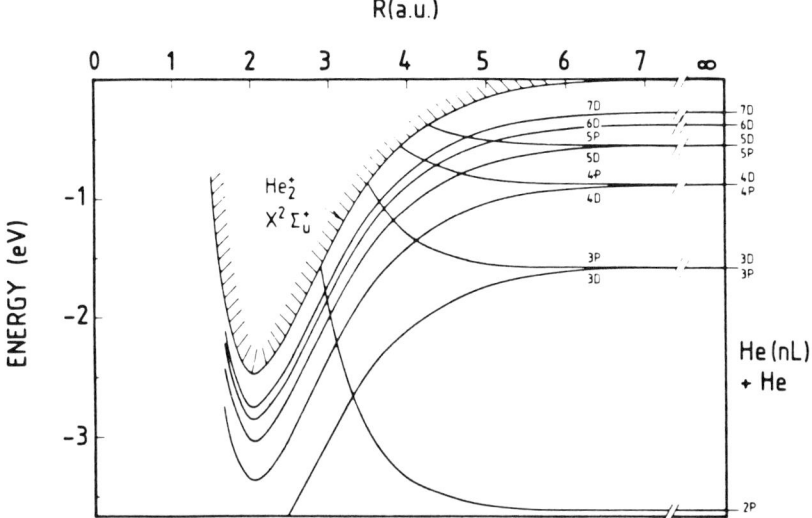

FIG. 20. Typical form of the diabatic curves used by Cohen (1976) to develop the MSCC model. Curves shown here are taken from Runge *et al.* (1985) and used to analyze process (15).

the help of Eq. (59) the probability of curve changing; (3) count all the pathways to the ionizing region and sum over the probabilities of following each one. Since the Landau-Zener expession is velocity dependent, the net result is a calculation of AI probability as a function of collision velocity. The total probability is simply related to the cross section through Eq. (58), and therefore the MSCC model is capable of predicting the excitation function (cross section versus collision velocity) for the ionizing collision. Cohen (1976), only had thermally averaged cross section results (Wellenstein and Robertson, 1972) with which to compare his calculations, and he found experiment and theory to be in reasonable accord. It was not until nine years later (Runge *et al.*, 1985) that the more stringent prediction of the excitation function itself could be compared with experiment; and, as discussed in Sec. II, the success of the theory is not unqualified.

In contrast to Cohen's theory which concerns itself entirely with Rydberg state diabatic crossings, Nielsen and Dahler (1979) have formulated a semiclassical treatment of AI in which the nuclei move primarily on valence-type adiabatic Born-Oppenheimer potential curves, and they neglect Rydberg interactions altogether. Their theory was motivated by the experimental results of Ringer and Gentry (1979) on the system

$$N(^4S, \,^2D) + (O \,^3P) \rightarrow NO^+ + e,$$

exhibiting a threshold for AI at about 0.4 eV collision energy. The essential elements of the treatment are (1) classical nuclear motion along adiabatic Born-Oppnheimer (ABO) potentials, and (2) the definition of a probability of ionization per unit length $A(R, E)$ with subsequent integration over the path to obtain the total ionization probability $P(R, E)$ [see Eqs. (55)-(58)]. It is worthwhile noting that the R dependence of Γ, estimated by an exponential expression,

$$\Gamma(R) = \Gamma_c \{\exp[-\gamma(R - R_c)]\}, \qquad R \leq R_c \tag{60}$$

implies a collision energy dependence which will be strongly reflected in the AI excitation function. The energy dependence comes from the classical turning point R_c—more energetic collisions will penetrate further into the ionizing region with a consequent increase in the size of Γ. In fact Nielsen and Dahler use γ as a variable parameter to fit theory to experimental results. Cohen (1976), on the other hand, treats the ionization event as a "black hole"—once the relative particle has found its way to the ionizing region, the probability of AI taking place is unity, independent of total energy. An interesting feature of Nielsen and Dahler's theory is the development of a threshold law for the endoergic AI cross section,

$$\sigma_{AI}(E) \sim (E - E_{AI})^{3/2}, \tag{61}$$

where E_{AI} is the collision energy threshold for associative ionization. However, they find the energy range of validity to be narrow (only about 0.1 eV above threshold) after which a linear dependence takes over. The slope in the linear part of the threshold behavior appears to be greater than in the experiment (Ringer and Gentry, 1979) by about a factor of two. The absolute values of the cross sections over an energy range of about 10 eV are in good agreement with experiment if γ is assigned a value of 4 (Eq. 60), although there is no *a priori* criterion for judging what the proper value of γ should be or even if the R dependence of Γ is exponential as expessed in Eq. (60).

Cohen (1976) as well as Nielsen and Dahler (1979) have both formulated semiclassical theories, but where Cohen has considered diabatic Rydberg crossings as the controlling physical phenomenon, Nielsen and Dahler have concentrated on adiabatic classical paths on *valence states* and have paramaterized the behavior of the ionizing width. Both theories calculate about the right values for the AI cross section, but neither treatment yields an excitation function in impressive agreement with experiment.

C. Quantum Mechanical Treatments

Two types of quantum mechanical treatments have been used for describing the dynamics of AI. The first one, developed by Nakamura (1971, 1972) and Bieniek (1978) is based on the extension of the configuration mixing treatment of atomic autoionization (Fano, 1961) to the molecular case. Briefly, the nuclear Schrödinger equation is solved for a *complex* potential, whose imaginary part is the electronic width $\Gamma(R)$ [Eq. (49)] measuring the ionization rate of the entrance neutral state. A refined treatment takes account of the *nonlocal* character of the ionization process. The second and more recent approach extends the multichannel quantum defect theory (MQDT) to include both ionization and dissociation channels. First introduced for dissociative recombination (Giusti, 1980) this approach is characterized by a symmetric treatment of the two half-collisions (electron–ion and atom–atom) and by the natural inclusion of the resonances due to the Rydberg levels below the ionization threshold.

The two classes of treatments will be reviewed below with an analysis of the approximations currently used for each of them in the actual calculations. We first define the notations and the general framework common to the two methods.

1. Notation

In both methods the total wave function is written as a superposition of two types of channel wave functions, associated with either the association or the ionization process. Each of them consists of the produce of a bound and a continuum part. For the entrance channel the bound part is the electronic wave function $\Phi_d(Q, R)$ already introduced in Eq. (50) (Q denotes collectively the N-electron coordinates) and the continuum part is the nuclear wave function $\chi_d(\kappa, R)$ for the incoming collision partners. For the ionization channels involved in the AI process (below the PI threshold), the bound part consists in the ion electronic wave function $\Phi^+(q, R)$ with q the $N - 1$ electron coordinates, and the nuclear wave functions

$$\chi_{v^+ J^+}(\mathbf{R}) = \frac{1}{R} Y_{J^+ M}(\mathbf{R}) \chi_{v^+ J^+}(R) \tag{62}$$

of the bound nuclear motion in the ion potential well. The continuum part is here the external electron wave function $\varphi(\mathbf{k}, \mathbf{r})$, $\mathbf{k}$ being the electron wave vector with the associated kinetic energy $\epsilon = \hbar^2 k^2 / 2m$.

The bound parts of the wave functions are considered as known or are calculated in a preliminary step. The "unknown" of the dynamical problem are the continuum components, electronic and nuclear. They are expanded in spherical harmonics in order to separate the radial and angular problems:

$$\phi(\mathbf{k}, \mathbf{r}) = \sum_{l,m} Y_{lm}^*(\hat{k}) Y_{lm}(\hat{r}) \frac{1}{r} \phi_l(\epsilon, r), \tag{63}$$

$$\chi_d(\mathbf{K}, \mathbf{R}) = \sum_{J,M} Y_{JM}^*(\kappa) Y_{JM}(\hat{R}) \frac{1}{R} \chi_d^J(E, R). \tag{64}$$

Analogous to the radial electronic wave function ϕ_l, already introduced in Eq. (63), the nuclear continuum wave functions χ_d^J are energy normalized

$$\int \chi_d^{*J}(E, R) \chi_d^J(E', R) dR = \delta(E - E')$$

and have also the dimension [length × energy]$^{-1/2}$. We recall that the electronic coupling V_l of Eq. (50) has the dimension [energy]$^{1/2}$ such that the effective channel interaction matrix elements

$$V_{dv^+}^{lJJ^+} = \int \chi_{v^+J^+}(R) V_l(\epsilon, R) \chi_d^J(E, E) dR \tag{65}$$

which play an important role in the quantum mechanical treatments, are *dimensionless* quantities.

2. The complex potential method

The resonance theory of Fano (1961) was first extended to molecules by Bardsley (1967, 1968) for the treatment of dissociative attachment and recombination processes. Adaptation to AI is mainly due to Nakamura (1971) and has been described in great detail and consolidated by Bieniek (1978). We summarize here the main steps of the treatment and refer the reader to the presentation of Bieniek for details.

The complex potential method is based on an atomic-like resonance treatment, within the Born-Oppenheimer representation. One starts from a linear superposition of electronic wavefunctions at fixed R, with a resonance corresponding to the neutral entrance state which may autoionize into the electronic continuum. In the atomic case, the Schrödinger equation leads to a system of *algebraic* equations for the unknown coefficients of this linear combination, which can only be satisfied for a complex energy characterizing the resonance energy and width. Here the same procedure leads to a system of coupled *differential* equations since the R-dependent coefficients are just the nuclear wave functions. By elimination of the ionization components, the system is transformed into a single *integro-differential* equation for the nuclear wave function of the neutral state, with a complex potential analogous to the complex energy of the atomic case.

a. The basic equations and their approximations. The most elegant way to translate this procedure into mathematical terms is to define a pair of Feshbach projection operators P and Q (Feshbach, 1962), which project the electronic wave function onto the orthogonal subspaces P and Q introduced in Sec. III,B,6 [we assume here that Q is restricted to a single "doubly excited" state (see Fig. 20) since we want to treat only one associative state at a time]. The operators P and Q may be written

$$P = \int d\epsilon |\Phi_\epsilon\rangle\langle\Phi_\epsilon|, \qquad Q = |\Phi_d\rangle\langle\Phi_d|,$$

where $\epsilon = (\epsilon, \hat{\epsilon})$ specifies the energy and direction of the ejected electron and $\Phi_\epsilon = \varphi_\epsilon \Phi_+$ is the total electronic wave function of the continuum part. The usual projector relations ($P^2 = P$, $Q^2 = Q$, $PQ = QP = 0$, and $P + Q = 1$) apply; the last one meaning that the neutral entrance and the continuum exit states are assumed to approximate adequately a complete set for the process under consideration.

Projection of the time-independent Schrödinger equation $(H - E)\psi = 0$, where H is the total Hamiltonian and ψ the total wave function, leads to the coupled equations

$$(PHP - E)P\psi = -PHQ\psi, \tag{66a}$$

$$(QHQ - E)Q\psi = -QHP\psi. \tag{66b}$$

A single equation for the neutral state component $Q\psi$ is obtained by elimination of $P\psi$ from the right-hand side of Eq. (66b), using the formal solution of Eq. (66a):

$$P\psi = -G_P^+ PHQ\psi. \tag{66c}$$

The complex Green operator, defined as $G_P^+ = \lim_{\eta \to 0} [P(E + i\eta - H)P]^{-1}$, yields the formal expression of the *outgoing wave* solution of Eq. (66a), corresponding to the ionization process. Substituting (66c) into (66b) and using $P^2 = P$,

$$(QHQ - E)Q\psi = -QHPG_P^+ PHQ\psi \tag{66d}$$

yields an expression restricted to the Q space with coupling to the ionization channels represented by the propagator $PG_P^+ P$ on the right-hand side.

This formal equation becomes more familiar in a coordinate representation. First, we note that the *homogeneous equations* associated with Eqs. (66a), (66b) and (66d) become, respectively, after left-multiplication by $\langle \Phi_\epsilon |$ and $\langle \Phi_d |$ and electronic integration,

$$\left(-\frac{\hbar^2}{2M}\nabla_R^2 + U_+(R) + \epsilon - E\right)\chi_+(\mathbf{R}) = 0, \tag{67a}$$

$$\left(-\frac{\hbar^2}{2M}\nabla_R^2 + U_d(R) - E\right)\chi_d(\mathbf{R}) = 0. \tag{67b}$$

One recognizes the nuclear Schrödinger equations with the ionic and neutral potentials. They are transformed into purely radial equations by separating angular and radial components, as in Eqs. (63) and (64) of Sec. IV,C:

$$\left(-\frac{\hbar^2}{2M}\frac{d^2}{dR^2} + U_+(R) + \frac{\hbar^2 J(J+1)}{2MR^2} + \epsilon - E\right)\chi_{J^+}(R) = 0, \tag{68a}$$

$$\left(-\frac{\hbar^2}{2M}\frac{d^2}{dR^2} + U_d(R) + \frac{\hbar^2 J(J+1)}{2MR^2} - E\right)\chi_d^J(R) = 0. \tag{68b}$$

The second step is to introduce for each value of J^+ the resolvant of the homogeneous Eq. (68a), i.e., the Green functions $G^+_{PJ^+}(R, R')$ which represent the partial wave decomposition and the coordinate representation of the Green operator G^+_P. The nonlocal character of these Green's functions, which involve simultaneously two radial distances $R \neq R'$, has important consequences for the coordinate representation of Eq. (66d) and is made manifest by their spectral decomposition into the solutions $\chi_{v^+J^+}$ of Eq. (68a):

$$G^+_{PJ}(E - \epsilon', R, R') = \lim_{\eta \to 0} \sum_{v^+} \frac{\chi_{v^+J^+}(R)\chi_{v^+J^+}(R')}{E - \epsilon + i\eta - E_{v^+J^+}} \tag{69}$$

with $E_{v^+J^+}$ the energy of the ion rovibrational levels (v^+, J^+).[7]

The remaining operators QHP and PHQ in Eq. (66d) correspond to the coupling terms between the P and Q spaces. Separation of radial and angular variables requires here partial wave expansion of the continuum electronic wave function Eq. (63) and introduction of the partial coupling terms V_l defined in Eq. (50). After averaging over the electron angular coordinates, the

---

[7] Rigorously, the v^+ summation over the discrete spectrum in Eq. (69) should be completed by an integration over the continuous spectrum above the ion dissociation limit. This part has negligible contributions for the low energies considered here (see Figs. 1a and 1b).

J partial wave component of Eq. (66d) reads in coordinate representation

$$\left(-\frac{\hbar^2}{2M}\frac{d^2}{dR^2} + U_d(R) + \frac{\hbar^2 J(J+1)}{2M R^2} - E\right)\tilde{\chi}_d^J$$

$$= -\int_0^\infty dR' \left[\sum_{lJ^+}(2J^+ + 1)\begin{pmatrix}l & J^+ & J\\0 & 0 & 0\end{pmatrix}^2\right.$$

$$\left.\times \int_0^\infty d\epsilon\, V_l(\epsilon, R)G_{PJ^+}^+(E - \epsilon, R, R')V_l(\epsilon, R')\right]\tilde{\chi}_d^J(R'). \quad (70)$$

This is the basic equation for the present treatment of AI dynamics. We show later (Sec. IV,C,2,c) how the partial and total cross sections are obtained from the neutral-state nuclear wave functions $\tilde{\chi}_d^J$ [distinct from the solutions χ_d^J of the associated *homogeneous* Eq. (68b)]. First, we comment on the physical meaning of Eq. (70) and indicate several approximations currently performed:

(i) The three J symbols on the right-hand side correspond to the conservation of total angular momentum: the initial momentum J of the nuclei is the sum of the two angular momenta of the final products, J^+ for the ion nuclear rotational motion and l for the ejected electron. Since the high-l electron partial waves usually contribute very little to the process due to the centrifugal barrier, J^+ and J have very similar values and one may average with respect to the different J^+ values on the right-hand side of Eq. (70), which reduces to

$$-\sum_l \int_0^\infty dR' \int_0^\infty d\epsilon\, V_l(\epsilon, R)G_{PJ}^+(E - \epsilon, R, R')V_l(\epsilon, R')\tilde{\chi}_d^J(R'). \quad (71)$$

This approximation amounts to neglect of any change in nuclear angular momentum during the transition. It has been numerically tested by Bieniek (1978) in a very severe case [He*(2 ^{3}S) + H] where electronic partial waves up to $l = 9$ contribute. The deviation from the exact computation was typically 10% and is certainly much smaller when only s, p or d waves play a role, as in the sodium or hydrogen cases.

(ii) Equation (70) is an *integro-differential* equation since the unknown wave function $\tilde{\chi}_d^J$ is included in the R' integral on the right-hand side. This is due to the *nonlocal* character of the Green function $G_{PJ^+}^+(R, R')$, which is also *complex* due to the outgoing wave form. The real and imaginary parts can be separated using the spectral decomposition in Eq. (69) and the relation

$$\lim_{\eta \to 0} \frac{1}{E + i\eta - E'} = P\left(\frac{1}{E - E'}\right) - i\pi\delta(E - E'),$$

where p represents the principal part distribution. The energy integrals in the simplified form [expression (71)] of the right-hand side of Eq. (70) may thus be written $\Delta_E(R, R') - i\pi\Gamma_E(R, R')/2$, where the two nonlocal operators,

$$\Delta_E(R, R') = \sum_{l, v^+} p \int d\epsilon \, \frac{V_l(\epsilon, R)\chi_{v^+ J}(R)\chi_{v^+ J}(R')V_l(\epsilon, R')}{E - \epsilon - E_{v^+ J}}, \tag{72a}$$

$$\Gamma_E(R, R') = 2\pi \sum_{l, v_{\text{op}}^+} V_l(E - E_{v^+ J}, R)\chi_{v^+ J}(R)\chi_{v^+ J}(R')V_l(E - E_{v^+ J}, R'), \tag{72b}$$

have the dimension of energies. In (72b), the v^+ summation is now restricted to the vibrational levels v_{op}^+, open for ionization ($E_{v_{\text{op}}^+} < E$) as required by the function $\delta(E - E_{v^+ J} - \epsilon)$, with $\epsilon > 0$.

At this stage, two more approximations are performed in most of the AI studies using this approach. The first one is the *local approximation* which replaces the nonlocal operators (72) by the local forms

$$\Delta_E(R) = \sum_l p \int d\epsilon \, \frac{|V_l(\epsilon, R)|^2}{E - \epsilon - \bar{E}}, \tag{73a}$$

$$\Gamma_E(R) = 2\pi \sum_l |V_l(E - \bar{E}, R)|^2, \tag{73b}$$

where $\bar{E}$ represents a mean value of the vibrational energies $E_{v^+ J}$. Equations (72) have been reduced to (73) by using the closure relation

$$\sum_{v^+} \chi_{v^+ J}(R)\chi_{v^+ J}(R') = \delta(R - R'). \tag{74}$$

The validity of this transformation will be discussed later. First we comment on the physical interpretation of Eqs. (73). The local forms of $\Gamma_E(R, R')$ and $\Delta_E(R, R')$ being proportional to $\delta(R - R')$, the R' integration in the right-hand side of Eq. (70) disappears and one is left with a homogeneous differential equation:

$$\left(-\frac{\hbar^2}{2M}\frac{d^2}{dR^2} + U_d(R) + \Delta_E(R) - \frac{i}{2}\Gamma_E(R) + \frac{\hbar^2 J(J+1)}{2M R^2} - E\right)\tilde{\chi}_d^J = 0. \tag{75}$$

The entrance neutral state potential is thus replaced by an effective complex potential (from which the method derives its name). The term $\Delta_E(R)$ represents an *energy shift* due to coupling with the ionization continuum, while the imaginary part $\Gamma_E(R)$ appears as a *width* for the neutral potential, measuring the loss of flux into the ionization channels. Note that $\Gamma_E(R)$ is just the electronic width $\Gamma(\epsilon, R)$ introduced in Eq. (49) (Sec. IV,A), with $\epsilon = E - \bar{E}$.

In most of the actual calculations, the shift $\Delta_E(R)$ is neglected (this is the second approximation mentioned above). The main justification is the very slow energy dependence of the electronic coupling $V_l(\epsilon, R)$. Since these electronic coupling terms act at short-range where the external electron is strongly accelerated by the Coulomb attraction, they are insensitive to small variations of the asymptotic kinetic energy ϵ of the electron, and the principal part integral in Eq. (73a) is thus very small.

b. Nonlocal effects. The physical interpretation of the local energy terms (73) may be extended to the nonlocal terms (72): they represent the shift and width of the entrance state, when the quantization of the nuclear motion in the bound ion state is taken into account. Each interaction with the final continuum involves a vibrational wave function χ_{v^+J} over the whole range of nuclear distances, and not only at a specific distance R as in the local approximation or in the semiclassical description. Physically, this means that the rate of electron emission at any particular nuclear separation is not purely determined by the local electronic interaction but also depends on the nuclear motion, which is quantized in the final state.

The validity of the local approximation has been discussed by Bieniek (1980) and by Lam and George (1984). The first requirement for using the closure relation (74) which transforms the nonlocal relations (72) into local ones, is that the v^+ summation in (72) bears only on the vibrational wave functions χ_{v^+J}. The ionization process must thus be insensitive to the vibrational energy of the final ion levels. As already noted, the electronic coupling depends very weakly on the kinetic energy $\epsilon = E - E_{v^+J}$ of the ejected electron such that this requirement is usually fullfilled, especially when the total energy E is large compared to the vibrational spacing. The second requirement for the validity of closure relation (74) is that the ion levels included in the summation form a nearly complete basis set. While this is true for the shift (72a), or even for the width (72b) in the domain of Penning ionization with total energy E well above the ion dissociation limit (see Fig 1c), it is certainly not so for the cases of Figs. 1a and 1b, with only a few open final states in the summation of Eq. (72b).

Inclusion of nonlocal effects requires solving the integro-differential Eq. (70) instead of the simplified differential form (75). The method derived by Bardsley (1968a) for the dissociative recombination process transforms the integro-differential equation into a system of algebraic coupled equations, with complex coefficients. It can be easily applied to AI with different boundary conditions. Using this method within a very simple model of *laser-induced* AI, Lam and George (1985) have predicted multiple resonances (as a function of laser frequency) and non-Franck-Condon transitions which demonstrate the importance of nonlocality in this process. Bieniek (1980) has also made a numerical evaluation of the nonlocal effects for the associative detachment process $H + H^- \rightarrow H_2 + e$ and found them substantial at low energies, especially for the partial cross section of individual rovibrational final states. To our knowledge, no other calculations of AI cross sections have yet been performed using the complex potential method without a local approximation. An alternative to this approach is the recent work on low-energy AI processes which has been performed with the MQDT approach described below (see Sec. IV,C,3).

c. Cross-section calculations. The energy-normalized solution $\tilde{\chi}_d^J$ of Eq. (70) or one of its simplified versions behaves asymptotically as

$$\tilde{\chi}_d^J \sim \sqrt{\frac{2M}{\hbar^2 \pi \kappa}} \sin(\kappa R - J\pi/2 + \delta_J),$$

where δ_J is a complex phase-shift due to the complex effective potential of the associative state. The J partial wave contribution to the partial cross section $\sigma_{v^+ J^+}$ may be expressed as (Bieniek, 1978);

$$\sigma_{v^+ J^+, J} = g_i \frac{4\pi^3}{\kappa^2} \sum_l |e^{i\delta_J} \langle \chi_{v^+ J^+} | V_l(E - E_{v^+ J^+}, R) | \tilde{\chi}_d^J \rangle|^2$$

$$= g_i \frac{4\pi^3}{\kappa^2} e^{-2\,\mathrm{Im}\,\delta_J} \sum_l |\langle \chi_{v^+ J^+} | V_l(E - E_{v^+ J^+}, R) | \tilde{\chi}_d^J \rangle|^2, \quad (76)$$

where g_i is the statistical weight of the initial associative state. This expression, obtained from the definition of the transition matrix elements, is exact in as much as the exact wave function $\tilde{\chi}_d^J$ is used. An obvious approximation is to use instead the real solution χ_d^J of the homogeneous Eq. (68b), retaining nevertheless a complex phase shift δ_J in Eq. (76). The cross section appears then as the product of two terms with clear physical interpretations:

$$\sigma_{v^+ J^+}^J = S_J \cdot \sigma_{v^+ J^+, J}^{\mathrm{Born}} \quad (77)$$

with $S_J = e^{-2\,\mathrm{Im}\,\delta_J}$. The second term is the first Born approximation to the cross section for transition to the final ion level (v^+, J^+). The first term S_J is the "survival factor" (Bardsley, 1968) representing the integrated autoionization probability of the entrance neutral state. Bieniek (1978) gives an approximate (local) expression for $\mathrm{Im}\,\delta_J$:

$$\mathrm{Im}\,\delta_J \simeq \frac{\pi}{2} \int_0^\infty \chi_d^J \Gamma(R) \chi_d^J \, dR \quad (78)$$

which relates the survival actor S_J to the same quantity derived below from the MQDT equations. The approximation (77) to the AI cross section, has been shown by Bieniek to give reasonable results, compared to the full expression (76), as long as the electronic width $\Gamma(R)$ is small compared to the real part $U_d(R)$ of the neutral potential.

3. The MQDT treatment

The complex potential method described above concentrates the computational effort on the neutral state nuclear wave function, since from the beginning the electronic continuum wave functions are eliminated and play only the role of a flux sink, represented by the imaginary part of the potential. Moreover, the Rydberg states converging to the ion ground state, have not yet been included in any AI calculation using the complex potential method. Although such an extension has been achieved for the analogous treatment of dissociative recombination (Giusti *et al.*, 1982; Hickman, 1986), the MQDT approach seems more suitable for a realistic calculation of AI cross sections at low energy, including nonlocal and resonance effects as well as couplings between the final ionization channels. We note, however, that several approximations are also made in the existing calculations using this second quantal method, as we describe below.

a. The basic concepts of the theory. The quantum defect approach, first developed by Seaton for atoms (see his review paper, Seaton, 1983) and extended to molecules mainly by Fano (1970), Jungen (see Greene and Jungen, 1985) and Mies (1984), rests on two basic ideas. First, it unifies the treatment of discrete and continuum states by using the collisional concept of *channel*, characterized by the bound or free motion of a relative particle in a long-range potential (Coulomb potential for an electron–ion channel, molecular potential for an atom–atom channel; see Fig. 21). If the relative motion is bound ($E < 0$) the channel is said to be closed, if the two interacting particles may escape from the potential well into the continuum ($E > 0$) the channel is said to be open.

Open and closed channels can be treated on the same footing at short range only, in the region of the potential well where the kinetic energy is large enough to be insensitive to the asymptotic conditions. Generally there is a zone at very short range (the *reaction zone* of Fig. 22) where a distinction between electron–ion and atom–atom channels becomes unrealistic, because the electrons and the nuclei are all close together and strongly interact. Thus the second important idea for MQDT is a partition of the configuration space into two regions corresponding to different physical situations and to different theoretical treatments: structure calculations in the inner region, collisional treatment in the outer one.

We illustrate briefly these concepts in the simplest case of an atomic ionization channel (or a molecular one at fixed R, in the Born-Oppenheimer representation). The "reaction zone" is the atomic or molecular core where electron exchange and correlation take place. The outer region beings when the Rydberg electron escapes the core electron cloud ($r > r_c$ in Fig. 22) and moves in a Coulomb potential. Its radial wave function in this outer region is

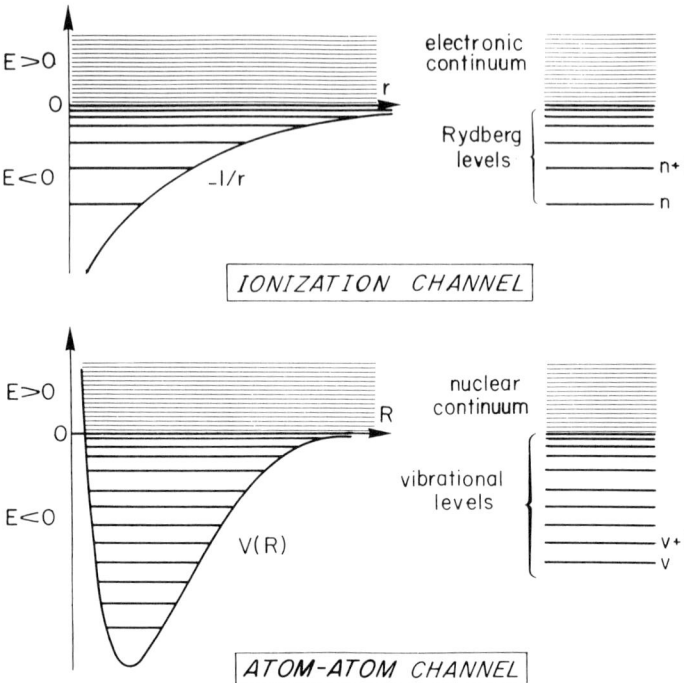

FIG. 21. Discrete and continuum potential states in an electron–ion channel (top) and an atom–atom channel (bottom) and their schematic energy-level representation (right-hand side).

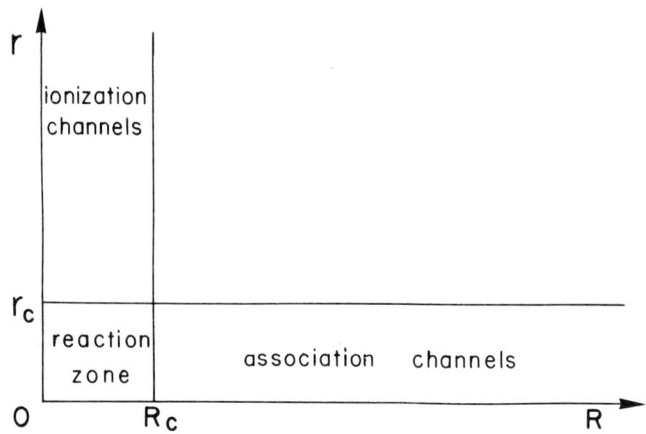

FIG. 22. Different regions of the two-dimensional configuration space (r, R) involved in the AI process.

a solution of the Coulomb equation, phase-shifted with respect to the hydrogen wave function due to the additional interactions experienced at short range. It may be written

$$\phi_l(\epsilon, r)_{r > r_c} = f_l(\epsilon, r)\cos \delta_l - g_l(\epsilon, r)\sin \delta_l, \tag{79}$$

where f_l is the regular Coulomb function, g_l the irregular one lagging in phase by $\pi/2$ with respect to f_l, and δ_l the short-range phase shift. The origin of δ_l is short-range interaction and therefore δ_l is almost independent of the asymptotic kinetic energy of the electron, on both sides of the ionization threshold. At positive energies the asymptotic behavior of the radial wave functions in Eq. (79) is

$$f_l \underset{r \to \infty}{\sim} a_k \sin(kr - l\pi/2 + \sigma_l), \qquad g_l \underset{r \to \infty}{\sim} -a_k \cos(kr - l\pi/2 + \sigma_l),$$

$$\phi_l \underset{r \to \infty}{\sim} a_k \sin(kr - l\pi/2 + \sigma_l + \delta_l), \tag{80}$$

where σ_l is the Coulomb phase and $a_k = (2m_e/\hbar^2\pi k)^{1/2}$ the energy normalization factor. At negative energies, f_l and g_l diverge exponentially except for certain discrete energies ϵ_n,

$$\epsilon_n = \frac{\text{Ry}}{(n - \delta_l/\pi)^2} \qquad (n = l + 1, l + 2, \ldots). \tag{81}$$

Comparison with the Rydberg formula (Eq. 28) yields the relation $\delta = \pi\mu$, where μ, or $\mu(R)$ in the molecular case, is the quantum defect.

Extension to multichannel situations and to non-Coulomb interactions requires the definition of a short-range *reaction matrix* K which generalizes the single-channel phase shift or quantum defect. As already indicated in Sec. III,A, the diagonal terms may be interpreted as phase shifts (more precisely $\pi K_{ii} = \tan \delta_i$, according to the usual definition of a reaction matrix) while the off-diagonal elements correspond to effective interchannel couplings. One often introduces the diagonal form ($\tan \delta_\alpha$) of the K matrix and the eigenvectors $|\alpha\rangle$ which define a new set of channels, the "eigenchannels" of the reaction zone. Uncoupled at short range, the $|\alpha\rangle$ channels combine into the asymptotic $|i\rangle$ channels, the frame transformation coefficients $\langle i|\alpha\rangle$ satisfying the diagonalization relation,

$$\pi K_{ii'} = \sum_\alpha \langle i|\alpha\rangle \tan \delta_\alpha \langle \alpha|i'\rangle. \tag{82}$$

This frame transformation technique (Fano, 1970) is very useful when the physical reactions prevailing at short and long range correspond to different coupling schemes between the particles or their angular momenta. The coefficients $\langle i|\alpha\rangle$ are then energy independent and easily determined, and knowledge of the eigen phase shifts δ_α [or eigen quantum defects $(\mu_\alpha = \delta_\alpha/\pi)$] is sufficient for obtaining the whole reaction matrix. Such an approach has been applied to vibrational interactions in the AI process, as we describe below.

b. The short-range reaction matrix. The treatment of AI involves two types of channels, schematized in Fig. 21: a set of ionization channels and a set of atom–atom (association or dissociation) channels, which in most calculations is reduced to a single one. According to the diabatic representation one usually proceeds in two steps (Giusti, 1980) studying first the couplings between channels of the same type, and then between the two types of channels. In this part we assume, for the sake of clarity and because this assumption has been made in all the existing calculations, that the nuclear angular momentum J is conserved during the reaction (see the discussion in Sec. IV,B,2).

Up to the present time, theoretical work has only progressed to the point of considering *vibrational* interaction among the ionization channels, neglecting the couplings induced by molecular rotation (l uncoupling) or the molecular anisotropy (l mixing). In scattering language, vibrational interaction may induce inelastic transitions between two channels with $v^+ \neq v'^+$ either open or closed. The MQDT approach avoids the introduction of the derivative operator d/dR to describe this non-adiabatic interaction. Instead, the *vibrational* reaction matrix is obtained using the frame-transformation technique defined above: at short range, the electron may be treated in the Born-Oppenheimer representation at fixed R, and the eigen quantum defects are represented by the function $\mu_l(R)$. At greater distance r, the electron is decoupled from the molecular ion vibration and the asymptotic i channels correspond to different v^+ levels. The reaction matrix elements are thus given by the relation (82), which reads in the present case

$$\pi K^{J,l}_{v^+v'^+} = \int dR\, \chi_{v^+J}(R)\tan \pi\mu_l(R)\chi_{v'^+J}(R), \tag{83}$$

where the discrete summation in (82) is replaced by an integral since $\alpha = R$ is a continuous index. The vibrational wave functions $\chi_{v^+J}(R) = \langle \chi_{x^+J}|R\rangle$ play the role of the frame-transformation matrix elements.

The second step of the method consists in introducing the *electronic* couplings between the ionization channels and the entrance associative one. At the Born-Oppenheimer level they are represented by the energy normalized functions $V_l(\epsilon, R)$, Eq. (50). We recall that these couplings may be obtained either from electron scattering or from bound Rydberg state calculations using the scaling law given by Eq. (52). The corresponding *electronic* reaction matrix $K^{el}(R)$ may be obtained from a fit to the adiabatic curves as in the case of H_2 (Ross and Jungen, 1987)—see Sec. III,A—or from a Lippmann-Schwinger equation (Ivanov and Golubkov, 1986), which lends itself well to a perturbation expansion, usually truncated to first-order $K^{el} = V$ in the case of weak coupling. Passage into the external zone as in (83) yields now the matrix elements

$$K^{el\,J}_{v^+d} = \int dR\ \chi_{v^+J}(R) K^{el}(R) \chi^J_d(R) \tag{84a}$$

and

$$K^{el\,J}_{v^+v'^+} = \int dR\ \chi_{v^+J}(R) K^{el}(R) \chi_{v'^+J}(R). \tag{84b}$$

To first order only the matrix elements (84a) are not zero, representing the direct coupling between ionization and neutral channels. At higher order the elements (84b) represent indirect electronic couplings between ionization channels.

The result of the different types of channel interactions are finally summarized in a global short-range reaction matrix, which combines the vibrational and electronic K matrices (83) and (84). We do not report here the complicated expressions of the resulting reaction matrix but rather show how it is used to calculate the AI cross sections.

c. Boundary conditions and resonances. In the external region ($r > r_c$, $R > R_c$) open and closed channels must be distinguished since they correspond to very different asymptotic behaviors. The AI process involves at least one open channel for the nuclear motion (the entrance channel) and one for the electronic motion (ionization). The closed ionization channels correspond to vibrational levels of the ion higher in energy than the collision energy E.

Several methods have been used for this last step, the aim being to eliminate the closed channels from the total reaction matrix K defined above. It is thus reduced to a smaller matrix $\bar{K}$ involving the open channels only, from which the scattering matrix and cross-sections may be calculated. The most direct method is due to Seaton and has been first used for dissociative recombination by Nakashima, et al. (1987). The K matrix is partitioned into four blocks

$$\bar{K} = \begin{Bmatrix} K_{oo} & K_{oc} \\ K_{co} & K_{cc} \end{Bmatrix},$$

where o and c denote the set of open and closed channels, respectively. From the asymptotic behavior of the wavefunctions, imposing convergence on the closed channel contributions, one gets the relation

$$\bar{K} = K_{oo} - K_{oc}(\tan \pi v I_{cc} + K_{cc})^{-1} K_{co}, \tag{85}$$

where $v = (-\text{Ry}/\epsilon)^{1/2}$ is defined for closed channels only ($\epsilon < 0$) and I_{cc} is the unit matrix in the closed channel space. In Eq. (85) the matrix in brackets is singular for each root of the determinantal equation

$$\det|\tan \pi v I_{cc} + K_{cc}| = 0$$

which determines the discrete level energies in the closed channels. In the vicinity of these discrete energies the effective matrix $\bar{K}$ varies rapidly, as well as the associated scattering matrix

$$S = (1 + i\pi \bar{K})(1 - i\pi \bar{K})^{-1}. \tag{86}$$

The partial cross sections

$$\sigma_{v^+ J^+, J} = g_i \frac{\pi}{\kappa^2} |S_{v^+ J^+, d}|^2 \tag{87}$$

where $J^+ = J$ in the present approximation, exhibit a series of *resonances* due to this rapid energy dependence.

These resonances correspond to the excited vibrational levels of the Rydberg states, embedded in the electronic continuum into which they can decay by vibrational autoionization (see Fig. 23). The resonance structure, which is shown in specific examples in the following paragraph, has been studied in detail for the inverse process, the dissociative recombination (Hickman, 1987), (Giusti-Suzor, 1988) with conclusions applying to AI as well. They correspond to an "indirect" AI process, in which the discrete Rydberg levels are populated via an electronic interaction with the entrance state, and then decay either by vibrational autoionization (in which case they contribute to the AI process) or by electronic predissociation back to the initial atomic products (in which case they cause a loss in the ionization yield). Since the autoionizing vibrational coupling is usually much weaker

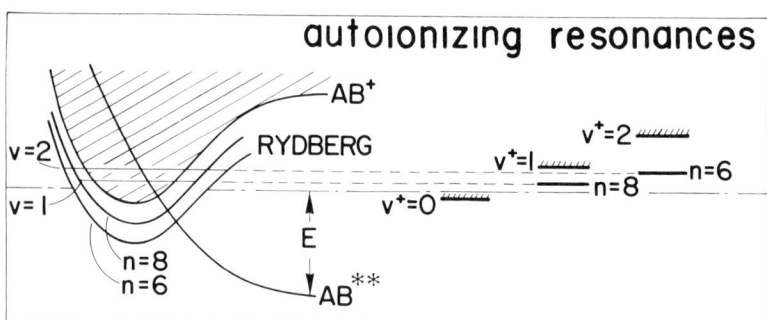

FIG. 23. Autoionizing discrete Rydberg levels giving rise to resonances in the AI cross sections.

than the predissociating electronic one, the resonances are more likely to appear as *dips* in the AI cross sections rather than as peaks. Electronic coupling between open and closed ionization channels due to high order terms in the electronic reaction matrix may nevertheless induce electronic autoionization of the Rydberg states, leading to peaks or Fano-profile shapes for the resonances which then may increase the ionization yield. Note finally that these resonances are the quantum mechanical manifestation of the numerous curve-crossings between the entrance state and the Rydberg series below the ion ground state potential. The semiclassical treatment of this multistate problem by Cohen (1976) has been discussed in Sec. IV,B. It is difficult to compare the results of the two treatments since in the present stage they rest on very different assumptions: the semiclassical treatment assumes isolated crossings and localized couplings, neglecting any interference effects and ignoring the vibrational structure of the Rydberg states. On the other hand, loss of ionization flux due to competition between AI and excitation transfer has been considered in the semi-classical model but not in the quantal one which has not yet included Rydberg series of repulsive states.

Before reporting on the existing AI studies performed with the MQDT approach, we show how the cross section (87) compares with the result of the complex potential method Eqs. (76) and (77) at the same level of approximations. Neglecting vibrational coupling and high-order terms in the electronic K-matrix a calculation restricted to *open* ionization channels leads to the approximate analytic formula (Giusti, 1980),

$$\sigma_{v^+ J, J} \simeq g_i \frac{4\pi^3}{\kappa^2} \frac{\sum_l |\langle \chi_{v^+ J} | V_l(R) | \chi_d^J \rangle|^2}{\left(1 + \pi^2 \sum_{v^+, l} |\langle \chi_{v^+ J} | V_l(R) | \chi_d^J \rangle|^2 \right)^2}, \quad (88)$$

where the integrations represented by brackets are over the nuclear radial coordinate, and the v^+ summation in the denominator includes only open vibrational thresholds. As for Eq. (76) this expression may be written as a product:

$$\sigma_{v^+ J, J} = S'_J \cdot \sigma_{v^+ J}^{\text{Born}},$$

where

$$S'_J = \left(1 + \pi^2 \sum_{v^+, l} |\langle \chi_{v^+ J} | V_l(R) | \chi_d^J \rangle|^2 \right)^{-2} \quad (89)$$

appears as the survival factor in the present formalism, to be compared with $S_J = e^{-2 \text{Im} \delta_J}$ in Eq. (77). Both S and S' are less than but close to unity in the weak-coupling case to which we are restricted here, and can be expanded as

$$S'_J \simeq 1 - 2\pi^2 \sum_{v^+, l} |\langle \chi_{v^+ J}|V_l(R)|\chi_d^J\rangle|^2, \tag{90a}$$

$$S_J \simeq 1 - \pi \langle \chi_d^J|\Gamma(R)|\chi_d^J\rangle, \tag{90b}$$

where the approximate expression (78) has been used for Im δ_J. One sees easily that to the extent that the closure relation $\sum_{v^+} \langle R|\chi_{v^+ J}\rangle\langle \chi_{v^+ J}|R'\rangle = \delta(R - R')$ may be used in (90a) as is required by the local approximation, these two expressions are identical since $\Gamma(R) = 2\pi \sum_l |V_l(R)|^2$. Thus within the local approximation, both theories coincide for weak coupling. When only a few vibrational thresholds are available, numerical comparison between (90a) and (90b) allows an estimation of the nonlocal effects in the direct AI process. The local approximation leads to *underestimated* cross sections ($S_J < S'_J$) as Bieniek (1980) already noticed from numerical tests on the H + H$^-$ associative detachment process.

d. Examples. The extension of MQDT to include both electron–ion and atom–atom channels has been first applied to dissociative recombination (Giusti 1980; Giusti-Suzor et al., 1982; Nakashima et al., 1987). The first application to AI is due to Urbain et al. (1986) who studied H$_2^+$ formation from H$^+$-H$^-$ collisions, a process previously measured at Louvain-la-Neuve (Poulaerts et al., 1978). The long-range dynamics is complex because the H$^+$, H$^-$ ionic curve crosses many covalent states as the internuclear distance decreases (see Fig. 24). The same dynamics governs the H$^+$-H$^-$ mutual neutralization process, for which coupled equation calculations have determined the path indicated by arrows on Fig. 24 as the most probable at short distances. The HH̄ state, with the dominant configuration (1s σ_g, 3d σ_g) has thus been selected as the main entrance channel into the autoionization region, being electronically coupled to the lower doubly excited state (2p σ_u)^2 of same symmetry $^1\Sigma_g^+$.

The study of the AI process itself begins at shorter distances ($R \leq 5$ a.u.). An MQDT calculation has been performed using quantum defects and electronic couplings extracted from the *ab initio* calculations of Wolniewicz and Dressler (1977), extrapolated beyond the ionization threshold where they join smoothly the values derived from electron scattering calculations. The electronic K matrix is represented by its first order approximation which has been shown, from H$_2^+$ dissociative recombination studies (Hickman, 1987), to be valid when many ion vibrational levels ($v \leq 9$ in the present case) are available for ionization. Two sets of calculations have been performed. The first one includes two *associative* channels, the entrance one (HH̄) and the doubly excited state (2p σ_u)^2, and only *open ionization* channels corresponding to s and d electronic partial waves ($l = 0, 2$). In the second calculation, all

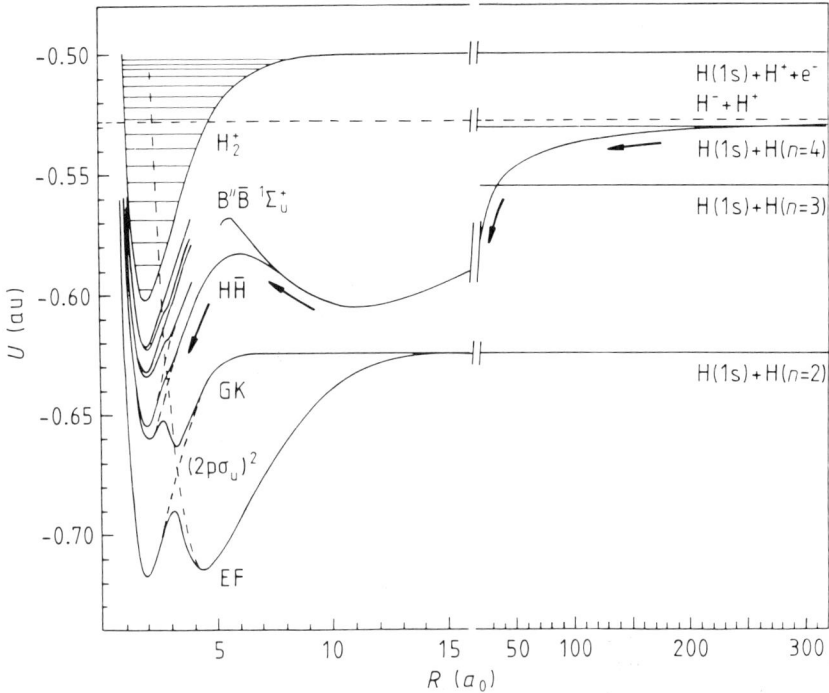

FIG. 24. Potential energy curves of H_2 and H_2^+ relevant to AI in H^+ -H^- and $H + H^*$ collisions. Left: adiabatic curves of Wolniewicz and Dressler (1979) (see Fig. 17). Letters refer to $^1\Sigma_g^+$ states unless stated otherwise. Broken curves show diabatic states. Right: Curve-crossing representation of the ionic and covalent states. Arrows indicate the principal route for the H^+, H^- collision [from Urbain et al. (1986)].

the singly excited states including exit channels open for dissociation ($n \leq 4$) are introduced, together with the closed ionization channels leading to numerous resonances in the partial cross sections. The results are shown in Fig. 25. The experimental energy dependence is correctly reproduced by both types of calculations, being essentially due to the π/κ^2 factor (proportional to E^{-1}) in the cross section, Eq. (87). The squared S matrix factor on the right side of (87) is slowly varying because the reaction is very exoergic and the effective kinetic energy in the reaction zone is so large that the vibrational overlap between the doubly excited and ion states is an insensitive function of energy. The resonance structure present in the second type of calculation (full line) is almost totally washed out by the partial-wave and vibrational level summations. Only the global effect of the resonances appears from comparison with the open-channel calculation (dashed line). The reduction by about 20% is partly due to loss of flux into the additional dissociative channels, and

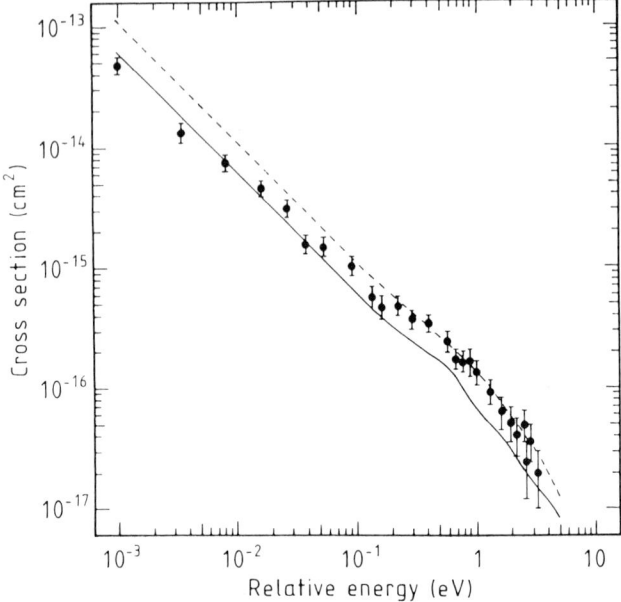

FIG. 25. Total cross section for $H^+ + H^- \to H_2^+ + e$ reaction. Dashed line: MQDT calculation including only the open ionization channels. Solid line: MQDT calculation including both open and closed ionization channels. Solid points: experimental results of Poulaerts et al. (1978).

partly to the effect of Rydberg states for which predissociation dominates over autoionization, as indicated above.

This averaging of resonance structure does not happen in the $H + H^*(2s)$ AI process, studied by Takagi and Nakamura (1988). The $n = 2$ atomic limit lies well below the first ionization threshold (see Fig. 24) and the energy dependence of this endoergic reaction differs radically from the previous one (see Fig. 26). Starting from negligible values at the reaction threshold, the cross section increases due to the successive opening of rotational and vibrational ion levels, while in the previous exoergic case a large number of thresholds were available from the beginning. However, the step-like structure expected from the *direct* AI process is obscured by the resonance structures, which here dominate the partial wave summation for two reasons: (1) many more channels are closed, and (2) fewer partial waves contribute. Using the same MQDT treatment as for the H^+-H^- case, this calculation is based on a different set of quantum defects and electronic couplings previously obtained from scattering calculations by the same authors, who found a negligible contribution of the s partial wave to the ionization process. The

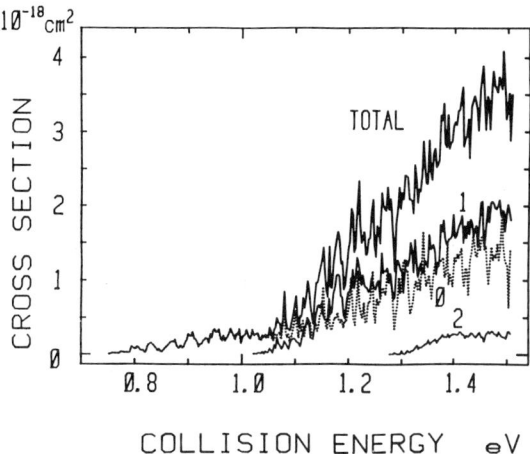

FIG. 26. Cross section for $H + H^*(2s) \to H_2^+ + e$. Integer numbers labeling the curves indicate the vibrational quantum number v^+ of the final ion level, and "total" indicates the cross section summed over v^+. From Takagi and Nakamura (1988).

entrance channel consists here of two asymptotically degenerate $^1\Sigma_g^+$ states correlated to the same $n = 2$ limit (Fig. 24) and the flux is simply assumed to be equally divided between these two paths.

New developments, both experimental and theoretical, are in progress for these hydrogen studies, as described by Urbain (1989). On the theoretical side, two main improvements have been achieved compared to the MQDT calculations described above:

(i) The electronic reaction matrix obtained by Ross and Jungen (1987) has been used in place of the first-order K matrix of the previous calculations.

(ii) Coupled equations are solved to determine more precisely the division of flux between the two entrance channels in the $H + H^*(2s)$ collision. Using a "half-collision" point of view (Mies and Julienne, 1985) the effective nuclear wave functions for both entrance states are obtained from the close-coupling system and then used for calculating the effective reaction matrix integrated over the nuclear motion. Interference effects between the two paths induce oscillations in each partial-wave cross section, but here also the sum over many partial waves washes out this oscillatory behavior which changes from one J value to another. Although the net effect of this refinement is weak in the present case, it is a first step toward an accurate treatment in more complex cases of the long range dynamics which determines the respective weight of each neutral state entering into the reaction zone.

We list two other notable applications of the MQDT formalism to associative ionization. The first one is due to Golubkov and Ivanov (1988) who studied the endoergic $N(^2D) + O(^3P) \to NO^+ + e$ reaction near threshold, using a variant of the MQDT approach in which the reaction matrix is obtained by solving a Lippmann-Schwinger equation, (Ivanov and Golubkov, 1986). Three main entrance channels ($B\ ^2\Pi$, $A'\ ^2\Sigma^+$, $B\ ^2\Delta$) have been selected. The A' state which correlates to the lower $N(^4S)$ limit is assumed to be populated by a Landau-Zener transition at the crossing point with the $I^2\ \Sigma^+$ state, much as in the semiclassical work of Nielsen and Dahler (1979) described in Sec. IV,B. The main difference with respect to the semiclassical calculation is the threshold behavior. A linear increase of the cross section at threshold is found here, mainly due to the centrifugal barrier: the cross section is roughly proportional to J_{max}^2, where J_{max}, the largest partial wave contributing to the AI process (with threshold energy E_{AI}) varies as $(E - E_{AI})^{1/2}$. The linear threshold law is in accord with the experimental results of Ringer and Gentry (1979) and differs from the semiclassical results of Nielsen and Dahler (1979) who found a threshold law of $(E - E_{AI})^{3/2}$ [see Eq. (61)]. It should be remembered that the range of Nielsen and Dahler's threshold law was quite short (~ 0.1 eV) and that their calculations also showed essentially linear behavior at higher energies. Above threshold, the quantal results overestimate the cross section, which could be due to inclusion of only some of the Rydberg resonances since only $\Delta v = \pm 1$ vibrational couplings are introduced between open and closed ionization channels. Adjustment of some parameters leads to an agreement with experiment similar to the one obtained by Nielsen and Dahler (1979). Unfortunately this system appears too complex to gauge accurately the importance of quantal effects. Nitric oxide associative ionization involves a large number of neutral states and the quality of the molecular data appears uncertain.

Finally, the MQDT treatment has been recently applied to the dynamics of the AI process $Na^*(3p) + Na^*(3p) \to Na_2^+ + e$, to which large parts of Sec. II and III have been devoted. Using the diabatic potential curves and electronic couplings described in Sec. III,B.6, Henriet et al. (1989) have estimated the cross section for this reaction at thermal energy. The cross section for the direct process (involving open channels only) is shown to be very small for the $^1\Sigma_g^+$ symmetry: the relevant curve does not cross the ground state ion curve and the overlap between the X_{v+J} and X_d^J vibrational functions in (84a) is too small to give a substantial contribution. In contrast, the interaction matrix elements $V_{dv^+}^{lJ}$ (Eq. 65) are so large in the case of a $^3\Sigma_u^+$ doubly excited state that the use of the weak coupling approximation is not valid. Quantitative calculation going beyond the weak coupling approximation are presently in progress.

V. Summary, Conclusions, and Perspectives

The experimental results show that in general AI occurs with near unit probability. Collision between two excited sodium atoms is a notable exception [process (16)] in which the probability of ionization has been shown to be 10% or less. Studies of orientation and alignment in process (16) show the Σ states are favoured over other molecular symmetries, but Π and Δ states also make significant contributions. At collision energies low enough so that only the long range quadrupole interaction predominates, Π states are the only open channels. In crossed-beam, circular polarization experiments laser excitation prepares a population of pure atomic states, $^2P_{3/2}$ (F = 3, M_F = 3), and collisions between them will produce (at large internuclear separation) only $^3\Delta_u$ states. Preparation of *initial* atomic states has advanced our understanding of the collision dynamics, but is not sufficient to provide a complete picture. Experiments yielding detailed information on the distribution of *final* molecular ion states are needed before theory can link the initial atomic preparation to the final molecular products. First attempts have been undertaken in this direction by measurement of the emitted electron energy spectra and the recoil velocity distributions of photodissociated ion products, but the results are not nearly detailed enough to provide reliable distributions of vibrational and rotational state populations. A true spectroscopy of the ion product is required.

Associative ionization will continue to be the collision of choice for experiments investigating the very long range, quantal behavior of weakly interacting particles at ultra-cold temperatures. In this new, low kinetic energy environment, the strength of the light-field–atom interaction becomes comparable to the collisional interaction even at the low intensities characteristic of cw lasers, and therefore the experimenter is provided a rare opportunity to study the effect of light-field "dressing" on collision dynamics under very well-defined conditions.

The problem of the calculation of the potential curves in the molecular region seems to be practically solved for systems with two active electrons. The correct treatment of core polarization effects has produced a large improvement in the accuracy of both the *ab initio* and the model potential (or pseudopotential) calculations. The quality of the results for the lighter alkali dimer cations and for the ground state of the alkali dimers is not far from the excellent accuracy of the hydrogen molecule potential curves. Concerning the excited states, it has been realized that the theoretical treatment should provide a correct representation of the wave function when one Rydberg electron moves far from the molecular ion core. Methods should be developed which improve the quality of this electronic asymptotic wavefunction.

Nevertheless, satisfactory results have already been obtained for the Na_2 molecular potential curves up to the 3p + 3p dissociation limit. An interpretation in terms of molecular quantum defects has been given, and the associative ionization process appears to be qualitatively understood in the framework of a diabatic representation of the two-electron wave function. One considers separately the Rydberg series converging to the Na_2^+ ground state in the singly excited configuration space and to the Na_2^+ first excited state in the space of the doubly excited configurations. Such a treatment has to be developed for other systems.

In contrast the connection between this representation and the separated atom limit is not yet well understood. Indeed, the diabatic representation assumes that one of the two electrons stays either on a bonding or on an antibonding orbital of the molecular ion, so that a chemical bond is already presupposed. Future work should investigate, at greater internuclear distances, how this bond is formed, considering in particular, the uncoupling of the spin of the two electrons.

At even greater internuclear distances, the relative importance of exchange forces compared to dispersion forces is not yet firmly established. This effect, together with the influence of the hyperfine coupling and of the possible coupling by laser fields (dressed states) needs to be investigated in view of the emerging results of ultra-cold-atom collisions, where the associative ionization reaction is controlled by the long-range part of the potential curves. The importance of accurate potentials in the *intermediate* and *long-range* regimes to the successful application of dynamical theories cannot be overestimated.

Several developments are required for the treatment of the AI dynamics, even in the quantum mechanical approach which has to be used for most of the cases reviewed here.

(i) In the complex-potential method described in Sec. IV,C,2, nonlocality and Rydberg state resonances must be included since MQDT and model calculations have shown they may have substantial effects on the cross sections.

(ii) Furthermore the electronic reaction matrix required in the MQDT approach has to be determined accurately enough for a proper account of the indirect electronic couplings between the ionization channels, due to their individual electronic interaction with the same doubly excited neutral state. The type of scattering calculations performed by Ross and Jungen (1987) for the case of H_2 must be applied to other molecules, in particular to the sodium case for which accurate adiabatic potential curves are now available.

(iii) Most of the treatments reported in Sec. IV focus on the case where the entrance channel is electronically coupled to the ionization continuum. Other physical mechanisms such as *vibrationally* induced ionization (never reconsidered since the early work of Nielsen and Berry, 1971) or Coriolis coupling between bound and continuum electronic states with different Λ symmetries require further theoretical development. The molecular frame-transformation technique (Greene and Jungen, 1985) seems well adapted to this purpose.

(iv) Most importantly, the long- and medium-range dynamics, outside the chemi-ionization region, have to be properly described. The post-collision interactions among the ionization channels (rovibrational couplings) have been—or may be, for the rotational effects—included in the MQDT treatment, but consideration of the pre-collisional couplings among the reactant channels has not yet been undertaken, although this problem is the key to understanding the collision physics at "ultracold" temperatures (see Sec. II,C). The case of radial or Coriolis non-adiabatic couplings has been elegantly approached by Jones and Dahler (1988) who used a representation for molecular states based on the *total* orbital angular momentum of the system, instead of the orbital momentum of the nuclei only. This representation is well suited to the derivation of selection rules for the analysis of AI experiments involving atoms excited in oriented hyperfine states (see Sec. II,B). Electronic, spin orbit, and hyperfine couplings between crossing states of same symmetry requires the solution of sets of coupled equations from which the "half-collision" matrix, describing the branching between the different neutral channels entering the ionization region, may be determined as in the work of Urbain (1989) on $H + H^*(2s)$. This half-collision formalism [see Band and Mies (1988) and references cited therein] will be quite useful for analyzing the atomic-to-molecular transformation of the entrance associative channels and will complement the MQDT analysis of the ionizing exit channels.

If the net result of the work reviewed here can be summarized in a single sentence, it would be that the job is about half done: experiments need to characterize *final* as well as initial states, accurate potentials are needed at *long range* as well as short range, and *entrance atomic channels* need to be treated by half-collision methods, paralleling the MQDT approach to the ionizing branch of the collision. Would that next decade be as fruitful as the last.

ACKNOWLEDGEMENTS

The authors would like to thank X. Urbain, A. Henriet, Ch. Jungen, P. Julienne, and F. H. Mies for valuable comments and discussion during the course of this work. A. Henriet kindly provided data prior to publication. The

1987 "Workshop on Reactive Collisions of Laser-Excited Na Atoms" organized by R. Morgenstern and H. A. J. Meijer at Utrecht contributed greatly to the clarification of many problems. J. W. gratefully acknowledges support from the National Science Foundation and the Scientific Computing Center of the University of Maryland.

REFERENCES

Aymar, M. (1978). *J. Phys. B: At. Mol. Phys.* **11**, 1413.
Aymar, M., Combet-Farnoux, F., and Luc-Koenig, E. (1976). *J. Phys. B: At. Mol. Phys.* **9**, 1279.
Bagus, P. S., Liu, B., McLean, A. D., and Yoshimine, M., ALCHEMY system of programs.
Band, Y. B., and Mies, F. H. (1988). *J. Chem. Phys.* **88**, 2309.
Barbier, L., and Cheret, M. (1987). *J. Phys. B: At. Mol. Phys.* **20**, 1229.
Bardsley, J. N. (1967). *Proc. Phys. Soc.* **91**, 300.
Bardsley, J. N. (1968a). *Proc. Phys. Soc. J. Phys. B.* **1**, 349.
Bardsley, J. N. (1968b). *Proc. Phys. Soc. J. Phys. B* **1**, 365
Bardsley, J. N. (1974). In "Case studies in Atomic Physics" (E. W. McDaniel and M. R. C. McDowell, eds.) Vol. 4, pp. 229–368. North Holland, Amsterdam.
Bardsley, J. N., Junker, B. R., and Norcross, D. (1976). *Chem. Phys. Lett.* **37**, 502.
Baylis, W. E. (1977). *Can. J. Phys.* **55**, 1924.
Bernheim, R. A., Gold, L. P., and Tipton, T. (1983). *J. Chem. Phys.* **78**, 2625.
Berry, R. S. (1970). In "Proceedings of the International School of Physics Enrico Fermi XLIV, Molecular Beams and Reaction Kinetics" (Ch. Schlier, ed), pp. 193–221. Academic Press, New York.
Bethe, H. A. (1964). In "Intermediate Quantum Mechanics," p. 29. Benjamin, New York.
Bezuglov, N. N., Klyucharev, A. N., and Sheverev, V. A. (1987). *J. Phys. B: At. Mol. Phys.* **20**, 2497.
Bieniek, R. J. (1978). *Phys. Rev. A.* **18**, 392.
Bieniek, R. J. (1980). *J. Phys. B: At. Mol. Phys.* **13**, 4405.
Bonanno, R., Boulmer, J., and Weiner, J. (1983). *Phys. Rev. A.* **28**, 604.
Bonanno, R., Boulmer, J., and Weiner, J. (1985). *Comments At. Mol. Phys.* **16**, 109.
Borodin, V. M., Klyucharev, A. N., and Sepman, V. Yu (1975). *Opt. Spectrosc.* **39**, 231.
Bottcher, C., and Dalgarno, A. (1974). *Proc. R. Soc. Lond. A.* **34D**, 187.
Bottcher, C., Dalgarno, A., and Wright, E. L. (1973). *Phys. Rev. A.* **7**, 1606.
Boulmer, J., Bonanno, R., and Weiner J. (1983). *J. Phys. B: At. Mol. Phys.* **16**, 3015.
Brencher, L., Nawracala, B., and Pauly, H. (1988). *Z. Phys. D.* **10**, 211.
Broyer, M., Chevaleyre, J., Delacretaz, G., Martin, S., and Wöste, L. (1983). *Chem. Phys. Lett.* **99**, 206.
Bussery, B., and Aubert-Frecon, M. (1985a), *J. Chem. Phys.* **82**, 3224.
Bussery, B., and Aubert-Frecon, M. (1985b), *J. Phys. B.* L379.
Carlson, N. W., Taylor, A. J., and Schawlow, A. L. (1980). *Phys. Rev. Lett.* **45**, 18.
Carré, B., Spiess, G., Bizau, J. M., Dhez, P., Gerard, P. Wuilleumier, F., Keller, J. C., LeGouet, J. L., Picque, J. L., Ederer, D. L., and Koch, P. M. (1984). *Opt. Commun.* **52**, 29.
Carré, B., Roussel, F., Spiess, G., Bizau, J. M., Gerard, P., and Wuilleumier, F. (1986). *Z. Phys. D* **1**, 79.

Caves, T. C., Dalgarno, A. (1972). *J. Quant. Spectrosc. Radiat. Transfer* **12**, 1539.
Cerjan, C. J. (1975), *Chem. Phys. Lett.* **36**, 569.
Cerjan, C. J., Docken, K. K., and Dalgarno, A. (1976). *Chem. Phys. Lett.* **38**, 401.
Cheret M., Spielfiedel, A., Durand, R., Deloche, R., (1981). *J. Phys. B: At. Mol. Phys.* **14**, 3953.
Chu, S., Hollberg, L., Bjorkholm, J. E., Cable, A., Ashkin, A. (1985). *Phys. Rev. Lett.* **55**, 48.
Cohen, J. S. (1976), *Phys. Rev. A* **13**, 99.
Dalgarno, A. (1967), *Adv. Chem. Phys.* **12**, 143.
Dalgarno, A. (1975). *Atomic Physics*, Vol. 4, pp. 325-334. Plenum Press, New York.
Dalgarno, A., and Black, J. H. (1976). *Rep. Prog. Phys.* **39**, 573
Dalgarno, A., Bottcher, C., and Victor, G. A. (1970). *Chem. Phys. Lett.* **7**, 265.
Das, G., and Wahl, A. C. (1972). *BISON M.C. Manual*, Argonne National Laboratory Report ANL-7955.
de Jong, A., and van der Valk, F. (1979). *J. Phys. B: At. Mol. Phys.* **12**, 1561.
Devdariani, A. Z., Klyucharev, A. N., Lazarenko, A. V., and Sheverev, V. A. (1978). *Sov. Tech. Phys. Lett.* **4**, 408.
Djerad, M. T., Harima, H., and Cheret, M. (1985). *J. Phys. B: At. Mol. Phys.* **18**, L815.
Djerad, M. T., Cheret, M., and Gounand, F. (1987). *J. Phys. B: At. Mol. Phys.* **20**, 3789.
Dobrolezh, B. V., Klyucharev, A. N., and Sepman, V. Yu (1975). *Opt. Spectrosc.* **38**, 630.
Dressler, K., Galluser, R., Quadrelli, P., and Wolniewicz, L. (1979). *J. Mol. Spectrosc.* **75**, 205.
Eisel, D., Demtröder, W., Müller, W., and Botschwina, P. (1983). *Chem. Phys.* **80**, 329.
Fano, U. (1961). *Phys. Rev.* **124**, 1866.
Fano, U. (1970). *Phys. Rev. A.* **2**, 353.
Feshbach, H. (1962). *Ann. Phys. (N.Y.)* **19**, 287.
Fontijn, A. (1985). In "Gas-Phase Chemiluminescence and Chemi-ionization" (A. Fontijin, ed), pp. 1-10, and references cited therein. North-Holland, Amsterdam.
Fort, J., Laucagne, J. J., Pesnelle, A., and Watel, G. (1976). *Phys. Rev. A.* **14**, 658.
Fort, J., Laucagne, J. J., Pesnelle, A., and Watel, G. (1978). *Phys. Rev. A.* **18**, 658.
Fort, J., Dalzinger, T., Corno, D., Ebding, T., and Pesnelle, A. (1978). *Phys. Rev. A* **18**, 2075.
Fuentealba, P., Preuss, H., Stoll, H., and von Szentpaly, L. (1982). *Chem. Phys. Lett.* **89**, 418.
Garver, W. P., Pierce, M. R., and Leventhal, J. J. (1982). *J. Chem. Phys.* **77**, 1201.
Geltman, S. (1988). *J. Phys. B: At. Mol. Phys.* **21**, L735.
Giusti, A. (1980). *J. Phys. B: At. Mol. Phys.* **13**, 3867.
Giusti-Suzor, A. (1989). In "Dissociative Recombination: Theory, Experiment, and Application" (J. B. A. Mitchell and S. L. Guberman, eds.). World Scientific, Singapore, p. 14.
Giusti-Suzor, A., Lefebvre-Brion, H. (1977). *Ap. J. Lett.* **214**, L101.
Giusti-Suzor, A., Bardsley, J. N., and Derkits, C. (1983). *Phys. Rev. A* **28**, 682.
Golubkov, G. V., and Ivanov, G. K. (1988), *J. Phys. B: At. Mol. Phys.* **21**, 2049.
Gould, P. L., Lett, P. D., and Phillips, W. D. (1987). In "Laser Spectroscopy VII" (S. Svanberg and W. Persson, eds), p. 64. Springer-Verlag, Berlin.
Gould, P. L., Lett, P. D., Julienne, P. S., Phillips, W. D., Thorsheim, H. R., and Weiner, J. (1988). *Phys. Rev. Lett.* **60**, 788.
Greene, C. H., and Jungen, Ch. (1985), *Adv. At. Mol. Phys.* **21**, 51.
Hameed, S., Herzenberg, A., and James, M. G. (1968). *J. Phys. B* **2**, 822.
Henriet, A. (1985). *J. Phys. B: At. Mol. Phys.* **18**, 3085.
Henriet, A., and Masnou-Seeuws, F. (1983). *Chem. Phys. Lett.* **101**, 535.
Henriet, A., and Masnou-Seeuws, F. (1987). *J. Phys. B: At. Mol. Phys.* **20**, 671.
Henriet, A., and Masnou-Seeuws, F. (1988). *J. Phys. B: At. Mol. Phys.* **21**, L339.
Henriet, A., and Masnou-Seeuws, F. (1989). *Z. Phys. D.* (Submitted).
Hickman, A. P. (1987). *J. Phys. B: At. Mol. Phys.* **20**, 2091.
Holstein, T. (1947). *Phys. Rev.* **72**, 1212.

Holstein, T. (1951). *Phys. Rev.* **83**, 1159.
Hornbeck, H. A., and Molnar, J. P. (1951). *Phys. Rev.* **84**, 621.
Huennekens, J., and Gallagher, A. (1983a). *Phys. Rev. A* **27**, 771.
Huennekens, J., and Gallagher, A. (1983b). *Phys. Rev. A* **28**, 1276.
Ivanov, G. K., and Golubkov, G. V. (1986). *Z. Phys. D* **1**, 199.
Janev, R. K., and Mihajlov, A. A. (1979). *Phys. Rev. A.* **20**, 1890.
Janev, R. K., and Mihajlov, A. A. (1980). *Phys. Rev. A* **21**, 819.
Jeung, G. H. (1983). *J. Phys. B: At. Mol. Phys.* **16**, 4289.
Jeung, G. H. (1987). *Phys. Rev. A* **35**, 26.
Jeung, G. H., Malrieu, J. P., and Daudey, J. P. (1982). *J. Chem. Phys.* **77**, 3571.
Jeung, G. H., Malrieu, J. P., and Daudey, J. P. (1983). *J. Phys. B.* **16**, 699.
Jeung, G. H., and Ross, A. J. (1988). *J. Phys. B: At. Mol. Phys.* **21**, 1473.
Johnson, B. C., Wang, M.-X., and Weiner, J. (1988). *J. Phys. B: At. Mol. Phys.* **21**, 2599.
Jones, D. M., and Dahler, J. S. (1987). *Phys. Rev. A* **35**, 3688.
Julienne, P. S. (1988). *Phys. Rev. Lett.* **61**, 698.
Keller, J., Bonanno, R., Wang, M.-X., deVries, M. S., and Weiner, J. (1986). *Phys. Rev. A* **33**, 1612.
Kircz, J. G., Morgenstern, R., and Nienhuis, G. (1982). *Phys. Rev. Lett.* **48**, 610.
Klapisch, M. (1969). Thesis Orsay (unpublished).
Klapisch, M. (1971). *Comput. Phys. Commun.* **2**, 239.
Klyucharev, A. N., and Ryazanov, N. S. (1972). *Opt. Spectrosc.* **33**, 230.
Klyucharev, A. N., Sepman, V. Yu, and Vuinovich, V. (1977). *Opt. Spectrosc.* **42**, 336.
Koch, M. E., Verma, K. K., Bahn, J. T., and Stwalley, W. C. (1982). Proc. Int. *Conf. Lasers, 82*, 119-123.
Koike, F., and Nakamura, H. (1972). *J. Phys. Soc. Jpn.* **33**, 1426.
Kolos, W. (1978). *J. Mol. Struct.* **46**, 73.
Kolos, W., and Wolniewicz, L. (1965). *J. Chem. Phys.* **43**, 2429.
Kolos, W., and Wolniewicz, L. (1969). *J. Chem. Phys.* **50**, 3228.
Konowalow, D. D., and Fish, J. L. (1983). *Chem. Phys.* **77**, 435.
Konowalow, D. D., and Fish, J. L. (1984). *Chem. Phys.* **84**, 463.
Konowalow, D. D., and Olson, M. L. (1979). *J. Chem. Phys.* **71**, 450.
Konowalow, D. D., and Rosenkrantz, M. E. (1979). *Chem. Phys. Lett.* **61**, 489.
Konowalow, D. D., and Rosenkrantz, M. E. (1982a). *J. Phys. Chem.* **86**, 1099.
Konowalow, D. D., and Rosenkrantz, M. E. (1982b). In "Metal Bonding and Interactions in High Temperature Systems," Am. Chem. Soc. Symposium Series 179 (J. L. Cole and W. C. Stwalley, eds), pp. 1-77. American Chemical Society.
Konowalow, D. D., Rosenkrantz, M. E., and Olson, M. L. (1980). *J. Chem. Phys.* **72**, 2612.
Kowalczyk, P. (1979). *Chem. Phys. Lett.* **68**, 203.
Kowalczyk, P. (1980). *Chem. Phys. Lett.* **74**, 80.
Kowalczyk, P. (1984). *J. Phys. B: At. Mol. Phys.* **17**, 817.
Kushawaha, V. S., and Leventhal, J. J. (1980). *Phys. Rev. A* **22**, 2468.
Kushawaha, V. S., and Leventhal, J. J. (1982). *Phys. Rev. A* **25**, 346.
Lam, K. S., and George, T. F. (1984). *Phys. Rev. A* **29**, 492.
Lam, K. S., and George, T. F. (1985). *Phys. Rev. A* **32**, 1650.
Lefebvre-Brion H., and Field, R. W. (1986), "Perturbations in the Spectra of Diatomic Molecules" Academic Press, Orlando.
Leutwyler, S., Hermann, A., Wöste, L., and Schumacher, E. (1980). *Chem. Phys. Lett.* **48**, 253.
Lucatorto, T. B., and McIlrath, T. J. (1980). *Appl. Opt.* **19**, 3948, and references cited therein.
Martin, S., Chevaleyre, J., Valignat, S., Perrot, J., Broyer, M., Cabaud, M., and Hoareau, A. (1982). *Chem. Phys. Lett.* **87**, 235.
Maynau, D., and Daudey, J. P. (1981). *Chem. Phys. Lett.* **81**, 273.

Meijer, H. A. J., v. d. Meulen, H. P. Morgenstern, R., Hertel, I. V., Meyer, E., Schmidt, H., and Witte, R. (1986). *Phys. Rev. A* **33**, 1421.
Meijer, H. A. J., v. d. Meulen, H. P., Morgenstern, R. (1987a). *Z. Phys. D* **5**, 299.
Meijer, H. A. J., Pelgrim, T. J. C., Heideman, H. G. M., Morgenstern, R., and Anderson, N. (1987b). *Phys. Rev. Lett.* **59**, 2939.
Meijer, H. A. J., Pelgrim, T. J. C., Dijkerman, H.A. Heideman, H. G. M., Nienhuis, G., Morgenstern, R., and Andersen, N. (1988). In "Workshop on Reactive Collisions of Laser-Excited Na Atoms," 11–13 January, Fysisch Laboratorium, Rijksuniversiteit Utrecht (unpublished).
Meijer, H. A. J., Zeegers, T. H., Pelgrim, T. J. C., Heideman, H. G. M., and R. Morgenstern (1989a). *J. Chem. Phys.* **90**, 729.
Meijer, H. A. J., Zeegers, T. H., Pelgrim, T. J. C., Heideman, H. G. M., Morgenstern R., and Andersen, N. (1989b). *J. Chem. Phys.* **90**, 738.
Mies, F. H. (1984). *J. Chem. Phys.* **80**, 2514.
Mies, F. H., and Julienne, P. S. (1985). *In "Spectral Line Shapes" (F. Rostas, ed)* p. 393. W. DeGruyter, Berlin.
Milne, E. (1926). *J. London Math. Soc.* **1**, 1.
Mihajlov, A. A., and Janev, R. K. (1981). *J. Phys. B: At. Mol. Phys.* **14**, 1639.
Miller, W. H. (1970). *J. Chem. Phys.* **52**, 3563.
Mohler, F. L., and Boeckner, C. (1930). *J. Res. Nat'l. Bur. Stds.* **5**, 831.
Müller, W., and Meyer W. (1984). *J. Chem. Phys.* **80**, 3311.
Müller, W., Flesch, J., and Meyer, W. (1984). *J. Chem. Phys.* **80**, 3297.
Müller, M. W., Meijer, H. A. J., Kraft, T., Ruf, M.-W., and Hotop, H. (1989). Abstract of paper presented at ECAMP 1989, April 3–7, Bordeaux, France.
Muschlitz, E. E. (1966). *In* Advances in Chemical Physics (John Ross, ed.) Vol. X pp. 171–193.
Nakamura, H. (1971). *J. Phys. Soc. Jpn.* **31**, 574, and references cited therein.
Nakashima, K., Takagi, H., and Nakamura, H. (1987). *J. Chem. Phys.* **86**, 726.
Niehaus, A. (1981). In "The Excited State in Chemical Physics, Part 2", (J. Wm. McGown, ed.), Advances in Chemical Physics XLV, pp. 399–486.
Nielsen, S. E., and Berry, R. S. (1971). *Phys. Rev. A* **4**, 865.
Nielsen, S. E., and Dahler, J. S. (1979). *J. Chem. Phys.* **71**, 1910.
Nienhuis, G. (1982). *Phys. Rev. A* **26**, 3137.
Norcross, D. W. (1973). *Phys. Rev. Lett.* **32**, 192.
Olson, M. L., and Konowalow, D. D. (1977). Chem. Phys. **21**, 393.
Olson, M. L., and Konowalow, D. D. (1979). Chem. Phys. **22**, 29.
Oppenheimer, M., and Dalgarno, A. (1977). Astrophys. J. **212**, 683.
Partridge, H., Bauschlicher, C. W., Jr., Siegbahn, P. E. M. (1983). *Chem. Phys. Lett.* **97**, 198.
Peach, G. (1978). *J. Phys. B.* **11**, 2107.
Pluvinage, Ph. (1950). Ann. Phys. (Paris) **5**, 145.
Pluvinage, Ph. (1951). *J. Phys. Rad.* **12**, 789.
Poulaert, G., Brouillard, F., Claeys, W., McGowan, J. W., and Van Wassenhove, G. (1978). *J. Phys. B: At. Mol. Phys.* **11**, L671.
Prentiss, M., Cable, A., Bjorkholm, J. E., Chu, S., Raab, E. L., and Pritchard, D. E. (1988). *Opt. Lett.* **13**, 452.
Prodan, J., Migdall, A., Phillips, W. D., So, I., Metcalf, H., and Dalibard, J. (1985). *Phys. Rev. Lett.* **54**, 992.
Reeh, H. (1960). *Z. Naturforsch., Teil A* **15**, 377.
Rosmus, P. and Meyer, W. (1976). *J. Chem. Phys.* **65**, 492.
Ross, A. J., and Jeung, G. H. (1986). *Chem. Phys. Lett.* **132**, 44.
Ross, S., and Jungen, Ch. (1987). *Phys. Rev. Lett.* **59**, 1297.

Runge, S., Pesnelle, A., Perdrix, M., Watel, G., and Cohen, J. S. (1985). *Phys. Rev. A.* **32**, 1412.
Schmidt-Mink, I., Müller, W., and Meyer, W. (1985). *Chem. Phys.* **92**, 263.
Seaton, M. J. (1983). *Rep. Prog. Phys.* **46**, 167.
Smirnov, B. M. (1981). *Sov. Phys. Usp.* **24**, 251.
Smith, F. T. (1969). *Phys. Rev.* **179**, 111.
Stevens, W. J., Hessel, M. M., Bertoncini, P. J., and Wahl, A. C. (1977). *J. Chem. Phys.* **66**, 1477.
Stwalley, W. C., Uang, Y. H., and Pichler, G. (1978). *Phys. Rev. Lett.* **41**, 1164.
Takagi, H., and Nakamura, H. (1988). *J. Chem. Phys.* **88**, 4552.
Taylor, A. J., Jones, K. M., and Schawlow, A. L. (1981). *Opt. Commun.* **39**, 47.
Thorsheim, H. R., Weiner, J., and Julienne, P. S. (1987). *Phys. Rev. Lett.* **58**, 2420.
Thorsheim, H. R., Wang, Y., and Weiner, J. (1989) (unpublished).
Urbain, X. (1989), *In* "Dissociative Recombination: Theory, Experiment, and Applications" (J. B. A. Michell, and S. L. Guberman, eds). World Scientific, Singapore, p. 84.
Urbain, X., Giusti-Suzor, A., Fussen, D., and Kubach, C. (1986). *J. Phys. B: At. Mol. Phys.* **19**, L273.
Uzer, T., Watson, D. K., and Dalgarno, A. (1978). Chem. Phys. Lett. **55**, 6.
Valance, A. (1979). *J. Chem. Phys.* **69**, 355.
Valance, A., and Nguyen Tuan, Q. (1982). *J. Phys. B.* **15**, 17.
Valiron, P. (1979). Thése de 3e cycle Université de Bordeaux (unpublished).
Veselov and Bersuker (1958). *Izv. Akad. Nauk SSSR Ser. Fiz* **22**, 662.
Veselov and Schtoff (1967). *Opt. Spectrosc.* **22**, 457.
Wang, M. X., deVries, M. S., Keller J., and Weiner, J. (1985). *Phys. Rev. A* **32**, 681
Wang, M.-X., deVries, M. S., and Weiner, J. (1986a). *Phys. Rev. A* **33**, 765.
Wang, M.-X., deVries, M. S., and Weiner, J. (1986b). *Phys. Rev. A* **34**, 1869.
Wang, M.-X., Keller, J., Boulmer, and Weiner, J. (1986). *Phys. Rev. A* **34**, 4497.
Wang, M.-X., and Weiner, J. (1987a). *J. Phys. B: At. Mol. Phys.* **21**, L15.
Wang, M.-X., and Weiner, J. (1987b). *Phys. Rev. A* **35**, 4424.
Wang, M.-X., and Weiner, J. (1989). *Phys. Rev. A* **39**, 405.
Wang, M.-X., Keller, J., Boulmer, J., and Weiner, J. (1987). *Phys. Rev. A* **35**, 934.
Watson, D. K. (1977). *Chem. Phys. Lett.* **51**, 513.
Watson, D. K., Cerjan, C. J., Guberman, S., and Dalgarno, A. (1977), *Chem. Phys. Lett.* **50**, 181.
Watson, W. D. (1978). *Ann. Rev. Astron, Astrophys.* **16**, 585.
Weiner, J., and Boulmer, J. (1986). *J. Phys. B: At. Mol. Phys.* **19**, 599.
Weisheit, J. C. (1972). *Phys. Rev. A* **5**, 1621.
Wolniewicz, L., and Dressler, K. (1977). *J. Mol. Spectrosc.* **67**, 416.
Wolniewicz, L., and Dressler, K. (1985). *J. Chem. Phys.* **82**, 3292.
Yan, G. Y., Duffey, T. P., Du, W. M., and Schawlow, A. L. (1987). *J. Opt. Soc. Am. B* **4**, 1829.
Zagrebin, S. B., and Samson, A. V. (1985a). *J. Phys. B: At. Mol. Phys.* **18**, L217.
Zagrebin, S. B., and Samson, A. V. (1985b). *Sov. Tech. Phys. Lett.* **11**, 283.

ON THE β-DECAY OF ^{187}RE: AN INTERFACE OF ATOMIC AND NUCLEAR PHYSICS AND OF COSMOCHRONOLOGY

ZONGHUA CHEN, LEONARD ROSENBERG, and LARRY SPRUCH

Department of Physics
New York University
New York, New York

I. Introduction 297
II. The Relative Production Rates of Different Isotopes of Re and of Os . . . 303
III. Refinements and Improvements 308
    A. Effects of Branching and of the Nuclear Excited State of ^{187}Os on the S-Process 308
    B. Effects of the Bound-State β Decay of ^{187}Re 309
IV. Conclusion 318
    Acknowledgements 318
    References 319

I. Introduction

There have been many excellent review papers on nuclear cosmochronology. These include, in particular, analyses of how a study of the ratios of various isotopes of Re and Os, combined with a knowledge of the half-life $T_{1/2}$ of the β decay of ^{187}Re, can be helpful in estimating the age of our galaxy (e.g., see Burbidge et al., 1957; Clayton, 1964; Fowler, 1984; Meyer and Schramm, 1986; Yokoi et al., 1983; Clayton, 1988). This paper has been written in the belief that it might nevertheless be useful to have available a review article on Re and Os as a chronometric pair written specifically for atomic physicists, a review which defines or expands on astrophysical terminology and concepts (isochron, astration, s-process, r-process, etc.) and which emphasizes the broad view of the problem rather than the details. We can only hope that the advantages associated with the relatively fresh view of authors who are rather new to the subject are not outweighed by the disadvantages.

One might well ask how atomic physics enters a problem which would seem to be one in nuclear cosmochronology. The answer is the following.

The rest energy of a ^{187}Re nucleus which decays is *less* by an amount $\Delta \approx 12$ keV than the sum of the rest energies of its decay products, a ^{187}Os nucleus, an electron, and an antineutrino. The continuum β decay of a neutral ^{187}Re atom is therefore possible only because the atomic binding energy of a neutral ^{187}Os atom exceeds that of a neutral ^{187}Re atom by more than Δ; the excess is $\Delta + E_0$, where the end-point energy E_0 of the electron spectrum is only 2.64 keV. The very small value of E_0 is one reason for the very large value of $T_{1/2}$, and might suggest that the bound-state β decay (bound-state β decay was first considered by Daudel et al., 1947a, 1947b; many of the details of the theory were worked out by Bahcall, 1961) of a ^{187}Re atom (with $Z = 75$), in which the β particle is captured into a bound state of ^{187}Os (with $Z = 76$), might be significant. The calculation of the bound-state decay rate requires the determination of atomic wave functions of Re and Os. In fact, bound-state β decay of a *neutral ^{187}Re atom* is not a significant effect, the rate being of the order of one percent of the continuum decay rate. [The atomic aspects of the bound-state β decay of a neutral ^{187}Re atom are covered in some detail in Williams et al. (1984) and Chen et al. (1987), and most of the detail will not be repeated in this article.][1] However, at a high temperature Θ characteristic of some stellar interiors, ^{187}Re is very highly ionized and continuum decay may not be energetically possible. (For sufficiently great ionization, the difference in atomic binding energies will not be enough to compensate for the energy difference Δ.) On the other hand, when ^{187}Re is sufficiently highly ionized, bound-state β decay may not only be energetically possible but the decay rate may be *very* much greater, of the order of 10^9 times greater, than the continuum decay rate of a neutral ^{187}Re atom! (See Takahashi et al., 1983, 1987.) This is a consequence of the considerable

---

[1] The normal procedure for determining the β-decay rate, detection of the emergent β particles, seemed for some experiments (Brodzinski and Conway, 1965) to give a rate different from that obtained by studying the ratio of ^{187}Re and ^{187}Os in geochemical deposits (Luck et al., 1983); in the latter analysis, one uses an age of the solar system of about 4.6 Gyr (1 Gyr $= 10^9$ years). Since the former measurement detects only continuum decay while the latter measurement encompasses continuum and bound state decay, it was sometimes argued that bound-state decay must play a role (Perrone, 1971). However, a recent measurement (Lindner, et al., 1986; see also references therein) of the number of ^{187}Os atoms present in a liquid four years after a known number of ^{187}Re atoms were placed in the initially nearly ^{187}Os-free liquid, a measurement of the total decay rate, agrees with some measurements of the continuum rate alone (Payne, 1965), as well as with some determinations based on geochemical data (Luck et al., 1983). The possibility that bound-state β decay is terribly significant would seem to be ruled out, but there is no experimental confirmation of the theoretical prediction (Williams et al., 1984; Chen et al., 1987) that the bound-state decay rate is about one percent of the continuum decay rate. Note that since $T_{1/2} \approx 4 \times 10^{10}$ years, only about 10^{-10} of the ^{187}Re nuclei initially present will have decayed in 4 years. The initial impurity level of ^{187}Os must therefore be exceedingly low.

binding energy of an inner electron of Os, of the nuclear properties of ^{187}Re and ^{187}Os, and of the selection rules. We will return to this point later.

Estimates of the age of the Earth based on nuclear half-lives were made by Rutherford much earlier, but nuclear cosmochronology may be said to have begun also by Rutherford (1929). Burbidge *et al.* (1957) used the chronometric pairs ^{232}Th/^{238}U and ^{235}U/^{238}U to estimate the age of our galaxy. Later Clayton (1964) introduced the ^{187}Re/^{187}Os chronometric pair as a tool; his method of analysis remains the basic approach for that pair. To be useful in estimating the age of our galaxy, the parent member of the chronometric pair must have a half-life roughly comparable to the age of our galaxy, believed to be of the order of 10 Gyr. ^{187}Re has a very long half-life, $T_{1/2} \approx 43$ Gyr, under laboratory conditions. The uncertainties in the estimate of the age of our galaxy when using ^{235}U, ^{238}U, and ^{232}Th are primarily the uncertainties in the relative production rates of those nuclei (e.g., see Clayton, 1964; Meyer and Schramm, 1986). The uncertainties arising in the use of the ^{187}Re/^{187}Os chronometric pair are rather different. As we will see, they include uncertainties in the neutron capture cross section ratio for ^{186}Os and ^{187}Os under "slow process" (defined below, and normally written as s-process) conditions, and in the Os to Re abundance ratio (e.g., see Clayton, 1983).[2] Furthermore, as noted above, the β-decay rate of neutral ^{187}Re is only of the order of 10^{-9} times the rate for ^{187}Re which is sufficiently highly ionized, an ionization which can occur for the ^{187}Re in the very hot environment in certain stars. The rate of β decay of highly ionized ^{187}Re is itself somewhat uncertain, but the greatest uncertainty may possibly be the fraction of time the ^{187}Re spends in a sufficiently hot environment. In the long run, the ^{187}Re/^{187}Os pair may well be the pair best suited for providing an estimate of the age of our galaxy. At the very least, it provides an estimate of the age of the galaxy which can be compared with the estimates obtained by using the ^{232}Th/^{238}U and ^{235}U/^{238}U pairs and by other methods.

To discuss the use of the ^{187}Re/^{187}Os pair in broad outline, we make the following simplifying assumptions—some will be relaxed later, and some additional assumptions will be noted later:

(1) No very heavy nuclei were synthesized during the Big Bang. The ^{187}Re found in meteorites and Earth samples was synthesized in the time interval T between the formation of our galaxy and the formation of the solar system. The synthesis occurred during supernovae, with the ^{187}Re therefore ejected immediately; the rate of synthesis is then dependent on the rate at which supernovae occur. The number of ^{187}Re nuclei *produced* per unit volume of our galaxy in the time interval $d\tau$ is $P(\tau)d\tau$, where τ is measured

---

[2] When an element is given without a superscript, its abundance is the sum of the abundance of all of its isotopes.

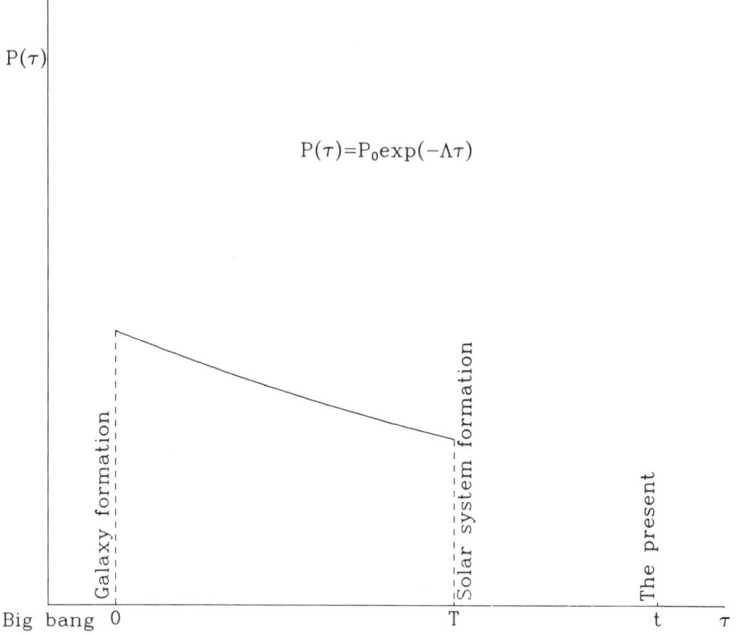

FIG. 1. The production rate per unit volume of ^{187}Re versus time in an exponential synthesis model. T is the time of formation of the solar system, and t, the present time, is the age of our galaxy, both measured from the beginning of nucleosynthesis. In the figure, we have ignored the short period, believed to be about 0.1 Gyr, for the solidification of the solar system.

from the time of formation of our galaxy. (See Fig. 1.) The time from the Big Bang to the formation of our galaxy, presumably of the order of 1 Gyr, plays no role in the present analysis of the production of ^{187}Re. It is the value of T, the time interval from the formation of our galaxy to the formation of the solar system, which is the primary question we consider here. For simplicity, we used a two-parameter model (Burbidge et al., 1957) to characterize $P(\tau)$,

$$P(\tau) = P_0 e^{-\Lambda \tau}. \qquad (1)$$

with $\Lambda \geq 0$. According to this "exponential synthesis" model, the number of ^{187}Re nuclei produced per unit volume in the time interval T—some will of course have decayed by the time T—is given by

$$^{187}\text{Re(produced by time } T) = (P_0/\Lambda)[1 - \exp(-\Lambda T)]. \qquad (2)$$

$P(\tau)$ reduces to a "constant synthesis" model for $\Lambda = 0$, and approaches a "sudden synthesis" model for $\Lambda \sim \infty$ and P_0/Λ finite. (As will be seen later, an estimate of the age of our galaxy can be expressed in a form which is independent of P_0 and which is not too sensitive to the value of Λ which one chooses. Some basis for the choice made in Eq. (1) for the form of $P(\tau)$ will be given in Sec. II).

(2) The ^{187}Os nuclei are synthesized by two independent processes. One component is synthesized within stars and is then ejected into interstellar space; it is designated by ^{187}Os$_s$, where the subscript s refers to the "slow process" (to be described in Sec. II), by which it is formed. A second component, formed by the β decay of ^{187}Re in the interstellar medium, is designated by ^{187}Os$_c$, with the subscript c denoting its cosmoradiogenic origin.

(3) Once in the interstellar medium, ^{187}Re and ^{187}Os atoms are not captured by stars.[3]

(4) The isotopes ^{186}Os, ^{187}Os and ^{187}Re are each uniformly mixed in the interstellar medium at the time the solar system was formed (and presumably at earlier times, but that is here irrelevant).

(5) The various meteorites for which isotope ratios have been measured, and the Earth's mantle, were formed at nearly the same time, though not necessarily under the same conditions (local density, temperature, etc.). Since ^{186}Os and ^{187}Os are chemically almost identical their ratio at the time of formation of the solar system should be almost the same, for all of the meteorites and Earth samples, independent of local conditions; differences associated with the slight mass difference can for present purposes be largely ignored. Because of chemical differences, the ratio of ^{187}Re to ^{187}Os (or to ^{186}Os), on the other hand, can vary from meteorite to meteorite to Earth sample.

The validity of assumptions 4 and 5 is very strongly attested to by the results of an isochron plot. (See Fig. 2; Luck and Allegre, 1983. These authors assume the age of the solar system to be known, and determine the half-life of ^{187}Re. Given an independent determination of $T_{1/2}$, the method can also be used in reverse, as we have, to determine the age of the solar system.) The terminology isochron, or "same time," is used because the plot is based upon the assumption—part of assumption 5—that the solar system was formed at a rather well-defined time. Thus, once the solar system forms, there is no further input of ^{187}Re or ^{187}Os, and the ^{187}Re undergoes "free decay" to ^{187}Os. (An end-point energy E_0 of a few keV is low by β-decay standards but is high enough for the β-decay rates of an atom of ^{187}Re within a meteorite and in free space to be almost identical.) Letting the free decay rate be $\lambda = \ln 2/T_{1/2}$, and letting t represent the present time, the relationship between ^{187}Os(t), ^{187}Re(t) and ^{187}Os(T) follows immediately. We then divide by ^{186}Os(t) and note that, since ^{186}Os is stable and is not being produced, we

---

[3] It should be reemphasized that assumption (3) is one of the weak points in the argument. Some fraction of the ^{187}Re and ^{187}Os may be recaptured by other stars and reprocessed.

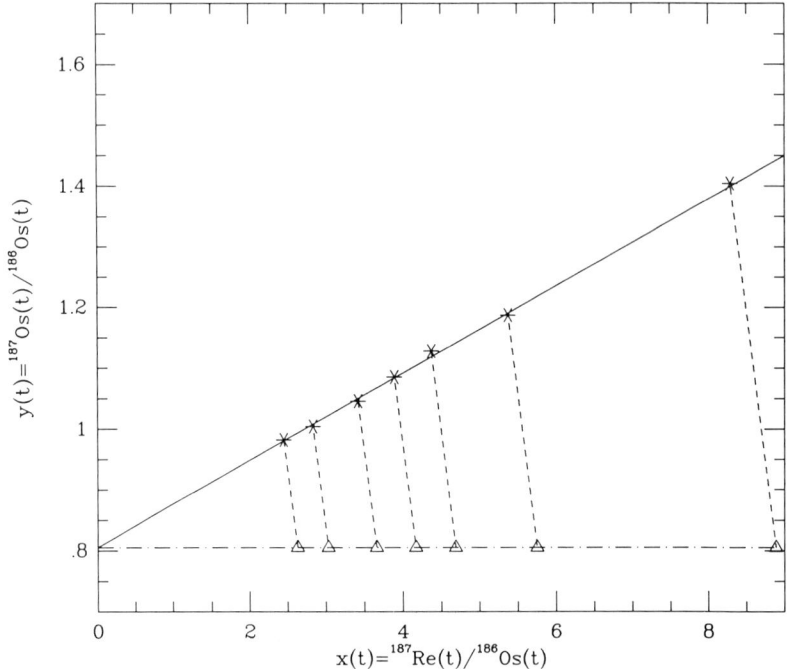

FIG. 2. A schematic plot of $y(t)$ versus $x(t)$ for various meteorites. $x(t)$ and $y(t)$ are the present abundance ratios of $^{187}\text{Re}/^{186}\text{Os}$ and $^{187}\text{Os}/^{186}\text{Os}$, respectively. The experimental values, each denoted by a $*$, lie very close to a straight line as is expected from the form of Eq. (3). The abundance ratios at the time T the solar system formed, each indicated by a Δ, are obtained from the experimental values by using the value of T and the experimental value of the decay rate λ of ^{187}Re. The Δ's lie very close to a horizontal dashed line, consistent with the assumption that all meteorites have the same abundance ratio of $^{187}\text{Os}/^{186}\text{Os}$ at time T. If we had used the same scale for $x(t)$ and $y(t)$, each dashed line would make an angle of 45° with the horizontal since the sum of the ^{187}Re and ^{187}Os atoms is time independent, the former decaying into the latter.

have $^{186}\text{Os}(t) = {}^{186}\text{Os}(T)$. A connection between abundance ratios at the present time t and at the earlier solar system formation time T is then easily found to be (Luck and Allegre, 1983)

$$\frac{^{187}\text{Os}(t)}{^{186}\text{Os}(t)} = [e^{\lambda(t-T)} - 1] \frac{^{187}\text{Re}(t)}{^{186}\text{Os}(t)} + \frac{^{187}\text{Os}(T)}{^{186}\text{Os}(T)}. \quad (3)$$

For purposes of discussion, we write this as

$$y(t) = m(t)x(t) + y(T).$$

A plot of $y(t)$ vs $x(t)$ (see Fig. 2), using the experimental values of $y(t)$ and $x(t)$ for the different meteorites and an Earth sample, shows an excellent fit to a straight line. The intercept is thereby well defined, validating assumptions 4 and 5; further, the slope $m(t) = 0.0716$ gives the value of $\lambda(t - T)$. The use of the laboratory value $T_{1/2} = 43.5 \pm 1.3$ Gyr (Lindner et al., 1987), which corresponds to $\lambda = (1.59 \pm 0.05) \times 10^{-11}$ yr^{-1}, then provides the estimate, $t - T = 4.34 \pm 0.13$ Gyr, for the age of the solar system. [The age of the solar system is known more accurately to be 4.55 ± 0.03 Gyr by other methods, which gives $\lambda = 1.52 \pm 0.04$ (Luck and Allegre, 1983). The inaccuracy in the determination of the slope and therefore of $\lambda(t - T)$, and the uncertainty in λ, provide the estimate of roughly 0.1 Gyr for the time interval during which the solar system was formed (Luck and Allegre, 1983).] The value of the intercept is

$$y(T) \equiv {}^{187}\text{Os}(T)/{}^{186}\text{Os}(T) = 0.805.$$

Thus, knowing the value of λ from experiments on ^{187}Re β decay, we have determined very good estimates of the age $(t - T)$ of the solar system and of the Os isotope ratio $y(T)$ at the time T of formation of the solar system by working *backward* from values of Os and Re isotope ratios at the present time t. The next step will be to work *forward* from the time 0 of the formation of the galaxy. Having determined various isotope ratios at the unknown time T, it will be possible to determine the value of T which gives those isotopes ratios, and in particular the ratios $y(T)$, if we can obtain the relative rates at which the isotopes are formed. We turn our attention now to the question of the relative formation rates.

II. The Relative Production Rates of Different Isotopes of Re and of Os

The production of ^{187}Re at a rate given in Eq. (1), was assumed to occur only during supernovae. The synthesis then proceeds via the so-called "rapid process", often written simply as "r-process", and briefly described just below. (See Fig. 3; the r-process path is indicated by a dashed line. See also the discussion which follows.) It should be understood that Eq. (1) has at least some rough justification. Thus, studies of the abundances of heavy elements (which include Re) in various stars of known ages suggest that the rate of production of Re has been constant or decreasing, but has not been increasing. Apart from imposing that constraint, we have chosen an explicit form (Burbidge et al., 1957), an exponential; the constraint, for the exponential form, is that Λ must be non-negative. That choice is based on the reasonable but not compelling assumption that the rate of production is linearly proportional to the density of the elements which produced ^{187}Re atoms. [The assumption encompassed in Eq. (1) has its strength as well as its

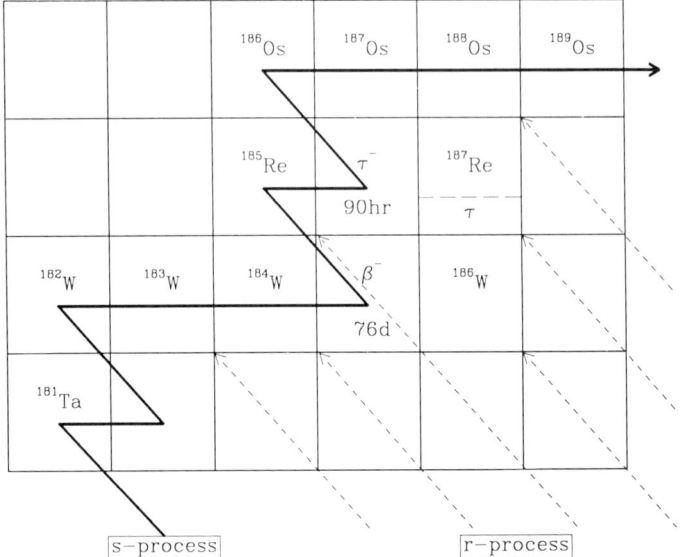

FIG. 3. The synthesis of Re and Os. The s-process path, shown as the heavy line, results in the synthesis of stable species lying on the line of beta-stability. The diagonal dashed lines indicate isobaric β-decays starting from the neutron-rich isobars synthesized in the r-process. The nuclei ^{186}Os and ^{187}Os are shielded from r-process nucleosynthesis by ^{186}W and ^{187}Re, respectively.
(Reprinted courtesy of D. D. Clayton and *The Astrophysical Journal*, published by the University of Chicago Press, © 1964 The American Astronomical Society.)

weaknesses. Thus, since P_0 and Λ are taken to be open parameters, Eq. (1), in some average sense, allows not only for the production of ^{187}Re via supernovae—all that we will consider for the moment (see assumption 3 in Sec. I)—but also, at least to some extent, for supernovae followed by capture by stars of ^{187}Re and ^{187}Os, with subsequent reprocessing.]

Once one has assumed the form given in Eq. (1), the details of the r-process play no role, but it is important to understand the nomenclature. The term "rapid" originates in the assumption that neutron rich isotopes are generated by a sequence of neutron captures, the neutron capture by a given nuclide in the sequence occurring at a much more *rapid* rate than the β-decay rate of that nuclide. Roughly speaking, there occurs a sequence of neutron captures and β decays, with the former predominating, in a very short time (a few seconds), along the "r-process path;" this is then followed by a sequence of β decays, terminating at a stable or near-stable isotope, in this case at ^{187}Re. The reader should be alerted to the fact that figures (such as Fig. 3) show only the latter stage, a series of β decays in a process defined as one in which neutron capture dominates. (For a more detailed discussion, see Clayton, 1983.)

Consider now an s-process path, which can occur in a number of types of stars, including red giants, but not during supernovae. The s-process path is indicated by a solid line in Fig. 3, the path by which ^{186}Os and ^{187}Os$_s$ nuclei, among others, are produced. (We need not put a subscript s on ^{186}Os since this nucleus is not produced by a second route.) In contrast to a rapid process, a slow process (Burbidge et al., 1957) is a neutron capture process in which the neutron capture rate is slower than the β decay rate for the radioactive isotopes. We are here concerned with a path containing steps in which neutron capture occurs only by stable nuclides; unstable nuclides undergo β decay, typically with half-lives of hours to years, the neutron capture rate for one nucleus being *slower*, taking of the order of thousands of years. (As for the r-process, there is an element of possible confusion in Fig. 3 showing an s-process. Although by definition an s-process is one in which neutron capture rates are slower than β-decay rates, an s-process path can show many more neutron captures than β decays. The point is that an s-process path proceeds in the valley of β-stability, with most of the nuclides along the path therefore stable. For unstable nuclei, for which neutron capture or β decay can occur, the β-decay rate is greater.) In this discussion of the s-process, we have (for the moment) made an assumption.

(6) There is no branching in an s-process path in the Re-Os region—each step is either β decay *or* neutron capture.

We will (again for the moment) make two further assumptions with regard to the s-process path:

(7) Every ^{186}Os and ^{187}Os nucleus undergoing neutron capture is initially in its ground state.

(8) Each step in the Re-Os region occurs under equilibrium conditions, the number of nuclei of a given nuclide being independent of time.

Assumption 8 has some marvelously simplifying consequences. Given that two stable isotopes with mass numbers A and $A-1$ each undergoes neutron capture, as is the case for ^{187}Os and ^{186}Os (see Fig. 3), with cross sections σ_A and σ_{A-1}, respectively, we have

$$\langle \sigma v \rangle_A n N_A = \langle \sigma v \rangle_{A-1} n N_{A-1},$$

where angular brackets indicate velocity averaging, n is the number density of neutrons (of any velocity), and N is an isotope number density. Now the s-process is assumed to occur at temperatures corresponding roughly to 30 keV. (This temperature is chosen because it leads to the distribution of nuclear abundances found in nature. The s-process is assumed to occur in an interval not of seconds but of perhaps 1000 years, give or take three powers of 10; this is one element of the argument allowing one to assume equilibrium.) At 30 keV, for heavy nuclei, it is a good approximation to set $\langle \sigma v \rangle = \langle \sigma \rangle v_\Theta$, where v_Θ is the average velocity for the given temperature. With the $\langle \ \rangle$ on σ henceforth understood, we thereby arrive at

$$\sigma_A N_A = \sigma_{A-1} N_{A-1}, \tag{4}$$

the so called "local approximation." The ratio N_A/N_{A-1} thereby follows from a knowledge of σ_{A-1}/σ_A; it is *independent* of (i) the neutron flux in the star, and (ii) the time of formation of the star, the time at which the s-process in the star began, and the time relative to the onset of the s-process. [The ratio does depend somewhat on the local temperature Θ, through σ, but Θ is presumably moderately well known. The fact that one can choose a Θ which leads to a distribution in the abundances of heavy elements in good agreement with the data is a strong argument in favor of the local approximation. For the particular case of Os, it is known experimentally that in the neighborhood of 30 keV, the cross sections σ_{186} and σ_{187} have about the same energy dependence (Winters et al., 1980; Browne and Berman, 1981), so that the ratio is relatively insensitive to the choice of Θ.]

When applied to the Os isotopes, it is the ratio $^{187}\text{Os}_s/^{186}\text{Os}$ which is obtained,

$$^{187}\text{Os}_s/^{186}\text{Os} = \sigma_{186}/\sigma_{187}; \tag{5}$$

the cross section ratio $\sigma_{186}/\sigma_{187}$ at 30 keV was recently measured to be 0.478 ± 0.022 (Winters and Macklin, 1982). The fact that this abundance ratio is reasonably reliable,—see (i) and (ii), just above, but also see Sec. III.A—represents a great advantage of the Os pair over other chronometric pairs, which, not being generated in an s-process, can depend upon fluxes, absolute densities, and/or times, as well as upon complicated branching ratios.

We must of course also determine the rate of production of $^{187}\text{Os}_c$, but that is relatively straightforward for the assumed production rate of ^{187}Re given by Eq. (1). We have

$$\frac{d\ ^{187}\text{Re}}{d\tau} = -\lambda\ ^{187}\text{Re} + P_0 e^{-\Lambda \tau}, \qquad ^{187}\text{Re}(0) = 0,$$

where λ is assumed to be time independent, since ^{187}Re is assumed to be in the interstellar medium, and therefore

$$^{187}\text{Re}(T) = \frac{P_0}{\lambda - \Lambda}(e^{-\Lambda T} - e^{-\lambda T}). \tag{6}$$

The number density of $^{187}\text{Os}_c$ at time T is just the difference between the number of ^{187}Re nuclei produced per unit volume by the time T, given in Eq. (2), and the ^{187}Re nuclei per unit volume which actually exist at time T, as just given. We therefore have

$$^{187}\text{Os}_c(T) = P_0\left(\frac{1 - e^{-\Lambda T}}{\Lambda} - \frac{e^{-\Lambda T} - e^{-\lambda T}}{\lambda - \Lambda}\right), \tag{7}$$

so that, using Eq. (6), we have

$$\frac{^{187}\text{Os}_c(T)}{^{187}\text{Re}(T)} = \frac{\lambda - \Lambda}{\Lambda}\frac{1 - e^{-\Lambda T}}{e^{-\Lambda T} - e^{-\lambda T}} - 1. \tag{8}$$

We also have the simple result

$$^{187}\text{Os}_c(T) = {}^{187}\text{Os}(T) - {}^{187}\text{Os}_s(T). \tag{9}$$

Dividing Eq. (9) through by $^{187}\text{Re}(T)$, and multiplying and dividing the right-hand side of Eq. (9) by $\text{Os}(T)$ and by $\text{Re}(T)$ for convenience, and using Eq. (5), we arrive at an alernative expression for the ratio given in Eq. (8),

$$\frac{^{187}\text{Os}_c}{^{187}\text{Re}} = \frac{(^{187}\text{Os}/\text{Os}) - (\sigma_{186}/\sigma_{187})(^{186}\text{Os}/\text{Os})}{^{187}\text{Re}/\text{Re}}\frac{\text{Os}}{\text{Re}}, \tag{10}$$

where each abundance is at time T and each appears within a ratio. The ratios on the right hand side of Eq. (10) can all be derived rather easily, and with reasonable precision, from the present measured ratios within meteorites and from a knowledge of the various half-lives. [The results are listed in Table I. We note that on the one hand there is some uncertainty in the Os/Re

TABLE I

THE ABUNDANCE RATIOS OF VARIOUS ISOTOPES OF Re AND Os, AND OF Re AND Os, AT THE PRESENT TIME AND AT THE TIME T OF THE FORMATION OF THE SOLAR SYSTEM.[a]

$t - T = 4.55$ Gyr, $\lambda = 1.52 \times 10^{-11}$ yr^{-1}, $\sigma_{186}/\sigma_{187} = 0.478$.

	Present	At time T
$^{185}\text{Re}/\text{Re}$	0.37[b]	0.315
$^{187}\text{Re}/\text{Re}$	0.63[b]	0.65
$^{186}\text{Os}/\text{Os}$	0.016[c]	0.016
$^{187}\text{Os}/\text{Os}$	0.016[c]	0.0129
Os/Re	14.1[c]	13.4

[a] Re is the sum of ^{185}Re and ^{187}Re; the latter decays into ^{187}Os. The total abundance of Os can be approximated as a constant, even though the abundance of ^{187}Os changes, since ^{187}Os is only a small component of Os.
[b] Clayton (1964).
[c] Winters (1984).

ratio, but that on the other it is an advantage of the approach that the value of the ratio seems to be relatively constant from one meteorite to another (Clayton, 1983). One finds a value of

$$^{187}Os_c(T)/^{187}Re(T) = 0.108.$$

Setting this value equal to the right-hand side of Eq. (8) gives an estimate of the value of T for an assumed value of Λ. In particular, one has

$$T = 6.75 \quad \text{Gyr} \quad \text{for sudden synthesis,}$$

$$T = 13.7 \quad \text{Gyr} \quad \text{for uniform synthesis.}$$

Adding the age of the solar system $t_{ss} = 4.55$ Gyr, one obtains the age of the galaxy as about 11 Gyr for the sudden synthesis model and about 18 Gyr for the uniform synthesis model (Clayton, 1964).

III. Refinements and Improvements

Eight explicit assumptions were listed in Secs. I and II. We now reconsider assumptions 6, 7 and 3.

A. Effects of Branching and of the Nuclear Excited State of ^{187}Os on the S-Process

With regard to assumption 6, we note that there *is* some branching in the s-process path in the W-Os region (Arnould, 1974; Beer *et al.*, 1981). As one example, consider ^{185}W. As well as undergoing β decay, which is the dominant process, ^{185}W can undergo neutron capture to ^{186}W followed by neutron capture to ^{187}W, which then β decays. One introduces a correction factor F_s to account for branching. The value of F_s follows easily from a knowledge of paths (the path is no longer unique) and the neutron capture and β decay rates (e.g. see Clayton, 1983, p. 576).

Turning to assumption 7, it is important to recall that the s-process takes place at a very high temperature, believed as noted above to correspond to about 30 keV. Using the Boltzman factor and taking into account the statistical weight factors, it is found that at that temperature the ratio of ^{187}Os nuclei in the 9.8 keV first excited state to those in the ground state is 48 to 33. The relevant cross section for neutron capture by ^{187}Os is a weighted average of the neutron capture cross sections for the ground state and all the excited states; as it happens, the contribution from states above the first excited state is negligible. A correction factor F_σ is introduced to account for the difference between the cross section ratio with the neutrons *and* nuclei in an ambient temperature of 30 keV, and under laboratory conditions, where 30 keV neutrons are incident on ground state nuclei. (The determination of F_σ must contain some theoretical input. One cannot measure the neutron capture cross section for a very short-lived state.)

The net effect is that one must replace Eq. (5) by

$$\frac{^{187}Os_s}{^{186}Os} = F_\sigma F_s \frac{\sigma_{186}}{\sigma_{187}}, \tag{11}$$

where the σ's are the laboratory values. Woosley and Fowler (1979) estimated F_σ to lie in the range 0.8 to 1; more recently, Hershberger *et al.* (1983) narrowed the range to 0.8 to 0.83. A likely range for F_s of 0.9 to 1.0 has been suggested by Arnould *et al.* (1984).

B. Effects of the Bound-State β Decay of ^{187}Re

We noted above that the half-life $T_{1/2}$ of a free ^{187}Re atom for continuum decay, or for continuum and bound-state decay, is almost the same according to some theoretical calculations and some experimental results (see Williams *et al.*, 1984; Chen *et al.*, 1987; and references therein). In other words, bound-state β decay of a free ^{187}Re atom seems to play no significant role. However, if a ^{187}Re atom is multiply ionized, the bound-state β decay rate may increase greatly. This is the case when we relax assumption 3; thus there is a possibility (Clayton 1969; Yokoi *et al.*, 1983) that some ^{187}Re or ^{187}Os in the interstellar medium will help form a new star, and will be re-ejected, having been reprocessed while in the interior of the star (so-called "astration"). We noted above that if one treats the P_0 and Λ in Eq. (1) as open parameters, various refinements, including astration, are partially accounted for. However, in that approach, one must allow Λ to range from 0 to ∞, and this introduces a range, by about a factor of 2, in the value of T. Certain constraints on the value of Λ may be placed by estimating the age of the galaxy using the "exponential model" for other chronometric pairs, such as $^{232}Th/^{238}U$, which are also produced by the r-process. As an alternative to somewhat arbitrarily choosing a model for the production rate of ^{187}Re and using data of various kinds to constrain the values of the parameters in that model one can attempt to proceed from first principles. Such an attempt, of course, requires a rather good understanding of astration. One form of reprocessing of ^{187}Os is through electron capture, and there may also be neutron capture (s-process). Some detailed discussions of these effects have been given (e.g. see Yokoi *et al.*, 1983). We will here consider in some detail the enhanced β decay of ^{187}Re when in a star in an environment of high temperature ($10^6 < \Theta < 10^8$ K—this temperature range is chosen from stellar evolution models and from the requirement that the temperature is high enough for the enhanced β decay to be important), and at various possible free electron densities.

In the following, we will sketch the major properties of bound-state β decay of a neutral ^{187}Re atom and of a nearly completely ionized ^{187}Re. At the price of being able to obtain only rough estimates—much more exact calculations have been carried out (Perrone, 1971; Takahashi et al., 1987; Chen et al., 1987)—we will present the argument in as simple and physical a manner as we can.

Generally, the *bound-state β-decay rate* for the nuclear ground state to ground state transition, $\lambda_{\text{bd},i}$, is proportional to the product of the antineutrino phase-space factor ρ_ν, the square of the relevant radial wave function components (determined by the selection rules) at the nuclear surface R of the electron in its final bound state i and of the antineutrino, $|\phi_{ei}(R)|^2$ and $|\phi_\nu(R)|^2$, respectively, and the square of the nuclear ground state to ground state matrix element for the transition, $|M|^2$. Using the symbol $\sim$ for the order of magnitude and ignoring some constant factors, we have

$$\lambda_{\text{bd},i} \sim \rho_{\bar{\nu}} |\phi_{ei}|^2 |\phi_{\bar{\nu}}|^2 |M|^2. \tag{12}$$

We will later use this expression to estimate the relative order of magnitude of various bound-state β-decay rates of ^{187}Re.

1. β Decay of Neutral ^{187}Re Atom

The total angular momenta and the parities of the initial and final nuclear states in the β decay of neutral ^{187}Re to ^{187}Os are $\frac{5}{2}^+$ and $\frac{1}{2}^-$, respectively. The transition therefore represents a nuclear change of parity and a nuclear change of total angular momentum of 2 or 3. Within the normal approximation of β-decay theory, the dominant contribution is that in which the electron-antineutrino ($e^- - \bar{\nu}$) pair carries off one unit of spin and one unit of orbital angular momentum, so aligned that the total angular momentum carried off is 2. The one unit of orbital angular momentum can be carried off either by the e^- or by the $\bar{\nu}$; which ever particle carries off the orbital angular momentum carries off a total angular momentum of $\frac{3}{2}$. In a non-relativistic description one would therefore say that the β particle will end up in a p$_{3/2}$ or s$_{1/2}$ state. Turning to relativistic theory one notes that the large and small components of a Dirac wave function differ by one unit in orbital angular momentum. The small component of the state labeled d$_{3/2}$ behaves as a p$_{3/2}$

state, and the small component of the state labeled $p_{1/2}$ behaves as an $s_{1/2}$ state. Therefore, in a relativistic description one would say that the β particle can end up in an $s_{1/2}$, $p_{1/2}$, $p_{3/2}$, or $d_{3/2}$ state. For a neutral ^{187}Re atom, it turns out that the dominant contribution for bound state β decay is that for which the β particle is emitted into a $p_{3/2}$ state, and—not surprisingly, though it is by no means obvious—this is then also true for the continuum β decay of a neutral ^{187}Re atom. Considering for the moment the continuum case, we can see by the following simple argument that the β particle will almost always carry off the orbital angular momentum. Temporarily ignoring the Coulomb field, let p and q represent electron and antineutrino momenta, and let E_0 be the kinetic energy available to the e^- and the $\bar{v}$. Then the probabilities for the e^- and the $\bar{v}$ carrying off orbital angular momentum $l = 1$ are proportional to $\overline{(pr)^2}$ and $\overline{(qr)^2}$, respectively, where the bars represent averages over the momenta and the coordinates, and we have retained the square of the leading nonvanishing terms of $\exp(i\mathbf{p}\cdot\mathbf{r})$ and $\exp(i\mathbf{q}\cdot\mathbf{r})$, respectively. The ratio will not be much affected if we replace p by p_{max} and q by q_{max}, where

$$p_{max}^2/(2m_e) = E_0; \quad q_{max}c = E_0.$$

The ratio of the decay rate for the β particle captured into the $s_{1/2}$ state to that into the $p_{3/2}$ state is then the ratio of the phase-space factors, namely,

$$\frac{q_{max}^2}{p_{max}^2} = \frac{E_0}{2m_e c^2} \approx 3 \times 10^{-3} \ll 1.$$

The effect of the attractive Coulomb field seen by e^- is to further enhance the likelihood that it is the e^- which carries off $l = 1$. Thus, for a final momentum p of the e^-, the momentum of the e^- at R—the relevant momentum—must be much greater than p, since the e^- must escape from the Coulomb field. We return now to the question of bound-state β decay. The contributions to the bound-state decay rate associated with the small components of the Dirac wave function can also be neglected, for they do not provide a larger $\bar{v}$ phase space for the decays under consideration and the small components are generally smaller than the corresponding large components by a factor $\alpha Z'$, where α is the fine structure constant, and Z', the effective charge seen by the captured β particle, is not much greater than unity for the bound-state decay of a neutral atom, the β particle having been captured into an outer state since the inner states are filled.

Because the nuclear charge Z changes from $Z = 75$ to $Z = 76$ during the β transition, there is an imperfect overlap between initial and final atomic wave functions. Since the overlap integral is less than unity, the effect of the overlap will be to decrease the decay rate. There also exists an exchange effect, in which an atomic electron is excited to an outer shell during the β transition, leaving a vacant inner orbital, and the β particle is captured into that vacant state.[4] It is found (Chen et al., 1987) that the exchange effect increases the bound-state decay rate by a factor of roughly 2 for a neutral ^{187}Re atom, but the bound state contribution remains relatively small. Calculations with a modified relativistic Thomas-Fermi potential (Williams et al., 1984) and with a relativistic Hartree-Fock model (Chen et al., 1987) give bound state to continuum contributions for the free ^{187}Re atom of only about 1%.

2. β Decay of Highly Ionized ^{187}Re

The situation is entirely different for highly ionized states. For simplicity, we begin with the case of complete ionization. (The probability of complete ionization as a function of temperature, for a few different free electron densities, is given in Fig. 4.) Continuum decay is now energetically forbidden. (See Fig. 5.) Bound state decay from the nuclear ground state of ^{187}Re to the nuclear ground state of ^{187}Os is energetically possible for any ^{187}Os electronic state bound by more than $\Delta \approx 12$ keV, where Δ/c^2 is the amount by which the mass of a ^{187}Re nucleus is less than the sum of the masses of a ^{187}Os nucleus and an electron. As in the case of a neutral ^{187}Re atom, the selection rules dictate that the β particle can end up in an $s_{1/2}$, $p_{1/2}$, $p_{3/2}$, or $d_{3/2}$ state. The decay rates to different atomic bound states for a completely ionized ^{187}Re ground state, relative to the continuum decay rate for the neutral ^{187}Re atom, are listed in Table II. The results in Table II are relatively simple to obtain because there is only one electron (in a Coulomb field) in the problem, but it may be useful to show that the *relative* decay rates can be obtained, if only roughly, by simple physical arguments. The $\bar{\nu}$ phase-space factor in Eq. (12) is

$$\rho_{\bar{\nu}} \sim q^2, \tag{13}$$

where q, the momentum with which the $\bar{\nu}$ is emitted, is given by

$$qc = \Delta - E_i, \tag{14}$$

---

[4] The "excitation" from an inner electronic state j to an outer state j' need not have been caused by an interaction with the emerging β particle. For simplicity, we restrict our discussion to a single configuration analysis, from which it will follow that there would be excitation even if there were no interaction between electrons. (Apart from simplicity, there is an additional reason for this restriction. The question of which state the β particle occupies is not very well defined when one uses multi-configurations.) The point is that the nuclear charge seen by the electrons changes suddenly from $Z = 75$ to $Z = 76$ by virtue of the β transition, and it is this sudden change, and the Pauli principle, which can cause the excitation; with the number of bound electrons fixed at 75 the initial wave function $\Psi_j(Z = 75)$ has (small) components $\Psi_{j'}(Z = 76)$ for $j' \neq j$.

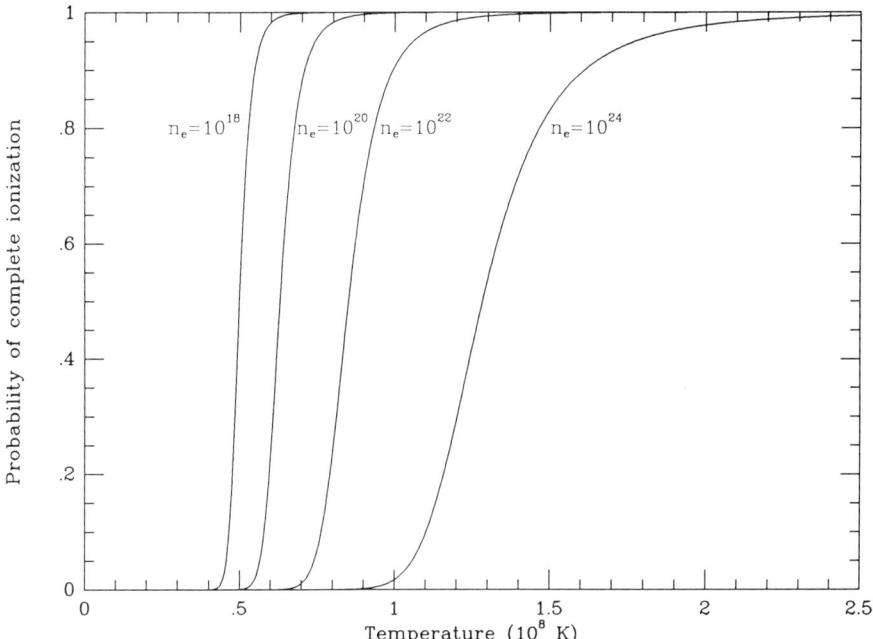

FIG. 4. The probability of the ^{187}Re nucleus being completely ionized, as a function of the temperature in units of 10^8 K, for four values of n_e, the number of free electrons per cubic centimeter. The Saha equation was used.

where Δ (rounded off) is about -12 keV, and E_i is the electron energy for bound state i. Ignoring the (fine structure) energy difference between the $n = 2$ states, the energies of the $n = 1$ and $n = 2$ states are—a rough relativistic value will suffice for our purposes—

$$E_1 \sim -86 \quad \text{keV},$$
$$E_2 \sim -20 \quad \text{keV}. \tag{15}$$

The $\bar{\nu}$ wave functions at the nuclear surface R can be approximated by

$$\phi_{\bar{\nu}} \sim 1 \quad \text{for an } s \text{ state},$$
$$\phi_{\bar{\nu}} \sim qR \quad \text{for a } p \text{ state}. \tag{16}$$

Whether the appropriate electron wave function ϕ_{ei} in Eq. (12) is the large component $g_i(R)$ or the small component $f_i(R)$ of a Dirac wave function is determined by the selection rules. For $j = l + \frac{1}{2}$, large components can be approximated by the corresponding Schrödinger solutions if $\alpha Z \ll 1$. In our case αZ is not so small, but, for $j = l + \frac{1}{2}$, that approximation will

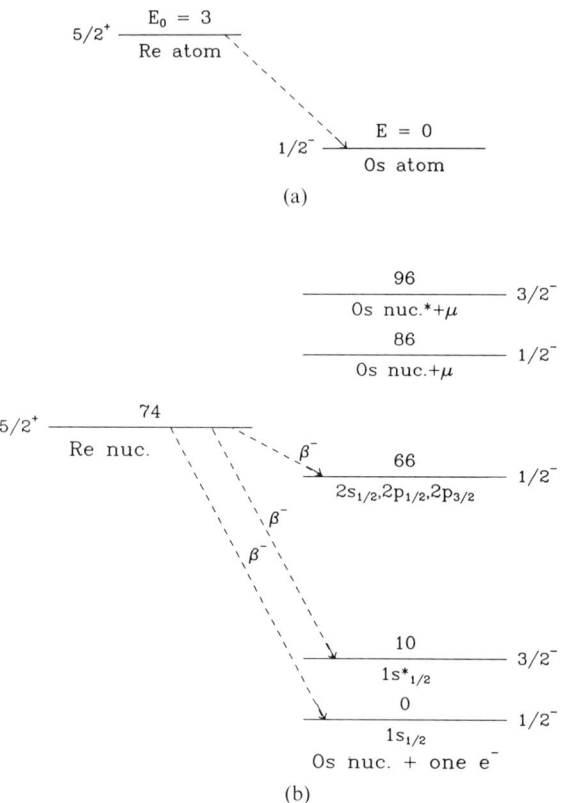

FIG. 5. Energy levels of neutral atoms of ^{187}Re and ^{187}Os (a), and of the bare nucleus ^{187}Re and the bare nucleus ^{187}Os with one electron (b). The values of $J^\pi(=\frac{5}{2}^+, \frac{3}{2}^-,$ and $\frac{1}{2}^-)$ denote the total nuclear angular momenta and the nuclear parities. The $\frac{3}{2}^-$ state, the first excited state of ^{187}Os, is at $E^* = 9.75$ keV, which we round off to 10 keV. Atomic states of the neutral atoms are those with the electrons in the ground state configuration, $\mu = 511$ keV is the rest energy of the electron. $E_0 = 2.64$ keV (≈ 3 keV) is the end-point energy of the β particle in the decay of the ^{187}Re atomic ground state to the atomic ground state of ^{187}Os. *In (a) and (b), and only in these figures*, Re represents ^{187}Re and Os represents ^{187}Os. All energies are in keV. nuc. represents nucleus, and nuc.* indicates that the ^{187}Os nucleus is in its first excited state. We have ignored the fine structure of the n = 2 hydrogen-like energy levels of an ^{187}Os nucleus with one electron. The β decay of the neutral Re atom is first forbidden unique—the decay must be to the nuclear ground state of Os—and is predominantly to the continuum. The β decay of completely ionized Re is first forbidden but not unique—the nuclear state of Os can be the ground state or the first excited state—and must be to a low-lying atomic bound state.

TABLE II

ACCURATE ESTIMATES OF THE RATIO OF THE BOUND-STATE DECAY RATE FOR A BARE ^{187}Re NUCLEUS TO ONE OF THE ENERGETICALLY ALLOWED INNER SUBSHELLS OF ^{187}Os TO THE CONTINUUM DECAY RATE OF A NEUTRAL ^{187}Re ATOM.[a]

State	$1s_{1/2}$	$2s_{1/2}$	$2p_{1/2}$	$2p_{3/2}$
Ratio	14400	0.624	0.0439	118

[a] The relativistic Hartree-Fock atomic binding energy is used here, while Thomas-Fermi estimates were used in Chen et al. (1987).

nevertheless give very meaningful results. In the Schrödinger small argument approximation, we have

$$g_{1s_{1/2}}(R) \approx 2Z^{3/2},$$

$$g_{2s_{1/2}}(R) \approx \frac{1}{\sqrt{2}} Z^{3/2}, \quad (17)$$

$$g_{2p_{3/2}}(R) \approx \frac{1}{2\sqrt{6}} Z^{5/2}(R/a_0).$$

(The ratio $|g_{2s_{1/2}}(R)/g_{1s_{1/2}}(R)|^2 = \frac{1}{8}$ is of course consistent with the nonrelativistic result that $|\Psi_{ns}(0)|^2 \propto 1/n^3$.) Generally for $\alpha Z \ll 1$, the ratio $(f/g)^2$ is of order $(\alpha Z)^2$. However, for states with $j = l - \frac{1}{2}$ and for $ZR/(na_0) \ll 1$ (which is the case for the β decay of ^{187}Re, with $R \approx 10^{-4} a_0$, for all n) where n is the principal quantum number, the ratio $(f/g)^2$ is of order

$$(Z\alpha)^2/(ZR/na_0),$$

which can be rather larger than one. Not independently, the large component of a $j = l - \frac{1}{2}$ state may also be quite different from the corresponding Schrödinger wave function for αZ not very small. In our case, the $2p_{1/2}$ state is an example. We therefore use the Dirac solution for the $2p_{1/2}$ state; while $(\alpha Z)^2 \approx 0.3$, the Dirac solution gives (e.g., see Bethe and Salpeter, 1957)

$$[f_{2p_{1/2}}(R)/g_{2s_{1/2}}(R)]^2 \approx 0.073. \quad (18)$$

In a comparison of decay rates we can choose the decay rate to the $2s_{1/2}$ state to be unity (this choice is based on the fact that the ratio of the $2s_{1/2}$ bound-state decay rate of a bare ^{187}Re nucleus to the continuum decay rate of a neutral ^{187}Re atom is close to unity); the relative decay rates to other states is given by Eq. (12) and by the rough approximations just outlined in Eqs. (13)–(18). The rough relative estimates listed in Table III compare reasonably well with the almost exact ratios given in Table II.

TABLE III

Rough "Physical" Estimates of the Relative Orders of Magnitude of Various Bound-State β-Decay Rates of a Bare ^{187}Re Nucleus.[a]

Final atomic state	Characteristic factors	Relative order of magnitude
$1s_{1/2}$	$q_1^2 \lvert g_{1s_{1/2}}(R) \rvert^2 (q_1 R)^2 \lvert M \rvert^2$	10^4
$2s_{1/2}$	$q_2^2 \lvert g_{2s_{1/2}}(R) \rvert^2 (q_2 R)^2 \lvert M \rvert^2$	1
$2p_{1/2}$	$q_2^2 \lvert f_{2p_{1/2}}(R) \rvert^2 (q_2 R)^2 \lvert M \rvert^2$	10^{-1}
$2p_{3/2}$	$q_2^2 \lvert g_{2p_{3/2}}(R) \rvert^2 \lvert M \rvert^2$	10^2
$1s_{1/2}^*$	$q_1^{*2}(\alpha Z)^2 \lvert g_{1s_{1/2}}(R) \rvert^2 \lvert M^* \rvert^2$	10^9

[a] The rate for capture to the $2s_{1/2}$ state has been set equal to unity. (That rate happens to be close to the continuum decay rate of a neutral ^{187}Re atom.) $1s_{1/2}^*$ represents the $1s_{1/2}$ atomic state for the first excited state of the ^{187}Os nucleus. M and M^* represent the nuclear matrix elements for the ground state to ground state ($\tfrac{5}{2}^+$ to $\tfrac{1}{2}^-$) and ground state to first excited state ($\tfrac{5}{2}^+$ to $\tfrac{3}{2}^-$) transitions. We assumed that $M^*/M \approx 1$. q_1 and q_2 are the anti-neutrino momenta for capture to the $n = 1$ and $n = 2$ atomic states, respectively, for the $\tfrac{5}{2}^+$ to $\tfrac{1}{2}^-$ transition; q_1^* is the $\bar{\nu}$ momentum for capture to the $n = 1$ state for the $\tfrac{5}{2}^+$ to $\tfrac{3}{2}^-$ transition. g and f are large and small components.

The huge ratio 1.4×10^4 shows that bound-state effects can be very significant if ^{187}Re nuclei are in stars under conditions for which there is a reasonable probability of complete ionization. But there is an even more—indeed far more—significant effect, first noted by Takahashi et al. (1983), and that is that a completely ionized ^{187}Re nucleus can undergo β decay not only to the ground state but to the first excited state of the ^{187}Os nucleus, which lies about 10 keV above the nuclear ground state of ^{187}Os. The rest energy of the ground state ^{187}Re nucleus, less by about 12 keV than the sum of the rest energies of an e^- and of a ^{187}Os nucleus in its ground state, is therefore less by about 22 keV than the sum of the rest energies of an e^- and of a ^{187}Os nucleus in its first excited state. The β decay of a bare ^{187}Re nucleus in its ground state to the first excited state of ^{187}Os is therefore energetically possible if and only if the β particle ends up in a bound state whose binding energy exceeds 22 keV. (The value 22 keV also follows from the energy level diagram of Fig. 5.) The bound state must therefore be the $1s_{1/2}$ state. Since we are concerned now with a $\tfrac{5}{2}^+$ to $\tfrac{3}{2}^-$ transition, rather than the $\tfrac{5}{2}^+$ to $\tfrac{1}{2}^-$ transition to the ground state, the total angular momentum carried off by the $e^- - \bar{\nu}$ pair can be 1 or 2. It is a nonunique first forbidden transition for it involves more than one nuclear matrix element. Since the parities of the nuclear states are different, in a non-relativistic description the $e^- - \bar{\nu}$ pair has to carry off one unit of orbital angular momentum. There are therefore two possibilities: (i) The β particle can emerge in a $p_{1/2}$ bound state with the antineutrino in an $s_{1/2}$ continuum state, or (ii) the β particle can emerge in an $s_{1/2}$ bound state with the $\bar{v}$ in a $p_{1/2}$ state. However, since in Dirac theory the small component of the wave function of an $s_{1/2}$ state behaves like a $p_{1/2}$ state, there is a third possibility, namely, (iii) the β particle can *also* emerge in an $s_{1/2}$ bound state with the $\bar{v}$ in an $s_{1/2}$ state, and this corresponds to a larger $\bar{v}$ phase space volume. Possibility (iii) has the advantage over (i) that the $\bar{v}$ has more energy and therefore more phase space, and the advantage over (ii) that $|\phi_{\bar{v}}(R)|^2$ is much larger. The disadvantage of (iii), that it involves a small component, is rather minor since $(\alpha Z)^2 \approx 0.3$. (The β particle and the $\bar{v}$ cannot both emerge in $s_{1/2}$ states for the $\frac{5}{2}^+$ to $\frac{1}{2}^-$ transition, for which at least two units of angular momentum must be carried off.) The third possibility is therefore the dominant one. In a comparison with the decay rate for a transition to the ^{187}Os ground state, the decay rate λ^*_{1s} for the transition to the first excited nuclear state of ^{187}Os can then be written as

$$\lambda^*_{\text{bd; 1s}} \sim q_1^{*2}(\alpha Z)^2 |g_{1s}|^2 |M^*|^2, \qquad (19)$$

where M^* is the nuclear matrix element for the ground state of ^{187}Re to the first excited state of ^{187}Os, and q_1^* is the $\bar{v}$ momentum for the transition to the first excited nuclear state. Assuming, as suggested by theory (e.g., see Konopinski, 1966), that the ratio of nuclear matrix elements $|M|/|M^*|$ for the transitions $\frac{5}{2}^+ \to \frac{1}{2}^-$ and $\frac{5}{2}^+ \to \frac{3}{2}^-$ is of order unity, we have

$$\lambda^*_{\text{bd, 1s}}/\lambda_{\text{bd, 1s}} \approx \left(\frac{q_1^*}{q_1}\right)^2 \frac{(\alpha Z)^2}{(q_1 R)^2} \approx 10^5.$$

It is interesting to note that while bound-state decay of a bare ^{187}Re nucleus to a $2p_{3/2}$ state is energetically forbidden, decay to a $2p_{3/2}$ state is possible for a ^{187}Re ion with two electrons in $1s_{1/2}$ states and with up to 9 bound electrons altogether (Takahashi *et al.*, 1987); the configuration for 9 electrons would be $(1s_{1/2})^2(2s_{1/2})^2(2p_{3/2})^3(2p_{1/2})^2$. (Additional energy is gained by the binding energy differences of atomic electrons.) The decay rate for a few bound electrons will be much greater than for a neutral atom, though much smaller than for a completely ionized ^{187}Re system, but can be relevant at much lower temperatures than for the completely ionized case (Yokoi *et al.*, 1983). For an almost totally ionized atom, the exchange effect is less important than for a neutral atom, because fewer exchange orbitals are involved. For example, using the relativistic Hartree-Fock approximation, we found that exchange effects increase the bound-state β-decay rate by about 13% for ^{187}Re with 18 atomic electrons, and by about 6% for ^{187}Re with 10 electrons.

IV. Conclusion

From the above discussion, we see that the bound-state β decay of ^{187}Re is the dominant contribution at high temperature. The nonunique first forbidden transition of ^{187}Re to the first excited nuclear state of ^{187}Os plays a very important role when the temperature is in excess of about 10^7 K. A detailed discussion of the effects on nuclear cosmochronology has been given (Yokoi et al., 1983). Orbital electron capture is also possible by the thermally populated 9.75 keV excited state of ^{187}Os at high temperature; it is energetically possible if the degree of ionization is not too high. Continuum-state electron capture is not negligible as long as the free electron density and energy are high. Including both the effect of enhanced β decay of ^{187}Re and the effect of electron capture, Yokoi et al (1983) calculated the age of the galaxy to be 11-15 Gyr in their chemical evolution model. In order to obtain a more accurate age of the galaxy using the Re/Os pair, it will be necessary to make more complete measurements and theoretical calculations for the uncertainties in the present solar system Os/Re abundance ratio, and the half-life of ^{187}Re in the environment appropriate to galaxy evolution. Some thought should also be given to the question of the neutron capture cross section ratio at 30 keV, though this problem would at least seem to have been resolved. These uncertainties were noted by Clayton 20 years ago (Clayton, 1964). In view of the enormously enhanced decay rate of ^{187}Re when it is very highly ionized, the most serious uncertainty is probably the amount of time ^{187}Re spends in very high temperature interiors of stars. This question (and more generally the question of the time various elements spend in stellar interiors) is a very important one in astrophysics. We therefore note that it may be possible to proceed in the opposite direction. If the age of the galaxy is determined by other means—the use of other chronometric pairs, or the study of the properties of globular clusters, for example—the requirement that the Re/Os pair lead to that galaxy age would provide information on the time ^{187}Re spends in stellar environments. That information could, in turn, provide some insight into the amount of recycling elements in general undergo.

ACKNOWLEDGEMENTS

We wish to express our sincere thanks to Professor David N. Schramm for a very helpful conversation. The work was supported in part by the National Science Foundation under Grant No. PHY87-06114.

REFERENCES

Arnould, M. (1974). *Astron. Astrophys.* **31**, 371.
Arnould, M., Takahashi, K., and Yokoi, K. (1984). *Astron. Astrophys.* **137**, 51.
Bahcall, J. N. (1961). *Phys. Rev* **124**, 495.
Beer, H., Kappelar, F., Wisshak, K., and Ward, R. A. (1981). *Astrophys. J. Suppl.* **46**, 295.
Bethe, H. A., and Salpeter, E. E. (1957). "Quantum Mechanics of One and Two Electron Atoms," Springer-Verlag, Berlin.
Brodzinski, R. L., and Conway, D. C. (1965). *Phys. Rev. B* **138**, 1368.
Browne, J. C., and Berman, B. L. (1981). *Phys. Rev. C* **23**, 1434.
Burbidge, E. M., Burbidge, G. R., Fowler, W. A., and Hoyle, F. (1957). *Rev. Mod. Phys.* **29**, 547.
Chen, Z., Rosenberg, L., and Spruch, L. (1987). *Phys. Rev.* **35**, 1981.
Clayton, D. D. (1964). *Astrophys. J.* **139**, 637.
Clayton, D. D. (1983). *In* "Principles of Stellar Evolution and Nucleosynthesis," Chap. 7. University of Chicago Press, Chicago.
Clayton, D. D. (1969). *Nature* **224**, 56.
Clayton, D. D. (1988). *Mon. Not. R. Astr. Soc.* **234**, 1.
Daudel, R., Jean, M., and Lecoin, M. (1947a). *J. Phys. Radium* **8**, 238.
Daudel, R., Jean, M., and Lecoin, M. (1974b). *C. R. Acad. Sci.* **224**, 1427.
Fowler, W. A. (1984). *Rev. Mod. Phys.* **56**, 149.
Hershberger, R. L., Macklin, R. L., Balakrishnan, M., Hill, N. W., and McEllistrem, M. T. (1983). *Phys. Rev. C* **28**, 2249.
Konopinski, E. J. (1966). "The Theory of beta Radioactivity." Oxford, London.
Lindner, M., Leich, D. A., Borg, R. J., Russ, G. P., Bazan, J. M., Simons, D. S., and Date, A. R. (1986). *Nature* **320**, 246.
Luck, J. M., and Allegre, C. J. (1983). *Nature* **302**, 130.
Meyer, B. S., and Schramm, D. N. (1986). *Astrophys. J.* **311**, 406.
Payne, J. A. (1965). Ph.D. thesis, University of Glasgow.
Perrone, F. A. (1971). Ph.D. thesis, Rice University.
Rutherford, E. (1929). *Nature* **123**, 313.
Takahashi, K., and Yokoi, K. (1983). *Nucl. Phys. A* **404**, 578.
Takahashi, K., Boyd, R. N., Mathews, G. J., and Yokoi, K. (1987). *Phys. Rev. C* **36**, 1522.
Williams, R. D., Fowler, W. A., and Koonin, S. E. (1984). *Astrophys. J.* **281**, 363.
Winters, R. R., Macklin, R. L., and Halperin, J. (1980). *Phys. Rev. C* **21**, 563.
Winters, R. R., and Macklin, R. L. (1982). *Phys. Rev. C* **25**, 208.
Winters, R. R. (1984). *In* "Neutron-Nucleus Collisions: A Probe of Nuclear Structure," *AIP Conf. Proc. No. 124* (J. Rapaport, R. W. Finlay, S. M. Grimes, and F. S. Dietrich, eds.), pp. 495–497. AIP, New York.
Woosley, S. E., and Fowler, W. A. (1979). *Astrophys. J.* **233**, 411.
Yokoi, K., Takahashi, K., and Arnould, M. (1983). *Astrophys. J.* **117**, 65.

PROGRESS IN LOW PRESSURE MERCURY–RARE-GAS DISCHARGE RESEARCH

J. MAYA and R. LAGUSHENKO

GTE Electrical Products
Danvers, Massachusetts

I. Introduction 321
II. Modelling of Low Pressure Mercury–Rare-Gas Discharge 323
   A. General Formulation 323
   B. Electron Energy Distribution Function (EEDF) in a Low Pressure Discharge 326
   C. Radiation Transport in a Low Pressure Discharge 332
   D. Theoretical Models of the Positive Column of the Mercury–Rare-Gas Discharge 336
III. Altered Low Pressure Mercury–Rare-Gas Discharge 342
   A. Isotope Effect 343
   B. Magnetic Fields and Striations 346
   C. High Frequency (HF) and Surface Waves 350
   D. Nonequilibrium and Altered EEDF 353
IV. Diagnostics 356
   A. Laser Diagnostics 357
   B. Probes and Optical Fibers 364
V. Summary 369
   References 370

I. Introduction

The low pressure Hg–rare gas discharge (LPD) is a well studied electrical phenomenon, partially due to its wide applicability in fluorescent lamps. Typically, a low pressure discharge in a Hg–rare-gas mixture implies an electrical discharge in an elongated tube containing several hundred pascals (133.3 Pa = 1 torr) of a rare gas and a few tenths of a pascal of mercury vapor. When the inside walls of such a tube are coated with phosphor material, the tube constitutes a fluorescent lamp. What differentiates a low pressure discharge from a high pressure discharge (other than the fact that high pressure implies about 10^3 times higher pressure) is in fact that the electron energy distribution function (EEDF) of such a discharge is not in thermodynamic equilibrium or even local thermodynamic equilibrium (LTE) with the gas temperature; while in a high pressure discharge, this is typically the case. We shall say more about this later. In this connection it may be mentioned that about 70% of all artificial lighting in the world is by fluorescent lamps (Koedam, 1985).

The fundamentals of the operation of a (LPD) have been covered in the books by Ellenbaas (1972) and Waymouth (1971). In this article we will not go into the details of the operation of the LPD. What we will try to do is cover some of the highlights in the physics of the LPD that have taken place primarily over the last decade. We will cover fundamental physical processes insofar as they are necessary to explain the progress of the last decade. Furthermore we will cover the topics from the point of view of physics research and more particularly as it reflects on atomic and molecular physics.

There are several areas of LPD research where significant activity has taken place recently. One area where a significant amount of research has taken place over the last decade, is discharge modeling by way of additional processes, more precise electron energy distribution functions (EEDF), better treatment of the radiation transfer, and introduction of new calculational techniques such as Monte Carlo simulations.

The second broad area of activity has been what might loosely be called attempts to increase the efficiency (electrical to radiation conversion efficiency) and to obtain more radiation per unit length. These have primarily been driven by the technological desire of a more compact and more efficient radiation source for energy conservation purposes. Topics such as high current densities, isotopically altered discharge, application of magnetic fields, recombination structures, altered electron energy distribution function (EEDF), high frequency operation, etc., are among the research subjects that fall into this category.

Finally, the third broad area of activity in LPD research during the last decade or so has been in diagnostics. The increased sophistication of the modeling coupled with the intense desire to increase the efficiency has generated the need to understand the finer details of the discharge in the hope of discovering novel approaches to modify it. Some of the diagnostics have been applications of relatively well known techniques to Hg–rare-gas discharge; e.g., Fabry Perot interferometry, Hook's method, Langmuir probes, etc. Some other diagnostics have more modern roots, such as fiber optics, lasers, and, in particular, very narrow bandwidth ($\sim$ several hundred KHz) tunable dye lasers. These diagnostics have generated considerable amount of information that is likely to be substantially useful to future research in this area.

In the following, we shall elaborate on these three broad areas of activity that have dominated the scene in low pressure mercury–rare-gas discharge

(LPD) during the last decade or so. (Note that inasmuch as LPD generally implies any low pressure discharge, for the purposes of this article, it will be used to denote low pressure discharge in mercury–rare-gas mixtures exclusively.) We should point out that recently there have been a number of very good articles reviewing various aspects of fluorescent lamps and low pressure discharges. The interested reader is referred to Polman, et al. (1975), Waymouth (1982), Jack and Vrehen (1986), and Jack (1986).

II. Modeling of Low Pressure Mercury-Rare-Gas Discharge

A. GENERAL FORMULATION

In this section, for the sake of convenience, we will use the word "discharge," meaning the positive column discharge. Up-to-date efforts to model low pressure discharges cover a very vast area. Because of that we restrict ourselves to modeling only the positive column discharge. In this section, we describe the general formulation of the problem applicable to a collision-controlled positive column of a low pressure discharge. We keep the time derivative term in the equations, so they are applicable both to steady state and time dependent conditions until the electron inertia and displacement current effects become important. When formulating the basic equations, we also mention simplifications which can be done without significant loss of generality.

A low pressure discharge as mentioned above is typically far from local thermodynamic equilibrium (LTE). Therefore, a theory of a low pressure discharge has to be based on the system of balance equations for the number densities of ions and the population densities of the atomic states involved in the ionization and energy balance.

The ion balance equations can be represented in the form

$$\frac{\partial N_{ps}}{\partial t} - \text{div}\left(\frac{\mu_{ps} N_{ps} \nabla p_e}{e N_e}\right) = C_{ps},$$

$$\sum_s N_{ps} = N_e. \tag{1}$$

Here e is the elementary charge; N_{ps} and μ_{ps} are ion number density and mobility, respectively, for an ion of the gas species s; N_e and p_e are electron number density and pressure, respectively; C_{ps} is the collisional ionization–recombination rate. Equations (1) follow from the equations of the multi-fluid hydrodynamics for ions and electrons when one neglects the inertia terms, ion pressure gradient terms, and electron thermal diffusion terms. The balance equations for the population densities of atomic states can be written as

$$\frac{\partial N_{sj}}{\partial t} - \text{div}\left(\frac{D_{sj}\nabla p_{sj}}{kT_g}\right) = C_{sj} - R_{sj}, \qquad (2)$$

where N_{sj} is the population density of the atomic state j of the gas species; k and T_g are Boltzmann constant and gas temperature, respectively; p_{sj} and D_{sj} are the partial pressure of atoms of the sort sj and relevant diffusion coefficient, respectively; C_{sj} is collisional population–depopulation rate; R_{sj} is radiative depopulation–population rate. The form of the diffusion term in Eq. (2) again follows from the multifluid hydrodynamics.

For radiative states, the diffusion effects can usually be neglected, setting $D_{sj} = 0$. For the metastable states and the ground state, the diffusion term often has to be considered. For instance, depopulation of metastables by the process of quenching on the wall is controlled by diffusion. For the ground state atoms ($j = 0$), the main contribution to the collisional depopulation is usually due to ionization. Therefore, one may set in Eq. (2) for $j = 0$ $C_{s0} = -C_{ps}$ and keep the diffusion term (3). Equation (2) for $j = 0$ describes the ionization depletion of atoms in the central part of discharge due to diffusion followed by recombination on the wall or the cathode. This effect may be very important in the case of a mixture of a low ionization potential species and an inert buffer gas.

The form of the radiative rate R_{sj} in Eq. (2) requires special discussion. If the discharge medium is optically thin for the depopulating radiation, the term R_{sj} in Eq. (2) may be represented in the form

$$R_{sj} = N_{sj}/\tau_{sj} - \sum_n A_{snj} N_{sn}. \qquad (3)$$

Here τ_{sj} is the natural lifetime of the state sj; A_{snj} and N_{sn} are Einstein coefficients for the $n \to j$ transitions and populational densities of states involved in the cascade radiative population of the state nj respectively. The situation is much more complex if the discharge medium is optically thick. In that case, an additional term has to be added to Eq. (3). This term considers the change of the populational density N_{sj} at the point (**r**) due to reabsorption of radiation reaching the point (**r**) from all other points (**r**$_v$). The form of the reabsorption term will be discussed further. Here we want to emphasize that the discharge medium is optically thick for the resonance radiation corresponding to transitions from the first excited states to the ground state. In turn, the first excited states are usually involved in the balance of energy and ionization. This situation is typical for a wide range of discharge conditions. Thus, the problem of discharge modeling is linked to the problem of radiation transport in an optically thick medium. Usually, these two problems have been decoupled. The problem of radiation transport has been analyzed for a simplified situation when all the processes of excitation and quenching were switched off. As a result, the decay time of an excited state which takes into account multiple reabsorptions and reemissions of photons is obtained.

This decay time is typically longer than the natural one. It is called the imprisonment time since multiple reabsorptions and reemissions of photons result in the imprisonment of radiation. For discharge modeling, this imprisonment time is inserted in Eq. (3) instead of the natural lifetime τ_{sj}. The accuracy of the above approach will be discussed further in Section II,C.

Generally, the gas is heated up by the discharge current. Therefore, the equation of the gas energy balance has to be considered. This equation may be written in the form

$$C_g \frac{\partial T_g}{\partial t} - \text{div}(\xi \nabla T_g) = Q_g, \tag{4}$$

where C_g and ξ are the gas thermal capacity and thermal conductivity, respectively; Q_g is the gas heating rate expressed in units of energy per volume.

Significant fraction of the electrical energy deposited in the discharge can be transferred to the wall by radiation and ions. Therefore, a good way to express the term Q_g is

$$Q_g = H_{el}, \tag{5}$$

where H_{el} is the elastic energy loss of electrons.

Equation (5) is usually very accurate for the positive column, since the ion current is much less than the electron current. However, the energy of ions and the radiation energy have to be considered in the energy balance of the wall if the wall temperature is not stabilized by external means.

One more equation has to be added to the system of Eqs. (4) and (5) to connect the discharge current and the electric field intensity in the discharge plasma. If the displacement current and the inertia terms may be neglected, this equation may be written as

$$j_e = eE(\mu_e N_e + \sum \mu_{ps} N_{ps}), \tag{6}$$

where j_e is the discharge current density, E is the absolute value of the electric field intensity, and μ_e is the electron mobility.

Actually, the ion current terms in Eq. (6) may be neglected in most cases. If the EEDF is known, Eqs (1)–(6) represent a closed system. Solution of this system allows one to calculate all the parameters of the discharge. However, the specifics of a low pressure discharge do not allow one to suggest any universal form of the EEDF. Therefore, the strict approach is to find the

EEDF by solving the Boltzmann equation simultaneously with Eqs. (1)–(6). However, this system of discharge equations, described above, is very complex and the strict approach is usually very hard to follow. In practice, one usually reduces the problem to more manageable tasks by proceeding along the following lines.

(1) Specify the elementary processes and atomic states to be considered for the adequate description of a discharge.
(2) Specify the appropriate form of the EEDF or the way to find it by solving the Boltzmann equation.
(3) Specify the form of radiative terms in Eq. (2), e.g., the treatment of the radiation transport problem.
(4) Specify an approach to solving the reduced system of the discharge Eq. (1)–(6).

The progress of the discharge modeling can be viewed as the progress along the four points above. This progress will be discussed in the next several sections. As far as methods and approaches are concerned, many results discussed below are applicable to any low pressure discharge. However, due to space considerations our review of calculational results will be confined to the discharge in a mixture of mercury vapor and rare gases.

B. Electron Energy Distribution Function (EEDF) in a Low Pressure Discharge

The full description of a discharge requires knowledge of the electron velocity distribution function $f_e(\mathbf{v}, \mathbf{r}, t)$. At collision-controlled conditions, the value of the electron drift velocity is usually much less than that of random velocity. This means the function $f_e(\mathbf{v}, \mathbf{r}, t)$ depends weakly on the direction of the velocity $\mathbf{v}$. This enables one to express $f_e(\mathbf{v})$ as the sum of the first four terms of the expansion into the spherical function series (Davydov, 1937). This sum can be represented as

$$f_e(\mathbf{v}, \mathbf{r}, t) = f_0(\varepsilon, \mathbf{r}, t) + \mathbf{v} \cdot \mathbf{f}_1(\varepsilon, \mathbf{r}, t)/v, \qquad (7)$$

where $\varepsilon = mv^2/2$ is the electron energy and $f_0, f_{1x}, f_{1y}, f_{1z}$ are coefficients of expansion. The average value of any quantity $\psi(\varepsilon)$ dependent on the electron energy ε can be expressed as

$$\langle \psi(\varepsilon) \rangle = \int_0^\infty f_0(\varepsilon, \mathbf{r}, t) W(\varepsilon) d\varepsilon. \qquad (8)$$

The average value of the flux density of the quantity $\psi(\varepsilon)$ can be found as

$$\langle \mathbf{v}\psi(\varepsilon)\rangle = \tfrac{1}{3}\int_0^\infty v\psi(\varepsilon)\mathbf{f}_1 W(\varepsilon)d\varepsilon. \qquad (9)$$

The function $W(\varepsilon) = 4\pi(2\varepsilon)^{1/2}/m^{3/2}$ has the meaning of volume in the velocity space corresponding to the unit interval of energy. The function f_0 will be called the EEDF because it directly enters the kinetic equation. Actually, the exact definition of the EEDF is Wf_0.

Equations for the functions f_0 and $\mathbf{f}_1$ are obtained by inserting Eq. (7) into the Boltzmann equation and equating to zero coefficients of spherical functions 1, v_x/v, v_y/v, v_z/v in the resulting expression. For the case of a steady state cylindrical positive column being uniform in the axial direction, the equations take the form (Davydov, 1937).

$$\frac{v}{3\rho}\frac{\partial}{\partial\rho}(\rho f_{1\rho}) - \frac{e}{3v}\frac{\partial}{\partial\varepsilon}[v^2(E_z f_{1z} + E_\rho f_{1\rho})] = \frac{\delta f_0}{\partial t}, \qquad (10)$$

$$q_t N f_{1z} = eE_z \frac{\partial f_0}{\partial \varepsilon}, \qquad (11)$$

$$q_t N f_{1\rho} = eE_\rho \frac{\partial f_0}{\partial \varepsilon} - \frac{\partial f_0}{\partial \rho}, \qquad (12)$$

where E_z and E_ρ are axial and radial component of the electric field, respectively, and q_t is the total cross section of the electron momentum transfer. The functions f_{1z} and $f_{1\rho}$ describe the axial and radial components of electron flux. The term $\delta f_0/\delta t$ represents the collisional rate of change of the function f_0.

In general, the collisional rate has to include the terms corresponding to the electron–electron interaction, and elastic, inelastic, and superelastic collisions with atoms and the ionization. Expressions for these particular terms are well described in the literature and may be found, for example, in Davydov (1937), Shkarofsky et al. (1966), Holstein (1946), and Ginzburg and Gurevich (1960).

Equations (10)–(12) are very complex to solve. Therefore, many models have accepted a simplified approach, which can be called the test function method. With this approach, one assumes a certain expression for the EEDF which depends on a number of parameters. These parameters are determined by satisfying conditions of the electron energy balance and some other conditions, which follow from Eq. (10). From a mathematical point of view, the test function method is equivalent to satisfying Eq. (10) on the average for certain intervals of energy rather than at every point ε. In the early stage of modeling (Waymouth and Bitter, 1956; Cayless, 1962; Polman et al., 1972;

Drop and Polman, 1972) the Maxwellian EEDF with a position independent temperature T_e had been chosen by setting

$$f_0(\varepsilon) = (m/2\pi k T_e)^{3/2} N_e(\rho) \exp(-\varepsilon/k T_e). \qquad (13)$$

The electron temperature T_e usually had been determined by satisfying the electron energy balance equation averaged over the discharge cross section. Clearly, the closer a test function simulates the real shape of the EEDF, the better it should work.

In parallel with the test function method, many workers have attempted to obtain the solution of Eq. (10), both analytically (Kagan and Lyagushchenko, 1961; 1962; 1964; Lyagushchenko, 1972; Kagan et al., 1971 and numerically (Atajew et al., 1972; Rockwood 1973; Saelee 1982; Winkler et al., 1983a, b). For a review of numerical solutions to the Boltzman equation, the reader is referred to Pitchford (1983) and Waters (1983). Usually, this was done with the assumption $f_{1p} = 0$, which is justified at not too low pressures. The most important result of these studies has been the realization that the high energy tail of the EEDF ($\varepsilon \gg k T_e$) is significantly depleted by electrons at many discharge conditions. Qualitative explanation of this phenomenon is plausible.

The population of the excited states in a low pressure discharge is typically much lower than the Boltzmann population with the electron temperature T_e. This is due to escape of radiation and quenching of metastables on the wall. The tail electrons lose their energy by inelastic collisions, but this loss cannot be compensated by superelastic collisions due to a low population density of the excited states. Thus, inelastic collisions result in a depletion of the tail. This effect can be greater or less, dependent on the electron density, since the electron–electron interaction tends to make the EEDF Maxwellian. As a result, the shape of the tail explicitly depends on the electron density.

Furthermore, the decline of the electron density from the axis to the wall leads to increasing tail depletion. Thus, the shape of the tail may vary with the radial coordinate. The effect of the tail depletion has been incorporated in the test function method by developing the two electron group model (2-EGM) (Vriens 1973; 1974; Vriens and Lighthart, 1977; Lighthart and Keijser, 1980). According to the 2-EGM the energy axis is divided into two domains: $\varepsilon < \varepsilon_1$ and $\varepsilon \geq \varepsilon_1$. These two domains are called the bulk and tail regions, respectively. Here, the energy ε_1, corresponds to the threshold of inelastic collision. The EEDF is represented as

$$\begin{aligned} f_0(\varepsilon) &= (m/2\pi k T_b)^{3/2} N_e \exp(-\varepsilon/k T_b) \quad \text{for} \quad \varepsilon < \varepsilon_1, \\ f_0(\varepsilon) &= S(m/2\pi k T_t)^{3/2} N_e \exp(-\varepsilon/k T_t) \quad \text{for} \quad \varepsilon > \varepsilon_1, \end{aligned} \qquad (14)$$

where T_b and T_t are electron temperatures for the "bulk" and "tail," respectively, and S is a parameter. For a continuous EEDF, the parameter S has to be taken as (Lighthart and Keijser, 1980)

$$S = (T_t/T_b)^{3/2} \exp(\varepsilon_1/kT_t - \varepsilon_1/kT_b), \qquad (15)$$

whereas in Vriens (1973, 1974) and Vriens and Lighthart (1977) this parameter was taken equal to unity. The bulk and tail temperature are determined by satisfying simultaneously the bulk energy balance and the tail energy balance. An alternative way is to use the total energy balance and the tail energy balance as demonstrated by Morgan and Vriens (1980).

The energy balance equations may be obtained as usual by multiplying Eq. (10) by the factor $W(\varepsilon)d\varepsilon$ and integrating over the relevant domains of energy. The 2-EGM calculations have been compared to calculations based on numerical solution of the kinetic equation by Morgan and Vriens (1980). These authors have derived an empirical formula for the factor S, which allowed them to obtain a good agreement.

The version of the 2-EGM described above has dealt with the constant bulk and tail temperatures. Such an approach is incapable of accounting for the radial variation of the EEDF in a positive column, which has been mentioned previously. Further development of the 2-EGM has been achieved by Dakin (1986) by allowing for the radial variation of T_b and T_t. This has been done by assuming the bulk and tail temperatures in the form

$$\begin{aligned} T_b(\rho) &= T_b(0)(1 - C_b \rho^2/R^2), \\ T_t(\rho) &= T_t(0)(1 - C_t \rho^2/R^2), \end{aligned} \qquad (16)$$

where R is the tube radius and C_b and C_t are constants for bulk and tail electrons, respectively. Parameters $T_b(0)$, C_b, $T_t(0)$, C_t have been determined from the position dependent equations of the bulk and tail energy balance. Small variations of the bulk temperature have been observed due to a considerable mixing effect of the electron thermal conductivity. However, variation of the tail temperature turned out to be quite significant. This result is consistent with results obtained by solving the kinetic equation (10). We believe incorporation of the spatially dependent bulk and tail temperatures considerably widens possibilities for the 2-EGM.

The 2-EGM offers a simple way to account for the depletion of the EEDF tail. Nevertheless, this model has natural limitations. In the first place, solutions of the kinetic equation have shown the shape of the EEDF tail to be fundamentally a non-Maxwellian one. Therefore, a Maxwellian approximation of the tail is good enough only for a relatively narrow range of energy, which contributes to the tail energy balance. Extension of this approximation to higher energies may result in a significant divergence from the real shape of the EEDF.

In addition to the above, the 2-EGM approximation has discontinuity of the derivative or both the function and the derivative at $\varepsilon = \varepsilon_1$. A real EEDF has to satisfy conditions of continuity for the function and the derivative. This is dictated by condition of continuity of the electron flux in the energy space as shown by Lyagushchenko (1972). Therefore, the 2-EGM discontinuity represents a built-in deviation from the real situation. The flaws of the 2-EGM just mentioned may have greater or lesser practical importance depending on conditions. However, any step to eliminate these flaws can only improve the model.

An alternative expression for the EEDF has been suggested by Lagushenko and Maya (1984). Derivation of this expression is briefly reviewed below. We introduce a new variable u and new function $f(u)$ defined as

$$u = \varepsilon/e, \qquad f = 2\pi(2e/m)^{3/2} f_0/N_e, \qquad \int_0^\infty u^{1/2} f \, du = 1. \tag{17}$$

The energy axis is divided into two regions: the bulk region $u < u_1$, and the tail region $u \geq u_1$. Here u_1 is the lowest excitation potential for a mixture of gases.

The Maxwellian shape of the EEDF is assumed for the bulk region.

$$f(u) = C_N \exp(-u/u_e), \tag{18}$$

where C_N and u_e are the normalization constant and effective temperature, respectively. The expression for the EEDF in the tail region is to be established on the basis of the kinetic equation (10). Neglecting the $f_{1\rho}$ terms and some small terms in the inelastic collision rate, Eq. (10) for $u \geq u_1$ may be written in the form

$$\frac{dF}{du} - \sum_{s,j} \frac{N_s u q_{sj}(u) f(u)}{N}$$

$$= \sum_{s,j} \frac{N_{sj} u q_{sj}(u) f(u - u_{sj})}{N g_{sj}}, \quad \text{where } N = \sum_s N_s \tag{19}$$

$$F = d(u) \frac{df}{du} + b(u) f, \tag{20}$$

$$d(u) = \frac{(E/N)^2 u}{3q_t} + \frac{2u_e Q_c N_e}{N}, \tag{21}$$

$$b(u) = 2 \sum_s \frac{m N_s q_{1s}(u) u^2}{M_s N} + \frac{2 Q_c N_e}{N}. \tag{22}$$

Here F is the electron flux in the energy space due to the electric field $E = E_z$, electron–electron interaction and elastic collision. The electron–electron interaction term has been simplified assuming $u_e \ll u_1$. Other notations are as follows: q_{1s} and M_s are elastic cross section of the electron momentum transfer and atomic mass, respectively, for the s-species; q_{sj} and u_{sj} are inelastic cross section and electron energy loss, respectively, for the j state of the s species. Summation over j in the inelastic term includes ionization; g_{sj} is the relevant statistical weight. The total cross section of the electron momentum transfer is defined as

$$q_t = \sum_s N_s q_{1s}/N + 1.72 q_c N_e/N. \tag{23}$$

Here $q_c = Q_c/u^2$ is the Coulomb cross section and coefficient 1.72 takes into account contribution of the electron–electron interaction. The right-hand-side term in Eq. (19) corresponds to superelastic collisions. We further make the assumption (for derivation of the test function only)

$$f(u - u_{sj}) = \exp(u_{sj}/u_e) f(u). \tag{24}$$

With this assumption, Eq. (19) is reduced to the form

$$\frac{dF}{du} - h(u) f(u) = 0$$

$$h(u) = \sum_{sj} N_s (1 - R_{sj}) u q_{sj}(u)/N \tag{25}$$

The value $R_{sj} = N_{sj} \exp(u_{sj}/u_e)/N_s g_{sj}$ is the ratio of actual to Boltzmann population. To evaluate the solution of Eq. (25) one can use the WKB method. Application of this method to solution of the kinetic equation has been discussed in detail in Kagan and Lyagushchenko (1962, 1964) and Lyagushchenko (1972).

The WKB solution of Eq. (25) may be written in the form

$$f(u) = C_N \exp\left(-\int_0^u G_w\, du\right), \tag{26}$$

where

$$G_w(u) = \frac{g(u)}{2} + [h(u)/d(u) + g^2(u)/4]^{1/2}, \tag{27}$$

$$g(u) = b(u)/d(u). \tag{28}$$

Strictly speaking, Eq. (26) may not be applicable in close proximity to $u = u_1$, where total inelastic cross section is close to zero. However, this, may be disregarded for derivation of the test function. If the electron–electron terms dominate the field and elastic terms in Eqs. (21) and (22), then $g(u) = 1/u_e$. We obtain the final expression for the test function by substituting $g(u) \to 1/u_e$ in Eq. (26). Since the inelastic cross section is equal to zero at $u = u_1$, then $h(u_1) = 0$. This permits one to combine expressions (17) and (25) into only one expression for the EEDF applicable to the whole range $0 \le u \le \infty$:

$$f(u) = C_N \exp\left(-\int_0^\infty G(u)du\right) \qquad (29)$$

where

$$G(u) = \frac{1}{2u_e} + [h(u)/d(u) + \tfrac{1}{4}u_e^2]^{1/2} \qquad (30)$$

Since $h(u_1) = 0$, Eq. (29) is continuous and it has continuous derivative at $u = u_1$.

The function G^{-1} may be regarded as an energy dependent tail temperature for the tail region. The tail part of the EEDF, Eq. (29), explicitly depends on the cross sections, particle number densities, and the electric field. The effective temperature u_e has to be determined by solving the electron energy balance equation. In general, this should be done by considering a position dependent temperature and energy balance. A more simple approach is to use a position independent temperature and determine it from the energy balance equation averaged over discharge cross section. Note, even in that case the radial dependence of the EEDF is taken into account via parametrical dependence on the particle number densities. Application of the EEDF Eq. (29) will be discussed in Section II,D.

C. Radiation Transport in a Low Pressure Discharge

A convenient way to treat the problem of radiation transport in a low pressure discharge is to use the Holstein-Biberman equations for the population densities of radiative states (Biberman, 1947; Holstein, 1947, 1951). For a multicomponent line radiation these equations may be represented in general form as follows:

$$\frac{dn_i(\mathbf{r},t)}{dt} = \alpha_i(\mathbf{r},t) - n_i(\mathbf{r},t)/\tau_i + \sum_j B_{ij} + \sum_j C_{ij}. \qquad (31)$$

In Eq. (31) and further in this section, the indexes i and j refer to one of the hyperfine structure (hfs) components of the line; $n_i(\mathbf{r},t)$ is the population density of the ith excited sublevel. The ith sublevel corresponds to an excited atomic state of a certain species. The quantities τ_i and α_i represent the natural lifetime and the population-depopulation rate due to any transitions between the ith sublevel and other atomic states, respectively. Usually, the term α_i has to include electron impact transitions and cascade population. The terms B_{ij} represent reabsorption terms which have been mentioned in the first section when discussing radiative rates in Eq. (2). These terms may be expressed as follows:

$$B_{ij} = \frac{1}{4\pi\tau_j} \int_0^\infty \varepsilon_j(v) k_i(v) dv$$

$$\times \int n_j(\mathbf{r}_v, t) \exp[-K(v)\Delta r]/(\Delta r)^2 d^3 r_v, \qquad (32)$$

where $\Delta r = |\mathbf{r} - \mathbf{r}_v|$ and v is the radiation frequency.

The quantities $\varepsilon_i(v)$, $k_i(v)$ represent emission and absorption coefficient for the ith hfs component, respectively; $K(v)$ is the total absorption coefficient defined as $K(v) = \sum k_i(v)$.

Note the use of K in Eq. (32) corresponds to the inclusion of overlap of atomic absorption profiles of the various hfs components. The terms C_{ij} describe collisional mixing between sublevels corresponding to a certain atomic state.

These terms may be expressed as

$$C_{ij} = (\langle v\sigma_{ji}\rangle N_i n_j - \langle v\sigma_{ij}\rangle N_j n_i)$$
$$+ N_g(\langle vq_{ji}\rangle n_j - \langle vq_{ij}\rangle n_i) \qquad (33)$$

Here, the first term represents transitions between sublevels due to energy transfer in the process of collision of excited and ground state atoms of the same species. The second term characterizes hyperfine mixing by collisions with buffer gas atoms. In the case, where a radiating gas is composed of several isotopes, the first term involves transitions between all isotopes, but the second term involves only transitions between sublevels corresponding to the same isotope.

The quantities σ_{ij}, q_{ij}, represent cross sections of relevant processes, N_i and N_g represent ground state density corresponding to the ith hfs component and number density of buffer gas, respectively. Equations (31) and (32) are strictly valid on two conditions.

(1) Emission and absorption coefficients are position independent.
(2) There is no correlation between frequencies of absorbed and emitted photons.

The first condition requires spatially uniform distributions of the gas temperature and number density of absorbing atoms. The second condition poses many complicated questions. Discussion of these questions may be found in the original papers of Holstein (1947, 1951). Holstein has derived, by variational method, expressions for the imprisonment time of radiation corresponding to the fundamental decay mode of Eq. (31). He has considered the cases of either Doppler or Lorentzian line shape and a single component line. These results have been applied to the decay of the mercury 253.7-nm radiation. The line has been treated as composed of five nonoverlapping components of equal intensity. Walsh (1950) has included effects of collisional line broadening and component overlap. To simplify the problem, Walsh (1950) has treated the line as composed of five equidistant, equal intensity components. The line shape of a component has been expressed as the sum of the Doppler core and the Lorentzian wing. The collisional mixing between the hfs sublevels has been neglected.

In spite of all the simplifications, the model of Walsh (1950) has given a very good agreement with experimental data of radiation decay. The imprisonment time of Walsh (1950) has been widely applied to calculations of the steady state discharges.

It can be shown that the use of the fundamental decay time is equivalent to the exact solution of the steady state Eq. (31) for one special situation. This situation is achieved, if the collisional rate α_i depends on the coordinate r exactly in the same manner as the function $n_{io}(r)$ corresponding to the fundamental decay mode. Functions $n_{io}(r)$ have typically a parabolic shape. Thus, if the radial profile of α_i is close to the parabolic one, the application of the decay imprisonment time to a steady state case can be a reasonable approximation and vice versa.

Grossman et al. (1986) have recently published an improved analysis of the transport of the 253.7-nm mercury radiation. Their calculations of the imprisonment time agree very well with recent experimental measurements performed by Van de Weijer and Cremers (1985). In their analysis, the authors have taken into account the following effects.

(a) Resonant energy transfer

$$^i\text{Hg } 6\ ^1S_0 + {}^j\text{Hg } 6\ ^3P_1 \rightarrow {}^i\text{Hg } 6\ ^3P_1 + {}^j\text{Hg } 6\ ^1S_0, \qquad (34)$$

where i and j are two different isotopes. This process has a cross section of 1.1×10^{-13} cm^2 (Lagushenko et al., 1985).

(b) Collisional line broadening by rare gases. The cross section for this process depending on the gas varies around 50–250 Å^2 (Grossman et al., 1986; Lagushenko et al., 1985).

(c) Overlap of the different isotopic component emissions.

(d) Hyperfine mixing due to collisions with rare gases.

By solving the appropriate equation with a series of mathematical details, these authors were able to calculate the amount of radiation emitted by each isotopic component and reconstruct the hyperfine structure.

In Fig. 1 the results of this calculation are shown. As one can see there is excellent agreement between calculations and experimental data. As one might expect, different parts of the hfs have different degrees of sensitivity to the collisional processes outlined above. For example, the depth of the valleys between different isotopic components is sensitive to the Lorentz broadening cross section, the depth of the self-reversal is most sensitive to the shape and details of the excited-state radial distribution. During the course of the calculations the authors realized that the envelope of the peaks of the hfs was sensitive only to the energy transfer cross section (σ_t) and very insensitive to the other parameters. This observation was exploited to obtain σ_t by varying it and finding the value for which the envelope of the calculated hfs best matches the experimental one. This cross section was found to be $(1.1 \pm 0.3) \times 10^{-13}$ cm^2 at 335 K. This value agrees well with the original order-of-magnitude estimate of $\sim 10^{-13}$ cm^2 of Holstein et al. (1952).

In addition to the analytical radiation transfer model, Anderson et al. (1985) have used the Monte Carlo technique to calculate the exit probability of photons for different concentrations of ^{196}Hg and the frequency distribution of the emitted spectra. The Monte Carlo technique essentially simulates

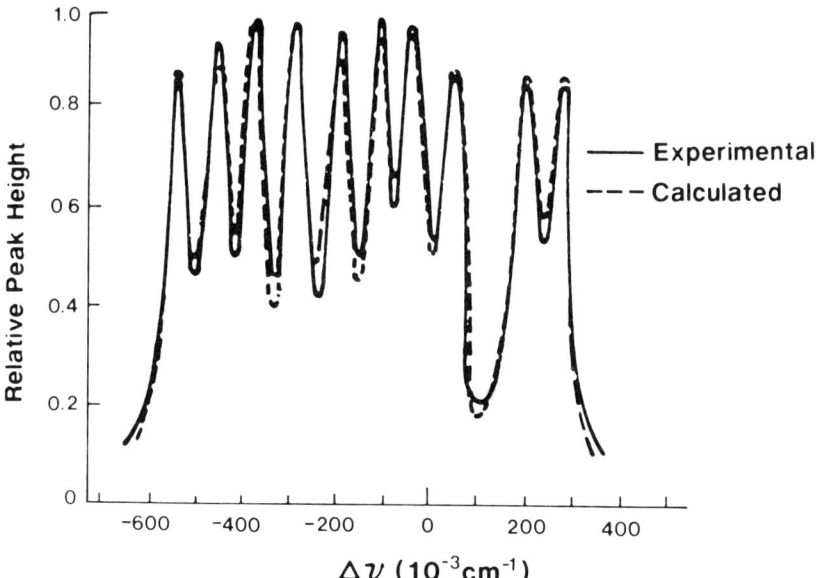

FIG. 1. Calculated and measured Hg 6 $^3P_1 \rightarrow$ 6 $^1S_0$ hfs. (Grossman et al, 1986).

the photon emission absorption, quenching and energy transfer processes that go on in the lamp. After an initial excitation, the energy of the mercury atom is tracked until the photon either leaves the tube or is quenched. The process is repeated thousands of times to obtain a reliable exit probability which is inversely proportional to trapping time.

There is very good agreement between calculated and experimental results. These results show that it is possible to use Monte Carlo techniques for radiation transfer in systems where there is a moderate amount of absorption. If the absorption is too strong, the computational time can get to be fairly substantial so as to make alternative techniques more economical.

Post (1986) has studied transport of the 185.0-nm mercury radiation. He has arrived at the conclusion that the condition of complete spectral redistribution is not satisfied for this radiation at typical discharge conditions. Applying a special formalism to the transport of 185.0-nm radiation, Post (1986) obtained calculational results that were in very good agreement with experimental data. Van Trigt (1969, 1970, 1971) has addressed questions of obtaining analytical solutions to Eq. (31) for cases of a single component line, optically thick conditions and slab geometry. In addition to these rigorous approaches Irons (1979a, b, c) discusses various approaches to approximate treatments of radiation transport.

D. Theoretical Models of the Positive Column of the Mercury–Rare-Gas Discharge

Fundamentals of modeling of the mercury–rare-gas positive column have been laid on in the early works of Kenty (1950), Waymouth and Bitter (1956), and Cayless (1962). These works have emphasized the important role of the excited Hg 6 ^3P states both for the energy and ionization balance of the discharge. Results obtained by these researchers and further modeling efforts have shown the following elementary processes to be important for an adequate description.

(1) Electron impact transitions between ground state Hg 6 ^1S and excited states Hg 6 ^3P, Hg 6 ^1P, Hg 6 ^3D, Hg 7 ^3S.

(2) Electron impact transitions between excited states Hg 6 $^3P_{0,1,2}$, Hg 6 ^1P, Hg 6 ^3D, Hg 7 ^3S.

(3) Radiative transitions Hg 6 $^3P_1$ and Hg 6 $^1P \to$ Hg 6 ^1S; Hg 7 ^3S and Hg 6 $^3D \to$ Hg 6 ^3P.

Production of ions has been considered to be due to electron impact ionization of mercury both from the ground state Hg 6 ^1S and the excited states Hg 6 $^3P_{0,1,2}$. It should be emphasized that stepwise ionization via the excited states Hg 6 $^3P_{0,1,2}$ is very important or even predominant under most discharge conditions. Excitation and ionization of rare gas and depletion of mercury have been neglected. The above scheme of elementary processes has been the basis for all further modeling.

Based on the above scheme, Waymouth and Bitter (1956) have developed the first calculational model of the mercury–argon positive column. Cayless (1962) has presented a model that is similar in many respects to the model of Waymouth and Bitter, but used more rigorous mathematical procedures for the solution of the model equations. Calculational results and details of these models are described in Waymouth (1971). Finding cross-section data involved in the above scheme of processes has always been a problem. Kenty (1950) has derived and compiled the first set of these cross sections and first models have used these cross sections. In further modeling, these data were updated along with progress of atomic physics or by efforts of modelers.

References to the updated cross section data may be found in relevant papers on modeling cited in this review. Calculations of the mercury–argon discharge based on the first models have shown an agreement with experimental observations, which may be regarded as being very good, considering the complexity of the problems. Nevertheless, significant discrepancies have been observed at many discharge conditions. This was a driving force to seek improvement.

Significant improvement has been achieved by Vriens *et al.* (1978) along two lines: incorporation of the two-electron group EEDF and the 6 ^3P–6 ^3P collisional ionization. This is the so-called two-electron group model (2-EGM). In addition to production of ions by the electron impact ionization, they have analyzed the role of ionization–recombination processes as follows:

$$\text{Hg } 6\,^3P_0 + \text{Hg } 6\,^3P_1 \to \text{Hg}_2^+ + e, \tag{35}$$

$$\text{Hg}_2^+ + e \to \text{Hg}^{**} + \text{Hg}, \tag{36}$$

$$\text{Hg } 6\,^3P_2 + \text{Hg } 6\,^3P_2 \to \text{Hg}^+ + \text{Hg } 6\,^1S + e, \tag{37}$$

$$\text{Hg } 6\,^3P + \text{Hg } 6\,^3P \to \text{Hg}^{**} + \text{Hg } 6\,^1S, \tag{38}$$

$$\text{Hg}^{**} + e \to \text{Hg}^+ + 2e. \tag{39}$$

Here symbols Hg_2^+ and Hg^{**} refer to the molecular ion and a high excited state of mercury, respectively.

Ionization rates due to processes (35)–(39) have been calculated for the mercury–argon discharge using the 2-EGM approximation of the EEDF and the literature data of the relevant cross sections and population densities of the Hg 6 ^3P states. Results obtained in this way have shown that contribution of processes (35)–(39) may be very significant, especially at elevated pressures of mercury vapor.

The 2-EGM approximation of the EEDF and the 6 ^{3}P–6 ^{3}P collisional ionization have been incorporated in the complete model of the mercury–rare-gas discharge (Lighthart and Keijser, 1980; Lighthart, 1979; Dennemann et al., 1980). These calculational results have shown a definite improvement as compared to previous modeling. It is essential to say that inclusion of the 2-EGM approximation of the EEDF and the 6 ^{3}P–6 ^{3}P collisional ionization has produced a good description of the mercury–rare-gas discharge using the Walsh imprisonment time without any further adjustment of the imprisonment time value. An example of the calculational improvement is shown in Fig. 2 where Dennemann et al. (1980) compare calculations for positive column efficiency using 2-EGM, collisional ionization, and Walsh's (1959) radiation transfer with the experimental data of Koedam et al. (1963).

As was mentioned previously, the most rigorous approach to modeling is the simultaneous solution of the kinetic equation for the EEDF and the

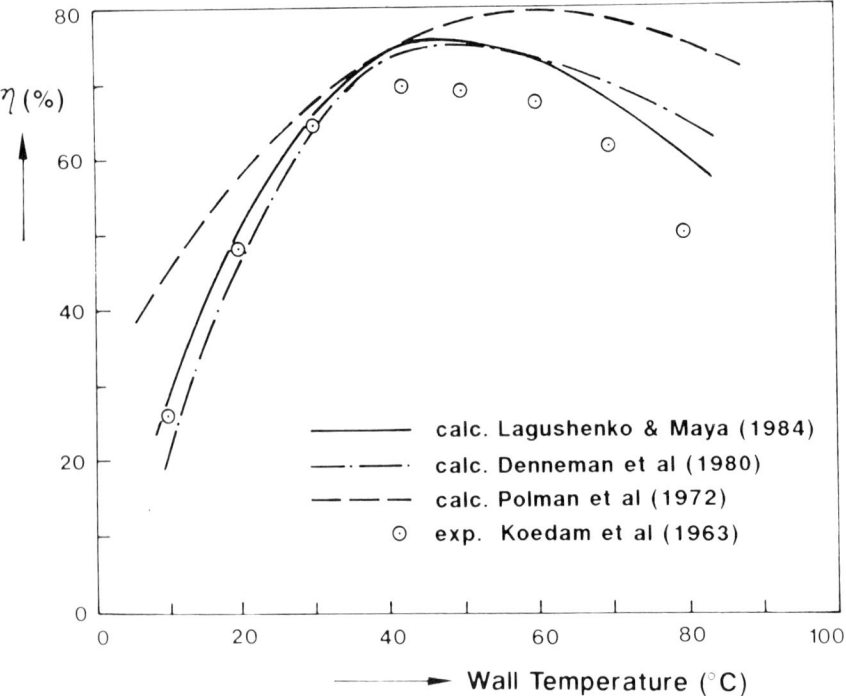

FIG. 2. Comparison of calculated and experimental results for conversion efficiency of electrical energy into both UV lines (253.4 nm + 185.0 nm) as a function of wall temperature.

particle balance equations. This approach has been attempted by Winkler *et al.* (1983a, b). To simplify the problem, the particle balance equations have been averaged over the discharge cross section assuming the Bessel radial distributions for all species. However, since the ionization balance equation was not solved, they could not calculate the discharge electric field, and these authors had to use experimental values of the electric field. The kinetic equation has been solved with the cross section averaged number densities of electrons and the Hg 6 ^{3}P atoms.

In spite of all the above simplifications, these authors have observed improved agreement with experimental data. Comparison of calculational results obtained with various models to experimental results of Koedam *et al.* (1962) for the mercury–argon discharge is given in Fig. 3. In this figure, the ultraviolet radiation efficiency is shown as a function of the mercury

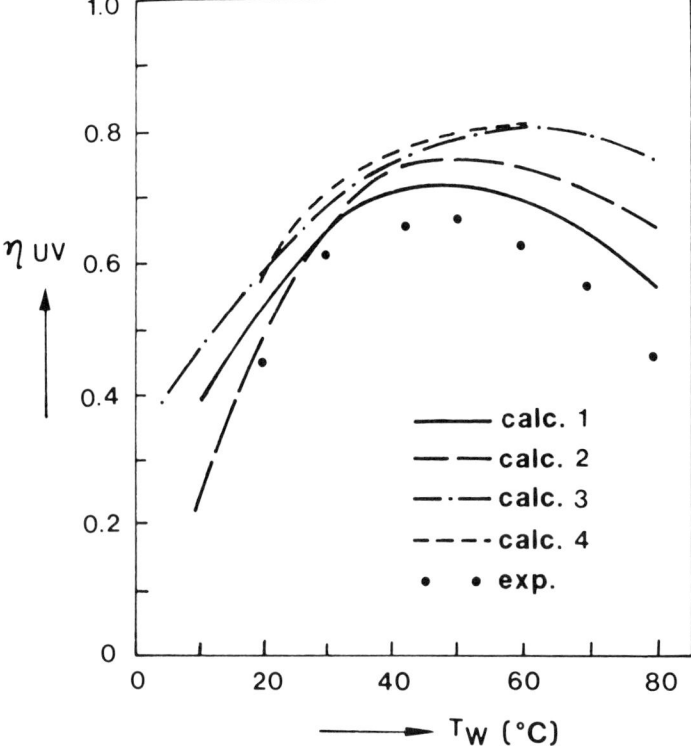

FIG. 3. Comparison of experimental and calculated results for conversion efficiency (see text) as a function of wall temperature. Here calc. 1 is Winkler *et al.* (1983a, b), calc. 2 is Denneman *et al.* (1980), calc. 3 is Polman *et al.* (1972), calc. 4 is Cayless (1962) and the points (●) are from Koedam, *et al.* (1963).

condensate temperature. The ultraviolet radiation efficiency is defined as the ratio

$$\eta_{uv} = (P_{253.7} + P_{185.0})/P \qquad (40)$$

Here, $P_{253.7}$ and $P_{185.0}$ are power outputs of the two mercury lines and P is the total power input.

Lagushenko and Maya (1984) have published calculations of the mercury–argon discharge based on a different expression for the EEDF. Salient features of their model are as follows.

(1) The EEDF has been approximated by Eq. (29) described in the previous section.

(2) Excitation and ionization of rare gas and Penning ionization of mercury have been included.

(3) Additional processes of the $6\,^3P$–$6\,^3P$ collisional ionization have been considered. For example,

$$\text{Hg } 6\,^3P_{2,1} + \text{Hg } 6\,^3P_{2,1} \rightarrow \text{Hg}_2^+ + e, \qquad (41)$$

$$\text{Hg}_2^+ + e \rightarrow \text{Hg}^{**} + \text{Hg} \qquad (42)$$

have been taken into account. Cross sections for the processes of Eq. (41) have been adjusted by fitting calculated to experimental values of the electric field (Verweij, 1961) at elevated pressures of mercury vapor and low currents. The fitting procedure resulted in cross sections for the process in Eq. (35).

(4) Equations of the population and energy balance have been averaged over the discharge cross section. The Bessel profile for electrons and ions and parabolic profile for excited atoms have been assumed.

(5) Radial nonuniformity of gas temperature T_g has been considered assuming the parabolic radial distribution in the form

$$T_g(\rho) = T_W + (T_C - T_W)(1 - \rho^2/R^2), \qquad (43)$$

where T_C and T_W are the center and wall temperature, respectively. The center temperature has been calculated from the gas energy balance.

(6) The radial distribution of the ground state mercury atoms have been expressed via the radial distribution of ions using the formula (Verweij, 1961)

$$N_{\text{Hg}}(\rho) + \frac{D_a}{D_{\text{Hg}}} N_{\text{Hg}}^+(\rho) = N_{\text{Hg}}(R) \qquad (44)$$

Here, D_a and D_{Hg} are the ambipolar and mercury atom diffusion coefficients respectively. Eq. (44) follows from equality of fluxes of Hg and Hg$^+$ at steady-state conditions.

Comparison of calculated (Lagushenko and Maya, 1984) to experimental data of UV efficiency as a function of wall temperature are presented in Fig. 2.

Unlike previous models, this model gives realistic values of the electric field down to the cold-spot temperature 10°C. We believe this is due to the use of the EEDF of Eq. (29) combined with the Penning ionization process. This is shown in Fig. 4.

A rigorous model of the mercury–argon discharge has been published recently by Dakin (1986). This model has used the scheme of elementary processes described in Sections II,A and II,C with the addition of the $6\,^3P$–$6\,^3P$ collisional ionization processes of Eqs. (35) and (37). In that respect, the model is similar to the model of Lighthart and Keijser (1980). However, significant improvements have been made along the following points.

(1) The improved version of the 2-EGM EEDF has been used. This allows for radial variation of bulk and tail temperatures as described in the previous section.

(2) Radial variation of the ground state mercury number density and gas temperature have been considered.

(3) Imprisonment of radiation has been accounted for by using position dependent expressions for the Biberman-Holstein net radiative brackets. (Irons, 1979a, b, c).

(4) Position dependent balance equations of population and energy have been solved numerically.

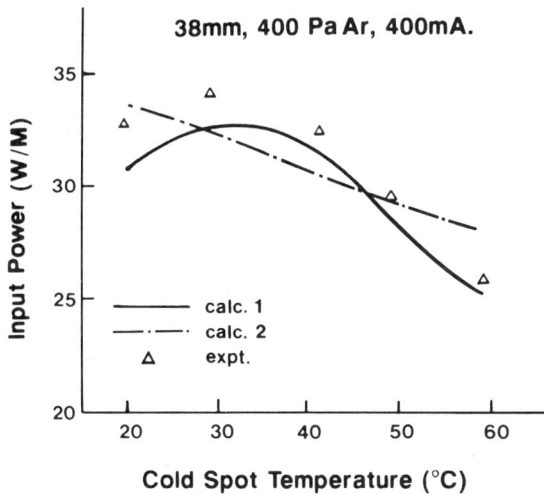

FIG. 4. Comparison of calculated and experimental results for input power (or E field since current is constant) as a function of cold-spot temperature, for a 38-mm-diameter discharge tube. Here calc. 1 is Lagushenko and Maya (1984), calc. 2 is Waymouth and Bitter (1956) and the triangles (△) are experimental data of Koedam, et al. (1963).

The model of Dakin has shown very good agreement with experimental data for conditions typical of fluorescent lamps. In particular, it has allowed for an accurate description of radial distributions of all species.

Most published calculational results have dealt with the Hg + Ar discharges. Calculations of the discharge in the Hg + Ar + Ne mixtures have been described by Kalyazin, et al. (1981). These calculations have been based on the authors' expressions of the EEDF obtained by an approximate analytical solution to the kinetic equation. In conclusion, one may say that modeling of the mercury–rare-gas discharge has attained an accuracy of 5%–30% range. As can be seen from Figs. (2) and (3), considerable progress in predicting the UV efficiency has been made over the years. However, still at higher temperatures ($\sim 80\,°C$), a discrepancy between calculations and experiments exists. Furthermore, all models suffer from lack of atomic data to calculate associative ionization rigorously. We believe further progress of modeling can be achieved by developing more realistic approaches to calculations of the EEDF.

III. Altered Low Pressure Mercury-Rare-Gas Discharge

During the last decade or so there have been numerous attempts and implementations to alter the low pressure Hg–rare-gas discharge. The driving forces behind these efforts are several. For one, energy conservation and the desire for a more energy efficient discharge has led to a number of approaches, some of which are still in the research and development stage, while some have already been incorporated in products. In the first part of this section we shall elaborate on some of these ideas. The second driving force has been the desire to have a more compact fluorescent lamp. This desire, again, fundamentally derives from the energy conservation force. (Rosenfeld and Hafemeister, 1988; Maya et al., 1984). However, the objective here is to have a low pressure compact discharge to replace the incandescent lamp which is notoriously inefficient compared to electrical discharge light sources and yet very practical. The relative merits of different light sources is outside the scope of this article and the reader is referred to Waymouth (1987) for an extensive discussion of this subject. One natural approach to making a more compact low pressure positive column discharge is clearly to bend it onto itself several times or spiral it and somehow twist it to fit into a small space! All of these have been tried and made into products; however, we shall not discuss any of these approaches in this article. What we shall discuss are approaches that inherently try to alter the positive column or some parameter in it. Subjects that we shall review include isotope and magnetic effects, introduction of recombination structures into the positive column, altering the EEDF in the discharge, altering the volt-ampere characteristic of the discharge, high frequency operation of the positive column, surface waves, etc., all of which have the common characteristic of attempting to alter the traditional positive column discharge, therefore providing some new insights into the physics of the discharge.

A final driving force behind much of the research and development in LPD, especially in the area of high frequency, RF and microwave (MW) radiation has been the gradually declining cost of increasingly available electronic components. The anticipation of the viability of an economic electronic ballast for the fluorescent lamp has been a strong incentive to study the high frequency, RF, and MW operation of the positive column.

A. Isotope Effect

One of the fundamental limitations to increasing the efficiency of fluorescent lamps is the phenomenon of resonance trapping. This is the phenomenon as discussed above where photons of a given frequency generated in the bulk of the discharge on their way out of the bulb are absorbed and reemitted many times by atoms whose absorption line profile overlaps the given frequency. In a typical fluorescent lamp photons of 253.7-nm wavelength are absorbed and reemitted on the average about 50–100 times before they emerge out of the glass bulb (Kenty, 1950). The larger the number of emissions and absorptions the longer the photons remain inside the tube and the higher the probability of a radiationless transition. Therefore, it is important to find ways of reducing this so called "imprisonment" of resonance radiation and its characteristic time, "imprisonment time." This characteristic time is nothing more than the observed lifetime of a photon inside the bulb, which, as discussed above, can be considerably longer than the natural lifetime depending on the amount of imprisonment it has undergone.

Increases in efficiency of fluorescent lamps by as much as 2%–5% due to isotopic altering of the Hg used in the lamps was reported by Maya and coworkers in a series of papers (Johnson *et al.*, 1983; Grossman *et al.*, 1983; Maya *et al.*, 1984; Anderson *et al.*, 1985; Grossman *et al.*, 1986). The reports indicated that by increasing the concentration of ^{196}Hg from 0.146% in natural Hg to about 3%, an increase in efficiency of visible light production of about 2%–5% was observed.

The effect was explained by the reasoning that ^{196}Hg offered a nonoverlapping distinct radiation channel and an easier way for the photons to escape, since the concentration of ^{196}Hg is still far lower than the other isotopes and each isotope absorbs only its own radiation. Therefore, by a strong resonant energy transfer, the other isotopes in their $6\,^3P_1$ state would unload part of their energy to the ^{196}Hg isotope and let that isotope "carry the ball," so to speak, to the outside world.

Subsequent measurements, by the same research team, of total light output in the integrating sphere, total 253.7-nm output using a circulating water bath, and hyperfine structure measurements of the 253.7-nm line using a Fabry Perot interferometer have verified the validity of this interpretation. Relative UV output per arc watt as a function of Hg cold-spot temperature (T_c, which determines the Hg pressure) measurements indicated that for about 2.6% ^{196}Hg, there is an increase in 253.7-nm production efficiency of about 6.8%. These measurements were conducted in circulating water bath. We should point out that the UV output and UV efficiency and total light output and efficacy (lumens/watt or LPW) measurements in an integrating sphere are all in good agreement when end losses (Faraday dark space and electrode losses) are taken into account.

Notice that increasing ^{196}Hg further does not help in efficiency or UV output. The reasons for this can be understood by looking at Fig. 5. Here the hyperfine structure (hfs) of the 253.7-nm line for natural Hg at two different temperatures is shown. As is well known, the low temperature structure has five distinct peaks. However, when one goes to higher temperatures, the peaks show a self-reversal due to the fact that the radial excited state distribution is not constant across the tube.

Now when one goes to a higher concentration of ^{196}Hg as shown in Fig. 6, a distinct sixth peak appears as an additional escape channel. However, as one increases the temperature, the back energy transfer from ^{196}Hg $6\,^3P_1$ to other isotopes and the overlap with the ^{199}Hg and ^{201}Hg peaks starts becoming appreciable such that one does not gain any more by increasing the ^{196}Hg concentration. The asymmetry in the self-reversed left peak of the ^{199}Hg/^{201}Hg line clearly demonstrates the overlap contribution of ^{196}Hg.

A calculation to demonstrate the effect of energy transfer was conducted as follows. The amount of radiation emerging from the tube in the ^{198}Hg and ^{196}Hg components as a function of ^{196}Hg and ^{198}Hg concentration was calculated and plotted versus the ^{196}Hg and ^{198}Hg concentrations (Grossman et al., 1986). What this plot showed was that due to strong energy transfer into a small concentration component (i.e., ^{196}Hg) the radiation coming out in this channel far exceeds the amount that would be expected from its concentration. This, however, is not the case for a large concentration component such as ^{198}Hg or when ^{196}Hg concentration becomes large. Clearly, this calculation is in support of why one reaches a saturation phenomenon in efficiency and light output as one increases the ^{196}Hg concentration.

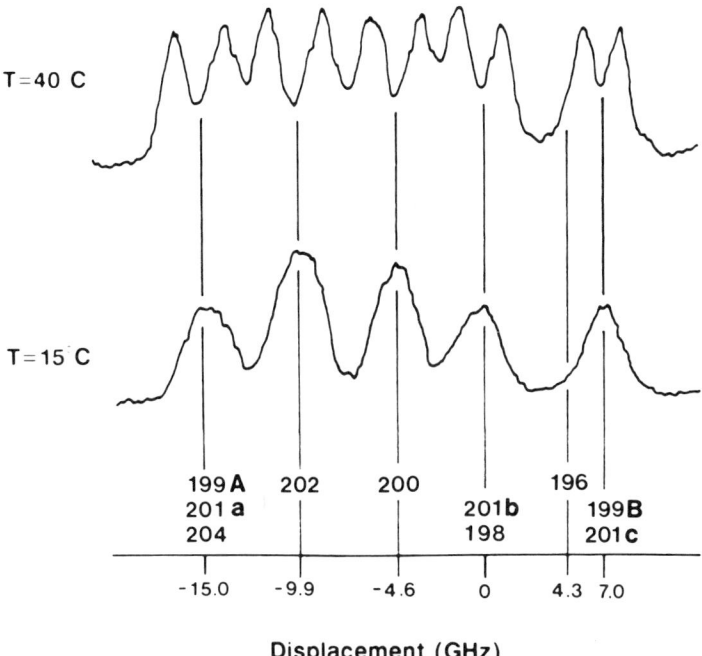

FIG. 5. Emission spectra of Hg 6 $^3P_1$–6 $^1S_0$ transition for natural Hg at 15 °C and 40 °C cold-spot temperatures. (Grossman et al. 1986.)

In Fig. 7 we show a summary of the LPW, UV per arc watt (P_{uv}/P_W) measurements, and τ_{rel} calculations reported by Grossman et al., 1986. As can be seen from the figure, the reduction in τ_{rel} reaches a plateau at around 8%–10% ^{196}Hg. In the same figure, the experimental 253.7-nm output per arc watt (P_{uv}/P_W) and measured integrated LPW (lumens/watt) curves as a function of ^{196}Hg are shown. Note that the experimental curves for LPW and P_{uv}/P_W reach saturation at about half the concentration of ^{196}Hg as reduction in τ_{rel}. Similar results were found in the Monte Carlo type calculations reported earlier (Anderson et al., 1985).

Here the photon exit probability, which is essentially inversely proportional to imprisonment time, reached a maximum at about 10% ^{196}Hg and decreased thereafter. This suggests that the efficiency and light output relationship to τ_{rel} is not a simple one to one direct relationship, as expected and discussed above. However, qualitatively, there is a definite trend and correlation between the reduction in τ_{rel} and LPW, as well as P_{uv}/P_W. Note further in the figure that the UV output reaches a plateau around 2.6 to 3% ^{196}Hg while the LPW reaches saturation around 5% ^{196}Hg. The reason for a

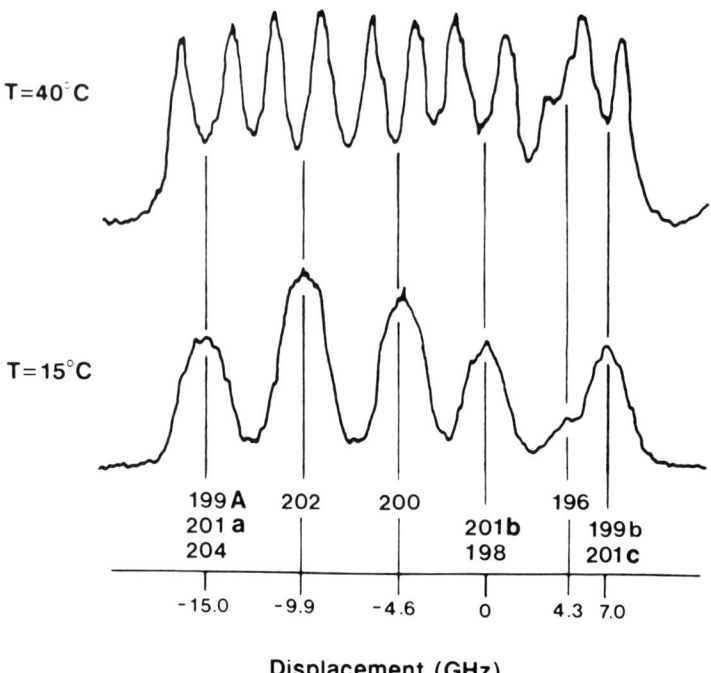

FIG. 6. Emission spectra of Hg $6\ ^3P_1-6\ ^1S_0$ transition for 3.9% ^{196}Hg enriched Hg at 15 °C and 40 °C cold-spot temperatures. (Grossman et al. 1986.)

later saturation in LPW may be due to the fact that the visible radiation includes about 6% visible Hg lines and about 6% phosphor emission due to 185.0-nm Hg radiation. Therefore, some of the trapped or quenched radiation from the Hg 6^3P manifold may end up as 185.0-nm and/or visible Hg lines resulting in a slight delay of saturation as ^{196}Hg concentration is increased.

The isotope effect has been investigated by Ingold and Roberts (1986) as well as by Sun et al., (1983). Ingold and Roberts conclude that there is an increase in efficiency of the positive column as a result of altered Hg isotopic distribution, based on their theoretical results. Similarly, Berman and co-workers (Sun et al., 1983) verify the increase due to altered isotopic distribution.

B. MAGNETIC FIELDS AND STRIATIONS

In this section we would like to briefly describe the work on the application of magnetic fields to LPD's. The use of an axial magnetic field for efficiency

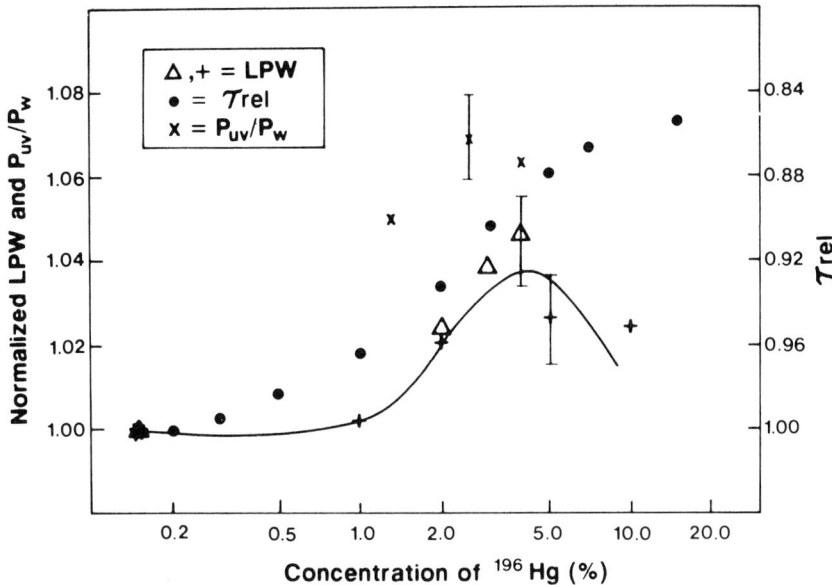

FIG. 7. Measured (P_{uv}/P_w), LPW and calculated τ_{rel} (see text) as a function of ^{196}Hg concentration. (Grossman et al., 1986.)

improvement has recently been suggested by Richardson and Berman (1983). Recent work of Zhou et al., (1987) and Ingold and Roberts (1984) indicate local efficiency improvements in LPD's due to the application of magnetic fields. While Richardson and Berman reported local radiation measurements with a magnetic field applied to a portion of the discharge, Ingold and Roberts, in a clever approach, immersed the lamp in an axial field inside an NMR machine in order to obtain a uniform axial magnetic field throughout the lamp. However, these researchers did not control the cold-spot temperature of the discharge. This is important because it determines the Hg number density, which subsequently determines, in a sensitive manner, the radiation output. Furthermore, due to alteration of the discharge cross section under magnetic field conditions it is important to run efficacy measurements inside the integrating sphere.

Moskowitz et al. (1987) have carried out a number of experiments to investigate, in a controlled manner, the effects of magnetic fields on lamp efficiency. Local emission measurements took place whereby radiation from only one section of the discharge and emission from phosphor coated lamps was measured in this way. In addition to local measurements, integrating sphere measurements were carried out which permitted detection over 4π steradians. Both axial and transverse (relative to positive column orientation) externally applied magnetic fields as well as a number of different length and diameter lamps were investigated. As will be shown below, significant improvement in lamp efficiency can be obtained with proper use of magnetic fields.

Experiments run outside the integrating sphere show substantially high increases in local efficiency of 253.7-nm generation in agreement with earlier reports. Here depending on a particular lamp one obtains local efficiency improvements of 10%–25%. However, when these lamps are inserted into an integrating sphere with temperature control the efficiency improvements are much more modest. The integrating sphere experiments were run for temperatures between 20–60°C of the water bath and the magnetic field was scanned up and down at any given temperature. The experiments were always run against a control lamp that had the exact same structure attached to it with identical reflectivity material but made out of dummy magnets. In Table I we summarize the results obtained.

As can be seen, efficiency improvements up to 5% are indicated. Moskowitz et al. (1987) found, however, that in these and many other experiments that transverse magnetic fields are more effective in bringing about a given increase in efficiency than axial fields. Since it is not exactly clear what is going on inside the discharge when magnetic fields are applied these researchers are not in a position to offer an explanation for this observation. Zeeman splitting of the odd components of the hfs has been offered as an explanation for the increase in efficiency by Richardson and Berman (1983). Moskowitz et al. (1987) have looked at the 253.7-nm line with a Fabry Perot interferometer and verified some splitting due to the magnetic field. As the field increases to substantial values, one can see some splitting taking place. However, at the lower field values, ~100 Gauss, one does not see any splitting. Therefore, they believe there is more than just Zeeman splitting going on. Clearly, one might expect a change in the EEDF, thereby resulting

TABLE I

SUMMARY OF MAGNETIC FIELD MEASUREMENTS BY MOSKOWITZ et al. (1987).

Lamp	B field	B field strength (Gauss)	ΔLPW (%)	Field generation
40 W, 38 mm	Transverse	40	+2.2	Permanent magnets
40 W, 38 mm	Transverse	110	+4.3	Permanent magnet strips
40 W, 38 mm	Transverse	180 Strips	~0.	Permanent magnet
20 W, 38 mm	Axial	395	+4.1	Helmholtz coils
8 W, 16 mm	Axial	430	+6.0*	Helmholtz coils

* The increase here was UV power/electrical power rather than LPW.

in a more efficient conversion of electrical energy into photons. The very-high-energy electrons of the EEDF which normally reach the walls of the container without doing any substantial excitation or ionization might be retained in the body of the discharge (due to $\mathbf{E} \times \mathbf{B}$ forces) and forced to do some work (excitation and ionization) thereby resulting in a more efficient discharge. Further EEDF or other diagnostic measurements certainly would help to clarify the observations.

While there are many forms of instabilities in LPDs, one form that has been studied for some time is striations. Most studies of striations however have taken place in noble gases, and there are only a handful of studies in LPD. Striations typically occur when either the temperature or the current falls below a certain value in the discharge and exhibit themselves as light fluctuations both in space and time. The same phenomenon occur in discharges of molecular gases as well and here the role of negative ions has been studied quite extensively. This subject, however, is outside the scope of our review. For further information on this subject, the reader is referred to Nighan and Wiegand (1974).

Recent studies by van den Heuvel (1988), van den Heuvel and Vrehen (1985), as well as Privalov and Fofanov (1987) have tried to answer the question of the origin of striations. The question has been whether striations are produced inherently by the plasma or are a response of the plasma to some local source of oscillation. The experimental and theoretical work of the above researchers determined that the current feedback of the external circuit plays a crucial role in the excitation of convective type of striations. Convective type striations are all types of striations which are not inherent in the plasma.

Furthermore, the picture that emerges from these studies is that striations are generated by current fluctuations in the cathode region leading to local inhomogenities which give rise to travelling waves. In a recent article, Dorleijn and Jack (1985) mention that Ar gas is less susceptible to striations than other gases in a LPD. Indeed further work would be necessary to elucidate what role the type of electrodes in convective instabilities in LPD plays as well as under what circumstances striations of the type that are inherent to the plasma are produced.

There may be practical considerations for studying the striation phenomena further. In the positive column, striations are regions of higher luminosity. This higher luminosity may be associated with higher electron density and possibly higher electron temperature (Grabec and Mikac, 1987). As mentioned earlier, higher electron temperature usually is associated with better conversion efficiencies as well as higher radiation output per unit length. Therefore further understanding of the microscopic processes going on inside striations may lead to some practical applications.

C. High Frequency (HF) and Surface Waves

High frequency operation of a LPD offers several distinct advantages. As we shall see below in more detail, it reduces the electrode losses and increases the positive column efficiency somewhat so as to result in a luminous efficacy increase of about 10% for a fluorescent lamp. From the viewpoint of discharge physics, this is the main advantage. However, from a technological point of view, there are several more advantages. For example, a high frequency operation of the discharge enables one to make a small, compact electronic ballast. The availability of such a ballast makes it possible to communicate and interact with the discharge light source remotely.

As mentioned above, it has been known for quite some time (Campbell, 1960) that HF operation results in an increased efficiency for the generation of light in a LPD. The primary reason for this as the experiments of Dorleijn and Jack show (1985) is a reduction in the cathode fall (V_c) and, therefore, end losses of the discharge (since $V_c \times I$ is basically a loss, where I is the discharge current). Part of the reason for the decrease in the V_c is the fact that ambipolar diffusion frequencies are on the order of 1 KHz, while electron relaxation rates are on the order of 100 KHz. Therefore, if the discharge is energized at frequencies above 1 KHz but below 100 KHz, there will always be some residual ionization in the vicinity of the electrodes, thereby requiring less of an electric field within the sheath to sustain the discharge. It may be fair to say that in the last decade the emphasis in HF operation has been on increasing the availability and types of circuitry (power supplies and ballasts) to drive the discharge. Most of the physics understanding has been accomplished prior to the last decade (e.g., Drop and Polman, 1972). The drive to bring HF operation of LPD (or fluorescent lamps) to the public has been slow. The use of HF ballasts is definitely on the rise and the costs are declining. We would, of course, point out that this trend also has its origins in energy conservation as the main driving force. Note that most HF applications in the marketplace use frequencies of 25–40 KHz.

Polman et al. (1975) in a review article discuss the effects of HF on electron temperature (T_e) and how variations in T_e effect the light distribution. They note that their experiments with pulsed discharges have exhibited fast variations in T_e in such a manner that a fast increase in high energy electrons resulted in the excitation of upper excited states more rapidly than in Hg $6\ ^3P_1$. In an elegant figure duplicated here as Fig. 8, Polman et al. (1985) discuss the effects of high frequency (up to 1 KHz) on T_e, concentration of excited states, and electron concentration and compare it to the corresponding DC values. As can be seen from the figure, even to as low a frequency as 1 KHz there are significant changes in all three parameters.

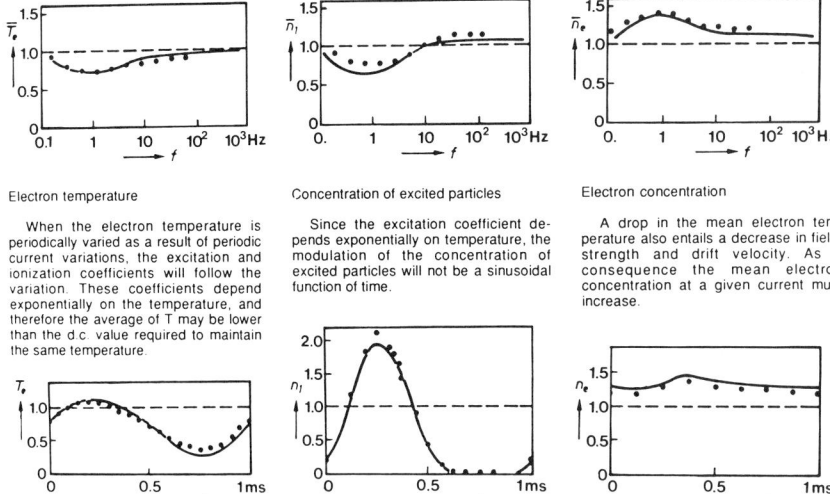

FIG. 8. Periodic variation of the electron temperature T_e and its consequences in a modulated mercury-neon discharge (modulation coefficient $\beta = 0.5$). The solid lines were calculated, the points give the results of measurements. Left: above, the average value T_e of the electron temperature as a function of the modulation frequency f of the discharge current; below, a period of T_e at a modulation frequency of 1 kHz. In both cases the dashed line as in the following figures give the value of the parameters for a DC discharge at the same mean current. Center: above, average concentration n_1 of the electron concentration as a function of the frequency; below, curve of the value n_1 of this concentration during one period at 1 kHz. Right: above, average value n_e of the electron concentration as a function of the frequency; below, curve of the value n_e of this concentration during one period at 1 KHz. The diameter of the discharge tube was 5.2 cm, the current in the DC discharge 0.4 A: the particle densities in this discharge were $n_0 Hg = 2 \times 10^{20}$ m^{-3}, $n_i = 6 \times 10^{17}$ m^{-3}, $nN_e = 2 \times 10^{23}$ m^{-3} and for each of the various $^3P_1$ states about 5×10^{17} m^{-3}. (Polman et al. 1975.)

During the last decade or so considerable research has been carried out on the properties of surface-wave excitation of plasmas (Moisan and Zakrzewski, (1986); Ferreira, 1986, and references in these two articles). As is well known, a surface-wave discharge utilizes a dielectric boundary (in many cases the walls of the container) to guide the travelling wave that produces the discharge. Unlike other MW or HF discharges that utilize a standing-wave pattern, the fact that a travelling wave produces the discharge enables one to excite a long plasma column from one end. Moisan and co-workers, for example, have studied the properties of rare-gas plasmas in the frequency range of 1–915 MHz using devices such as surfaguides and surfatrons (Moisan et al (1979, 1984) (for >300 MHz) and Ro-boxes (Moisan and Zakrzewski, 1987 (for <300 MHz). In addition, Ferreira and co-workers in a series of papers (e.g., Ferreira, 1986, and references therein) have shed considerable light on the theoretical aspects of surface wave excitation of weekly ionized plasmas.

Surface-wave excitation is of interest to the researchers in LPD activity because it has the potential of a new mode of excitation of a fluorescent lamp with several advantages. As is well known, one of the fundamental limitations to higher efficiency of photon generation in LPD is self-absorption. The majority of the excited states in the traditional thermionically emitting electroded positive columns are produced at the center of the discharge. These photons, by the time they get to the edge of the vessel are emitted and reabsorbed about 80–100 times depending on conditions. Therefore, during their journey to the walls they are susceptible to quenching and some of them never make it to the outside world. Clearly, one way of avoiding this trap is to create the majority of the excited states next to the walls of the discharge thereby offering the photons a much reduced path to travel. Surface-wave excitation offers this possibility since the electric field in this mode of excitation is strongest next to the walls of the discharge. This is shown in Fig. 9 for a discharge in Ar. Here the relative emission intensity of ArI peaks towards the edges as opposed to the center, which would be the case in a regular positive column excitation. One additional advantage, of course,

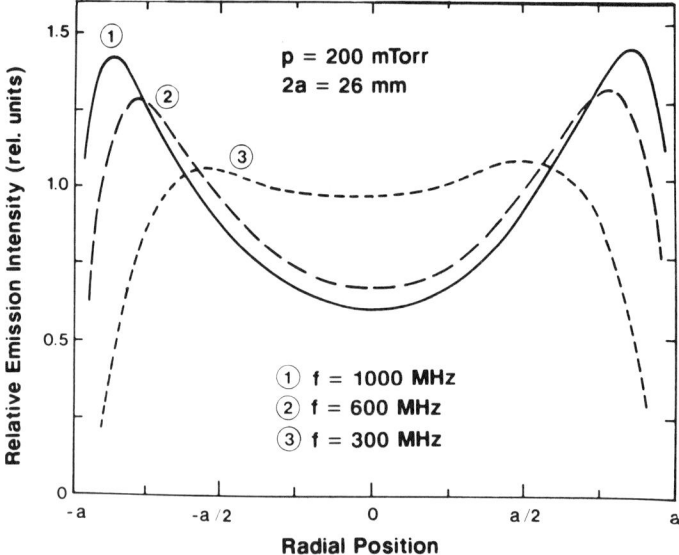

FIG. 9. Observed dependence of the emission intensity of the ArI, 545.6-nm thin line on wave frequency (Moisan et al., 1982). Note the tendency of peak emission to be closer to the walls as frequency increases.

which is common to all electrodeless modes of operation is the fact that electrode losses associated with the traditional mode of excitation are eliminated.

These advantages have been recognized by researchers in the field of LPD. Recently Beneking and Anderer (1988) reported studies of surface-wave excitation using a surfatron type of device to produce a low pressure plasma in a Hg–rare-gas mixture. One practical problem with surface waves is due to the fact that energy is absorbed by the plasma, as the wave travels less and less energy is available to be deposited on the column at points further away from the source. This is likely to cause a nonuniform appearance of light distribution in the axial direction. It may be fair to say, however, that further studies would be necessary to understand what is going on inside the plasma. In particular, spatial distribution of excited states, EEDF, and other plasma parameter studies would be desirable. From a practical point of view, whether this form of excitation is a commercially viable mode of driving fluorescent lamps remains to be seen. The high cost of generating surface waves efficiently and the attenuation of light as one moves away from the launcher in a long positive column are some of the issues that would need to be tackled once the physics of the plasma is thoroughly understood.

D. Nonequilibrium and Altered EEDF

The importance of accurately knowing what the EEDF in LPD is cannot be overemphasized. Druyvesteyn (1930) some time ago formulated some of the fundamentals for probe measurements of EEDF in LPD. These and subsequent measurements by other workers (e.g., Waymouth, 1966; Kagan et al., 1971) in the field have demonstrated that the fraction of high energy electrons, (i.e., electrons that can accomplish a single-step excitation of the Hg $6\,^3P_1$ state in a LPD) is rather small. In other words, if one is to describe the EEDF by a two-electron group model (2-EGM) as discussed above, the majority of the electrons are below the threshold for excitation. This realization, of course, has led to efforts to raise the fraction of high energy electrons. Recently, Godyak et al. (1988) reported a novel approach to raising the electron temperature in a LPD. In this work, a sudden constriction was introduced in the positive column transverse to the discharge. This sudden constriction leads to the development of a sharp electron-accelerating potential drop, which in turn gives rise to significant changes in the EEDF. The authors studied both experimentally and by calculations the spatial evolution of the EEDF in the vicinity of this sudden constriction or orifice. Their measurements of the EEDF using an RF-modulated double probe

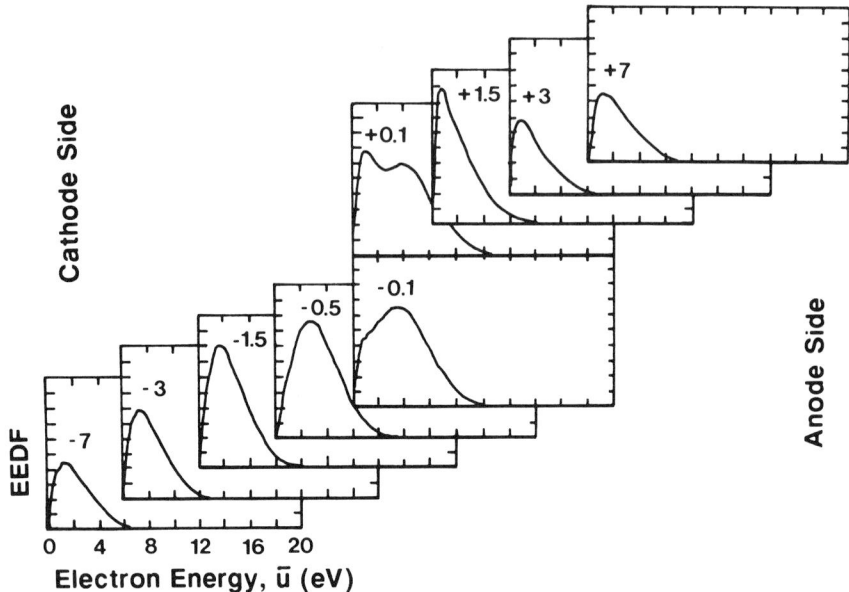

FIG. 10. Spatial evolution of the EEDF along the discharge (the numbers inside each rectangular box indicate the distance in cm from the orifice). Here $p_{Ne} = 133$ Pa, $T_c = 40$ °C, D = tube diameter = 5 cm, d_0 = orifice diameter = 1.25 cm, $I = 0.2$ A. The EEDF is expressed in units of 10^{10} eV^{-1} cm^3. (Godyak et al., 1988.)

technique showed a substantial increase in the fraction of high energy electrons as one approaches the orifice. This is shown in Fig. 10. They also found that the plasma density N_e and mean electron energy obtains a jump at the orifice and as the orifice diameter to tube diameter ratio decreases the jumps become more pronounced. These results are shown in Figs. 11 and 12.

From the absence of direct correlation between the electric field and the mean electron energy, as well as other observations of the EEDF, the authors conclude that there is significant nonequilibrium between the electric field and electrons under their experimental conditions in the vicinity of the orifice.

By using a one-dimensional profile of the measured electric field and assuming nonequilibrium, and also assuming the only energy loss mechanism to be inelastic collisions with Hg atoms, the authors modeled the discharge by accepting an electron accelerating region in front of the orifice. This approach appears to give calculational results that are in good agreement with the experimental determinations of EEDF and mean electron energy. It would appear that the sudden sharp constriction in the discharge offers a simple way of controlling the EEDF and also of separating the energy input and energy dissipation (photon production) regions in a LPD. One other result of a

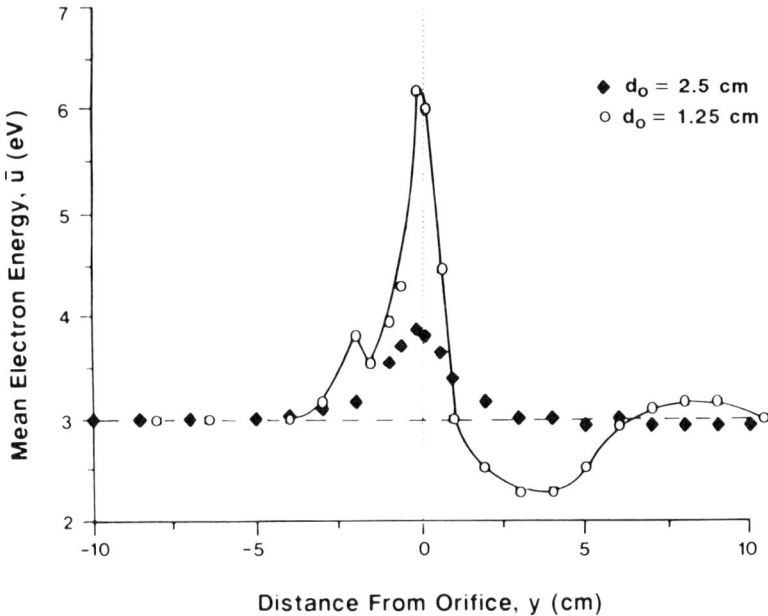

FIG. 11. Mean electron energy $\bar{u}$ for two different orifice diameters (d_0) as a function of distance from the orifice y. Here $p_{Ne} = 133$ Pa, D = tube diameter = 5.0 cm, $T_c = 20$ °C and I = 0.1 A. (Godyak et al. 1988.)

discharge with a sharp constriction is the higher light saturation level due primarily to a higher electron temperature. As is well known (Waymouth, 1971) the light obtained from a positive column in LPD saturates as power into the column is increased. The saturation level, however, tends to be higher for higher T_e. A similar effect is observed when one goes to a lighter rare gas, say from Ar to Ne. In this case, since ambipolar diffusion losses of ions and electrons is higher for Ne, the electric field together with T_e goes up somewhat, resulting in a higher saturation level. This also results in somewhat higher efficiency which typically is wiped out due to higher electrode losses in lighter gases, especially for short plasma columns (lighter gases have higher cathode falls which result in greater end losses).

In an attempt to increase the luminous flux intensity per unit length of a positive column in LPD, Hasker (1976) a number of years ago introduced solid recombination structures such as glass wool or glass fiber into the discharge. The issue here was, again, to increase the flux intensity without a substantial penalty in the electrical to photon conversion efficiency. The idea was to increase the ambipolar recombination losses for ions and electrons by offering solid recombination surfaces in the volume. This resulted in an in-

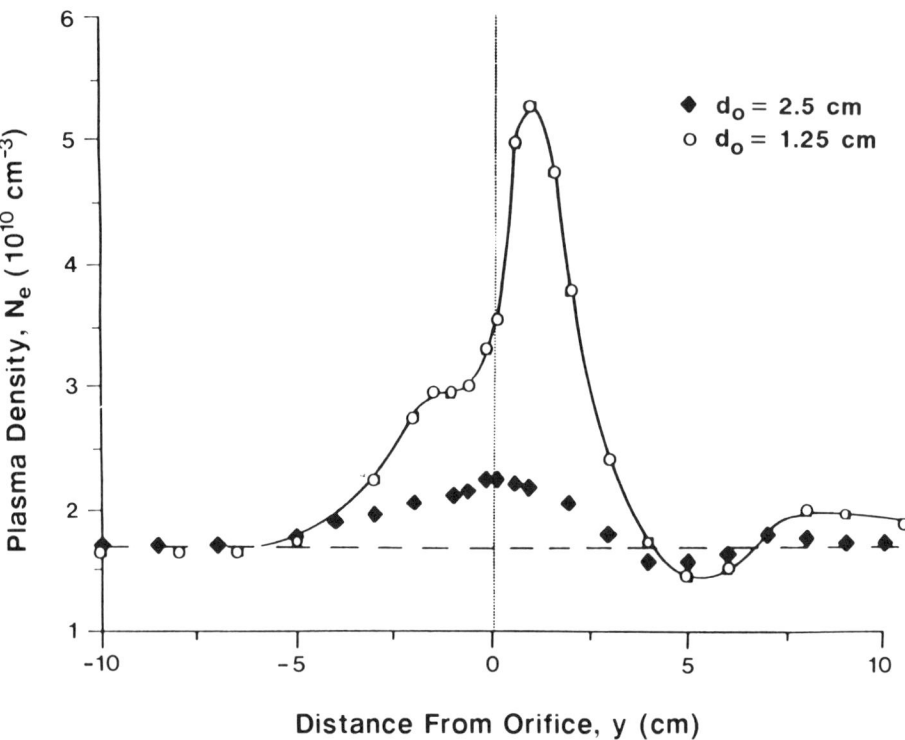

FIG. 12. Plasma density for two different orifice diameters (d_0). Here $P_{Ne} = 133$ Pa, D = tube diameter = 5.0 cm, $T_c = 20$ °C, $I = 0.1$ A. (Godyak, et al., 1988.)

crease of the electric field by as much as a factor of two depending on the density and configuration of the wool and in addition increased T_e by 20–25%. Both of these increases, of course, were expected, since, as mentioned above, these quantities are determined by the ionization balance in the discharge. However, what was somewhat surprising was the magnitude of the increase in the electric field, which probably is indicative of the importance of volume recombination in these discharges.

IV. Diagnostics

Generally speaking, diagnostics are an integral part of lamp research development and engineering. Among the many benefits derived from diagnostics of light sources, one can cite the discovery of new phenomena in the discharge, understanding of the processes that contribute to or subtract from light generation, accurate data for more accurate modeling of light sources, and possibilities of early detection of lamp failure mechanisms, etc. It is only natural that from very early on lamp research and development employed optical nondestructive, nonintrusive techniques such as emission and absorption spectroscopy. Over the years additional optical techniques such as fluorescence, interferometry, and Hook's method, which measures species densities by relying on the large index of refraction change in the vicinity of an absorption line, have been added to the arsenal of light source detectives. In addition to these optical nondestructive, nonintrusive techniques which yield very valuable information about the miscoscopic processes going on inside the electrical discharge, there are a number of plasma techniques which view the electrical discharge as a weakly ionized plasma rather than a photon generating atomic and molecular medium. Among these, probably the most useful and widely used technique is the Langmuir probe and its various improvements. Probes are simply small ($\sim$ a few mm diameter and few mm long) wires which are inserted into the plasma and collect a current from the plasma without substantially disturbing its characteristics. As we shall see below in somewhat greater detail, probe techniques yield valuable information on plasma parameters such as EEDF, plasma density N_e, mean electron energy u, and electron field E in weakly ionized LPD's.

More recently, the wide spread availability of tunable dye lasers has opened up a whole new area of diagnostic research in light sources. The very narrow bandwidth, (~ 0.5 MHz), wide-range tunability (~ 200-900 nm), very low divergence, long coherence length, and high power density of the lasers have opened up many possibilities to increase the accuracy and extend the range of information that can be obtained from operational light sources.

In the last decade or so the availability of both visible and UV grade optical fibers has enabled researchers to come up with several additional diagnostics to supplement the techniques mentioned above.

In the following we shall concentrate primarily on laser, probe and fiber-optic techniques reported in the literature in the last decade or so and some of the information relevant to LPD obtained through them. In as much as the traditional diagnostic methods of emission and absorption spectroscopy are widely used in LPD research, we shall not review them here since they have already been reviewed in an excellent article by van den Hook (1983).

A. Laser Diagnostics

A few years ago van den Hook (1983) summarized some of the diagnostic techniques used in lamp research in general. While the article primarily concentrated on high pressure discharges, some of the laser diagnostics techniques mentioned in the article are applicable to LPD laser diagnostics. More recently van de Weijer and Cremers (1987) summarized their laser diagnostics work in LPD over the prior several years.

Van de Weijer and Cremers (1984) used tunable pulsed dye lasers to determine the effective lifetimes of excited states. (As mentioned earlier, due to repeated absorptions and emissions in a LPD, the natural lifetime of an excited state is elongated. This elongated lifetime is of importance since it determines the radiation output of the low pressure discharge and it is called the "effective lifetime"). Van de Weijer and Cremers accomplished this by pumping a state higher than the state whose effective lifetime they wanted to measure and which was optically linked to it, thereby overpopulating it. By monitoring the time dependent fluorescence signal from the state of interest, these researchers determined the effective lifetimes of Hg 6 $^3P_1$ and Hg 6 $^1P_1$ both in pure Hg as well as in Hg-Ar and Hg-Kr discharges as a function of K_0R. Here R is the tube radius and K_0 is the absorption coefficient for the 253.7-nm radiation, basically a measure of Hg pressure. Their results for Hg, Hg-Ar, and Hg-Kr are shown in Fig. 13. In the same figure, the calculational results for effective lifetimes are also included. Their experimental results are in good agreement with earlier determinations of effective lifetimes as well as with calculations, which gives further legitimacy to the use of lasers in this type of measurements. Furthermore, since their excitation profile was substantially different than the excitation profile of other researchers, they support the expectation that effective lifetimes for the fundamental decay mode are independent of excitation profiles. Since their experimental results for Hg-Ar and Hg-Kr are almost identical, within experimental error of about 10%, they conclude that the line broadening by Kr and Ar for the Hg 6 $^3P_1$–Hg 6 $^1S_0$ transition is the same.

The measurement of the effective radiation lifetime of the Hg 6 $^1P_1$ state (natural lifetime 3 ns) and its comparison to calculation have pointed to an important conclusion. While the assumption of complete spectral redistribution is valid for Hg 6 $^3P_1$, (natural lifetime 120 ns) it is not valid for Hg 6 $^1P_1$. The reason apparently is that the natural lifetime of the 6 $^3P_1$, state is long enough for collisions to take place and cause the state to emit a photon any place within the doppler profile, while this is not the case for 6 $^1P_1$. The 6 $^1P_1$ state natural lifetime is short enough not to be perturbed by collisions during its natural lifetime, therefore, photons are not emitted at the wings of the doppler profile and the effective lifetime is longer than the one calculated with the assumption of complete spectral redistribution. (Note that photons escape faster at the wings of the profile because K_0R is smaller.)

Van de Weijer and Cremers (1983) and (1987) used a pulsed dye laser as a source for the Hook's method to determine the density of the Hg 6 ^3P and

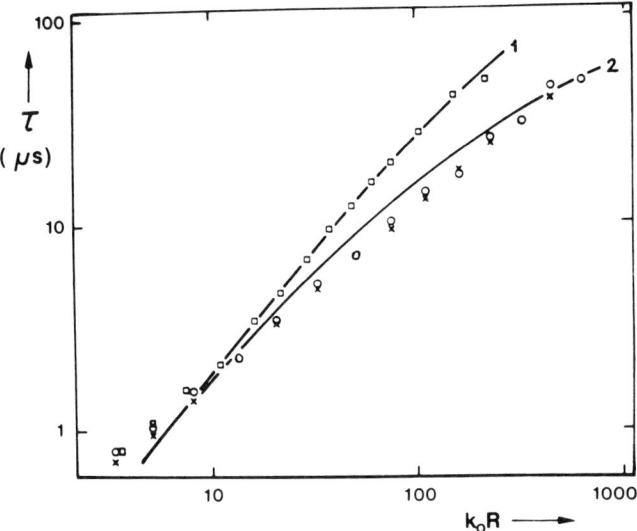

FIG. 13. The effective radiation lifetime (τ) of the $6\ ^3P_1$ level of mercury atoms in a mercury discharge (squares □) a mercury-argon discharge (circles ○) and a mercury-krypton discharge (crosses ×) as a function of k_0R. The pressure of the inert gas is 400 Pa in both cases and the diameter of the tube is 36 mm. The continuous lines relate to the lifetime calculated from Walsh (1959) for a mercury discharge (1) and for a mercury-argon discharge (2). (van de Weijer and Cremers, 1987.)

Hg $6\ ^1P_1$ states. While their results agreed well with previous absorption measurements for these states by Koedam et al. (1963), they disagreed significantly (especially $6\ ^3P_0$ and $6\ ^3P_2$ but not $6\ ^3P_1$) with previous non-laser Hook method results by Uvarov and Fabrikant (1965). Once they knew the density of excited states (N) and the effective lifetimes (τ) Van de Weijer and Cremers (1987) were able to calculate the UV radiation output (N/τ) of the Hg and Hg-Ar discharges as a function of Hg pressure or coldspot temperature and compared it to previous direct measurements of UV output. The agreement of the two different approaches for UV output is good for pure Hg discharge but not so good for Hg-Ar discharge. The reasons for this are not entirely clear, however, one may speculate that perhaps to calculate UV output by N/τ, which may be correct for a pulsed discharge, may be too simplistic an approach for steady state UV output calculation. Nevertheless, one, on the whole, can conclude that the laser diagnostic measurements are fairly relevant and augment our understanding of the microscopic processes inside the LPD.

Recently, Moskowitz (1987) has invented an elegant technique for measuring species concentrations point by point which may have far reaching applications. As is well known, in order to obtain spatial distribution of excited states, one can use laser absorption spectroscopy (LAS) whereby a certain absorption across the discharge tube at various spatial positions is measured and then Abel inverted in order to obtain the density at each point. Unfortunately, Abel inversion tends to propagate and amplify experimental errors and uncertainties in such a manner as to yield large error bars at any given spatial point. The modulated laser absorption (MLA) technique invented by Moskowitz eliminates the necessity of Abel inverting by obtaining the density at each point via two intersecting laser beams originating from the same laser. The laser beam originating from a ring tunable dye laser at the wavelength of interest is split into a strong "pump" and weak "probe" beam. This scheme is schematically shown in Fig. 14. The idea is to monitor the weak "probe" beam after it traverses the discharge and chop the pump beam. When the pump beam is *off*, the probe beam traverses the entire discharge, while when the pump beam is *on* absorption of the probe beam is everywhere in the discharge path except at the crossing point of the two beams. This is because the transition is already saturated by the pump beam. Therefore, by

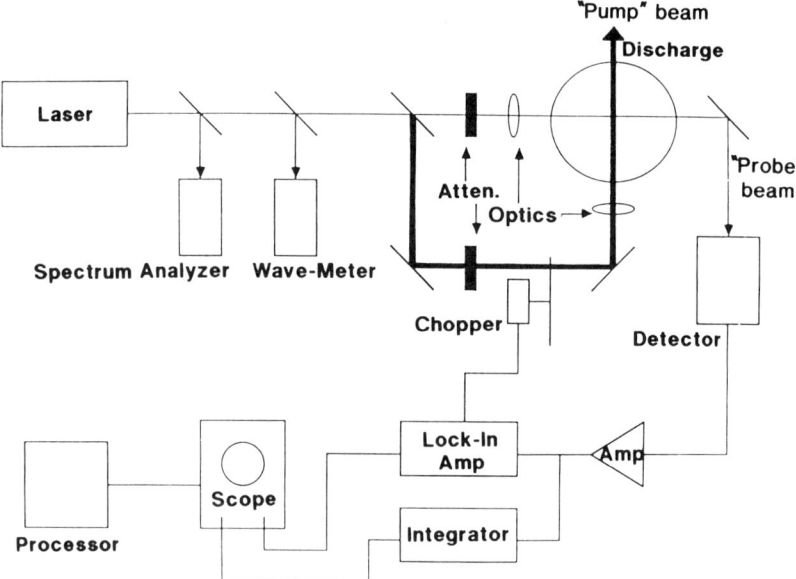

FIG. 14. Experimental setup for the modulated laser absorption (MLA) technique, indicating the pump and probe beam. Typically the discharge vessel is moved in the direction of the pump beam, while both beams remain fixed to obtain the pinpoint data (Moskowitz, 1987.)

lock-in techniques the information contained at the crossing point can be retrieved. By moving the point of intersection of the two beams one obtains the relative distribution of the state of interest. This in turn can be integrated and normalized against a total cord integrated absorption measurement to obtain an absolute spatial distribution of density for the state of interest without resorting to any inversion procedure such as Abel. Clearly, the technique could be expanded to measure other parameters in LPD's as well as to high pressure discharges.

In another laser diagnostic experiment the radial distribution of Hg 6 ^{3}P states were measured by Bigio (1987), again with a technique that avoids Abel or some other type of mathematical inversion. Here the laser was shone in the axial direction of a discharge tube where the electrodes were removed to the side. By utilizing saturated laser spectroscopy Bigio was able to obtain the radial distribution of the Hg 6 ^{3}P manifold for various discharge conditions. These data were later incorporated into a discharge model by Dakin and Bigio (1988). They conclude that the tail electrons having energy above the threshold for Hg inelastic scattering near the walls of the discharge have a temperature lower than in the center of the discharge.

Recently Bhattacharya and Awadallah (1988) reported on laser induced fluorescence (LIF) measurements of Ba (I) in the vicinity of the triple oxide thermionic cathode normally found in fluorescent lamps. In these experiments the authors used the strength of the LIF signal as the measure of evaporation of Ba from the cathode which ultimately controls the life of the lamp. From their measurements they concluded that Ba loss mainly takes place by thermal evaporation during the anode half cycle due to hot spot formation resulting from electron bombardment. During the cathode half cycle the hot spot runs somewhat cooler (20–50 K) due to thermionic emission and thermal evaporation is less. Whatever Ba is removed by sputtering during the cathode half cycle due to ion bombardment presumably could be ionized and attracted back to the cathode.

Another powerful technique introduced to the field of LPD diagnostics is laser optogalvanic (LOG) or optogalvanic effect (OGE). As is well known, this effect is the observation of a change in the electrical property of a gas discharge (usually external to the discharge vessel) when a species participating in the ionization mechanism of the discharge is illuminated with a wavelength corresponding to a transition of that species. Loosely speaking, the alteration of the ionization balance typically leads to a voltage or current spike in the external circuit which can be easily detected. While LOG signals are fairly easy to obtain experimentally, their interpretation and understanding their origin is far more complicated. The first observation of LOG in a gas discharge was reported by Green et al. (1976) in a hollow-cathode Ne lamp. Later on the technique had been used by Lawler and co-workers (e.g., Den

362   *J. Maya and R. Lagushenko*

Hartog et al., 1988, and references therein) to obtain a wealth of information in the cathode region of electrical discharges. In a series of papers, the authors reported spatial variations of electric field, absolute gas temperature and species concentrations using optogalvanic diagnostics. Since most of these results are in cold-cathode rare-gas discharges, we shall not review them here in any detail. The interested reader is referred to the above reference for further information.

More recently van de Weijer and Cremers (1985) applied the LOG technique to low pressure mercury discharge in an attempt to further understand the mechanisms of excited state production. These researchers conducted a series of experiments where they excited the $6\ ^3P$ manifold to $7\ ^3S_1$, as well as the $6\ ^1P$ state to a number of higher levels with a pulsed dye laser (Fig. 15). Subsequently, they watched the optogalvanic signal (electrical), the fluorescence signal and a long term fluorescence signal which is of the same time scale as the LOG signal. The latter signal which resulted primarily

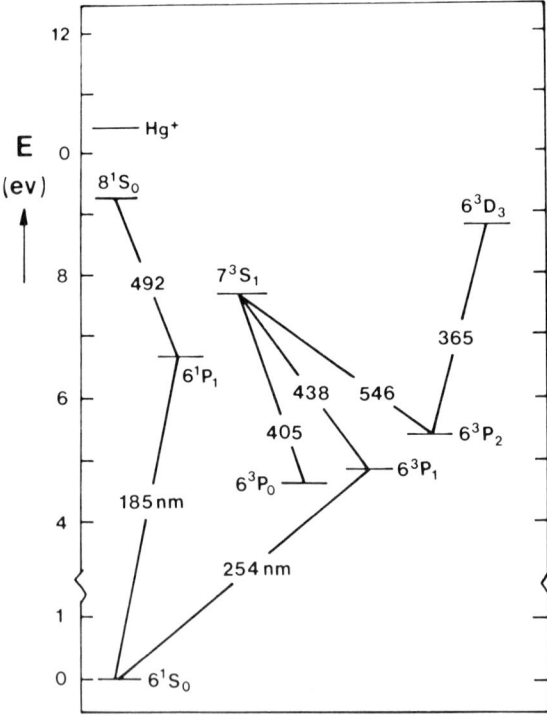

FIG. 15.   Energy level diagram of Hg showing various excited states relevant to LOG (van de Weijer, 1986; © 1986 IEEE.)

from the rearrangement of the excited states was called the indirect fluorescence. When low pressure Hg is irradiated with a pulse of 405 nm the resulting LOG and fluorescence signals are shown in Fig. 16. The authors argue that the sign of the LOG signal depends on the ballast resistance used in the circuit. Their argument goes as follows [van de Weijer (1986)].

"The negative OGE induced by the 405-nm pulse corresponds to an increase in electron density and a decrease in electron temperature. The increase in electron density results in an increase in excitation rate from the ground state to the $6\,^3P_1$ level, whereas the decrease in electron temperature results in a decrease of this excitation rate. The ratio of the relative change in

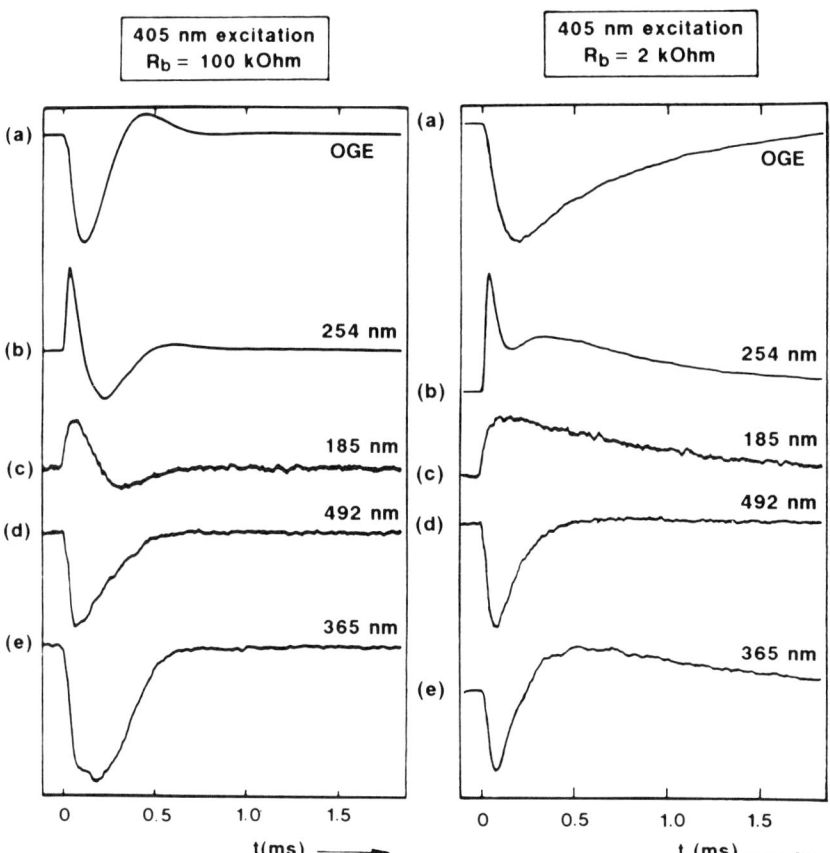

FIG. 16. The LOG and fluorescence signals on four Hg transitions induced by optical pumping on the 405-nm line. Figure on left is for R_b = ballast resistance = 100 kΩ while the figure on the right is for R_b = 2 kΩ (van de Weijer, 1986; © 1986 IEEE.)

discharge voltage V, which is a measure of the electron temperature, to the relative change in electron density n depends on the value of the ballast resistance R_b in the discharge circuit

$$\frac{\Delta V/V}{\Delta n/n} (:)(1/R_b + 1/R_d)$$

where R_d is the dynamic resistance of the discharge. As a consequence, the 254 nm long-term fluorescence is positive for low values of the ballast resistance, due to the increase in electron density, and negative for high values due to the decrease in electron temperature."

As can be seen by exciting different levels and monitoring the LOG signal, as well as the fast and slow parts of the fluorescence of various levels, it is possible to draw qualitative conclusions about the population mechanisms and their relative importance and interrelationships. These clearly are preliminary results and undoubtedly more work needs to be done to further unravel the microscopic information staring at us in this ubiquitous light source.

Before closing this section, we should mention that we left out many fine techniques and experiments used in other type of discharges and which are applicable to LPD. It is only a matter of time for this to happen. For example, the utilization of Stark spectroscopy (e.g., Garscadden, 1986) to measure electron densities and the applications of resonance ionization spectroscopy (RIS) to measure very low concentrations (Morgan, 1986, and references therein) are along these lines. We feel the whole field of laser diagnostics in LPD research is only in its infancy and much more useful and detailed information is likely to be generated by greater utilization of these techniques.

B. Probes and Optical Fibers

It is hardly possible to give a review on low pressure mercury–rare-gas plasmas without touching, at least briefly, on Langmuir type probes. After the pioneering works of Langmuir (1926) and Druyvestein (1930), probes became highly versatile tools especially for weakly ionized, plasma research. Over the years, however, the techniques had to be improved in order to obtain reliable data. The general field of electrostatic probes has recently been reviewed by Cherrington (1980) and Godyak (1988). For an overview of the utilization of probes in electrical discharges the reader is referred to these articles and the references therein. As far as the use of probes in LPD, Verweij (1961) has laid out many of the fundamentals in his by now almost classical treatise. For the fundamental properties and theory of operation of probes the reader is referred to this work. Probes have been used by various researchers over the last ten years or so in one form or another to primarily obtain EEDF's. We should stress that probe techniques work best in low pressure plasmas (as opposed to high pressure or magnetized plasmas) and still is the most widely used technique to obtain EEDF's, plasma potentials, mean energies and other microscopic plasma parameters. One has to be extremely careful, however, in interpreting the probe measurements.

Two of the most severe problems with probe measurements in LPD's have been probe contamination by Hg, and, therefore, change in the work function and plasma noise which leads to erroneous results. The problem of probe contamination was successfully addressed by Waymouth (1959); by pulsing the probe Waymouth was able to reduce the contamination problem to manageable levels.

The second problem of noise had been addressed by Kagan et al. (1971) using a second reference probe. More recently Godyak et al (1988) utilized the technique in a satisfactory manner for Hg–rare-gas discharges. [For an extensive discussion of noise associated problems in probe measurements see Godyak (1988) and references therein.] Godyak et al. (1988) used a second probe in the vicinity of the main probe which was responsive essentially to the noise in the Hg–rare-gas discharge. By using a noise suppression circuit and referencing the signal of the first probe to the second one he was able to obtain EEDF's with more than three orders of magnitude dynamic range. An example of the kind of information obtained with this improved technique is shown in Fig. 17. As can be seen at high currents the EEDF is essentially Maxwellian as determined by a single slope or single temperature to characterize the distribution. However, at lower currents the depletion of high energy electrons is fairly clear and, therefore the EEDF tends to drift away from a Maxwellian distribution. The fact that there are two slopes for say $I_d = 0.05$ clearly demonstrates that one cannot assign a single temperature to the distribution. Details of the electronic circuitry to enable one to make these kinds of measurements are given in Godyak (1988). The same technique was used in the investigation of the constriction phenomenon discussed above. It would be fair to say that probes are still very useful tools in obtaining microscopic plasma parameters in LPD's provided one is aware of the pitfalls and is careful to avoid them.

Low loss optical fibers, originally developed for optical telecommunications, introduced an additional tool to the repertoire of the electrical discharge researchers. The availability of UV, visible, and IR grade fibers makes the optical fiber a versatile medium for radiation transfer as well as for diagnosing plasmas that might otherwise be difficult due to for example electromagnetic interference or other interfering noise. Optical fibers offer a

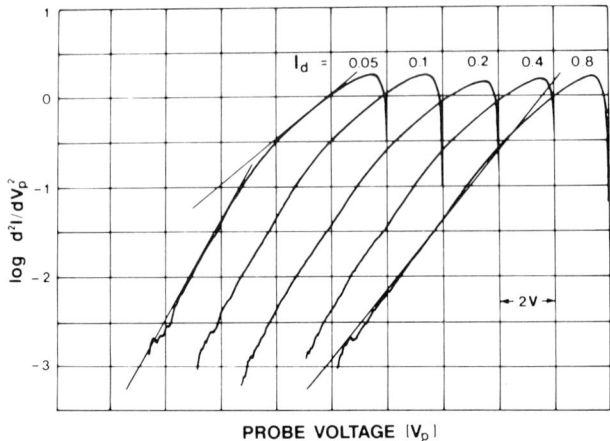

FIG. 17. Second derivative of probe current with respect to probe voltage versus probe voltage for Hg-Ar discharge. Data obtained by Godyak (unpublished) using the RF-modulated double probe technique described in Godyak et al. (1988). Note the two different slopes for I_d = discharge current = 0.05 A while a single slope can be drawn for I_d = 0.8 A (therefore Maxwellian distribution).

number of advantages for measurements. In addition to immunity from electromagnetic interference, they are insulating, safe as a passive medium, and they are low in weight, small, flexible, inexpensive, and easy to install. Furthermore, they provide a one-dimensional measurement medium allowing one to integrate or differentiate along a given path. We should point out that the use of optical fibers in LPD is in its infancy. While there have been many articles in the literature (e.g., Rogers, 1986, and references therein) on very inventive utilization of optical fibers for scientific and technological measurements, the applications in LPD during the last decade or so are only a handful.

For example, Ingold and Roberts (1984) utilized optical fibers to transport radiation from a fluorescent lamp inside an NMR device subjected to a uniform axial magnetic field. Recently Rogoff (1985a, b) suggested the use of an optical fiber probe for insertion into a plasma to obtain three dimensional spatial information, such as emission and absorption coefficients. The technique can be applied to plasmas of arbitrary shape and optical thickness.

Grossman et al. (1986) recently reported on a new application of optical fibers coupled with calculations to obtain radial distributions of different Hg isotopes in a LPD. Their investigations of Hg isotope separation involved the coupling of radiation from an excitation source to a Fabry-Perot interferometer using quartz fiber optics. Initial observations of the hfs led eventually to a measurement technique for radial distribution of excited states. The methodology used was as follow:

(a) Measure hfs at several radial points.
(b) Using modified Holstein-Biberman equations calculate the hfs for the same points.
(c) Determine radial distribution of excited states from the best fit of measured and calculated hfs.

The modified Holstein-Biberman equation was solved numerically so as to obtain the radial excited state distribution of each isotope. The authors calculated the effect of $K_0 R$ on the radial distribution of a single isotope. The difference in the distribution starts becoming appreciable after a reduced radius of about 0.7. Then they calculated the radial distribution of all the isotopes for natural Hg mixture. This is shown in Fig. 18. As can be seen, the only isotope that shows any substantial difference from the others is ^{196}Hg, which is consistent with the $K_0 R$ calculations. Further calculations of the distribution with 10% enriched ^{196}Hg showed that the difference in radial distribution between ^{196}Hg, and the other isotopes at this concentration disappears.

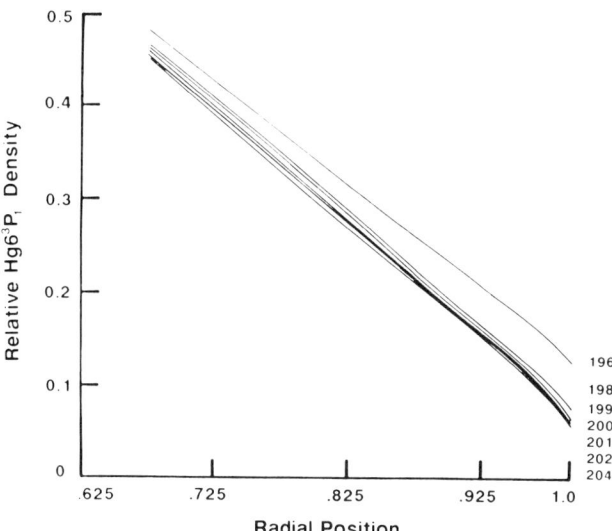

FIG. 18. Calculated relative Hg 6 $^3P_1$ density as a function of reduced radius for different isotopes using the modified Holstein-Biberman equation. Note the slower decline of ^{196}Hg as one approaches the wall.

In order to verify the expectations and see how well the coupling of the radiation transfer with hfs measurements via fiber-optic–Fabry-Perot works, Grossman et al. (1986) ran some experiments with an artificial Hg mixture that would be simpler to analyze than the natural Hg mixture. They measured the hfs structure of 87.8% ^{202}Hg and 10.5% ^{201}Hg mixture at several radial distances and noted the differences. Then they calculated the expected hfs for different excited state distributions and tried to obtain a best match to the measured hfs. This is shown in Fig. 19. As can be seen from the figure, they found that the minority isotope declines substantially less rapidly than the majority component, as expected. One can perhaps say that it is possible to obtain isotope specific radial excited state distributions in a low pressure electrical discharge, by combining optical measurements *external* to the discharge with a numerical solution to the modified Holstein-Biberman radiation transfer equation. One important conclusion of these studies is the demonstration that there is a substantial difference in the excited state radial distribution between a low concentration and high concentration isotope in a Hg-Ar discharge.

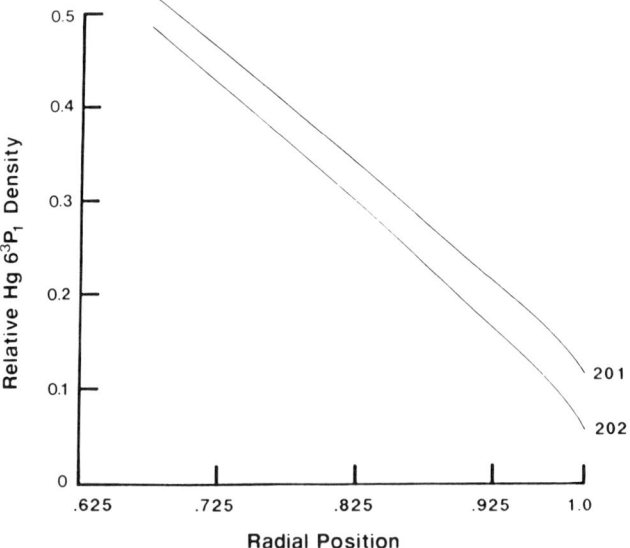

FIG. 19. Measured radial density of ^{201}Hg 6 $^3P_1$ and ^{202}Hg 6 $^3P_1$ states in a Hg-Ar discharge containing 87.8% ^{202}Hg and 10.5% ^{201}Hg. Here $P_{Ar} = 3$ torr, $I = 0.425$ A, tube diameter is 38 mm. (Grossman et al., 1986.)

V. Summary

In this chapter we have attempted to review the progress made in the last decade or so on low pressure mercury–rare-gas discharge research. As is well known, this discharge forms the basis of the fluorescent lamp and therefore much of the research activity takes place in industrial laboratories interested in this particular application. We have attempted to give a flavor of the relationships behind much of the theoretical and experimental activity in this area. Energy conservation and the desire to replace the incandescent lamp with a compact electrical discharge are fundamental underlying forces in this research. The three broad areas we covered in this review are (a) efforts on modeling the fairly complex discharge, (b) attempts to alter the discharge so as to make it either more efficient or increase the radiation density (i.e., more compact), and (c) deployment of increasingly sophisticated diagnostics to increase the accuracy of our understanding of the atomic, molecular, and plasma processes going on inside the discharge.

Broadly speaking one can say that considerable progress has been achieved in the modeling efforts during the last decade or so. The recognition of the non-Maxwellian electron energy distribution function, the Hg $6\,^3P$–$6\,^3P$ collisional ionization process and improvements in modeling techniques are areas that come immediately to mind. The primary limitations to better agreement between experimental and calculational results (which now stands at 5%–30% depending on the regime) probably lie in the lack of accurate atomic and molecular data and a far more accurate electron energy distribution function. Among the challenges that still lie ahead in modelling one can cite the simultaneous solution of spatially dependent population balance equations and radiation transport, a better treatment of fast electrons that might be escaping to the walls, incorporation of radially dependent electric field and radial dependence of electron energy distribution function in the Boltzman equation, etc.

In the area of altered discharges we mentioned efforts in isotopic alteration, magnetic fields, both of which increase the electrical to radiation conversion efficiency, recombination structures which increase the radiation density, constrictions which result in a higher proportion of high energy electrons, striations which have potential for an altered discharge and finally high frequency and surface wave operation. This list is by no means exhaustive, but it covers the bulk of the activities along the line of altering the positive column so as to accommodate the two fundamental forces mentioned above. These efforts have been primarily experimental and a better theoretical understanding of what goes on inside these "altered positive column" discharges would certainly be desirable. We expect these types of efforts to continue, especially since the forces are still in effect and as yet none of the attempts mentioned above have resulted in a product that has a dominant position in the marketplace.

In diagnostics the major activity during the last decade has been in the area of laser diagnostics. Techniques such as laser absorption spectroscopy (LAS), modulated laser absorption (MLA) which give pinpoint information directly, and application of lasers to obtain effective lifetimes, population densities and various cathode parameters have been reviewed. Finally, improvements in the classical Langmuir probe techniques and a couple of applications of the fiber optic diagnostics have been briefly touched upon.

REFERENCES

Anderson, J. B., Maya, J., Grossman, M. W., Lagushenko, R., and Waymouth, J. F. (1985). *Phys. Rev. A* **31**, 2968.
Atajew, A., Rutscher, A., and Winkler, R. (1972). *Beitr. Plasma Phys.* **12**, 239.
Beneking, C., and Anderer, P. (1988). *Bull. Am. Phys. Soc.* **33**, 134.
Bhattacharya, A. K., and Awadallah, A. (1988). *Bull. Am. Phys. Soc.* **33**, 138.
Biberman, L. M. (1947). *J. Exper. Theor. Phys* **17**, 416.
Bigio, L. (1987). *J. Appl. Phys.* **63**, 5259.
Campbell, J. H. (1960). *Illum. Eng.* **55**, 247.
Cayless, M. A. (1962). *J. Phys. D* **14**, 863.
Chalmers, A. G., and Pillow, M. E. (1971). *J. Phys. B* **4**, 1587.
Cherrington, B. E. (1982), *Plasma Chemistry and Plasma Process* **2**, No. 2, 113.
Dakin, J. T. (1986). *J. Appl. Phys.* **60**, 563.
Dakin, J. T., and Bigio, L. (1988). *J. Appl. Phys.* **63**, 5270.
Davydov, B. I. (1937). *Zh. Eksp. Teor. Fiz.* **7**, 1064.
Den Hartog, E. A., Doughty, D. A., and Lawler, J. E. (1988). *Phys. Rev. A* **38**, 2471.
Dennemann, J. W., de Groot, J. J., Jack, A. G., and Lighthart, F. A. S. (1980). *J. Illum. Eng. Soc.* **10**, 2.
Dorleijn, J. W. F., and Jack, A. G. (1985). *J. Illum. Eng. Soc.* **15**, 75.
Drop, P. C., and Polman, J. (1972). *J. Phys. D* **5**, 562.
Druyvestein, M. J. (1930), *Zeitschrift. Phys.* **64**, 781.
Ellenbaas, W. (1972), "Light Sources," Crane Russak and Co., New York.
Ferreira, C. M. (1986), *In* "Radiative Processes in Discharge Plasmas" Vol. 149 of *NATO Advanced Study Institute Series B: Physics* (J. M. Proud and L. H. Luessen eds.), pp. 431–466. Plenum Press, New York.
Garscadden, A. (1986). *In* "Radiative Processes in Discharge Plasmas" Vol. 149 of *NATO Advanced Study Institute Series B: Physics* (J. M. Proud and L. H. Luessen eds.), pp. 547–568. Plenum Press, New York.
Ginzburg, V. L., and Gurevich, A. V. (1960), *Sov. Phys. Usp.* **3**, 115.
Godyak, V., (1988). *In* "Plasma Surface Interaction and Processing of Materials" *NATO Advanced Study Institute Series*, lecture delivered at Alicante, Spain, Sept. 4–16 (unpublished).

Godyak, V., Lagushenko, R., and Maya, J. (1988), *Phys. Rev. A* **38**, 2044.
Grabec, I., and Mikac, S. (1974), *Plasma Physics* **16**, 1155.
Green, R. B., Keller, R. A, Luther, G. G., Schenck, P. K., and Travis, J. C. (1976), *Appl. Phys. Lett.* **29**, 727.
Grossman, M. W., Johnson, S. G., and Maya, J. (1983), *J. Illum. Eng. Soc.* **13**, 89.
Grossman, M. W., Lagushenko, R., and Maya, J. (1986), *Phys. Rev. A* **34**, 4094.
Grossman, M. W., Lagushenko, R. L., and Maya, J (1986), *Bull. Am. Phys. Soc.* **31**, 152.
Hasker, J. (1976a), *Appl. Phys. Lett.* **28**, 586.
Hasker, J. (1976b), *J. Illum. Eng. Soc.* **6**, 29.
Hollister, D., and Berman, S. M., (1983). *Third International Symposium on the Science and Technology of Light Sources* contributed paper 7, Toulouse, France (unpublished).
Holstein, T. (1946), *Phys. Rev* **70**, 367.
Holstein, T. (1947), *Phys. Rev.* **72**, 1212.
Holstein, T. (1951), *Phys. Rev.* **83**, 1159.
Holstein, T., Alpert, D., and McCoubrey, A. O. (1952), *Phys. Rev.* **85**, 985.
Ingold, J. H. (1986), *In* "Radiative Processes in Discharge Plasmas" Vol. 149 of *NATO Advanced Study Institute Series B: Physics* (J. M. Proud and L. H. Luessen eds.), p. 27–37. Plenum Press, New York.
Ingold, J. H., and Roberts, V. D. (1984), *Illum. Eng. Soc. National Tech. Conference*, St. Louis, MO (unpublished).
Irons, F. E. (1979a), *J. Quant. Spectr. Rad. Trans.* **22**, 1.
Irons, F. E. (1979b), *J. Quant. Spectr. Rad. Trans.* **22**, 21.
Irons, F. E. (1979c), *J. Quant. Spectr. Rad. Trans.* **22**, 37.
Jack, A. G. (1986). *In* "Radiative Processes in Discharge Plasmas" Vol. 149 of *NATO Advanced Study Institute Series B: Physics* (J. M. Proud and L. H. Luessen eds.), pp. 309–326. Plenum Press, New York.
Jack, A. G., and Vrehen, Q. H. F. (1986), *Philips Tech. Rev.* **42**, 342.
Johnson, S. G., Work, D. E., Maya, J., and Waymouth, J. F. (1983), *Third International Symposium on the Science and Technology of Light Sources*, contributed paper 66, Toulouse, France (unpublished).
Kagan, Yu. M., and Lyagushchenko, R. I. (1961), *Sov. Phys. Tech. Phys.* **6**, 321.
Kagan, Yu. M., and Lyagushchenko, R. I. (1962), *Sov. Phys. Tech. Phys.* **7**, 134.
Kagan, Yu. M., and Lyagushchenko, R. I. (1964), *Sov. Phys. Tech. Phys.* **9**, 627.
Kagan, Yu. M., Kolokolov, N. B., Lyagushchenko, R. I., Milenin, V. M., and Mirzabekov, A. M. (1971) *Sov. Phys. Tech. Phys.* **16**, 561.
Kalyazin, Yu. F., Milenin, V. M., and Timofeev, N. A. (1981), *Sov. Phys. Tech. Phys.* **26**, 925.
Kenty, C. (1950), *J. Appl. Phys.* **21**, 1309.
Koedam, M. (1985). *Lighting Design Appl.* **15**, No. 1, 18.
Koedam, M., Kruithof, A. A., and Riemens, J. (1963), *Physica* **29**, 565.
Lagushenko, R., and Maya, J. (1984), *J. Illum. Eng. Soc.* **14**, 306.
Lagushenko, R., Grossman, M. W., and Maya, J. (1985). *Chem. Phys. Lett.* **120**, 21.
Langmuir, I. (1926), *Phys. Rev.* **28**, 727.
Lighthart, F. A. S. (1979), *Bull. Am. Phys. Soc.* **24**, 126.
Lighthart, F. A. S., and Keijser, R. A. J. (1980), *J. Appl. Phys.* **51**, 5295.
Lyagushchenko, R. I. (1972), *Sov. Phys. Tech. Phys.* **17**, 901.
Maya, J., Grossman, M. W., Lagushenko, R., and Waymouth, J. F. (1984), Science **226**, 435.
Moisan, M., Pantel, R., and Ricard, A. (1982), *Can. J. Phys.* **60**, 379.
Moisan, M., Pantel, R., Zakrzewski, Z., and Leprince, P. (1984). *IEEE Trans. Plasma Sci.* **PS-12**, 203.

Moisan, M. and Zakrzewski, Z. (1986). *In* "Radiative Processes in Discharge Plasmas" Vol. 149 of *NATO Advanced Study Institute Series B: Physics* (J. M. Proud and L. H. Luessen eds.) pp. 381–430. Plenum Press, New York.
Moisan, M. and Zakrzewski, Z. (1987), *Rev. Sci. Instrum.* **58**, 1895.
Moisan, M., Zakrzewski, Z., and Pantel, R. (1979). *J. Phys. D.* **12**, 219.
Morgan, C. G. (1986). *In* "Radiative Processes in Discharge Plasmas" Vol. 149 of *NATO Advanced Study Institute Series B: Physics* (J. M. Proud and L. H. Luessen eds.) pp. 569–575. Plenum Press, New York.
Morgan, W. L. and Vriens, L. (1980), *J. Appl. Phys.* **51**, 5300.
Moskowitz, P. E. (1987), *Appl. Phys. Lettr.* **50**, 891.
Moskowitz, P. E., Whitney, F., and Maya, J. (1987), *J. Illum, Eng. Soc.* **16**, 105
Nighan, W. L., and Wiegand, W. J. (1974), *Phys. Rev.* **10**, 922.
Pitchford, L. C. (1983). *In* "Electrical Breakdown and Discharges in Gases", *NATO Advanced Study Institute Series B: Physics* (E. E. Kunhardt and L. H. Luessen eds.) pp. 313–329. Plenum Press, New York.
Polman, J., van der Werf, J. E., and Drop, P. C. (1972). *J. Phys. D* **5**, 266.
Polman, J., van Tongeren, H., and Verbeek, T. G. (1975), *Philips Tech. Rev.* **35**, 321.
Post, H. A. (1986), *Phys. Rev. A*, **33**, 2003.
Privalov, V. E., and Fofanov, Y. A. (1978), *Sov. Tech. Phys. Lett.* **4**, 116.
Richardson, R. W., and Berman, S. M. (1983), *Third International Symposium, on The Science and Technology of Light Sources*, contributed paper 67, Toulouse, (unpublished).
Rockwood, S. J. (1973), *Phys. Rev. A* **8**, 2348.
Rogers, A. J. (1986), *J. Phys. D* **19**, 2237.
Rogoff, G. L. (1985a), *Appl. Optics* **24**, 1733.
Rogoff, G. L. (1985b), *Appl. Optics* **24**, 2755.
Rosenfeld, A. H., and Hafemeister, D. (1988), *Scientific American*, **258**, No. 4, 56.
Saelee, H. T. (1982), *J. Phys. D* **15**, 873.
Shkarofsky, I. P., Johnston, T. W., and Bachynski, M. P. (1966). *In* "Particle Kinetics of Plasmas," Addison-Wesley, New York.
Sun, R. K., and Berman, S. M. (1983). *Third International Symposium on The Science and Technology of Light Sources*, contributed paper, 8, Toulouse, France. (unpublished).
Uvarov, F. A., and Fabrikant, V. A. (1965), *Opt. Spectr.* **18**, 433.
van de Weijer, P. (1986), *IEEE Trans. Plasma Sci.* **PS-14**, 464.
van de Weijer, P., and Cremers, R. M. M. (1983), *Appl. Optics* **22**, 3500.
van de Weijer, P., and Cremers, R. M. M. (1984), *J. Appl. Phys.* **57**, 672.
van de Weijer, P., and Cremers, R. M. M. (1985), *Optics Commun.* **53**, 109.
van de Weijer, P., and Cremers, R. M. M. (1985), *J. Appl. Phys.* **57**, 672.
van de Weijer, P., and Cremers, R. M. M. (1987). *Philips Tech. Rev.* **43**, 62.
van den Heuvel, F. C. (1988). *Philips Tech. Rev.* **44**, 89.
van den Heuvel, F. C., and Vrehen, Q. H. F. (1985), *Phys. Fluids* **28**, 3034.
van den Hook, W. J. (1983), *Philips J. Res.* **38**, 188.
van Trigt, C. (1969), *Phys. Rev.* **181**, 97.
van Trigt, C. (1970), *Phys. Rev. A* **1**, 1298.
van Trigt, C. (1971), *Phys. Rev. A* **4**, 1303.
Verweij, W. (1961), *Philips Res. Rep. Suppl.* **2**, 1.
Vriens, L. (1973), *J. Appl. Phys.* **44**, 3980.
Vriens, L. (1974), *J. Appl. Phys.* **45**, 1191.
Vriens, L., and Lighthart, F. A. S. (1977), *Philips Res. Rep.* **32**, 1.
Vriens, L., Keijser, R. A. J., and Lighthart, F. A. S. (1978), *J. Appl. Phys.* **49**, 3807.
Walsh, P. J. (1959), *Phys. Rev.* **116**, 511.

Waters, R. T. (1983), *In* "Electrical Breakdown and Discharges in Gases" *NATO Advanced Study Institute Series B: Physics* (E. E. Kunhardt and L. H. Luessen eds.) pp. 203–267. Plenum Press, New York.
Waymouth, J. F. (1959), *J. Appl. Phys.* **30**, 1404.
Waymouth, J. F. (1966), *J. Appl Phys.* **37**, 4493.
Waymouth, J. F. (1971), "Electric Discharge Lamps, "M.I.T. Press, Cambridge, MA.
Waymouth, J. F. (1982) *In* "Applied Atomic Collision Physics" (H. S. W. Massey, E. W. Daniel, and B. Bederson eds.), Vol. 5, pp. 331–347. Academic Press, New York.
Waymouth, J. F. (1987), *Encyclopedia of Physical Science and Technology* **7**, 224.
Waymouth, J. F., and Bitter, F. (1956), *J. Appl. Phys.* **27**, 122.
Winkler, R. B., Wilhelm, J., and Winkler, R. (1983a), *Annal. der. Phys.* **7**, 90.
Winkler, R. B., Wilhelm, J., and Winkler, R. (1983b), *Annal. der. Phys.* **7**, 119.
Zhou, T. M. Wang, L. Z., Hollister, D. D., Berman, S. M., and Richardson, R. W. (1987), *J. Illum. Eng. Soc.* **16**, 176.

INDEX

A

Ab initio
 calculations, 244–245, 250, 260
  inclusion of core valence correlation effects and of pseudo-potentials in, 253–255
  molecular treatment, 255
  treatment of the excited states of the Alkali dimers, 252–253
Absorption
 coefficient, 333
 effects, e^+-Ar elastic scattering, 15–16
Abundance ratios, 307
 branching, 308
Adiabatic
 assumption, 235
 behavior, 184, 193
 Born–Oppenheimer potentials, 268
 potential curves, 217, 242–243, 257–258
 representation, 97
  electronic, 168
 to diabatic, transformation, 168–170
 vibrational basis, 169, 184
 vibronic energy levels, 195–196
Adiabatic and nonadiabatic neutralization, 150
Age of the galaxy, 308, 318
Age of the solar system, 303
Alchemy program, 252
Alignment and orientation parameters, 113
Alkali
 associative ionization, 212–213
 dimer, calculation of the potential curves for, 244–261
 dimer ion
  effective Hamiltonian, 246
  two electron effective Hamiltonian, 247
Alkalis, *ab initio* pseudopotential calculations, 253

Altered low-pressure discharge, 343–357
Analytical model representation of the MO expansion method, 106
Angular momentum
 $\hbar J$, 264
 product of Na_2^+, 224–226
Antiproton
 collisions with H, 119
 scattering, by He, ionization (double-to-single), 9–10
AO + method, 102
Approximate AO methods, 103
Associative
 channels, 284–285
 ionization (AI), 210–211, 242–244, 259, 275–276, 280–281, 283, 286, 288–289
  at ultracold temperatures, 235–240
  between
   ground-state and Rydberg atoms, 212, 217–220
   He* and He($1\ ^1S$), 220–222
   Na(3p) atoms, 214–217
  calculations of cross sections, 262
  collisions, 230
  cross sections, 212–220
  dynamics, 261–289
  polarization, dependence of, 213
  rate coefficients, 212–220
  trap experiments, 235–240
  in ultracold collisions, 232
Astration, 309
Asymmetric case
 C^{6+} collisions with H, 133
 Na^+ collisions with Ne, 124
Asymptotes, 260–261
Atom–atom
 channels, 280
 coupling, 236
Atom–field coupling, 236

INDEX

Atom–light-field interactions, 227
Atom-like resonance treatment, 270
Atomic
   autoionization, 269
   dipole movement, 255
   effects in beta decay, 297–319
   ionization channel, 277
   molecular angular momentum, 235
   orbital electron translation factor (AO-ETF), 94
   photoionization, 227
   states expansion method (AO method), 99
Atomic and molecular processes in discharges, 334, 336–337, 340
Autoionization, 286
Axial–axial
   approach, 228
   collisions, 229

B

Balance equation, for population density of atomic state, 324
Basis sets, 167–170
Beta decay, 298, 309–317
   bound-state, 298, 309–310
   end-point energy, 298
   exchange effect, 311–312, 317
   excited-state, 316–317
   first forbidden
      non-unique, 314, 316–317
      unique, 314
   half life of Re(187), 299
   of highly ionized Re(187) ion, 312–317
   of neutral Re(187) atom, 310–312
   overlap effect, 311
Body-fixed reference frame, 165–166
Boltzmann
   constant, 213
   equation, 325–327
Born approximation, 110
Born–Oppenheimer, 280
   quasimolecular states, 234–235
   representation, 277
   separation, 245
   state, 262
Bound-continuum coupling $V(\epsilon, R)$, 262
Boundary conditions and resonances, 281–288
Bulk electron temperature, 328–329

C

Cell experiments, 214
   associative ionization, 217
   crossed beam, 214
   single beam, 214
Channel wave functions, 269
Channel exchange, *see also* Dissociative charge exchange, 161–166
Charge transfer, 161, 163, 174, 178, 186, 191, 196
Circular polarization, 233, 239
Classical, *see* Trajectory surface hopping
   double scattering, 57–58
   formation, 266
   quasi-, 162, 175
   turning points, 173
Classical common, path, *see also* Trajectory, 163
Close
   collisions, 84
   coupling
      calculations, 163, 188, 192, 198
      equations, 170–171
Closure relation, 275
Cold atom collisions, 235–240, 290
Cold beam, 239
Collision
   energy, 210
   of neutral atoms (lowly charged ions) with an atom, 114
   ionization, 210–211
   mixing, between hyperfine components, 333
   velocity, 227
Collisions of a highly charged ion with an atom, 127
Comparison of He^+ + Si collision with Si atoms and Si_7 cluster, 148
Complex
   Green operator, 271
   potential method, 270–277
Configuration interaction
   technique, 251–252
   treatments, 254
Continuum
   distorted wave approximation, 111
   wave functions, 263
Conversion efficiency, electrical to radiation, 338–340
Coordinate systems, 164–165

Core–core interaction, 247
Core-polarization
  effects, 245–253, 257, 289
  potentials, 254–256
Coriolis coupling, 290
Cosmoradiogenic chronology, 297–308
Coulombic functions, 256
Counterpropagating laser, 228
Coupled equations, 87
Coupling terms
  Coriolis, 166, 169, 171, 173
  electron, 181, 188–189
  non-adiabatic, 169
  nuclear-electronic, 167, 169
  rovibronic, 170
  vibronic, 188–189, 193
Cross section, 167, 175, 179, 204, 212
  calculations, 276
  differential, 172, 177, 185, 189
  integral (or total), 178, 181, 192, 196
  vs. collision energy, 222
Cross-beam
  associative ionization, 217
  collisions, 234
  experiment, 235
  studies in Na, 219
Cross-polarization correction V_3, 249
Crossing
  avoided, 168, 193, 197–198
  pseudo-, 169
  seam, 185, 188–189, 197
  vibronic, level, 180, 186, 191, 193–196

D

de Broglie wavelength, 240, 264–265
Decreasing solenoidal magnetic field, 237
Demkov medel, 108
Density matrix theory, 223
Depletion of electrons, 329, 365
Detection
  matrix analysis, 233
  operator, 232
Diabatic, *see* Crossing
Diabatic curve-crossing, 265
Diabatic curves, 221, 243, 259–260
  representation, 97, 280
  representation and states, electronic, 168–170, 187, 196
  Rydberg crossings, 268

vibrational basis, 160–170
vibronic, interaction matrix, 192
Diagnostics, 356–365
  electrostatic probe, 364–365
  laser absorption spectroscopy, 360
  modulated laser absorption technique, 360
  optical fiber, 365–367
  optogalvanic effect, 361
Diagonal terms, 244, 279
Dielectric correction V_{diel}, 248
  for asymmetric case, 134
Differential cross sections, 111
Dipole
  coupling, 217
  moment matrix elements, 252
Direct channel, 219
Direct numerical solution of the time-dependent Schrödinger equation, 111
Discharge, 321
  altered low-pressure, 342–356
  low-pressure, 321
  mercury-rare-gas, 321
  modelling of, 323–342
  positive column of, 323, 336–342
Dispersion forces, 261, 290
Dissociative
  charge exchange, 182, 195–202
  processes, 175, 204
  recombination, 242, 269, 275, 281–282, 284
Distant collisions, 85
Distorted wave approximation, 110
Double minimum
  potential curves, 242
  structure, 243

E

e^+ scattering
  Ps formation
    by Ar, 12
    by H_2, 24–25
    by He, 7–8
    by Kr, 16–17
    by Ne, 10–11
  Schumann–Runge excitation of O_2, 30
$e^{+,-}$ scattering
  channels, 3

$e^{+,-}$ scattering (Continued)
  comparisons of cross sections
    by CH_4, 36–37
    by CO and N_2, 28–29
    by H, 21–24
    by H_2, 24–26
    by He, 7–9
    by Na and K, 19–21
    by Ne, 9–11
  differential elastic, by Ar, 13–16
  interactions, 2
  ionization
    by H_2, 25–26
    by He, 7–10
    by O_2, 30
  total
    by atoms
      Ar, 11–13, 42, 44
      H, 21–24, 26–27, 42
      He, 6–9, 42
      K, 17–21, 42
      Kr, 16–17, 42
      Na, 17–21, 42
      Ne, 9–11, 42
      Xe, 16–18, 42
    by molecules
      C_2H_4, 36–38, 43
      C_2H_6, 36–38, 43–44
      C_3H_6, 36–38, 43
      C_3H_8, 36–38, 43
      C_4H_8, 36–38, 43
      C_6H_6, 37–39, 43
      CF_4, 40, 43
      CH_4, 35–38, 43–44
      CO, 26–29, 34–35, 43
      cycloC_3H_6, 36–38, 43
      H_2, 24–27, 43
      H_2O, 31–32, 43
      isoC_4H_{10}, 36–38, 43
      N_2, 26–28, 34–35, 43
      N_2O, 32–35, 43
      nC_4H_{10}, 36–38, 43
      NH_3, 35–36, 43
      O_2, 29–30, 43
      SF_6, 40–41, 43
      SiH_4, 39–40, 43–44
Effective
  channel interaction matrix elements, 270
  interaction parameter, 221

nuclear wave functions, 287
reaction matrix, 287
Eigen phase shifts (eigen quantum defects), 280
Eikonal
  approximation, 65–72
  phase factors, 70–72
Eikonal–Glauber approximation, 110
Electron capture at relativistic energies, 51–77
  classical double scattering, 57–58
  continuum distorted wave approximation (MCDW), 64–65
  eikonal approximations, 65–72
  eikonal phase factors, 70–72
  experimental data and comparisons with theory, 74–77
  first-order Born approximation with Coulomb boundary conditions (R1B), 72–74
  first-order relativistic OBK approximation (ROBK1), 53–57
  numerical solution of coupled equations, 74
  second-order relativistic OBK approximation (ROBK2), 58–64
  symmetric eikonal approximation, 66–69
  Thomas peak, 57, 62–64
Electron energy distribution from Na(3p) associative ionization, 226
Electron energy distribution function (EEDF), 321
  altered, 353–356
  in low-pressure discharge, 326–332
Electron translation factors, 85
Electronic coupling, 211, 280
  reaction matrix, 281
Electronic width, 263
Elementary process, in mercury-rare-gas, low-pressure discharge, 336
Emission coefficient, 333
Energy analysis of the ionizing electron, 226
Energy levels of Re(187) and Os(187), 314
Examples of close collisions, 141
Examples of distant collisions, 143
Exchange
  channel, 218
  effects, 261

forces, 290
Excitation function, 227
   for circular polarization, 229–230
Exponential terms, 242

F

Fano-profile shapes, 283
Feshbach projection operators, 271
Fiber, 364
   optical, 365–367
   recombination, 355
First order electrostatic energies, 260
Fourier expansion, 232
Frame-transformation technique, 280
Franck–Condon (FC)
   approximations, 181–187
   factors, 185–186
   principle, 181, 184
Frozen core approximation, 252–253

G

Galilean invariance, 92
Gaussian orbitals, 253–255
Ground state potential curves of Na_2^+, 249

H

H+H
   associative detachment process, 284
   mutual neutralization process, 284
Half collision, 287
Hartree–Fock
   potential, 248
   wave function, 253
Helium
   ground-state, 266
   metastable, 266
Heteronuclear
   collisions, 217
   rate coefficient interbeam, 219
High frequency, operation of low-pressure discharge, 350–353
Homonuclear
   collisions, 217
   rate intrabeam, 219
   systems, 218
Hondo package, 253
Hyperfine
   coupling, 290–291
   structure of mercury line, 335

I

Imprisonment time, of radiation, 325, 334
Infinite Order Sudden (IOS) approximation
   quantal, 163, 170–174
   semiclassical, 180, 192, 198, 204
Integral cross sections, 112
Integro-differential equation, 271
Interelectronic distance, 242
Interferences, *see also* Quantum beats, 176, 186, 191, 201, 203
Internuclear distances, inner and outer regions, 240
Intrabeam collisions, 226
Ion ground-state configuration, 261
Ion–atom and atom–atom collisions, 114
Ion–molecule and atom–molecule collisions, 138
Ion-pair formation, 188–191
Ion-surface collisions, 146
Ionization, *see* Antiprotons; $e^{+,-}$; Proton scattering
   channels, 280
      rovibrational couplings, 291
      spin orbit, 291
   complete, 313
   energies for the sodium atom levels, 256
   in mercury-rare-gas
      low-pressure discharge, 336–337, 340
      Hg 6^3p–6^3P collisional, 337, 340
   partial, 317
Isochron, 301–302
Isotope effect, in mercury-rare-gas, low-pressure discharge, 343–346

J

Janov model of associative ionization, 218
JM theory, 217, 219
JWKB, *see* Semiclassical
JWKB approximation, 266

K

K matrix, 279, 281
   electronic, 284
Klapisch potential, 255

L

Lamps, 321
   fluorescent, 321–322

Landau–Zener
  expression, 221
  formula, 266–267
  transition, 288
Landau–Zener–Stuckelberg (LZS) model, 108
Laser
  excitation, 213
  polarization, 223
  propagation axis, 228
Laser-induced fluorescence, 365
Laser-initiated, alkali vapor plasma ignition, 214
Lifetime, atomic, 333–336, 345, 358–359
Light-field-atom reaction, 289
Linear polarization, 232–233, 235, 239
  effect on the collision function of velocity, 227–228
Linear threshold law, 288
Lippman–Schwinger equation, 281, 288
Local
  approximation, 275, 284
  electronic coupling, 211
  (vertical) ionization process, 266
Locking radius model, 229
Long-range forces, 260–261
Lorentz
  distribution, 239
  width, 238
Lorentzian, 234
Low-pressure discharge, 321–322
  electrical to radiation conversion efficiency of, 340
  electron energy distribution function in, 326–332, 353–356
  in magnetic field, 346–349
  radiation transport in, 332–336
LTE, 323

M

Magnetic fields, in discharges, 346–349
Maxwell–Boltzmann distribution, 223, 227
Maxwellian distribution, 328–330, 365
Milne theory of radiation diffusion, 217
Model
  potential calculations for the intermediate Rydberg states of Na_2^+, K_2^+ and of Na_2, 255–258
  potentials, 244–251, 289
Modelling, 323–342

  electron energy distribution function of, 326–332, 354
  positive column of mercury-rare-gas discharge, 336–342
  radiation transport of, 332–335
Molecular
  ansiotropy, 280
  autoionization, 268–269, 281–286
  collisions, nonreactive, 161
  multichannel quantum defect theory (MQDT), 220, 269, 276–288, 290–291
  orbital electron translation factor (MO-ETF), 96
  orbitals, 229
  potentials, 240–261
  quantization axis, 224
  quantum defect, 240, 256, 290
  quantum defects, diabatic potential curves, 258–260
  rotation, 280
  states, symmetries of, 223
  states expansion method (MO method), 92
Monomode dye laser, 227
Monte Carlo simulation, 322, 335
Multicenter AO method (MC-AO), 102
Multichannel
  (molecular) quantum defect theory (MQDT), 220, 269, 276–288, 290–291
  quantum defect approach to dissociative recombination, 226
Multichannel-Vainstein-Presnyakov-Sobel'man (M-VPS) method, 104
Multiconfiguration self-consistent field method, 252
Multipole expansion
  coefficients, 260
  of the detection operator, 232
Multistate curve-crossing model (MSCC), 220–221

N

Na_2/Na_2^+ potential curves, 224
Neutral-state
  coupling, 210
  nuclear wave functions, 273
Neutron capture, 305–306
Nienhuis G matrix theory, 234

Nonadiabatic coupling, 261
Nonequilibrium, of electrons and electric field, 353–355
Non-Franck–Condon
  distributions, 242
  transitions, 275
Non-local effects in molecular dynamics, 275–276
Non-negative rate coefficients, 224
Nuclear Shrödinger equations, 272

O

Observables, 111
Off-diagonal
  elements, 279
  matrix elements, 233
  terms, 244
On-diagonal
  matrix elements, 233
One- and one-half center AO expansion method, 103
One-center AO method (OC-AO), 99
Open ionization channels, 284
Optical
  fibers, 365–367
  molasses, 236–238
  pumping, 227
  velocity selection and polarization, 223–235
Optogalvanic effect, 361–364

P

Particle beam axes, 228
Particle de Broglie wavelengths, 236
Pauli exclusion principle, 249
Penning ionization (PI), 210–211, 265, 275
Perturbation
  expansion, 281
  theory, 249, 258
Perturbed Stationary State (PSS) method, 85
Perturbed Stationary States, 167
Photofragmentation analysis, 224–226
Photoionization, 215–217
  cross sections, 255
Polarization effect in associative ionization, 223–224
Potential curves for the H_2 system, molecular quantum defect, 240–244

Potentials for Rydberg molecular states, 240–261
Power broadening, 227
Predissociation, 286
Principal quantum number n, 217
Probes, 364
  electrostatic, 364–365
  Langmuir type, 364
Proton scattering, by He, ionization (double-to-single), 9–10
Ps formation, see e^+ scattering
Pseudopotential calculations, 255
Pseudopotentials, 244–251, 289
Pseudostates, 101

Q

Quadrupole interaction, 289
Quadrupole–quadrupole
  interaction, C_5/R^5, 239
  term, C_5/R^5, 232
Quantum
  beats, 188–191, 203
  mechanical treatments, 268–269
  resonance effects, 236
Quasidiabatic curves, 258–259
Quasimolecule, 240
Quasiresonant
  collisions, 217
  heteronuclear associative ionization, 219
Quenching, 324, 326
  metastable states, 324
  photon, 326

R

r process, 303–304, 309
  exponential synthesis model, 300
  production rate, 300, 303
R-matrix method, 105
R1B approximation, 72–74
Rabi cycles, 230
Radiation
  transport, 332–336
  trapping, 214–215, 227
Ramsauer–Townsend effects
  in $e^{+,-}$ scattering, 5–6, 42
  by CH_4, 35–37
Rate coefficients
  in cells, 212–214
  crossed beams, 212–214
  single beams, 212–214

Rate coefficients and cross sections for Na($3p^2P_{3/2}$) + Na($3p^2P_{3/2}$), 216
Reaction
  matrix elements, 280
  zone, 277
Recombination
  in mercury-rare-gas, low-pressure discharge, 337, 355
  process of, 337
  structure, 355
Reionization process, 152
Relationships of close-coupling to perturbation methods, 109
Relative collision energy, 221
Relativistic electron capture, see Electron capture at relativistic energies
Resonance
  fluorescence, 215
  frequency, 227
  lamp excitation, 213
  structure, 285–286
  theory of Fano, 270
Resonances
  Feshbach, in $e^{+,-}$ scattering, 45
  shape
    in e^- scattering, 43
    by C_2H_4, 37–38
    by CO_2 and N_2O, 32–35
    by N_2 and CO, 26–29, 34–35
    by SF_6, 40–41
Resonant
  energy transfer, 339, 344
  excitation pulse, 217
ROBK1 approximation, 53–57
ROBK2 approximation, 58–64
Rosenthal oscillations, 125
Rotational energy, 225
Rovibrational
  coupling, 211
  final states, 276
Rydberg
  atom collisions, 126
  electron, 217
  orbital, 261
  state diabatic crossings, 267
  states, 212–213
    doubly excited, 262
    singly excited, 262
  wave function, 258
Rydberg-valence mixing, 243

S

s process, 301, 304–305, 308
  branching, 305, 308
  local approximation, 306
Second-order dispersion energies, 260
Seed electron producing mechanisms, 214
Semiclassical, see also IOS, 163, 264–268, 283
  multistate curve crossing (MSCC) 266–267
  vibronic excitation, treatment of, 176–181, 192, 203
Short-range reaction matrix, 280–281
Single-channel phase shift, 279
Single-mode laser, 223
Slab-geometry radiation diffusion theory, 215
Slater orbitals, 251–252
Sodium dimers, 260
Space-fixed reference frame, 165–166, 171
Spectrometer resolution, 226
Spectroscopic constants, 254
  for the $^1\Sigma_g^+$ intermediate Rydberg states of Na_2^a, 258
Spin-orbit
  coupling, 224, 233, 235, 260
  effects, 260–261
Spin-selected excitation functions, 230
Stationary phase, 185
Stray magnetic fields, effect on polarization, 227
Striation, in low-pressure discharge, 346, 349
Sturmian functions, 101
Sudden, see IOS
  approximation
    for rotation, 178, 182
    for vibration, 180–181, 185
  centrifugal, 171–172
  collision conditions, 203
  electronic transition, 184
  energy, 171–172, 179–180
Sum rule in $e^{+,-}$ scattering by atoms, 45–46
Surface-wave excitation, of plasmas, 351–353
Symmetric
  and near symmetric cases (Li^+ + Li, Na^+ + Na, or Li^+ + Na), 121
  case

H$^+$(p) collision with H, 115
He^{2+} collision with He, 130
eikonal approximation, 66–69

T

Tail electron temperature, 328–329
Thermal
  beam polarization effect, 228
  Boltzmann distribution, 234
Thomas peak, 57, 62–64
Three-channel R-dependent reaction matrix K, 244
Threshold
  behavior, 288
  law for endoergic AI cross-section, 268
Time dependent, *see* Semiclassical: Vibrational wavepacket
  description, 163, 176
  Schrödinger equation, 176, 180
Time-of-flight (TOF) mass spectroscopy, 225–226
Total
  electronic Hamiltonian, 243
  wave function, 269
Trajectory surface hopping (TSH), 162, 175–176, 188, 198, 203
Trajectory, *see also* Classical; Semiclassical
  common classical, 176–178, 180, 184, 186, 200
Translation factor, 53
Two electron group model, *see also* 2-electron group model, 328–329
2-Electron Group Model (2-EGM), 328–330
Two-electron
  integrals, 257
  interaction, 257
  model potential method, 255
  wave function, 257
Two-center AO method (TC-AO), 100

U

Ultracold-temperature collisions, 236
Ultraviolet radiation, of mercury, 339–340
Unified AO-MO matching method, 104
Unitarized distorted wave approximation (UDWA), 104

Utrecht atomic physics group, 213, 224, 229, 232–235
UV photon
  flux, 227
  radiation density flux, 217

V

Valence bond treatment, 260
Valence-electron wave function, 246
Valence-type adiabatic Born–Oppenheimer potential curves, 267
Velocity
  and polarization dependence of the AI cross section, 220–235
  dependence of the cross section, 214
  distribution, 212–213
  selection and polarization in atomic beams, 226–235
Vibrational
  autoionization, 282–283
  eigenfunctions, 169
  energy, 225–226
  excitation, 169, 174–175, 196, 203
  interaction, 280
  level summation, 285
  reaction matrix, 280
  wave function, 275
  wavepacket, 186, 204
Vibronic amplitudes, 179
Vibronic, *see* Basis sets; Charge transfer; Coupling terms; Crossing; Quantum beats; Sudden
Vibronic
  complex energies, 200
  cross sections, 179
  Hamiltonian, 177, 179–180
  interaction matrix, diabatic, 192
  phenomena, 161, 185, 195, 200
  transitions, 162–163, 172, 175, 187

W

Wave functions, 256–257
  bound parts, 269–270
  continuum parts, 269–270
Weak coupling, 284
Weak-field condition, *see* Optical molasses

Contents of Previous Volumes

Volume 1

Molecular Orbital Theory of the Spin Properties of Conjugated Molecules, *G. G. Hall and A. T. Amos*

Electron Affinities of Atoms and Molecules, *B. L. Moiseiwitsch*

Atomic Rearrangement Collisions, *B. H. Bransden*

The Production of Rotational and Vibrational Transitions in Encounters between Molecules, *K. Takayanagi*

The Study of Intermolecular Potentials with Molecular Beams at Thermal Energies, *H. Pauly and J. P. Toennies*

High-Intensity and High-Energy Molecular Beams, *J. B. Anderson, R. P. Andres, and J. B. Fenn*

Volume 2

The Calculation of van der Waals Interactions, *A. Dalgarno and W. D. Davison*

Thermal Diffusion in Gases, *E. A. Mason, R. J. Munn, and Francis J. Smith*

Spectroscopy in the Vacuum Ultraviolet, *W. R. S. Garton*

The Measurement of the Photoionization Cross Sections of the Atomic Gases, *James A. R. Samson*

The Theory of Electron–Atom Collisions, *R. Peterkop and V. Veldre*

Experimental Studies of Excitation in Collisions between Atomic and Ionic Systems, *F. J. de Heer*

Mass Spectrometry of Free Radicals, *S. N. Foner*

Volume 3

The Quantal Calculation of Photoionization Cross Sections, *A. L. Stewart*

Radiofrequency Spectroscopy of Stored Ions I: Storage, *H. G. Dehmelt*

Optical Pumping Methods in Atomic Spectroscopy, *B. Budick*

Energy Transfer in Organic Molecular Crystals: A Survey of Experiments, *H. C. Wolf*

Atomic and Molecular Scattering from Solid Surfaces, *Robert E. Stickney*

Quantum Mechanics in Gas Crystal-Surface van der Waals Scattering, *E. Chanoch Beder*

Reactive Collisions between Gas and Surface Atoms, *Henry Wise and Bernard J. Wood*

Volume 4

H. S. W. Massey—A Sixtieth Birthday Tribute, *E. H. S. Burhop*

Electronic Eigenenergies of the Hydrogen Molecular Ion, *D. R. Bates and R. H. G. Reid*

Applications of Quantum Theory to the Viscosity of Dilute Gases, *R. A. Buckingham and E. Gal*

Positrons and Positronium in Gases, *P. A. Fraser*

CONTENTS OF PREVIOUS VOLUMES

Classical Theory of Atomic Scattering, *A. Burgess and I. C. Percival*

Born Expansions, *A. R. Holt and B. L. Moiseiwitsch*

Resonances in Electron Scattering by Atoms and Molecules, *P. G. Burke*

Relativistic Inner Shell Ionization, *C. B. O. Mohr*

Recent Measurements on Charge Transfer, *J. B. Hasted*

Measurements of Electron Excitation Functions, *D. W. O. Heddle and R. G. W. Keesing*

Some New Experimental Methods in Collision Physics, *R. F. Stebbings*

Atomic Collision Processes in Gaseous Nebulae, *M. J. Seaton*

Collisions in the Ionosphere, *A. Dalgarno*

The Direct Study of Ionization in Space, *R. L. F. Boyd*

Volume 5

Flowing Afterglow Measurements of Ion-Neutral Reactions, *E. E. Ferguson, F. C. Fehsenfeld, and A. L. Schmeltekopf*

Experiments with Merging Beams, *Roy H. Neynaber*

Radiofrequency Spectroscopy of Stored Ions II: Spectroscopy, *H. G. Dehmelt*

The Spectra of Molecular Solids, *O. Schnepp*

The Meaning of Collision Broadening of Spectral Lines: The Classical Oscillator Analog, *A. Ben-Reuven*

The Calculation of Atomic Transition Probabilities, *R. J. S. Crossley*

Tables of One- and Two-Particle Coefficients of Fractional Parentage for Configurations $s^\lambda s'^\mu p^q$, *C. D. H. Chisholm, A. Dalgarno, and F. R. Innes*

Relativistic Z-Dependent Corrections to Atomic Energy Levels, *Holly Thomis Doyle*

Volume 6

Dissociative Recombination, *J. N. Bardsley and M. A. Biondi*

Analysis of the Velocity Field in Plasmas from the Doppler Broadening of Spectral Emission Lines, *A. S. Kaufman*

The Rotational Excitation of Molecules by Slow Electrons, *Kazuo Takayanagi and Yukikazu Itikawa*

The Diffusion of Atoms and Molecules, *E. A. Mason and T. R. Marrero*

Theory and Application of Sturmian Functions, *Manuel Rotenberg*

Use of Classical Mechanics in the Treatment of Collisions between Massive Systems, *D. R. Bates and A. E. Kingston*

Volume 7

Physics of the Hydrogen Master, *C. Audoin, J. P. Schermann, and P. Grivet*

Molecular Wave Functions: Calculation and Use in Atomic and Molecular Processes, *J. C. Browne*

Localized Molecular Orbitals, *Harel Weinstein, Ruben Pauncz, and Maurice Cohen*

General Theory of Spin-Coupled Wave Functions for Atoms and Molecules, *J. Gerratt*

Diabatic States of Molecules—Quasi-Stationary Electronic States, *Thomas F. O'Malley*

Selection Rules within Atomic Shells, *B. R. Judd*

Green's Function Technique in Atomic and Molecular Physics, *Gy. Csanak, H. S. Taylor, and Robert Yaris*

A Review of Pseudo-Potentials with Emphasis on Their Application to Liquid Metals, *Nathan Wiser and A. J. Greenfield*

CONTENTS OF PREVIOUS VOLUMES

Volume 8

Interstellar Molecules: Their Formation and Destruction, D. McNally

Monte Carlo Trajectory Calculations of Atomic and Molecular Excitation in Thermal Systems, James C. Keck

Nonrelativistic Off-Shell Two-Body Coulomb Amplitudes, Joseph C. Y. Chen and Augustine C. Chen

Photoionization with Molecular Beams, R. B. Cairns, Halstead Harrison, and R. I. Schoen

The Auger Effect, E. H. S. Burhop and W. N. Asaad

Volume 9

Correlation in Excited States of Atoms, A. W. Weiss

The Calculation of Electron–Atom Excitation Cross Sections, M. R. H. Rudge

Collision-Induced Transitions between Rotational Levels, Takeshi Oka

The Differential Cross Section of Low-Energy Electron–Atom Collisions, D. Andrick

Molecular Beam Electric Resonance Spectroscopy, Jens C. Zorn and Thomas C. English

Atomic and Molecular Processes in the Martian Atmosphere, Michael B. McElroy

Volume 10

Relativistic Effects in the Many-Electron Atom, Lloyd Armstrong, Jr. and Serge Feneuille

The First Born Approximation, K. L. Bell and A. E. Kingston

Photoelectron Spectroscopy, W. C. Price

Dye Lasers in Atomic Spectroscopy, W. Lange, J. Luther, and A. Steudel

Recent Progress in the Classification of the Spectra of Highly Ionized Atoms, B. C. Fawcett

A Review of Jovian Ionospheric Chemistry, Wesley T. Huntress, Jr.

Volume 11

The Theory of Collisions between Charged Particles and Highly Excited Atoms, I. C. Percival and D. Richards

Electron Impact Excitation of Positive Ions, M. J. Seaton

The R-Matrix Theory of Atomic Process, P. G. Burke and W. D. Robb

Role of Energy in Reactive Molecular Scattering: An Information–Theoretic Approach, R. B. Bernstein and R. D. Levine

Inner Shell Ionization by Incident Nuclei, Johannes M. Hansteen

Stark Broadening, Hans R. Griem

Chemiluminescence in Gases, M. F. Golde and B. A. Thrush

Volume 12

Nonadiabatic Transitions between Ionic and Covalent States, R. K. Janev

Recent Progress in the Theory of Atomic Isotope Shift, J. Bauche and R.-J. Champeau

Topics on Multiphoton Processes in Atoms, P. Lambropoulos

Optical Pumping of Molecules, M. Broyer, G. Gouedard, J. C. Lehmann, and J. Vigué

Highly Ionized Ions, Ivan A. Sellin

Time-of-Flight Scattering Spectroscopy, Wilhelm Raith

Ion Chemistry in the D Region, George C. Reid

CONTENTS OF PREVIOUS VOLUMES

Volume 13

Atomic and Molecular Polarizabilities—A Review of Recent Advances, *Thomas M. Miller and Benjamin Bederson*

Study of Collisions by Laser Spectroscopy, *Paul R. Berman*

Collision Experiments with Laser-Excited Atoms in Crossed Beams, *I. V. Hertel and W. Stoll*

Scattering Studies of Rotational and Vibrational Excitation of Molecules, *Manfred Faubel and J. Peter Toennies*

Low-Energy Electron Scattering by Complex Atoms: Theory and Calculations, *R. K. Nesbet*

Microwave Transitions of Interstellar Atoms and Molecules, *W. B. Somerville*

Volume 14

Resonances in Electron Atom and Molecule Scattering, *D. E. Golden*

The Accurate Calculation of Atomic Properties by Numerical Methods, *Brian C. Webster, Michael J. Jamieson, and Ronald F. Stewart*

(e, 2e) Collisions, *Erich Weigold and Ian E. McCarthy*

Forbidden Transitions in One- and Two-Electron Atoms, *Richard Marrus and Peter J. Mohr*

Semiclassical Effects in Heavy-Particle Collisions, *M. S. Child*

Atomic Physics Tests of the Basic Concepts in Quantum Mechanics, *Francis M. Pipkin*

Quasi-Molecular Interference Effects in Ion–Atom Collisions, *S. V. Bobashev*

Rydberg Atoms, *S. A. Edelstein and T. F. Gallagher*

UV and X-Ray Spectroscopy in Astrophysics, *A. K. Dupree*

Volume 15

Negative Ions, *H. S. W. Massey*

Atomic Physics from Atmospheric and Astrophysical Studies, *A. Dalgarno*

Collisions of Highly Excited Atoms, *R. F. Stebbings*

Theoretical Aspects of Positron Collisions in Gases, *J. W. Humberston*

Experimental Aspects of Positron Collisions in Gases, *T. C. Griffith*

Reactive Scattering: Recent Advances in Theory and Experiment, *Richard B. Bernstein*

Ion–Atom Charge Transfer Collisions at Low Energies, *J. B. Hasted*

Aspects of Recombination, *D. R. Bates*

The Theory of Fast Heavy Particle Collisions, *B. H. Bransden*

Atomic Collision Processes in Controlled Thermonuclear Fusion Research, *H. B. Gilbody*

Inner-Shell Ionization, *E. H. S. Burhop*

Excitation of Atoms by Electron Impact, *D. W. O. Heddle*

Coherence and Correlation in Atomic Collisions, *H. Kleinpoppen*

Theory of Low Energy Electron–Molecule Collisions, *P. G. Burke*

Volume 16

Atomic Hartree–Fock Theory, *M. Cohen and R. P. McEachran*

Experiments and Model Calculations to Determine Interatomic Potentials, *R. Düren*

Sources of Polarized Electrons, *R. J. Celotta and D. T. Pierce*

Theory of Atomic Processes in Strong Resonant Electromagnetic Fields, *S. Swain*

Spectroscopy of Laser-Produced Plasmas, *M. H. Key and R. J. Hutcheon*

Relativistic Effects in Atomic Collisions Theory, *B. L. Moiseiwitsch*

Parity Nonconservation in Atoms: Status of Theory and Experiment, *E. N. Fortson and L. Wilets*

Volume 17

Collective Effects in Photoionization of Atoms, *M. Ya. Amusia*

Nonadiabatic Charge Transfer, *D. S. F. Crothers*

Atomic Rydberg States, *Serge Feneuille and Pierre Jacquinot*

Superfluorescence, *M. F. H. Schuurmans, Q. H. F. Vrehen, D. Polder, and H. M. Gibbs*

Applications of Resonance Ionization Spectroscopy in Atomic and Molecular Physics, *M. G. Payne, C. H. Chen, G. S. Hurst, and G. W. Foltz*

Inner-Shell Vacancy Production in Ion-Atom Collisions, *C. D. Lin and Patrick Richard*

Atomic Processes in the Sun, *P. L. Dufton and A. E. Kingston*

Volume 18

Theory of Electron–Atom Scattering in a Radiation Field, *Leonard Rosenberg*

Positron–Gas Scattering Experiments, *Talbert S. Stein and Walter E. Kauppila*

Nonresonant Multiphoton Ionization of Atoms, *J. Morellec, D. Normand, and G. Petite*

Classical and Semiclassical Methods in Inelastic Heavy-Particle Collisions, *A. S. Dickinson and D. Richards*

Recent Computational Developments in the Use of Complex Scaling in Resonance Phenomena, *B. R. Junker*

Direct Excitation in Atomic Collisions: Studies of Quasi-One-Electron Systems, *N. Andersen and S. E. Nielsen*

Model Potentials in Atomic Structure, *A. Hibbert*

Recent Developments in the Theory of Electron Scattering by Highly Polar Molecules, *D. W. Norcross and L. A. Collins*

Quantum Electrodynamic Effects in Few-Electron Atomic Systems, *G. W. F. Drake*

Volume 19

Electron Capture in Collisions of Hydrogen Atoms with Fully Stripped Ions, *B. H. Bransden and R. K. Janev*

Interactions of Simple Ion–Atom Systems, *J. T. Park*

High-Resolution Spectroscopy of Stored Ions, *D. J. Wineland, Wayne M. Itano, and R. S. Van Dyck, Jr.*

Spin-Dependent Phenomena in Inelastic Electron–Atom Collisions, *K. Blum and H. Kleinpoppen*

The Reduced Potential Curve Method for Diatomic Molecules and Its Applications, *F. Jenč*

The Vibrational Excitation of Molecules by Electron Impact, *D. G. Thompson*

Vibrational and Rotational Excitation in Molecular Collisions, *Manfred Faubel*

Spin Polarization of Atomic and Molecular Photoelectrons, *N. A. Cherepkov*

Volume 20

Ion–Ion Recombination in an Ambient Gas, *D. R. Bates*

Atomic Charges within Molecules, *G. G. Hall*

Experimental Studies on Cluster Ions, *T. D. Mark and A. W. Castleman, Jr.*

CONTENTS OF PREVIOUS VOLUMES

Nuclear Reaction Effects on Atomic Inner-Shell Ionization, W. E. Meyerhof and J.-F. Chemin

Numerical Calculations on Electron–Impact Ionization, Christopher Bottcher

Electron and Ion Mobilities, Gordon R. Freeman and David A. Armstrong

On the Problem of Extreme UV and X-Ray Lasers, I. I. Sobel'man and A. V. Vinogradov

Radiative Properties of Rydberg States in Resonant Cavities, S. Haroche and J. M. Raimond

Rydberg Atoms: High-Resolution Spectroscopy and Radiation Interaction—Rydberg Molecules, J. A. C. Gallas, G. Leuchs, H. Walther, and H. Figger

Volume 21

Subnatural Linewidths in Atomic Spectroscopy, Dennis P. O'Brien, Pierre Meystre, and Herbert Walther

Molecular Applications of Quantum Defect Theory, Chris H. Greene and Ch. Jungen

Theory of Dielectronic Recombination, Yukap Hahn

Recent Developments in Semiclassical Floquet Theories for Intense-Field Multiphoton Processes, Shih-I Chu

Scattering in Strong Magnetic Fields, M. R. C. McDowell and M. Zarcone

Pressure Ionization, Resonances, and the Continuity of Bound and Free States, R. M. More

Volume 22

Positronium—Its Formation and Interaction with Simple Systems, J. W. Humberston

Experimental Aspects of Positron and Positronium Physics, T. C. Griffith

Doubly Excited States, Including New Classification Schemes, C. D. Lin

Measurements of Charge Transfer and Ionization in Collisions Involving Hydrogen Atoms, H. B. Gilbody

Electron–Ion and Ion–Ion Collisions with Intersecting Beams, K. Dolder and B. Peart

Electron Capture by Simple Ions, Edward Pollack and Yukap Hahn

Relativistic Heavy-Ion–Atom Collisions, R. Anholt and Harvey Gould

Continued-Fraction Methods in Atomic Physics, S. Swain

Volume 23

Vacuum Ultraviolet Laser Spectroscopy of Small Molecules, C. R. Vidal

Foundations of the Relativistic Theory of Atomic and Molecular Structure, Ian P. Grant and Harry M. Quiney

Point-Charge Models for Molecules Derived from Least-Squares Fitting of the Electric Potential, D. E. Williams and Ji-Min Yan

Transition Arrays in the Spectra of Ionized Atoms, J. Bauche, C. Bauche-Arnoult, and M. Klapisch

Photoionization and Collisional Ionization of Excited Atoms Using Synchrotron and Laser Radiation, F. J. Wuilleumier, D. L. Ederer, and J. L. Picqué

Volume 24

The Selected Ion Flow Tube (SIFT): Studies of Ion–Neutral Reactions, D. Smith and N. G. Adams

Near-Threshold Electron–Molecule Scattering, Michael A. Morrison

Angular Correlation in Multiphoton Ionization of Atoms, S. J. Smith and G. Leuchs

CONTENTS OF PREVIOUS VOLUMES

Optical Pumping and Spin Exchange in Gas Cells, R. J. Knize, Z. Wu, and W. Happer

Correlations in Electron-Atom Scattering, A. Crowe

Volume 25

Alexander Dalgarno: Life and Personality, David R. Bates and George A. Victor

Alexander Dalgarno: Contributions to Atomic and Molecular Physics, Neal Lane

Alexander Dalgarno: Contributions to Aeronomy, Michael B. McElroy

Alexander Dalgarno: Contributions to Astrophysics, David A. Williams

Dipole Polarizability Measurements, Thomas M. Miller and Benjamin Bederson

Flow Tube Studies of Ion-Molecule Reactions, Eldon Ferguson

Differential Scattering in He–He and He^+–He Collisions at KeV Energies, R. F. Stebbings

Atomic Excitation in Dense Plasmas, Jon C. Weisheit

Pressure Broadening and Laser-Induced Spectral Line Shapes, Kenneth M. Sando and Shih-I Chu

Model-Potential Methods, G. Laughlin and G. A. Victor

Z-Expansion Methods, M. Cohen

Schwinger Variational Methods, Deborah Kay Watson

Fine-Structure Transitions in Proton-Ion Collisions, R. H. G. Reid

Electron Impact Excitation, R. J. W. Henry and A. E. Kingston

Recent Advances in the Numerical Calculation of Ionization Amplitudes, Christopher Bottcher

The Numerical Solution of the Equations of Molecular Scattering, A. C. Allison

High Energy Charge Transfer, B. H. Bransden and D. P. Dewangan

Relativistic Random-Phase Approximation, W. R. Johnson

Relativistic Sturmian and Finite Basis Set Methods in Atomic Physics, G. W. F. Drake and S. P. Goldman

Dissociation Dynamics of Polyatomic Molecules, T. Uzer

Photodissociation Processes in Diatomic Molecules of Astrophysical Interest, Kate P. Kirby and Ewine F. van Dishoeck

The Abundances and Excitation of Interstellar Molecules, John H. Black

Volume 26

Comparisons of Positrons and Electron Scattering by Gases, Walter E. Kauppila and Talbert S. Stein

Electron Capture at Relativistic Energies, B. L. Moiseiwitsch

The Low-Energy, Heavy Particle Collisions—A Close-Coupling Treatment, Mineo Kimura and Neal F. Lane

Vibronic Phenomena in Collisions of Atomic and Molecular Species, V. Sidis

Associative Ionization: Experiments, Potentials, and Dynamics, John Weiner, Françoise Masnou-Sweeuws, and Annick Giusti-Suzor

On the β Decay of ^{187}Re: An Interface of Atomic and Nuclear Physics and Cosmochronology, Zonghau Chen, Leonard Rosenberg, and Larry Spruch

Progress in Low Pressure Mercury-Rare Gas Discharge Research, J. Maya and R. Lagushenko